PURE
MATHS

...rnrp'n ...nuing Professional
...velopment Programme

Front cover: Ian Giblin, University of Sussex. This image comes from the one-dimensional logistic map, based upon the mapping $x_{n+1} = x_n(1 - x_n)$, over a range of values in the (λ, x) plane. Each point is coloured by setting x_0 and counting the number of iterations before the system (i.e. the value of x) reaches a stable attractor, or cycle. In this case we have a period 1 attractor bifurcating to period 2 in the vicinity of $\lambda = 3$. By examining (with a computer) the ratio of the distances between successive bifurcations, it is possible to find experimentally the Feigenbaum number F, which, like e and π, arises in many different systems.

PURE
MATHS

R. C. SOLOMON

John Murray

First published 1995
by John Murray (Publishers) Ltd
50 Albermarle Street
London W1S 4BD

Reprinted 1996, 1998, 2000 (twice), 2001, 2002

Designed by Eric Drewery
Typeset by P&R Typesetters Ltd, Salisbury, Wiltshire
Printed and bound in Great Britain by Athenæum Press Ltd, Gateshead, Tyne & Wear

A catalogue entry for this title may be obtained from the British Library

ISBN 0-7195-5344-X

Contents

Introduction

This book covers all the Pure Mathematics necessary for the new core of A and AS level Mathematics. In particular it contains chapters covering the topics of 'the mathematics of uncertainty'.

Throughout the book the applications of theory are emphasised. Many of the exercises, especially the longer exercises at the end of each chapter, provide links between theory and practice. Applications from Physics, Biology, Economics and so on are introduced.

The National Curriculum

You will have begun A or AS level after Key Stage 4 of the National Curriculum. The starting point of this book is Level 7 of the National Curriculum, and it is assumed that you have at least met some of the material of Level 8.

It is hoped that this book will cater for a wide range of expertise. Much GCSE material is reviewed and developed beyond GCSE level in the first few chapters.

The exercises range from routine drill questions to reinforce the ideas of the text to questions which develop the theory further. Challenge questions are marked with an asterisk (*). There are a few very hard problems, marked with two asterisks (**), which to some extent require the reader to invent the means of solving them.

Computers and calculators

The use of calculators and computers is encouraged throughout this book, though there is no attempt to teach Computer Science. Computers are considered as an aid to the learning of mathematics. Used well, they can perform much of the repetitive numerical calculation, leaving a clear view of the underlying structure.

For many mathematical problems, no exact solution is possible, although there is often a numerical method to approximate the solution very closely. Computers are ideal for handling the calculations involved in these numerical methods, and this book contains investigations concerning approximations.

In recent years the new subject of **Chaos Theory** has captured the imagination of the general public. To provide an example of chaotic behaviour, it is often necessary to perform vast amounts of calculation, and it is doubtful whether Chaos Theory would have been developed so fully without computers. In this book we provide an investigation into the chaotic behaviour of a certain function.

There are many computer packages for learning mathematics. Many students are skilled at programming and can write programs to perform involved calculations. In this book though, the use of **spreadsheets** is encouraged. There are two chief reasons for this. Firstly, the spreadsheet is a tool widely used outside the academic world. Secondly, it is a general tool, rather than a special program which performs only one task. It can be thought of as an environment within which to do mathematics – it is possible to keep a spreadsheet running while doing mathematics, and to use it as an area in which to try out ideas and perform calculations.

Units

The units used are from the SI system. Speed, for example, is written as cm s^{-1} (centimetres per second) or as m s^{-1} (metres per second). Acceleration is written as cm s^{-2} or m s^{-2}.

Numerical answers are given correct to three significant figures, unless it is appropriate to give a greater or lesser degree of accuracy.

Structure of this book

The course material is covered in the 29 chapters. Each chapter contains the following:

- Introduction
- Chapter outline
- Explanation, examples, exercises
- Longer exercise
- Examination questions
- Summary and key points.

The chapters themselves are arranged into seven groups. At the end of each group there is a Consolidation section, which contains:

- Extra questions, similar to those in the preceding chapters
- A mixed exercise, containing a variety of problems, concentrating on those which link together the work of different chapters
- Longer exercises
- Examination questions.

There are three Computer Investigation sections in the book, containing suggestions on how to deepen your knowledge of the mathematics by computer exploration.

Halfway through the book there is a set of four Progress tests. At the end of the book there is a set of three Mock examinations.

Complete numerical answers are given at the end of the book.

Why study Pure Mathematics?

In this introduction we have mentioned the many uses of mathematics in other subjects. The exercises of the book contain applications of mathematics in many different contexts.

Pure Mathematics is more than simply a collection of techniques for other subjects. In all ages, and in all cultures, it has been a source of intellectual stimulation second to none. In his autobiography, the philosopher and mathematician Bertrand Russell writes of how, at the age of eleven, he was introduced to Euclid's geometry:

> *This was one of the great events of my life, as dazzling as first love. I had not imagined there was anything so delicious in the world.*

To describe the delight obtainable from mathematics, 'delicious' is not too strong a word.

Acknowledgements

I am very grateful to the following examination authorities for permission to reproduce examination questions.

Associated Examining Board (A)
University of Cambridge Local Examinations Syndicate (C)
Northern Examinations and Assessment Board
   (incorporating JMB) (J)

University of London Examinations and Assessment
Council   (L)
Northern Ireland Schools Examinations and Assessment
Council   (NI)
University of Oxford Delegacy of Local Examinations   (O)
Oxford and Cambridge Schools Examination Board   (O&C)
The Schools Mathematics Project   (SMP)
The Mathematics in Education and Industry project   (MEI)
Welsh Joint Education Committee   (W)

The source of each question is indicated by the appropriate letter. Questions from AS level papers are indicated by the letters AS. All other questions are from A level papers.

The photograph of a Babylonian cuneiform tablet on page 28 is reproduced with permission of the British Museum.

The photograph of Babbage's Difference Engine on page 171 is reproduced with permission of the Science Museum.

The photograph of a window in York Minster on page 376 is reproduced by kind permission of the Dean and Chapter of York.

The solutions to the examination questions are the work of the author, and have neither been provided nor approved by the boards concerned.

The University of London Examinations and Assessment Council accepts no responsibility whatsoever to the accuracy or method of working in the answers given.

Algebra I

In all areas of science, in order to advance beyond the most basic stages, it is necessary to be competent in algebra. This is especially true for mathematics itself. Throughout your A or AS level course you will have to use algebra.

This chapter includes many of the basic algebraic techniques which you will need throughout the course.

1.1 Algebraic expressions

In an algebraic expression, letters are used to represent numbers. We can treat algebraic expressions in a similar way to numbers. There are several operations with which we need to be familiar.

Substitution

If we have an expression involving x, then the process of replacing x by a number or another expression is called **substitution**.

EXAMPLE 1.1
Let $f(x)$ be the expression $3x^2 - 2x + 1$.
Find **a)** $f(2)$ **b)** $f(y)$ **c)** $f(x^2)$

Solution
a) Replace x by 2.
$$f(2) = 3 \times 2^2 - 2 \times 2 + 1 = 9$$

b) Replace x by y.
$$f(y) = 3y^2 - 2y + 1$$

c) Replace x by x^2 and then use the fact that $(x^2)^2 = x^4$.
$$f(x^2) = 3(x^2)^2 - 2x^2 + 1 = 3x^4 - 2x^2 + 1$$
$$f(x^2) = 3x^4 - 2x^2 + 1$$

EXERCISE 1A

1 Find the value of the formula $3x + 2$ in the following cases.

    a) $x = 2$ **b)** $x = -3$ **c)** $x = \frac{1}{2}$

2 Find the value of the formula $2x + 5y$ in the following cases.

    a) $x = 2$ and $y = 3$ **b)** $x = -4$ and $y = 2$

3 Let $f(x) = x^2 - x + 1$. Find the value of the following.

    a) $f(3)$ **b)** $f(z)$ **c)** $f(x^3)$

4 Let $g(x) = 2 - x - x^3$. Find the value of the following.

    a) $g(2)$ **b)** $g(-3)$ **c)** $g(y)$ **d)** $g(-x)$

5 Let $h(x) = x^2 + x + 1$. Evaluate the following.

    a) $h(x) + h(-x)$ **b)** $h(x) - h(-x)$

*6 $h(x)$ is an expression involving powers of x. For all x, $h(x) = h(-x)$. What can you say about h?

Expanding and simplifying

Expanding

Suppose we have an expression such as $4(x + y)$. We can **expand** the expression by multiplying both x and y by 4.

$$4(x + y) = 4x + 4y$$

Simplifying

Expressions which are similar, such as $4x$ and $5x$, can be added, to obtain $9x$.

EXAMPLE 1.2

Expand and simplify these expressions.

a) $2(2x - y) + 3(4x + 3y)$ **b)** $(2a + 3b)(a - b)$

Solution

a) First expand out the brackets.

$$2(2x - y) + 3(4x + 3y) = 4x - 2y + 12x + 9y$$

Now simplify by collecting the x and y terms.

$$2(2x - y) + 3(4x + 3y) = 16x + 7y$$

b) Expand the expression, multiplying both terms in the second bracket by both terms in the first bracket.

$$(2a + 3b)(a - b) = 2a^2 - 2ab + 3ba - 3b^2$$

Simplifying by collecting the ab terms.

$$(2a + 3b)(a - b) = 2a^2 + ab - 3b^2$$

EXERCISE 1B

1 Expand and simplify the following expressions.

a) $3(x + y)$

b) $4(-2x + 3y)$

c) $x(x + y)$

d) $2(x + y) + 3(x - y)$

e) $3(a - b) - 4(a + b)$

f) $4(2x + y) + 3(x - 2y)$

g) $2(p + 2q) - 3(2q - p)$

h) $x(x + y) + y(x - y)$

i) $a(2a + b) - b(3a - b)$

j) $(a + b)(2a + b)$

k) $(2x + y)(3x + y)$

l) $(p - q)(p + q)$

m) $(4a + 3b)(a - 3b)$

n) $(p + 2)(p - 3)$

o) $(2q + 3)(q - 3)$

p) $(x + y)^2$

q) $(2a + b)^2$

r) $(2r - 3s)^2$

2 Let $f(x) = 3x + 3$. Evaluate the following.

a) $f(x + 2)$

b) $f(x + y)$

3 Let $f(x) = x^2 + 2x - 1$. Evaluate the following.

a) $f(2x)$

b) $f(x + 1)$

c) $f(x + y)$

4 Let $f(x) = 1 - 2x - 3x^2$. Evaluate the following.

a) $f(x + 2)$

b) $f(1 - x)$

c) $f(x - y)$

1.2 The algebra of polynomials

A polynomial in x is a function built up from powers of x, which have been multiplied by constants and added. Here are some typical polynomials.

$$3x^3 - 4x^2 + 2.5x - 3 \qquad -x^{10} - x^5 + 3 \qquad 3x - 2$$

Note that the constants can be negative or fractional, but that the powers must be non-negative whole numbers.

The multiplying constants are called **coefficients**. For the first polynomial above, the coefficient of x^3 is 3, the coefficient of x is 2.5 and the constant term is -3.

The **degree** of a polynomial is the highest power of x. The degrees of the polynomials above are 3, 10 and 1 respectively. A constant polynomial has degree 0.

Addition and subtraction

Two (or more) polynomials can be added or subtracted, simply by adding or subtracting the corresponding terms. The result is another polynomial.

EXAMPLE 1.3
Let $P(x) = x^3 - 2x^2 - 3x + 4$ and $Q(x) = -2x^2 + 4x - 3$.
Find $P(x) + Q(x)$ and $P(x) - Q(x)$.

Solution
$Q(x)$ does not have an x^3 term. We can combine the x^2 terms by adding the coefficients, to obtain $-4x^2$. Similarly we can combine the x terms and the constants.

$$P(x) + Q(x) = x^3 - 2x^2 - 3x + 4 - 2x^2 + 4x - 3 = x^3 - 4x^2 + x + 1$$

$$P(x) + Q(x) = x^3 - 4x^2 + x + 1$$

and for $P(x) - Q(x)$, simply reverse all the signs in $Q(x)$.

$$P(x) - Q(x) = x^3 - 7x + 7$$

EXERCISE 1C

1 Write down the degrees of the following polynomials.

   a) $x^2 - 3x - 3$ **b)** $3x^4 - x - 2$ **c)** $4x - 1$

   d) $6 - x^3 + 5x^5$ **e)** $1 + x + x^2 + x^3$ **f)** 4

2 Letting $P(x) = 2x^2 + x - 3$ and $Q(x) = -x^2 - 3x + 1$, evaluate $P(x) + Q(x)$ and $P(x) - Q(x)$.

3 Letting $R(x)$ and $S(x)$ be the polynomials (a) and (e) from Question 1 above, evaluate $R(x) + S(x)$ and $R(x) - S(x)$.

4 Letting $F(x)$ and $G(x)$ be the polynomials (b) and (d) from Question 1 above, evaluate $F(x) + G(x)$ and $F(x) - G(x)$.

*5 Suppose $P(x)$ has degree n and $Q(x)$ has degree m. What can you say about the degrees of $P(x) + Q(x)$ and $P(x) - Q(x)$?

Multiplication of polynomials

Sometimes we need to multiply two polynomials together. Then every term of the first polynomial must be multiplied by every term of the second. The result will be another polynomial.

EXAMPLE 1.4
With $P(x)$ and $Q(x)$ as in Example 1.3 above, find $P(x) \times Q(x)$.

Solution
Write down the product, putting both expressions in brackets.

$$P(x) \times Q(x) = (x^3 - 2x^2 - 3x + 4)(-2x^2 + 4x - 3)$$

The x^3 term of P(x) must be multiplied by all the terms of Q(x). The same holds for the other terms. Write out all the terms.

$$\begin{aligned}
&-2x^5 + 4x^4 - 3x^3 &&\text{(multiplying } x^3 \text{ by Q}(x)) \\
&+4x^4 - 8x^3 + 6x^2 &&\text{(multiplying } -2x^2 \text{ by Q}(x)) \\
&+6x^3 - 12x^2 + 9x &&\text{(multiplying } -3x \text{ by Q}(x)) \\
&-8x^2 + 16x - 12 &&\text{(multiplying 4 by Q}(x))
\end{aligned}$$

Now simplify by collecting all the terms of each power of x.

$$\text{P}(x) \times \text{Q}(x) = -2x^5 + 8x^4 - 5x^3 - 14x^2 + 25x - 12$$

EXAMPLE 1.5

Express $(x + 3)^2$ as a polynomial.

Solution

Here $(x + 3)$ is multiplied by itself.

$$(x + 3)^2 = (x + 3)(x + 3)$$

Multiply out and simplify.

$$(x + 3)(x + 3) = x^2 + 3x + 3x + 9 = x^2 + 6x + 9$$
$$(x + 3)^2 = x^2 + 6x + 9$$

EXAMPLE 1.6

Express $(ax + 2b)(3ax - 4b)$ as a polynomial. (Here a and b are constants.)

Solution

Multiplying out, we obtain

$$(ax + 2b)(3ax - 4b) = 3a^2x^2 - ax4b + 2b3ax - 8b^2$$
$$= 3a^2x^2 - 4abx + 6abx - 8b^2$$

The middle terms can be combined, to obtain

$$(ax + 2b)(3ax - 4b) = 3a^2x^2 + 2abx - 8b^2$$

EXERCISE 1D

1 Evaluate the following products, simplifying your answers as far as possible.

a) $(2x - 3)(x + 7)$
b) $(4x - 3)(x^2 - 3x - 1)$
c) $(x^2 + 2x + 1)(x - 8)$

d) $(4x^2 - 3x - 1)(2x + 8)$
e) $(x^2 + x + 1)(x^2 - x + 1)$
f) $(x^3 + x)(x^2 - 1)$

g) $(x + 4)^2$
h) $(2x + 1)^2$
i) $(x - 3)^2$

j) $(2x - 3)^2$
k) $(ax + b)(ax - b)$
l) $(2ax - 3b)(3ax - 5b)$

m) $(ax + b)^2$
n) $(ax - b)^2$
o) $(x + 1)^3$

2 $(ax + b)(x^2 + 3x - 1) = 5x^3 + 14x^2 - 8x + 1$. Find a and b.

*3 The degrees of P(x) and Q(x) are n and m respectively. What is the degree of P(x) $\times$ Q(x)?

4 Expand and simplify these expressions.

a) $(x - a)(x - b)$ ***b)** $(x - a)(x - b)(x - c) \cdots (x - y)(x - z)$

Division of polynomials

The process is very similar to long division for ordinary numbers. Suppose we are dividing $(x + 3)$ into $x^3 + 7x^2 - 5x + 9$. Lay out the polynomial and its **divisor** as shown.

$$x + 3 \overline{)x^3 + 7x^2 - 5x + 9}$$

Divide the highest power by x. Dividing x^3 by x gives x^2, so put x^2 on the top line. Multiply $(x + 3)$ by x^2 and subtract it from the first two terms of the polynomial. Set it out as shown.

$$\begin{array}{r} x^2 \\ x + 3 \overline{)x^3 + 7x^2 - 5x + 9} \\ \underline{x^3 + 3x^2 } \\ 4x^2 \end{array}$$

Bring down the $-5x$, and now multiply $(x + 3)$ by $4x$ and subtract.

$$\begin{array}{r} x^2 + 4x \\ x + 3 \overline{)x^3 + 7x^2 - 5x + 9} \\ \underline{x^3 + 3x^2 } \\ 4x^2 - 5x \\ \underline{4x^2 + 12x } \\ -17x \end{array}$$

Bring down the 9, multiply $(x + 3)$ by -17 and subtract.

$$\begin{array}{r} x^2 + 4x - 17 \\ x + 3 \overline{)x^3 + 7x^2 - 5x + 9} \\ \underline{x^3 + 3x^2 } \\ 4x^2 - 5x \\ \underline{4x^2 + 12x } \\ -17x + 9 \\ \underline{-17x - 51} \\ 60 \end{array}$$

Since $(x + 3)$ cannot be divided into 60, the division stops here. The quotient is $x^2 + 4x - 17$ and the remainder is 60.

We can write the division in terms of quotient and remainder as follows.

$$x^3 + 7x^2 - 5x + 9 = (x + 3)(x^2 + 4x - 17) + 60$$

Notice that the remainder is a constant. In general, the remainder will always have at least one less power of x than the divisor, i.e. the degree of the remainder will be at least one less than the degree of the divisor. So if we are dividing by a quadratic expression, the remainder will be a linear expression. This will be seen in the following example.

EXAMPLE 1.7

Find the quotient and remainder when $x^3 - 4x^2 + x + 2$ is divided by $(x^2 - 3)$.

Solution

Set out the division as above. Notice that there is no x term in the divisor.

$$
\begin{array}{r}
x \\
x^2 - 3 \overline{)\,x^3 - 4x^2 + x + 2} \\
\underline{x^3 - 3x} \\
-4x^2 + 4x
\end{array}
$$

$$
\begin{array}{r}
x - 4 \\
x^2 - 3 \overline{)\,x^3 - 4x^2 + x + 2} \\
\underline{x^3 - 3x} \\
-4x^2 + 4x + 2 \\
\underline{-4x^2 + 12} \\
4x - 10
\end{array}
$$

$$x^3 - 4x^2 + x + 2 = (x^2 - 3)(x - 4) + 4x - 10$$

The quotient is $x - 4$ and the remainder is $4x - 10$.

EXERCISE 1E

1 In the following divisions, find the quotient and the remainder.

a) $(x^3 + 5x^2 + 4x - 17) \div (x - 4)$ b) $(x^3 - 3x^2 + 4x + 1) \div (x - 2)$

c) $(x^4 + x^3 + x^2 + x + 1) \div (x^2 + 1)$ d) $(x^3 + 4x - 3) \div (x - 2)$

e) $(8x^3 + 5x^2 - 3x - 1) \div (2x + 1)$ f) $(4x^4 + 2x^3 - 7x^2 + x - 3) \div (x - 2)$

g) $(16x^4 + x - 3) \div (2x - 3)$ h) $(x^5 + 1) \div (x - 1)$

2 When $x - 3$ is divided into $x^3 - 4x^2 + 5x + a$, the remainder is 7. Find a.

3 When $x + 4$ is divided into $2x^3 + 2x^2 - ax + 2$, the remainder is 8. Find a.

4 When $x^2 + x + 1$ is divided into $x^4 + 5x^3 - 3x^2 + ax + b$, the remainder is $2x - 3$. Find a and b.

1.3 Factorising

In Section 1.1 we dealt with the expansion of brackets. The opposite procedure to this is **factorising**. If all the terms of an expression are multiples of the same factor, that factor can be taken out.

EXAMPLE 1.8

Factorise $2ax^2 + 3a^2x - ax$.

Solution

All the terms can be divided by ax.

$$2ax^2 + 3a^2x - ax = ax(2x + 3a - 1)$$

EXERCISE 1F

Factorise the following expressions.

1   $3x + 9x^2$ 2 $5a + 15ab$ 3 $x^2 - x^3$

4   $4xy + 2x^2y - 6xy^2$ 5 $9p - 3pq^2$ 6 $z^2 + 2z^3 + 3z^4$

Expressions of the form $x^2 + bx + c$

An example of the expansion of two brackets is

$$(x - 3)(x + 5) \equiv x^2 + 2x - 15$$

Factorisation is the reverse procedure to expansion. Given $x^2 + 2x - 15$, we can factorise it to obtain $(x - 3)(x + 5)$.

Suppose we can factorise an expression of this form as $(x + \alpha)(x + \beta)$. When it is multiplied out we get

$$x^2 + (\alpha + \beta)x + \alpha\beta \equiv x^2 + bx + c$$

so we look for numbers α and β, for which $\alpha + \beta = b$ and $\alpha\beta = c$. If α and β are whole numbers, then they must be factors of c.

Note that it may not be possible to find α and β. In this case the expression cannot be factorised in terms of whole numbers.

EXAMPLE 1.9

Factorise the following expressions.
a) $x^2 + 7x + 10$ b) $x^2 + 3x - 18$

Solution

a) We want two numbers with product 10 and sum 7. The factors of

10 are either 10 and 1 or 2 and 5. The second pair is the one that works.

$$x^2 + 7x + 10 \equiv (x + 2)(x + 5)$$

b) We want two numbers with product -18 and sum 3. One of these numbers must be negative, so we look for factors of 18 with **difference** 3. The numbers are 6 and 3. Because the $3x$ term is positive, the factors must be $+6$ and -3.

$$x^2 + 3x - 18 \equiv (x + 6)(x - 3)$$

EXERCISE 1G

Factorise the following expressions.

1 $x^2 + 7x + 12$ **2** $x^2 + 8x + 15$ **3** $x^2 + 6x + 9$

4 $x^2 - 13x + 30$ **5** $x^2 - 7x + 12$ **6** $x^2 - 8x + 16$

7 $x^2 + 3x - 40$ **8** $x^2 - 5x - 50$ **9** $x^2 - 7x - 8$

10 $x^2 - 4$ **11** $x^2 - 16$ **12** $x^2 + 8x$

13 $x^2 - 12x$

Expressions of the form $ax^2 + bx + c$

These are more difficult to handle, as a is not necessarily equal to 1.

Suppose an expression of this form is factorised as $(\alpha x + \gamma)(\beta x + \delta)$.

Multiplying out gives

$$\alpha\beta x^2 + (\alpha\delta + \gamma\beta)x + \gamma\delta \equiv ax^2 + bx + c$$

So we have $\alpha\beta = a$, $\gamma\delta = c$, $\alpha\delta + \gamma\beta = b$. If α, β, γ and δ are whole numbers, then α and β will be factors of a, and γ and δ will be factors of c. If a and c have many factors there will be many different possibilities. A systematic way to proceed is as follows.

Note that $(\alpha\beta)(\gamma\delta) = (\alpha\delta)(\gamma\beta) = ac$ and that $(\alpha\delta) + (\gamma\beta) = b$. So we look for two numbers k and m with product ac and sum b. We then split the middle bx term into $kx + mx$, and separately factorise $ax^2 + kx$ and $mx + c$. These will have a common factor which can be taken out. The two following examples will illustrate the method.

Note that it may not be possible to find the numbers k and m. In this case we will not be able to factorise the expression in terms of whole numbers.

EXAMPLE 1.10
Factorise $6x^2 + 11x + 3$.

Solution
We look for two numbers with sum 11 and product $6 \times 3 = 18$. These numbers are 2 and 9. Rewrite the expression as

$$6x^2 + 2x + 9x + 3$$

and factorise the first two terms and the last two.

$$2x(3x + 1) + 3(3x + 1)$$

Note that $(3x + 1)$ is now a common factor. The full factorisation is

$$6x^2 + 11x + 3 \equiv (3x + 1)(2x + 3)$$

EXAMPLE 1.11
Factorise $2x^2 - 5x - 3$.

Solution
Here we look for two numbers with sum -5 and product $2 \times (-3) = -6$. These numbers are -6 and 1. Rewrite the expression as

$$2x^2 - 6x + x - 3$$

and factorise the first pair and the last pair.

$$2x(x - 3) + 1(x - 3)$$
$$2x^2 - 5x - 3 \equiv (2x + 1)(x - 3)$$

EXERCISE 1H

Factorise the following expressions.

1 $2x^2 + 7x + 3$ 2 $2x^2 + 5x + 2$ 3 $6x^2 + 19x + 3$

4 $6x^2 - 7x + 2$ 5 $10x^2 - 9x + 2$ 6 $2x^2 - 5x + 3$

7 $2x^2 + x - 15$ 8 $3x^2 - x - 2$ 9 $3x^2 + x - 10$

10 $2x^2 - 8$ 11 $3x^2 - 27$ 12 $2x^2 + 8x$

13 $3x^2 + 12x$

1.4 Rational functions

A **rational function** consists of one polynomial divided by another. It can be thought of as the ratio of two polynomials, or as an algebraic fraction. Here are some examples of rational functions.

$$\frac{x+3}{x-7} \qquad \frac{x-2}{2x^3+x^2-x} \qquad \frac{x^2+3x-2}{1}$$

The last example is the same as $x^2 + 3x - 2$, showing that an ordinary polynomial is itself a rational function.

The arithmetic rules for ordinary fractions also apply to these algebraic fractions.

Multiplying and dividing

If the numerator and denominator expressions of an algebraic fraction have a common factor, we can **simplify** the fraction by dividing top and bottom by the factor.

When two fractions are multiplied, their numerators are multiplied together and their denominators are multiplied together.

To divide one fraction by another, turn the second fraction upside-down and multiply.

EXAMPLE 1.12

Let $R(x) = \dfrac{3(x+2)}{x+1}$ and $S(x) = \dfrac{2x-1}{9(x+1)}$.

Find $R(x) \times S(x)$ and $R(x) \div S(x)$ and simplify your answers as far as possible.

Solution

Follow the rules for ordinary fractions.

$$R(x) \times S(x) = \frac{3(x+2)(2x-1)}{(x+1)9(x+1)}$$

Divide numerator and denominator by 3, and multiply out to obtain

$$R(x) \times S(x) = \frac{2x^2+3x-2}{3x^2+6x+3}$$

To find $R(x) \div S(x)$ turn $S(x)$ upside-down and multiply it with $R(x)$.

$$R(x) \div S(x) = \frac{3(x+2)}{x+1} \times \frac{9(x+1)}{2x-1}$$

The $(x + 1)$ terms cancel. Multiply the numerators and multiply the denominators.

$$R(x) \div S(x) = \frac{27x + 54}{2x - 1}$$

EXERCISE 1I

Evaluate the following multiplications and divisions. Simplify your answers where possible.

1 $\quad \dfrac{x+3}{x-1} \times \dfrac{x-1}{x-3}$

2 $\quad \dfrac{2x+1}{x-2} \times \dfrac{x-3}{x+1}$

3 $\quad \dfrac{x+2}{x-5} \times \dfrac{x-4}{x+2}$

4 $\quad \dfrac{2(x-3)}{x+1} \times \dfrac{x-5}{x-2}$

5 $\quad \dfrac{4(x+1)}{x+2} \times \dfrac{x-3}{2(x-1)}$

6 $\quad \dfrac{2x-3}{9(x+1)} \times \dfrac{3(x-1)}{x+2}$

7 $\quad \dfrac{x+2}{x-2} \div \dfrac{x-1}{x-4}$

8 $\quad \dfrac{2x+3}{x-1} \div \dfrac{x-2}{x+1}$

9 $\quad \dfrac{x+2}{x-4} \div \dfrac{x+2}{x-5}$

Addition and subtraction

When we add or subtract ordinary fractions we must express them both in terms of a **common denominator**. The same applies to algebraic fractions. The common denominator of two algebraic expressions is an expression that they will both divide into. This common denominator is often the product of the original denominators.

EXAMPLE 1.13

Evaluate the following expressions.

a) $\dfrac{1}{x+1} + \dfrac{2}{x+2}$ $\qquad$ **b)** $\dfrac{2}{(x+1)(x+2)} - \dfrac{1}{x+1}$

Solution

a) Put both expressions over a common denominator.

$$\frac{1}{x+1} + \frac{2}{x+2} = \frac{1(x+2) + 2(x+1)}{(x+1)(x+2)}$$

Now simplify the top.

$$\frac{1}{x+1} + \frac{2}{x+2} = \frac{3x+4}{(x+1)(x+2)}$$

b) Note that the denominator of the second expression is a factor of the denominator of the first expression. Hence a common denominator is the denominator of the first expression.

$$\frac{2}{(x+1)(x+2)} - \frac{1}{x+1} = \frac{2 - 1(x+2)}{(x+1)(x+2)}$$

Now simplify the top.

$$\frac{2}{(x+1)(x+2)} - \frac{1}{(x+1)} = \frac{-x}{(x+1)(x+2)}$$

EXERCISE 1J

Express the functions in Questions **1–8** as single fractions, simplifying your answers as far as possible.

1 $\dfrac{1}{x-2} + \dfrac{1}{x-3}$

2 $\dfrac{1}{x-2} + \dfrac{1}{2x+1}$

3 $\dfrac{2}{2x-1} - \dfrac{4}{x+2}$

4 $\dfrac{x+3}{x-1} + \dfrac{x-3}{x-1}$

5 $\dfrac{2x+1}{2(x-2)} + \dfrac{x-3}{x-2}$

6 $\dfrac{x+2}{x-5} - \dfrac{x-4}{x+2}$

7 $\dfrac{1}{(x-1)(x-3)} + \dfrac{1}{x-1}$

8 $\dfrac{1}{x-2} - \dfrac{x-3}{(x+1)(x-2)}$

9 Factorise $x^2 - 1$. Hence express $\dfrac{1}{x+1} + \dfrac{3}{x^2-1}$ as a single fraction.

10 Factorise $x^2 - 3x + 2$ and $x^2 - 4x + 3$. Hence express $\dfrac{1}{x^2-3x+2} - \dfrac{1}{x^2-4x+3}$ as a single fraction.

11 Express the following as single fractions.

a) $\dfrac{1}{x^2-4} + \dfrac{1}{x-2}$

b) $\dfrac{x+5}{x^2-3x+2} - \dfrac{1}{x-1}$

c) $\dfrac{1}{x^2-4} + \dfrac{1}{x^2-3x+2}$

1.5 The remainder and factor theorems

There is a quick way to find the remainder after division by a linear function.

The remainder theorem

When P(x) is divided by ($x - \alpha$), the remainder is P(α).

Proof
After division, there will be a quotient Q(x), and a remainder R which is a constant.

$$P(x) \equiv (x - \alpha)Q(x) + R$$

This equation is true for all values of x. In particular, it is true when x is equal to α. Put $x = \alpha$ into the line above.

$$P(\alpha) \equiv (\alpha - \alpha)Q(\alpha) + R = R$$

So we have $P(\alpha) = R$. This gives us the remainder theorem.

We can check this with the division on page 10, in section 1.2.

$P(x) = x^3 + 7x^2 - 5x + 9$ and $x - \alpha = x + 3$, which gives $\alpha = -3.$

$$P(-3) = (-3)^3 + 7(-3)^2 - 5(-3) + 9 = -27 + 63 + 15 + 9 = 60$$

This agrees with the result found by long division.

Note

If our factor had been of the form $(2x - 3)$, then we could rewrite it as $2(x - \frac{3}{2})$. In the proof above we would put $x = \frac{3}{2}$. So the remainder would be $P(\frac{3}{2})$.

EXAMPLE 1.14

When $8x^3 - 12x^2 + ax + 2$ is divided by $2x - 3$ the remainder is 5. Find a.

Solution

Let $\alpha = \frac{3}{2}$. Put this into the remainder theorem.

$$P(\tfrac{3}{2}) = 8 \times (\tfrac{3}{2})^3 - 12 \times (\tfrac{3}{2})^2 + a\tfrac{3}{2} + 2 = 5$$

$$\frac{3a}{2} = 3$$

$$a = 2$$

EXERCISE 1K

1 Use the remainder theorem to find the remainders in the following divisions.

a) $(x^3 - 8x^2 - 4x + 5) \div (x - 2)$ **b)** $(x^3 - 2x^2 + 7x - 3) \div (x + 1)$

c) $(x^4 - 4x^3 + 2x^2 + 3x + 2) \div (x - 3)$ **d)** $(x^6 - x^3 + 3) \div (x + 1)$

e) $(8x^3 + 4x^2 - 6x + 1) \div (2x + 1)$ **f)** $(x^3 + x^2 - x + 1) \div (3x - 2)$

2 When $x^3 - 6x^2 + 5x + a$ is divided by $(x + 2)$, the remainder is 3. Find a.

3 When $x^3 - 4x^2 - ax + 5$ is divided by $(x - 3)$, the remainder is 8. Find a.

4 When $x^4 - ax + 3$ is divided by $(2x - 1)$, the remainder is 2. Find a.

5 When $x^3 + 3x^2 + ax + b$ is divided by $(x + 1)$ the remainder is 5, and when it is divided by $(x - 2)$ the remainder is 8. Find a and b.

6 When $ax^3 + bx^2 + 3x - 4$ is divded by $(x - 1)$ the remainder is 3, and when it is divided by $(x + 2)$ the remainder is 6. Find a and b.

The factor theorem

$(x - \alpha)$ is a factor of P(x) if and only if P(α) = 0.

Proof

Suppose P(x) is divided by $(x - \alpha)$.

If $(x - \alpha)$ is a factor, then by the remainder theorem the remainder P(α) will be zero. Conversely, if P(α) = 0, then the remainder is zero. Hence $(x - \alpha)$ is a factor of P(x).

Let us take the polynomial P(x) = $x^3 + x^2 - 10x + 8$. Put $x = 2$, to obtain

$$P(2) = 2^3 + 2^2 - 10 \times 2 + 8 = 0$$

It follows that $(x - 2)$ is a factor of $x^3 + x^2 - 10x + 8$.

Note that 2 divides exactly into 8. When searching for factors of the form $(x - \alpha)$, we need only consider those α which divide into the constant term of P(x). For the example above, α could only be ± 1, ± 2, ± 4 or ± 8. We need not test to see whether $(x - 17)$, for example, is a factor.

EXAMPLE 1.15

If $(x + 1)$ and $(x - 2)$ are both factors of $x^3 + ax^2 + bx + 6$, find a and b.

Solution

Apply the factor theorem for $\alpha = -1$ and $\alpha = 2$.

$$(-1)^3 + a(-1)^2 + b(-1) + 6 = 0$$

Hence $a - b = -5$.

$$2^3 + a(2)^2 + b(2) + 6 = 0$$

Hence $2a + b = -7$.

Solve these simultaneous equations.

$$\left.\begin{array}{l} a - b = -5 \\ 2a + b = -7 \end{array}\right\} \rightarrow 3a = -12 \text{ and } 3b = 3$$

$$a = -4 \text{ and } b = 1$$

EXERCISE 1L

1 Which of the following are factors of $x^3 + 3x^2 - 5x - 10$?

    a) $x - 1$ **b)** $x - 2$ **c)** $x + 5$

2 Find factors of the following expressions.

    a) $x^3 - x^2 - 17x - 15$ **b)** $x^3 - 7x + 6$ **c)** $x^3 + 2x^2 + x + 2$

3 $(x + 1)$ is a factor of $x^3 - 8x^2 + ax + 4$. Find a.

4 $(x - 2)$ is a factor of $P(x) = x^3 + ax^2 + bx + 6$, and the remainder when $P(x)$ is divided by $(x + 3)$ is 30. Find a and b.

5 $x^2 - 3x + 2$ is a factor of $x^3 + ax^2 + bx + 4$. Find a and b.

Factorising

Once we have found one factor of a polynomial, we can divide through by that factor to obtain a simpler polynomial. There are two methods.

Method 1 Use the division procedure of Section 1.2 to find the quotient when $P(x)$ is divided by $(x - \alpha)$.

Method 2 Find the quotient by inspection of coefficients.

To illustrate Method 2, let us look at an example. By the factor theorem, we can show that $(x - 2)$ is a factor of $x^3 + x^2 - 10x + 8$. So let us write

$$x^3 + x^2 - 10x + 8 = (x - 2)(ax^2 + bx + c)$$

Here the letters a, b and c represent the coefficients. Now a must be 1, to ensure that $x^3 = x \times x^2$ and c must be -4, to ensure that $8 = (-2)(-4)$. We now have

$$x^3 + x^2 - 10x + 8 = (x - 2)(x^2 + bx - 4)$$

Consider the x^2 term. Multiply out the right hand side and equate the x^2 terms.

$$x^2 = -2x^2 + bx^2$$
$$\Rightarrow 1 = -2 + b$$

So b is 3.

$$x^3 + x^2 - 10x + 8 = (x - 2)(x^2 + 3x - 4)$$

Method 2 is quicker and less subject to error, but it takes confidence.

Whichever method we use, the quotient $x^2 + 3x - 4$ can now be factorised by the methods of section 1.3. We obtain the complete factorisation.

$$x^3 + x^2 - 10x + 8 \equiv (x - 2)(x + 4)(x - 1)$$

EXERCISE 1M

Factorise the following expressions.

1 $x^3 + 3x^2 - 4x - 12$ **2** $x^3 + x^2 - 4x - 4$ **3** $x^3 - 9x^2 + 26x - 24$

4 $x^3 - 6x^2 + 5x + 12$ **5** $x^3 - 4x^2 + 5x - 2$ **6** $x^3 + 5x^2 + 7x + 3$

7 $2x^3 - x^2 - 2x + 1$ **8** $3x^3 + 5x^2 - 4x - 4$ **9** $6x^3 - 13x^2 + x + 2$

1.6 Laws of proportionality

If two quantities increase at the same rate then they are **proportional** to each other. One quantity will be a constant multiple of the other.

The symbol for proportionality is $\propto$.

If y is proportional to x, then $y \propto x$.

If y is proportional to x, then $y = kx$ for some k.

Notice that the ratio of y to x is the constant k. This constant (k) is the **constant of proportionality**.

If we draw a graph of two quantities that are in proportion, the result will be a straight line going through the origin, as shown.

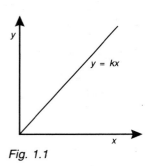

Fig. 1.1

Inverse proportion

If two quantities increase at opposite rates, then they are **inversely proportional** to each other. One quantity will be a constant divided by the other.

If y is inversely proportional to x, then $y = \dfrac{k}{x}$ for some k.

The graph of two quantities that are in inverse proportion is as shown in Fig. 1.2.

Notice that the product of x and y is constant. We still use the $\propto$ symbol.

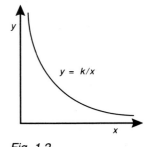

Fig. 1.2

If y is inversely proportional to x, then $y \propto \dfrac{1}{x}$.

EXAMPLE 1.16

y is proportional to x. When $x = 2$, $y = 6$. Find the equation giving y in terms of x.

Solution

Write the proportionality statement in terms of an equation.

$y = kx$, for some k

We know that $y = 6$ when $x = 2$. This gives $k = 3$.

The equation is $y = 3x$.

EXAMPLE 1.17

In an electrical circuit, if the voltage is constant then the current A amps is inversely proportional to the resistance R ohms. The current is 6 amps when the resistance is 20 ohms. Find an equation giving A in terms of R. Find the current if the resistance is raised to 60 ohms.

Solution

Write down the proportionality statement.

$$A = \frac{k}{R}$$

We know that $A = 6$ when $R = 20$. This gives $k = 120$.

The equation is $A = \dfrac{120}{R}$.

Now put $R = 60$.

$$A = \frac{120}{60} = 2$$

When the resistance is 60 ohms, the current is 2 amps.

EXERCISE 1N

1 y is proportional to x. When $x = 3$, $y = 15$. Find an equation giving y in terms of x, and find y when $x = 30$.

2 At small speeds, the air resistance experienced by a moving body is proportional to its speed. A body experiences 5 N when travelling at 20 m s^{-1}. What will the resistance be at 5 m s^{-1}?

3 We are told that P is proportional to Q and $P = a$ when $Q = b$. Find an equation giving P in terms of Q.

4 R is inversely proportional to S and $R = 5$ when $S = 4$. Find an equation giving R in terms of S. Find R when $S = 2$, and S when $R = 50$.

5 Parkinson's **Law of Triviality** states that the time a committee spends discussing an item on the agenda is inversely proportional to the cost of the item.

The governing committee of a college spends 50 minutes discussing the £500 loss made by the bar. How long will it spend on the £20 000 repairs to the roof?

6 We are told that P is inversely proportional to Q and $P = a$ when $Q = b$. Find an equation giving P in terms of Q.

Proportionality to powers

Often one quantity is proportional to a power of another. For example, in a wire of fixed resistance, the power generated by a current is proportional to the square of the current.

If y is proportional to the square of x, then $y \propto x^2$ or $y = kx^2$. We can then proceed as above.

If y is proportional to the square of x, the graph of y against x is as shown.

Similarly, if y is inversely proportional to the square of x, we write $y \propto \dfrac{1}{x^2}$ or $y = \dfrac{k}{x^2}$.

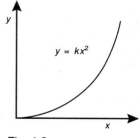

Fig. 1.3

EXAMPLE 1.18

The energy E joules stored in a stretched string is proportional to the square of the extension e cm. An extension of 2 cm stores 8 J of energy. Find the extension necessary to store 18 J.

Solution

The equation is of the form $E = ke^2$. Put in the known values.

$$8 = k2^2$$

Hence $k = 2$.

Hence $E = 2e^2$.

Now put $E = 18$.

$$18 = 2e^2$$

Hence $e^2 = 9$

$$e = 3 \text{ (since } e \geq 0)$$

An extension of 3 cm is required.

EXAMPLE 1.19

The acceleration due to gravity of a body is inversely proportional to the square of the distance of the body from the centre of the Earth. On the surface of the Earth the acceleration is 9.8 m s^{-2}. What is it 1000 km out to space? (Take the radius of the Earth to be 6400 km.)

Solution

Letting r km be the distance from the centre of the Earth, and a m s^{-2} the acceleration, we have

$$a = \frac{k}{r^2}$$

Put in the known facts.

$$a = 9.8 \text{ when } r = 6400$$

Hence $k = 9.8 \times 6400^2 = 4.014 \times 10^8$

In space, 1000 km high, the distance from the Earth's centre is 7400 km.

$$a = \frac{4.014 \times 10^8}{7400^2} = 7.33$$

The acceleration is 7.33 m s^{-2}.

EXERCISE 10

1 y is proportional to the square of x and $y = 100$ when $x = 5$. Find an equation giving y in terms of x. Find y when $x = 15$.

2 The weight of a circular disc is proportional to the square of its radius. A disc of radius 2 cm weighs 20 grams. What is the weight of a disc of radius 0.5 cm?

3 The weight of a metal sphere is proportional to the cube of its radius. A sphere of radius 4 cm weighs 128 grams. What is the weight of a sphere of radius 5 cm?

4 If a body is moving at high speed, the air resistance is proportional to the square of the speed. A body moving at 100 m s^{-1} experiences a resistance of 50 N. What is the resistance at a speed of 150 m s^{-1}?

5 The period of swing of a pendulum is proportional to the square root of its length. A pendulum of length 1.1 m takes 1.05 seconds to make a complete swing. How long will a pendulum of length 1.2 m take?

6 Y is inversely proportional to the square of X and $Y = 5$ when $X = 2$. Find an equation giving Y in terms of X. Find X when $Y = 80$.

7 R is inversely proportional to the square of S and $R = b$ when $S = a$. Find an expression giving R in terms of S.

8 Illumination is inversely proportional to the square of the distance from the light source. At a distance of 5 m, the illumination is 10 candelas. What is the illumination at a distance of 2 m?

*9 The product of x and the cube of y is constant. Express this in terms of a proportionality statement.

LONGER EXERCISE

Proportion in biology

> *The Emperor allowed me a quantity of meat and drink sufficient for the support of 1728 Lilliputians.* ... *his Majesty's mathematicians, having taken the height of my body with the help of a quadrant, and finding it to exceed theirs in the proportion of twelve to one, they concluded from the similarity of their bodies, that mine must contain 1728 of theirs.*
>
> *Gulliver's Travels*, Jonathan Swift, 1726

We see that the weight of a person is proportional to the cube of the height. Gulliver was 12 times as tall as a Lilliputian, but $12^3 = 1728$ times as heavy.

What other facts of proportionality are there? In these exercises we shall give very rough estimates. We shall also assume that all animals are the same shape, and only differ in size. Let h be the height of the animal.

1 Food as percentage of weight What fraction of your weight do you eat per day? What about a cat? As above, the weight of an animal is proportional to the cube of its height, h^3. So the weight of the animal is kh^3. The heat loss is proportional to its surface area, which is proportional to the square of its height, h^2.

If the main function of food is to keep the body warm, the amount needed will be proportional to the heat loss, i.e. to h^2. So the amount eaten will be $K'h^2$. What percentage of its weight does an animal need to eat per day? How does this depend on h?

2 Speed on the horizontal The power output of an animal is related to the amount of oxygen it can use, which is proportional to the surface area of its lungs. How is this related to h?

The drag felt by a running animal is proportional to h^2v^2, where v is the speed. At maximum speed, the power output is equal to the drag. How does the maximum speed depend on h? Is this result true, when we consider a rabbit, a horse and an elephant?

3 **Jumping** The force F that an animal can exert is proportional to the cross-sectional area of its muscles, which is proportional to h^2. If the animal exerts this force through a distance d, and jumps a height H, then we will have

$Fd = WH$

where W is the weight of the animal.

How does H depend on h? Is the result true when we consider a cat and a horse?

4 Are the bones of a mouse thicker or thinner, relative to its size, than those of an elephant? Can you explain why?

EXAMINATION QUESTIONS

1 Show that

$$\left(\frac{1+x^2}{1-x^2}\right)^2 - \left(\frac{2x}{1-x^2}\right)^2$$

has the same numerical value for all x ($x \neq \pm 1$) and determine this value.

JMB AS 1990

2 Express $\dfrac{1}{x+1} + \dfrac{3}{(x+1)(2x-1)}$ as a single algebraic fraction, simplifying your answer.

C AS 1992

3 The cubic function f is given by

$f(x) = x^3 + ax^2 - 4x + b$

where a and b are constants.

Given that $(x - 2)$ is a factor of $f(x)$ and that a remainder of 6 is obtained when $f(x)$ is divided by $(x + 1)$, find the values of a and b.

W AS 1992

Summary and key points

1 If we have an expression P(x), the result of replacing x by y is written P(y) where y could be a number or another algebraic expression.

The process of multiplying out brackets is called expansion.

2 A polynomial in x is built up from powers of x, multiplied by constants and added. The result of adding, subtracting or multiplying polynomials is another polynomial.

3 A rational function consists of one polynomial divided by another.

When rational functions are added or subtracted, they must be put over a common denominator.

4 If we express an algebraic expression as a product, we have factorised it. Quadratic expressions can sometimes be factorised.

5 The remainder after dividing a polynomial $P(x)$ by $(x - \alpha)$ is $P(\alpha)$.

$(x - \alpha)$ is a factor of $P(x)$ if and only if $P(\alpha) = 0$.

Be careful if α is negative.

6 If y is proportional to x we can write $y = kx$, where k is constant. In some cases y is inversely proportional to x $\left(\text{e.g. } y = \dfrac{k}{x} \right)$, or proportional to a power of x (e.g. $y = kx^2$).

CHAPTER 2

Quadratics

Chapter outline

2.1 Solution by factorising
2.2 Completing the square
2.3 The formula for solving quadratic equations
2.4 Other equations reducible to quadratic equations

A quadratic expression is one of the form $ax^2 + bx + c$, where a, b and c are constant and a is not zero. An example of a quadratic expression is $3x^2 + 2x - 5$. A quadratic equation can be formed by putting a quadratic expression equal to zero. So $3x^2 + 2x - 5 = 0$ is a quadratic equation.

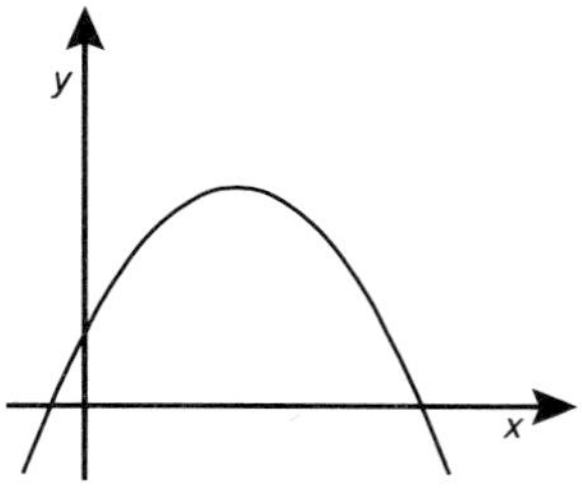

Fig. 2.1

The graph of a quadratic expression is a curve called a **parabola**, as shown.

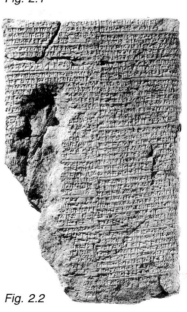

If a ball is thrown upwards, the path that it follows is that of a parabola. Its height h can be given approximately as a quadratic in time t. We can find when the ball reaches a certain height, by solving a quadratic equation. We can also find the greatest height reached by the ball.

Natural problems often give rise to quadratic expressions, and so they have been studied for many years. On the right is a cuneiform tablet from the British Museum, dating from about 2000BC. It contains mathematical problems, some of which involve quadratic equations. These will be discussed in Exercise 2F.

Fig. 2.2

2.1 Solution by factorising

If a quadratic expression is set equal to 0, then we have a quadratic equation. Suppose the quadratic expression can be factorised, by the methods covered in Chapter 1. Then the equation can be solved directly by putting each factor equal to zero.

Suppose the equation is $x^2 - 5x + 6 = 0$. The left-hand side factorises, to become

$$(x - 2)(x - 3) = 0$$

So either $x - 2 = 0$ or $x - 3 = 0$. The solutions of the equation are $x = 2$ and $x = 3$.

EXAMPLE 2.1
Solve this equation.

$$x^2 = 4 - 3x$$

Solution

First collect all the terms on one side.

$$x^2 + 3x - 4 = 0$$

Now factorise the left-hand side.

$$(x - 1)(x + 4) = 0$$
$$x = 1 \text{ or } x = -4$$

EXAMPLE 2.2
Solve this equation.

$$\frac{9}{x} + \frac{12}{x + 1} = 6$$

Solution

Multiply both sides of the equation by $x(x + 1)$.

$$9(x + 1) + 12x = 6x(x + 1)$$

Expand the brackets and rearrange.

$$9x + 9 + 12x = 6x^2 + 6x$$
$$0 = 6x^2 - 15x - 9$$

This factorises to

$$0 = 3(x - 3)(2x + 1)$$
$$x = 3 \text{ or } x = -\tfrac{1}{2}$$

EXERCISE 2A

1 Solve the following equations.

a) $x^2 - 8x + 15 = 0$ b) $x^2 - 7x + 12 = 0$ c) $x^2 + 3x - 28 = 0$

d) $x^2 - 4x - 45 = 0$ e) $2x^2 - 5x + 2 = 0$ f) $6x^2 + 11x + 3 = 0$

g) $x^2 - 25 = 0$ h) $x^2 - 5x = 0$ i) $2x^2 - 18 = 0$

2 Solve the following equations.

a) $x^2 = 5x - 6$ b) $x^2 - 8x = 20$ c) $7 - 3x^2 = -4x$

d) $x + \dfrac{6}{x} = 5$ e) $\dfrac{1}{x - 1} + \dfrac{6}{x + 1} = 3$ f) $\dfrac{3x + 4}{x} = \dfrac{x + 3}{x - 1}$

3 The length of a rectangle is 1 metre greater than the width. The area of the rectangle is 72 m². Letting the width be x m, show that $x^2 + x - 72 = 0$. Solve this equation.

4 The area of a triangle is 28 m². If the sum of the base and the height is 15 m, find the base.

5 A man takes £600 on holiday. If he had spent £5 less each day the holiday could have lasted four more days. How long did the holiday last?

Solving cubics

A **cubic** equation is of the form $ax^3 + bx^2 + cx + d = 0$, where a, b, c and d are constants, with a not equal to zero. We may be able to factorise the left-hand side by the methods covered in Chapter 1. Then we shall be able to solve the equation.

EXAMPLE 2.3
Solve this equation.

$$x^3 - 6x^2 + 3x + 10 = 0$$

Solution
By using the factor theorem (see page 19), we find that $(x + 1)$ is a factor of the left-hand side. We can then divide by $(x + 1)$, to find a quotient of $x^2 - 7x + 10$.

This quotient can be factorised to $(x - 2)(x - 5)$. Now the solutions can be written down.

The solutions are $x = -1$, $x = 2$ or $x = 5$.

EXERCISE 2B

Solve the following equations.

1 $x^3 - 7x + 6 = 0$

2 $x^3 - 6x^2 - x + 6 = 0$

3 $x^3 + 10x^2 + 31x + 30 = 0$

4 $x^3 - 6x^2 - x + 30 = 0$

5 $2x^3 + 5x^2 + x - 2 = 0$

6 $3x^3 - 5x^2 - 16x + 12 = 0$

2.2 Completing the square

Sometimes a quadratic expression can be factorised as a perfect square. For example

$$x^2 - 4x + 4 \equiv (x - 2)^2$$

Consider $x^2 - 4x + 3$. This is 1 less than the expression above.

$$x^2 - 4x + 3 \equiv x^2 - 4x + 4 - 1 \equiv (x - 2)^2 - 1$$

The procedure of writing a quadratic expression as a perfect square with a constant added is known as **completing the square**.

Because $(x-2)^2$ is a perfect square it cannot be negative. Therefore, because $(x-2)^2$ is always greater than or equal to 0, the expression $(x-2)^2 - 1$ is always greater than or equal to -1.

The minimum or least value of $x^2 - 4x + 3$ is -1, achieved at $x = 2$. Note that the graph of this function has a lowest point at $(2, -1)$.

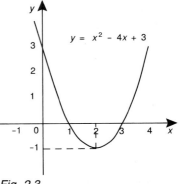

Fig. 2.3

Expressions of the form $x^2 + bx + c$

We complete the square for expressions like this by considering the expansion of $(x + \frac{1}{2}b)^2$, and then adjusting the constant. The expansion is

$$(x + \tfrac{1}{2}b)^2 = x^2 + 2 \times \tfrac{1}{2}bx + (\tfrac{1}{2}b)^2 = x^2 + bx + (\tfrac{1}{2}b)^2.$$

So we add c (which we *do* want) and subtract $(\frac{1}{2}b)^2$ (which we *don't* want).

$$x^2 + bx + c = (x + \tfrac{1}{2}b)^2 + c - (\tfrac{1}{2}b)^2$$

EXAMPLE 2.4
By completing the square, find the least value of $x^2 + 3x - 1$.

Solution
Consider the expansion of $(x + \frac{3}{2})^2$.

$$(x + \tfrac{3}{2})^2 \equiv x^2 + 3x + (\tfrac{3}{2})^2$$

This is $3\frac{1}{4}$ greater than the original expression.

$$x^2 + 3x - 1 \equiv (x + \tfrac{3}{2})^2 - 3\tfrac{1}{4}$$

This is least when $(x + \frac{3}{2})^2 = 0$, at $x = -\frac{3}{2}$ or $-1\frac{1}{2}$.

The least value of $x^2 + 3x - 1$ is $-3\frac{1}{4}$.

EXERCISE 2C

1 Complete the square for each of the following expressions.

  a) $x^2 + 2x - 3$ **b)** $x^2 - 4x + 2$ **c)** $x^2 - 8x + 11$

  d) $x^2 + 5x - 1$ **e)** $x^2 - 3x - 7$ **f)** $x^2 + x + 1$

2 Find the least values of the expressions in Question 1, and the corresponding values of x.

*3 Find the completed square form of some of the quadratic expressions in Exercise 2A. What do they have in common?

Expressions of the form $ax^2 + bx + c$

An expression of this form can be reduced to the simpler form above by taking out a factor of a.

This also applies when a is negative. In this case, the expression will have a maximum or greatest value instead of a least value. Note that the graph has a lowest point if $a > 0$, and a highest point if $a < 0$.

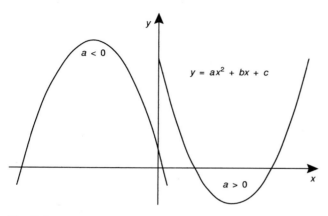

Fig. 2.4

EXAMPLE 2.5
Complete the square of $2x^2 + 4x + 3$.

Solution
Take out a factor of 2, to obtain

$$2(x^2 + 2x + 1.5)$$

Now complete the square of the expression inside the brackets.

$$2x^2 + 4x + 3 = 2[(x + 1)^2 + 0.5]$$

EXAMPLE 2.6
By completing the square, find the greatest value of $6 - 3x - x^2$.

Solution
Take the minus sign outside the brackets, to obtain

$$-(x^2 + 3x - 6)$$

Now complete the square of the expression inside the brackets.

$$6 - 3x - x^2 \equiv -[(x + 1.5)^2 - 8.25] \equiv 8.25 - (x + 1.5)^2$$

Since $(x + 1.5)^2$ is always greater than or equal to zero, the greatest value of this expression is when $x = -1.5$, when the square term becomes zero.

The greatest value of $6 - 3x - x^2$ is 8.25.

EXERCISE 2D

1 Complete the square for the following expressions.

a) $3x^2 + 6x - 5$ b) $3x^2 - 9x + 2$ c) $2x^2 - 4x - 7$

d) $2x^2 + 3x - 5$ e) $3x^2 - x + 3$ f) $5x^2 - 2x + 3$

g) $3 - 2x - x^2$ h) $5 - x - x^2$ i) $-7 + 3x - x^2$

j) $5 - 4x - 2x^2$ k) $5 + 3x - 3x^2$ l) $7 - x - 2x^2$

2 Find the least or greatest values of the expressions in Question 1.

*3 Take a general quadratic $ax^2 + bx + c$. Find an expression for its completed square form.

4 A stone is thrown up in the air. After t seconds, its height h m above ground is given by

$$h = 2 + 30t - 5t^2$$

By completing the square of this expression, find the greatest height it reaches.

*5 I am 50 m south of a crossroads, walking north at 2 m s^{-1}. A friend is at the crossroads, walking east at 4 m s^{-1}. Find expressions for our distances from the crossroads t seconds later. Use Pythagoras's theorem to find an expression for our distance apart d m, and show that it reduces to

$$d = \sqrt{20t^2 - 200t + 2500}$$

By completing the square of the expression inside the square root sign, find the least distance between us.

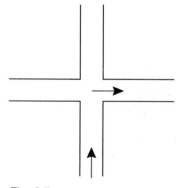

Fig. 2.5

*6 A farmer has 200 m of fencing with which to make a rectangular sheep pen. One side of the pen is a stone wall. If he uses x m for each of the sides perpendicular to the wall, find an expression for the other side.

Find an expression for the area A m^2, and show that it reduces to

$$A = 200x - 2x^2$$

By completing the square of this expression, find the greatest possible area the farmer can enclose.

Solution of equations

Another method of solving a quadratic equation is by completing the square. This method is more powerful than the factorisation method. It will allow us to find solutions even when the quadratic is not easily factorised.

EXAMPLE 2.7
Solve the equation $x^2 + 3x - 3 = 0$ by completing the square.
Give your answers correct to three decimal places.

Solution
Complete the square of the left-hand side.

$(x + 1.5)^2 - 5.25 = 0$

$(x + 1.5)^2 = 5.25$

Now take square roots of both sides. The square root of 5.25 could be
either positive or negative.

$x + 1.5 = 2.291$ or $x + 1.5 = -2.291$

Finally subtract 1.5 to obtain the solutions.

$x = 0.791$ or $x = -3.791$

EXERCISE 2E

Where relevant, give your answers correct to three significant figures.

1 Solve the following equations by completing the square.

a) $x^2 + 2x - 5 = 0$ **b)** $x^2 + 3x + 1 = 0$ **c)** $x^2 + 5x - 3 = 0$

d) $2x^2 - 8x + 1 = 0$ **e)** $3 - 8x - x^2 = 0$ **f)** $7 + 2x - 3x^2 = 0$

*2 I am 100 m due south of a crossroads, and walking north at 1 m s^{-1}. A friend is at the crossroads, and is walking east at 1 m s^{-1}. Find expressions for our distances from the crossroads t seconds later. Use Pythagoras's theorem to find an expression for our distance d m apart after t seconds, and show that it reduces to

$$d = \sqrt{2t^2 - 200t + 10\,000}$$

Find the times when we are exactly 80 m apart.

3 Find the value of p so that the equation $x^2 - 4x + p = 0$ has equal roots.

*4 The general quadratic equation is $ax^2 + bx + c = 0$. Find an expression for the solutions in terms of a, b and c.

2.3 The formula for solving quadratic equations

The method of completing the square gives a set of rules for
solving a quadratic equation. If these rules are applied to the
general quadratic equation, we obtain a formula.

Formula

If $ax^2 + bx + c = 0$, then $x = \dfrac{-b \pm \sqrt{b^2 - 4ac}}{2a}$

Proof

First divide the general equation, $ax^2 + bx + c = 0$, through by a.

$$x^2 + \frac{b}{a}x + \frac{c}{a} = 0$$

Complete the square of the left-hand side.

$$\left(x + \frac{b}{2a}\right)^2 + \frac{c}{a} - \frac{b^2}{4a^2} = 0$$

Hence,

$$\left(x + \frac{b}{2a}\right)^2 = \frac{b^2}{4a^2} - \frac{c}{a}$$

$$= \frac{b^2 - 4ac}{4a^2}$$

Take the square root of both sides. Note that the root could be positive or negative.

$$x + \frac{b}{2a} = \pm\sqrt{\left(\frac{b^2 - 4ac}{4a^2}\right)} = \frac{\pm\sqrt{b^2 - 4ac}}{2a}$$

Finally, subtract $\dfrac{b}{2a}$ from both sides and the formula is obtained.

$$x = -\frac{b}{2a} \pm \frac{\sqrt{b^2 - 4ac}}{2a} = \frac{-b \pm \sqrt{b^2 - 4ac}}{2a}$$

The formula involves taking the square root of the expression $b^2 - 4ac$. If this expression is negative we shall not be able to find the square root, and no solutions exist. This corresponds to the situation shown, where the graph of the quadratic expression does not cross the x-axis.

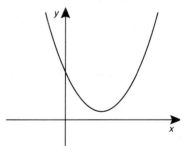

Fig. 2.6

EXAMPLE 2.8

Solve the equation $2x^2 - 3x - 7 = 0$, giving your answers correct to three decimal places.

Solution

Here $a = 2$, $b = -3$, $c = -7$. Put these values into the formula.

$$x = \frac{-(-3) \pm \sqrt{(-3)^2 - 4 \times 2 \times (-7)}}{2 \times 2}$$

$$x = \frac{3 \pm \sqrt{9 + 56}}{4}$$

$$= \frac{3 \pm \sqrt{65}}{4}$$

$$= \frac{3 \pm 8.062}{4}$$

$$x = 2.766 \text{ or } x = -1.266$$

EXERCISE 2F

Where relevant, give your answers correct to three significant figures.

1 Solve the following equations.

a) $3x^2 + 5x + 1 = 0$ b) $x^2 - 4x - 7 = 0$ c) $x^2 + x - 9 = 0$

d) $4x^2 + 5x - 3 = 0$ e) $x^2 + 7x + 3 = 0$ f) $x^2 - 5x + 2 = 0$

2 A stone is thrown in the air. After t seconds its height is given by $h = 3 + 40t - 5t^2$. Find t when it is 60 m high.

*3 This is another way of obtaining the quadratic formula. First write out the expansion of

$$(2ax + b)^2$$

Multiply the general quadratic equation $ax^2 + bx + c = 0$ by $4a$, to obtain

$$4a^2x^2 + 4abx + 4ac = 0$$

By combining the two expressions obtain the quadratic formula.

*4 Show that the solution of the equation $x^2 + 2bx + c = 0$ is

$$x = -b \pm \sqrt{b^2 - c}$$

A method for solving quadratics has been known for many years. The next four questions are concerned with problems which were written in Babylonia about 4000 years ago.

5 The area of a square less the side of the square is 870. Find the side.

Solve this problem by any of the methods of this chapter.

6 The solution to Question 5 is given as, 'Take half of 1, which is $\frac{1}{2}$, multiply it by itself, which gives $\frac{1}{4}$. Now add $\frac{1}{4}$ to 870, which is 870.25. This is the square of 29.5. Now add $\frac{1}{2}$ to 29.5. The result is 30, which is the side of the square.' Check the arithmetic of this solution. Does the answer agree with your answer for Question 5?

*7 Babylonian mathematicians seem to have used $x^2 - px = q$ as one standard form of the quadratic. In the example above $p = 1$ and $q = 870$. Follow through the method of Question 6 to find a general formula for the solution in terms of p and q. Check your formula on some other quadratics.

*8 Another Babylonian problem is, 'An area A consisting of the sum of two squares is 1000. The side of one square is 10 less than $\frac{2}{3}$ of the side of the other square. What are the sides of these squares?'

9 Show that $(x - 1)$ is a factor of $x^3 - 4x^2 + 4x - 1$. Find the other factor, and hence solve the equation $x^3 - 4x^2 + 4x - 1 = 0$.

10 Solve the following equations.

  a) $x^3 - 6x^2 - 2x + 7 = 0$ **b)** $x^3 - 3x^2 - 3x + 10 = 0$ **c)** $2x^3 + 3x^2 - 3x - 4 = 0$

2.4 Other equations reducible to quadratic equations

Sometimes other equations can be reduced to the form of a quadratic equation.

Concealed quadratics

Some equations which are not quadratic can be changed into a quadratic equation by means of a substitution.

EXAMPLE 2.9
Solve the equation $x^4 + 3x^2 - 5 = 0$.

Solution
This seems to be a **quartic** (fourth degree) equation, but notice there is no x^3 or x term. If we put $y = x^2$ the equation is reduced to a quadratic in y.

$$y^2 + 3y - 5 = 0$$

By completing the square or by the formula, we can solve this equation.

$$y = 1.193 \text{ or } y = -4.193$$

Now we can put it back in terms of x.

$$x^2 = 1.193 \text{ or } x^2 = -4.193$$

Now x^2 cannot be negative, so the only solutions come from the square root of 1.193.

$$x = 1.092 \text{ or } x = -1.092$$

EXERCISE 2G

1 Solve each of the following equations by the substitution indicated.

a) $x^4 - 4x^2 + 1 = 0 \ (u = x^2)$

b) $x^6 + 4x^3 - 2 = 0 \ (u = x^3)$

c) $(2x + 1)^2 - 3(2x + 1) - 7 = 0 \ (u = 2x + 1)$

d) $\dfrac{3}{x^2} - \dfrac{4}{x} + 1 = 0 \ \left(u = \dfrac{1}{x}\right)$

e) $x - 8\sqrt{x} + 6 = 0 \ (u = \sqrt{x})$

f) $(1 + 2x) - 4\sqrt{1 + 2x} + 3 = 0 \ (u = \sqrt{1 + 2x})$

2 Solve the following equations.

a) $x^4 + 7x^2 - 30 = 0$

b) $(1 - x)^4 - 5(1 - x)^2 + 6 = 0$

c) $\dfrac{2}{x + 1} = \dfrac{x + 1}{8}$

d) $\dfrac{1}{x^4} - \dfrac{9}{x^2} + 18 = 0$

e) $x + 3\sqrt{x} - 88 = 0$

f) $x + \sqrt{x} = 1$

*3 Solve the equation: $x(x + 1) + \dfrac{6}{x(x + 1)} = 5$

*4 Solve the equation $x^4 - 3x^3 + 2x^2 - 3x + 1 = 0$ by dividing through by x^2, and then making the substitution $u = x + \dfrac{1}{x}$.

*5 Solve the equation: $x^4 + 5x^3 + x^2 + 5x + 1 = 0$

Simultaneous equations

Suppose we have two simultaneous equations in x and y, such as $x^2 + y^2 = 3$ and $x + 2y = 1$. A procedure to solve them is as follows.

Use one of the equations to express y in terms of x (or x in terms of y).

Substitute into the other equation, to obtain an equation in x alone (or y alone).

Solve this equation. If it is quadratic then we can try to solve it by the methods of this chapter.

Put our solutions back in the first equation, to find the corresponding values of the other variable.

EXAMPLE 2.10

Solve the simultaneous equations $2x + y = 5$ and $xy = 3$.

Solution

We could use either of these equations to write y in terms of x. Using the second equation, we obtain $y = \dfrac{3}{x}$. Substitute this into the first equation.

$$2x + \frac{3}{x} = 5$$

Multiply through by x and rearrange.

$$2x^2 - 5x + 3 = 0$$

By factorisation or by the formula, solve this equation.

$$x = 1.5 \text{ or } x = 1$$

Find the corresponding values of y, from $y = \dfrac{3}{x}$.

Either $x = 1.5$ and $y = 2$, or $x = 1$ and $y = 3$.

EXERCISE 2H

1 Solve the following pairs of simultaneous equations.

a) $x - 3y = 4$ and $xy = 2$ **b)** $x^2 + y^2 = 3$ and $y = 2x + 1$ **c)** $y^2 = 3x$ and $2y + 3x = 4$

d) $y + 2x = 3$ and $x^2 + 3y^2 = 7$ **e)** $2y + 3x = 13$ and $xy = 2$ **f)** $\dfrac{x}{2} + \dfrac{y}{3} = 2$ and $x^2 + y^2 = 20$

g) $xy = 7$ and $\dfrac{1}{x} + \dfrac{1}{y} = 4$ **h)** $x + y = 3$ and $x^2 + 2xy - y^2 = 4$

*2 Solve the equations $x^2 + y^2 = 10$ and $yx^2 = 9$.

3 A line has equation $y = 2x + 1$ and a curve has equation $x^2 + 3xy + 3y^2 = 12$. Find the coordinates of the points of intersection.

*4 The equation $y = x^2$ and $y = mx + c$ represent a curve and a straight line respectively. Show that the line is a tangent to the curve if and only if $m^2 + 4c = 0$.

LONGER EXERCISE

Solution by iteration

This chapter has contained many methods of solving quadratic equations. Another method, which can be generalised to more complicated equations, is **iteration**.

The method obtains a succession of numbers x_1, x_2, x_3 and so on. The first number is the starting point x_1. Then we rewrite the equation to find a rule to give x_2 in terms of x_1, x_3 in terms of x_2, and in general x_{n+1} in terms of x_n.

Take the equation $x^2 + 3x - 2 = 0$. Two ways of rewriting this equation are

a) $x = -3 + \dfrac{2}{x}$ (by subtracting $3x - 2$ from both sides and dividing by x)

b) $x = -\dfrac{x^2}{3} + \dfrac{2}{3}$ (by subtracting $x^2 - 2$ from both sides and dividing by 3).

Either of (a) or (b) can be used for an iteration. For (a), start with $x_1 = 0.5$, and then use the iteration

$$x_{n+1} = -3 + \frac{2}{x_n}$$

So the next two values are

$$x_2 = -3 + \frac{2}{0.5} = 1$$

$$x_3 = -3 + \frac{2}{1} = -1$$

Continue the process. Do the numbers approach one of the solutions of the equation?

What happens with (b)? What happens if you change the starting point?

A computer can perform all the calculations involved in iteration. If you have access to a spreadsheet the exercise on page 171 will take the investigation further.

EXAMINATION QUESTIONS

1 Solve the simultaneous equations

$$4x^2 + y^2 = 25$$
$$xy = 6$$

giving all possible pairs of values of x and y.   *W AS*

2 Express $6 - 4x - x^2$ in the form $p - (q + x)^2$, where p and q are constants.

Hence, or otherwise, find the maximum value of $6 - 4x - x^2$.   *C AS*

3 The polynomial $x^5 - 3x^4 + 2x^3 - 2x^2 + 3x + 1$ is denoted by f(x).

(i) Show that neither $(x - 1)$ nor $(x + 1)$ is a factor of f(x).
(ii) By substituting $x = 1$ and $x = -1$ in the identity

$$f(x) \equiv (x^2 - 1)q(x) + ax + b$$

where q(x) is a polynomial and a and b are constants, or otherwise, find the remainder when f(x) is divided by $(x^2 - 1)$.

(iii) Show, by carrying out the division, or otherwise, that when f(x) is divided by $(x^2 + 1)$, the remainder is $2x$.

(iv) Find all the real roots of the equation f(x) = $2x$.   *C*

Summary and key points

1 If $(x - \alpha)(x - \beta) = 0$, then $x = \alpha$ or $x = \beta$.

2 To complete the square of $x^2 + bx + c$, consider the expansion of $\left(x + \frac{1}{2}b\right)^2$.

To complete the square of $ax^2 + bx + c$, take out a factor of a and proceed as above.

The least or greatest value of the expression is found by putting the squared term equal to zero.

3 The solutions of $ax^2 + bx + c = 0$ are

$$x = \frac{-b \pm \sqrt{b^2 - 4ac}}{2a}$$

Be careful with negative numbers. For example, when solving $x^2 - 3x - 5 = 0$, be sure to put $b = -3$ and $c = -5$.

When using an ordinary scientific calculator to evaluate the formula, be sure to press $=$ after entering the top line.

4 Some equations can be converted to quadratic equations by a substitution.

If we have two equations in x and y, use one equation to express y in terms of x (or x in terms of y) and then substitute into the other equation.

When you have two sets of answers, make plain which value of y goes with which value of x.

CHAPTER 3

Trigonometric functions

Chapter outline

3.1 The functions sine, cosine and tangent

3.2 Special angles

3.3 The functions cosecant, secant and cotangent

3.4 Trigonometric functions for all angles

3.5 Graphs of trigonometric functions

The trigonometric functions sine, cosine and tangent are used to find the sides and angles of right-angled triangles. A scientific calculator has buttons to evaluate these functions for specific angles.

Below there are several triangles. They are all right-angled, and their sides are all in the ratio 3:4:5

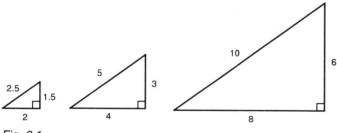

Fig. 3.1

All these triangles are similar, so they have the same angles. The angles of the triangles are determined by the ratios of the sides, which are called **trigonometric** (angle measuring) ratios or functions. These functions can be used to evaluate the angles.

In fact, these functions are useful in many areas besides the solution of triangles. Trigonometric functions can be used to represent the behaviour of the tides, alternating electric current, vibrating strings and many other phenomena, even though these may seem to have no connection with triangles and angles.

3.1 The functions sine, cosine and tangent

Look at the triangle on the right. The longest side, the side farthest from the right-angle, is always called the **hypotenuse**. For angle P, the side farthest away is its **opposite** side, and the side next to P is its **adjacent** side.

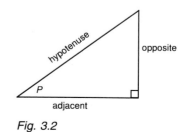

Fig. 3.2

Finding sides

The three basic trigonometric functions, which appear on any scientific calculator, are defined for this triangle as follows.

$$\sin P = \frac{\text{OPP}}{\text{HYP}} \qquad \cos P = \frac{\text{ADJ}}{\text{HYP}} \qquad \tan P = \frac{\text{OPP}}{\text{ADJ}}$$

Note that the words sine, cosine and tangent are usually abbreviated to sin, cos and tan.

If we know one side and the angles of a right-angled triangle, we can find the other sides.

Finding angles

If we know the sides of a right-angled triangle we can find the angles. For example, in the triangles of Fig. 3.1 the sides are all in the ratio 3:4:5. The smallest angle of the triangle can be found using the **arcsin** or **sin^{-1}** function.

$$P = \sin^{-1} \frac{3}{5} = 37°$$

cos^{-1} and **tan^{-1}** are defined similarly. They all give the same answer for P.

$$P = \sin^{-1} \tfrac{3}{5} = \cos^{-1} \tfrac{4}{5} = \tan^{-1} \tfrac{3}{4}$$

EXAMPLE 3.1

You need to find the height of a tree. You measure out 50 metres from its base, and from there you measure the angle of elevation of the top as 28°. How high is the tree?

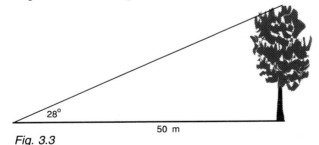

28°

50 m

Fig. 3.3

Solution

The situation is represented by the triangle of Fig. 3.3. The height of the tree is the length of the **opposite** side. The distance of 50 m is the length of the **adjacent** side. The ratio which connects opposite and adjacent is tan.

$$\tan 28° = \frac{\text{height}}{50}$$

The height is $50 \times \tan 28° = 26.6$ m.

EXAMPLE 3.2

In the triangle shown, $B = 90°$, $AC = 7$ cm, $AB = 4$ cm. Find A.

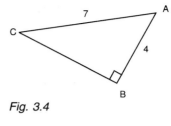

Solution

AC is the hypotenuse of ABC. Relative to angle A, AB is the adjacent side. The ratio which connects hypotenuse and adjacent side is cos.

$\cos A = \frac{4}{7}$

$A = \cos^{-1}\frac{4}{7} = 55°$

So A is 55°.

Fig. 3.4

EXERCISE 3A

1 Find the unknown side in each of the triangles below. Assume all lengths are in cm.

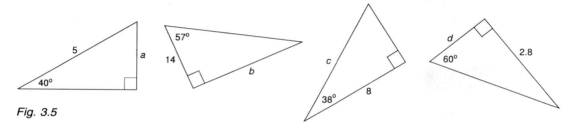

Fig. 3.5

2 Find the unknown angle in each of the triangles below. Assume all lengths are in cm.

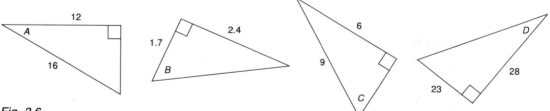

Fig. 3.6

3 Triangle ABC has $B = 90°$, $AC = 10$ cm and $A = 40°$. Find AB and BC.

4 Triangle PQR has $P = 90°$, $PR = 5$ cm and $QR = 7$ cm. Find Q.

5 I am 5000 m away from the base of a mountain, which is 1000 m higher than I am. Find the angle of elevation of its summit.

6 A 12 ft plank leans against a wall at an angle of 40° to the horizontal. How high up the wall does the plank reach?

7 In the diagram ABC is an isosceles triangle, with $AB = AC = 10$ cm, and $BC = 6$ cm. By splitting ABC into two right-angled triangles, find B.

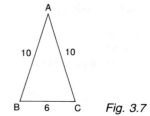

Fig. 3.7

8 Find the unknown sides and angles in the triangles below. Assume all lengths are in cm.

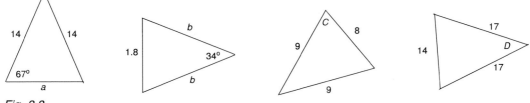

Fig. 3.8

9 By considering the definitions of sin, cos and tan, show that $\tan P = \dfrac{\sin P}{\cos P}$.

10 By considering the definitions of sin and cos, show that $\sin P = \cos(90° - P)$.

11 Find a relationship between $\tan P$ and $\tan(90° - P)$.

3.2 Special angles

There are some angles for which the trigonometric functions can be found without using a calculator or tables.

45°

Suppose a triangle is isosceles as well as right-angled. Then the two smaller angles are each 45°.

If the shorter sides are of length 1 unit, the hypotenuse will be, by Pythagoras's theorem,

$$\sqrt{1^2 + 1^2} = \sqrt{2} \text{ units}$$

We can now write down the trigonometric ratios for this triangle.

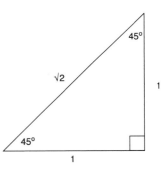

Fig. 3.9

$$\sin 45° = \frac{1}{\sqrt{2}} \qquad \cos 45° = \frac{1}{\sqrt{2}} \qquad \tan 45° = 1$$

30° and 60°

Consider an equilateral triangle ABC with each side of length 2 units. Convert it into two right-angled triangles by dropping a perpendicular AD.

By Pythagoras's theorem, $AD = \sqrt{2^2 - 1^2} = \sqrt{3}$ units. Also, $B = 60°$ and $B\widehat{A}D = 30°$. The trigonometric ratios for 60° and 30° can be written down.

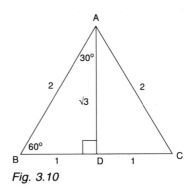

Fig. 3.10

$$\sin 30° = \frac{1}{2} \qquad \cos 30° = \frac{\sqrt{3}}{2} \qquad \tan 30° = \frac{1}{\sqrt{3}}$$

$$\sin 60° = \frac{\sqrt{3}}{2} \qquad \cos 60° = \frac{1}{2} \qquad \tan 60° = \sqrt{3}$$

0° and 90°

For the sake of completeness the trigonometric functions are defined for 0° and 90°, even though a figure with angles of 0°, 90° and 90° will consist of a line segment.

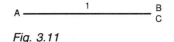

Fig. 3.11

In the 'triangle' of Fig. 3.11, $A = 0°$, $B = C = 90°$, $AB = AC = 1$ unit and $BC = 0$. The ratios can be written down.

$$\sin 0° = 0 \quad \cos 0° = 1 \quad \tan 0° = 0 \quad \sin 90° = 1 \quad \cos 90° = 0$$

Note that $\tan 90° = \frac{1}{0}$, which is not defined.

EXERCISE 3B

1 Use your calculator to verify the formulae above for the angles 0°, 30°, 45°, 60°.

2 Simplify the following expressions without the use of a calculator.

a) $\sin 60° - \cos 30°$ **b)** $\sin 60° \cos 30° + \cos 60° \sin 30°$ **c)** $(\sin 45°)^2 + (\cos 45°)^2$

3 In Questions 9, 10 and 11 of Exercise 3A, you found relationships between sin, cos and tan. Verify that they hold for the angles given in this section.

4 Try to find $\tan 90°$. What happens?

5 Enter a very large number into your calculator, and find its $\tan^{-1}$. How close is this to 90°?

Summary

	0°	30°	45°	60°	90°
$\sin x$	0	$\frac{1}{2}$	$\frac{1}{\sqrt{2}}$	$\frac{\sqrt{3}}{2}$	1
$\cos x$	1	$\frac{\sqrt{3}}{2}$	$\frac{1}{\sqrt{2}}$	$\frac{1}{2}$	0
$\tan x$	0	$\frac{1}{\sqrt{3}}$	1	$\sqrt{3}$	not defined

3.3 The functions cosecant, secant and cotangent

There are three other ratios between the sides of a right-angled triangle.

$$\text{cosec } P = \frac{\text{HYP}}{\text{OPP}} \qquad \sec P = \frac{\text{HYP}}{\text{ADJ}} \qquad \cot P = \frac{\text{ADJ}}{\text{OPP}}$$

The functions cosecant, secant and cotangent are usually abbreviated to cosec, sec and cot.

These functions do not appear on your calculator. Notice that in the definition of cosec, the sides involved are the same as for sin, but turned upside-down. A similar relationship holds for sec and cot.

$$\text{cosec } P = \frac{1}{\sin P} \qquad \sec P = \frac{1}{\cos P} \qquad \cot P = \frac{1}{\tan P}$$

So if we want to find the cosec of an angle, we find the sin and then find its reciprocal, by dividing it into 1. The $1/x$ button on a calculator is very useful for this.

EXERCISE 3C

1 Find the following.

  a) cot 56°　　　　　　　**b)** cosec 43°　　　　　　　**c)** sec 17°

2 In each of the following find the value of x.

  a) cosec $x = 3$　　　　**b)** cot $x = 0.7$　　　　**c)** sec $x = 1.8$

3 Find the cosec, sec and cot for the angles of 30°, 45° and 60°.

4 For what angles are cosec, sec and cot not defined?

3.4 Trigonometric functions for all angles

Angles greater than 90⁰

So far we have defined trigonometric functions in terms of ratios within a right-angled triangle. The angles in a right-angled triangle cannot be greater than 90°, but for many of the other applications of trigonometry we need to define the functions for all angles.

Figure 3.12 shows a wheel of radius 1 unit. Put the wheel on graph paper, with its centre at the origin.

The spoke turns through an angle P. The x-coordinate of the end of the spoke is cos P, and the y-coordinate is sin P.

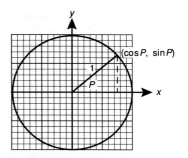

Trigonometric functions for angles bigger than 90° are defined to follow this pattern. When the wheel has turned through an angle P, however large, cos P is the x-coordinate and sin P is the y-coordinate. The ratio $\dfrac{y}{x}$ is tan P.

Fig. 3.12

Angles between 0° and 360°

Below are diagrams showing the position of the wheel when the end of the spoke is in the four quadrants. In each case we shall consider the angle between the spoke and the x-axis. This will be an acute angle, for which the trigonometric ratios are already defined.

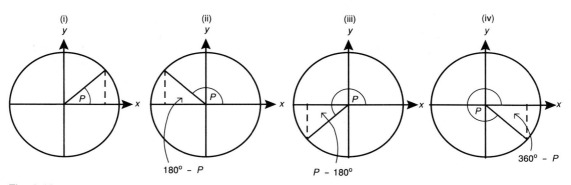

Fig. 3.13

For $90° < P < 180°$ (position (ii)), the angle between the spoke and the x-axis is $180° - P$.

This is an acute angle. The x- and y-coordinates of the end are given in terms of the cos and sin of $180° - P$. Note that the y-coordinate is positive and the x-coordinate is negative.

$$\sin P = \sin(180° - P)$$
$$\cos P = -\cos(180° - P)$$
$$\tan P = -\tan(180° - P)$$

For $180° < P < 270°$ (position (iii)), the angle between the spoke and the x-axis is $P - 180°$. Note that the y-coordinate is negative and the x-coordinate is also negative.

$$\sin P = -\sin(P - 180°)$$
$$\cos P = -\cos(P - 180°)$$
$$\tan P = \tan(P - 180°)$$

For $270° < P < 360°$ (position (iv)), the angle between the spoke and the x-axis is $360° - P$. Note that the y-coordinate is negative and the x-coordinate is positive.

$$\sin P = -\sin(360° - P)$$
$$\cos P = \cos(360° - P)$$
$$\tan P = -\tan(360° - P)$$

In the first quadrant (position (i)), all the functions are positive.
In the second quadrant (position (ii)), only sin is positive.
In the third quadrant (position (iii)), only tan is positive.
In the fourth quadrant (position (iv)), only cos is positive.

Figure 3.14 shows which functions are positive in which quadrant.

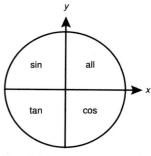

Fig. 3.14

Negative angles

The spoke of the wheel has been turning in an anticlockwise direction. If it turns clockwise, then the angle through which it turns is negative.

Notice that in the diagram, when the spoke has turned through angle $-P$, the y-coordinate is negative and the x-coordinate positive. Hence we have these rules.

$$\sin(-P) = -\sin P$$
$$\cos(-P) = \cos P$$
$$\tan(-P) = -\tan P$$

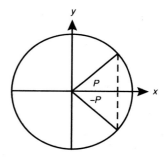

Fig. 3.15

Angles greater than 360°

After the wheel has rotated through a full circle, the coordinates repeat themselves. So we have

$$\sin P = \sin(P - 360°)$$
$$\cos P = \cos(P - 360°)$$
$$\tan P = \tan(P - 360°)$$

This process of subtracting 360° can be repeated as often as we please. We could also add 360°, by turning the wheel in the other direction, and again the functions would be unchanged. So for example

$$\sin P = \sin(P - 720°)$$
$$\cos P = \cos(P + 360°)$$
$$\tan P = \tan(P + 1080°)$$

Cosec, sec and cot

The functions cosec, sec and cot are the reciprocals of sin, cos and tan, respectively, so they will obey the same rules as sin, cos and tan respectively. For example, the following hold.

$$\csc P = \csc (180° - P)$$
$$\sec P = -\sec (P - 180°)$$
$$\cot P = -\cot (360° - P)$$

EXAMPLE 3.3
Express these in terms of the ratios of acute angles.
a) $\sin 210°$ **b)** $\cos (-100°)$ **c)** $\cot 700°$

Solution
a) Note that this angle lies between 180° and 270°. Apply the rule above, by subtracting 180°.

$$\sin 210° = -\sin 30°$$

b) First note that we have a negative angle. Apply the rule for negative angles.

$$\cos (-100°) = \cos 100°$$

Note that the angle of 100° lies between 90° and 180°. Apply the rule above, by subtracting from 180°.

$$\cos (-100°) = -\cos 80°$$

c) This angle is greater than 360°. Subtract 360°.

$$\cot 700° = \cot 340°$$

This angle now lies between 270° and 360°, and cot obeys the same rule as tan, so apply the rule above, by subtracting from 360°.

$$\cot 700° = -\cot 20°$$

EXAMPLE 3.4
Find two angles in the range $0° < x < 360°$ for which $\cos x = -0.5$.

Solution
From the results above, there will be solutions in the second and third quadrants. We know that $\cos 60° = +0.5$. Apply the rule for the second quadrant.

$$\cos x = -\cos (180° - x)$$

Putting $180° - x = 60°$, we obtain $x = 120°$. This is one of the solutions.

For the third quadrant, apply the rule

$$\cos x = -\cos (x - 180°)$$

Putting $x - 180° = 60°$, we obtain $x = 240°$. This is the second solution.

The solutions are 120° and 240°.

EXERCISE 3D

1 Express the following in terms of the ratios of acute angles.

a) sin 120°   b) cos 135°   c) tan 98°   d) cot 150°   e) sec 176°

f) cosec 160°   g) cos 250°   h) tan 310°   i) sin 190°   j) sec 340°

k) cosec 318°   l) cot 247°   m) sin 370°   n) cos 460°   o) tan 750°

p) sin (−32°)   q) sec (−30°)   r) cot (−70°)   s) sin (−300°)   t) tan (−500°)

u) cos (−120°)

2 Find the following without the aid of a calculator.

a) sin 135°   b) cos 150°   c) tan 120°   d) cosec 150°   e) cot 120°

f) sec 180°   g) sin 315°   h) tan 210°   i) sin 300°   j) cosec 225°

k) cot 315°   l) sec 300°

3 You are given that x is an angle between 0° and 360°, for which sin $x = -0.6$ and cos $x = 0.8$. Which quadrant is x in? Find the value of x.

4 Find the angles between 0° and 360° for which the following are true.

a) sin $x = 0.6$ and cos $x = -0.8$ b) tan $x = 1$ and cos $x = -\dfrac{1}{\sqrt{2}}$

5 You are given that sin $x = -$sin 20° and cos $x = -$cos 20°. What could x be?

6 For each of the following equations, find two solutions in the range 0° $< x <$ 360°.

a) sin $x = -0.5$ b) tan $x = -1$ c) cos $x = 0.5$

d) cosec $x = 3$ e) sec $x = -4$ f) cot $x = 0.3$

3.5 Graphs of trigonometric functions

Graphs of sin and cos

The graph of $y = \sin x$ is shown in Fig. 3.16. Note that the shape confirms the rules above.

sin x is positive for 0° $< x <$ 180°
sin x is negative for 180° $< x <$ 360°
sin x repeats itself after 360°.

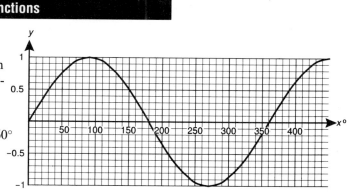

Fig. 3.16

The graph of $y = \cos x$ is shown in Fig. 3.17.

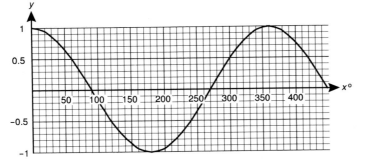

Fig. 3.17

Note that its shape is similar to that of $y = \sin x$. In fact, the cos graph can be obtained from the sin graph by shifting it 90° to the left.

$$\cos x = \sin(x + 90°)$$

Both graphs take the shape of a wave, so they can be used to illustrate the behaviour of anything that goes to and fro like a wave. For example

a) waves on the ocean
b) sound waves
c) the swing of a pendulum
d) vibration of a weight on the end of a spring
e) alternating electric current.

EXERCISE 3E

1 Verify that the graph of $\cos x$ obeys the rules of the previous section.

2 Use the graphs of $y = \sin x$ and $y = \cos x$ to solve the following inequalities, in the range $0° < x < 360°$.

a) $\sin x > 0$ **b)** $\cos x < 0$ **c)** $\sin x > 0.5$

3 Draw the graph of $y = \sin x + \cos x$. What is the greatest value of this function? For what value of x is this greatest value reached?

4 Draw the graph of $y = 3 \sin x - 4 \cos x$. What is the least value of this function? For what value of x is this least value reached?

5 On the same diagram, sketch the graphs of $y = \sin x$ and $y = \cos x$. How many times do these graphs cross in the range $0° < x < 720°$?

Graphs of tan, cot, sec and cosec

The graph of $y = \tan x$ is shown in Fig. 3.18. Note that the graph shoots up to infinity near $x = 90°$, $270°$ etc. and so the function is not defined at these values.

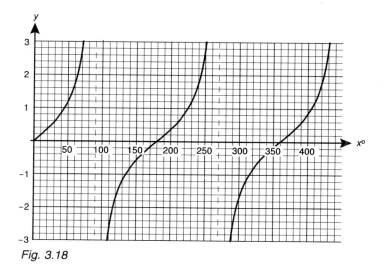

Fig. 3.18

The lines corresponding to these x values are shown dotted. They are called **asymptotes** of the curve.

Note also that the graph repeats itself every 180°.

$$\tan x = \tan (x + 180°)$$

Below are the graphs of $y = \operatorname{cosec} x$, $y = \sec x$ and $y = \cot x$.

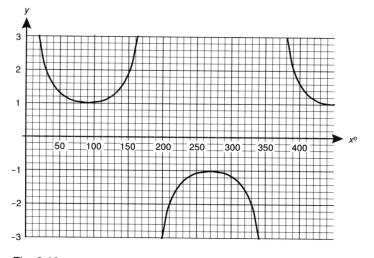

Fig. 3.19 $y = \operatorname{cosec} x$

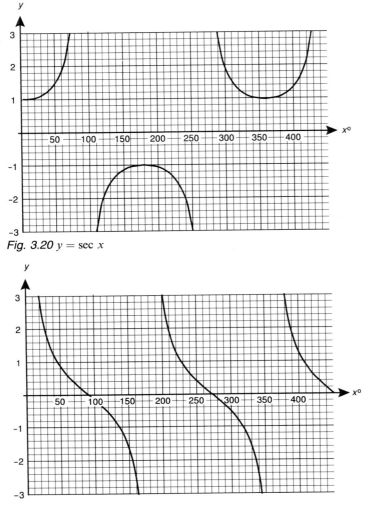

Fig. 3.20 $y = \sec x$

Fig. 3.21 $y = \cot x$

EXERCISE 3F

1 Show how the graphs of the functions tan, cosec, sec and cot confirm the rules given in Section 3.3.

2 For what values of x are the functions cot, cosec and sec not defined?

3 Use the graph of $y = \tan x$ to solve the following inequalities in the range $0° < x < 360°$.

   a) $\tan x < 0$ **b)** $\tan x > 1$

4 Solve the following inequalities in the range $0° < x < 360°$.

   a) $\cot x < 1$ **b)** $\sec x > 0$ **c)** $\mathrm{cosec}\, x < 2$

Period

The values of sin and cos repeat themselves after 360°. The value of tan repeats itself after 180°. These are called the **periods** of the functions.

We can find functions with any period we please, by multiplying x by a suitable constant. Consider $f(x) = \sin 4x$. When x varies between 0° and 90°, $4x$ will vary between 0° and 360°. So $\sin 4x$ will have period 90°, as it will repeat itself after every 90°.

EXERCISE 3G

1 Write down the periods of the following functions.

   a) $\sin 2x$ **b)** $\cos 3x$ **c)** $\tan \frac{1}{4}x$

2 Find functions of the form $y = \sin kx$ which have these periods.

   a) 20°   **b)** 720°   **c)** 300°

3 Figure 3.22 is the graph of $y = \sin x$, for x between 0° and 720°. Make a copy of this graph, and on it sketch the following graphs.

   a) $\sin 2x$ **b)** $\sin \frac{1}{2}x$ **c)** $\sin 2.5x$

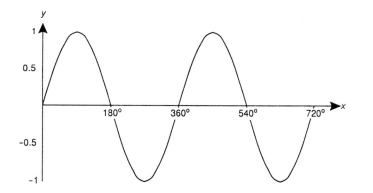

Fig. 3.22

4 Alternating electric current is given by the formula $I = A \sin 18\,000t$, where A amps is the size of the current and t is time in seconds. What is the period of this function?

5 When a musical instrument plays middle C, the air vibration could be given by $V = A \sin 93\,600t$, where A measures the loudness of the note and t is time in seconds. What is the period of this function?

6 The note an octave above middle C is C′. The period of this note is half that of middle C. What formula could give the vibration caused by C′?

LONGER EXERCISE

The chord function

The earliest trigonometric function was called **chord** (abbreviated to crd) used by Greek mathematicians before the invention of sine, cosine, etc. Tables of the crd function were included in the *Almagest*, which was written in about 150AD by the mathematician and astronomer Ptolemy.

Suppose the radius of a circle is r. Suppose a chord of length d subtends an angle of P at the centre. Then r, d and P are related to each other, by the relationship:

$$\text{crd } P = \frac{d}{r}$$

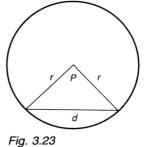

Fig. 3.23

In fact, with a table of values of the crd function you can find the modern functions, and with a table of sin you can find crd. See if you can find the following.

a) crd $60°$ **b)** crd $40°$ **c)** crd $100°$

Now try these.

d) If crd $P = 0.6$, what is sin P?   **e)** If crd $Q = 1.9$, what is sin Q?

f) If crd $P = 0.5$, what is P?   **g)** If crd $P = 1.2$, what is P?

h) Can you obtain general formulae, to express crd P in terms of sin P and vice versa?

Tides

Around the coasts of Britain, the sea level rises and falls throughout the day. This determines the times when a ship can dock at a harbour.

Suppose the depth of water in a harbour is given by

$$d = 20 + 6 \sin 30t$$

where d is the depth in feet, and t is the time (in hours) after midnight.

a) What is the period of this function?

b) When is high tide (when the depth is greatest)? What is this depth?

c) When is low tide (when the depth is least)? What is this depth?

d) At what times in the morning is the depth 23 feet?

e) The **draught** of a ship is the depth of water needed for it to float. What ships can dock at the harbour at all times of the day? What ships can never dock at the harbour?

f) The draught of a ship is 17 feet. When can the ship dock at the harbour?

EXAMINATION QUESTIONS

1 Sketch the graph of $y = \tan \frac{1}{3}x$ for $0° \leq x \leq 540°$. Indicate clearly any asymptotes.

Calculate the ranges of values of x in the interval $0° \leq x \leq 540°$ for which $-1 < \tan \frac{1}{3}x < 1$.

J AS 1991

2 The triangle ABC is equilateral with each side of length 6 cm. With centre A and radius 6 cm, a circular arc is drawn joining B to C. Similar arcs are drawn with centres B and C and with radii 6 cm joining C to A and A to B respectively, as shown in the figure. The shaded region R is bounded by the three arcs AB, BC and CA. Calculate, giving your answer in cm^2 to three significant figures,

Fig. 3.24

a) the area of triangle ABC   **b)** the area of R.

L 1993

3 The depth of water at the entrance to a harbour is y metres at time t hours after low tide. The value of y is given by

$$y = 10 - 3\cos kt$$

where k is a positive constant. Write down, or obtain, the depth of water in the harbour.

(i) at low tide   (ii) at high tide.

Show by means of a sketch graph how y varies with t between two successive low tides.

Given that the time interval between a low tide and the next high tide is 6.20 hours, calculate, correct to two decimal places, the value of k.

J 1992

4 The figure shows a triangle ABC with the usual notation.

a) Show that its area is $\frac{1}{2}ac\sin B$.

b) In a case when AB = 8 cm and BC = 6 cm, the area of the triangle ABC is 12 cm^2. Find the two possible values of

(i) angle B   (ii) the length of AC.

O 1990

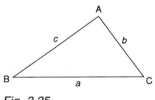

Fig. 3.25

Summary and key points

1 The functions sine, cosine and tangent for acute angles are defined in terms of the sides of a right-angled triangle.

2 The values of the functions can be found without using a calculator for 0°, 30°, 45°, 60° and 90°.

3 Three other functions, cotangent, secant and cosecant are the reciprocals of tangent, cosine and sine respectively.

4 The definitions can be extended to angles other than acute angles.

5 The graphs of the functions demonstrate their properties. In particular, the graphs of sine and cosine have a wave-like pattern.

Coordinate geometry

In coordinate geometry we use arithmetic and algebra to establish results about points, lines, circles and so on. This is often more systematic and accurate than finding the same results by construction and measurement.

Representation of points

The **coordinates** of a point give its position in terms of a pair of numbers. Sometimes we want to deal with particular points; for example, (2, 3) represents a fixed point with known coordinates.

Sometimes, however, we want to establish a result about **all** lines or **all** points. Fixed points with unknown coordinates will be represented by (a, b), (x_1, y_1), (x_2, x_2) and so on. A variable point, which may be anywhere in the plane, will be represented by (x, y). For example, suppose we want to establish a result about circles. The fixed centre of the circle might be represented by (a, b). A variable point on the circumference of the circle might be represented by (x, y).

4.1 Distances in terms of coordinates

The formula for the distance between two points

Suppose we have two points in a plane, with coordinates (x_1, y_1) and (x_2, y_2) relative to the x and y axes. The horizontal distance between them is the difference of their x-coordinates, $x_2 - x_1$. The vertical distance between them is the difference of their

y-coordinates, $y_2 - y_1$. Using Pythagoras's theorem to find the distance between them:

$$\text{distance} = \sqrt{(x_2 - x_1)^2 + (y_2 - y_1)^2}$$

With this formula we can prove facts about straight-line figures such as triangles, quadrilaterals and so on.

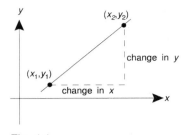

Fig. 4.1

Notes
1 It does not matter which order we take the points.

$$(x_2 - x_1)^2 = (x_1 - x_2)^2$$

2 The formula still holds if some or all of the coordinates are negative.

EXAMPLE 4.1
A, B and C are the points (1, 1), (8, 25) and (16, 21) respectively. Find the lengths of the sides of the triangle ABC. What can you say about triangle ABC?

Solution
Apply the formula.

$$AB = \sqrt{(8 - 1)^2 + (25 - 1)^2} = \sqrt{7^2 + 24^2} = \sqrt{625} = 25$$

$$BC = \sqrt{(16 - 8)^2 + (21 - 25)^2} = \sqrt{8^2 + (-4)^2} = \sqrt{80}$$

$$AC = \sqrt{(16 - 1)^2 + (21 - 1)^2} = \sqrt{15^2 + 20^2} = \sqrt{625} = 25$$

The sides are 25, 25 and $\sqrt{80}$.

Note that two sides are equal.

The triangle is isosceles.

EXERCISE 4A

1 Find the distances between the following pairs of points. Where relevant leave answers in square root form.

a) (1, 2) and (4, 7)   **b)** (3, 4) and (8, 9)   **c)** (5, 6) and (2, 12)

d) (3, 8) and (4, 1)   **e)** (−1, 3) and (0, 5)   **f)** (−3, −4) and (2, 1)

g) (4, −3) and (−3, 8)   **h)** (−3, 5) and (4, −10)   **i)** (1, 1) and (x, y)

j) (−2, −3) and (x, y)

2 The four vertices of a quadrilateral are (1, 1), (3, 4), (7, 1) and (5, −2). Show that the quadrilateral is a parallelogram.

3 The four vertices of a quadrilateral are (2, 3), (5, 7), (8, 6) and (7, 3). Show that the quadrilateral is a kite.

4 The three vertices of a triangle are (1, 1), (5, 3) and (−1, 5). Find the lengths of the sides and show that the triangle is right-angled.

5 Three points are A(1, 1), B(4, 5) and C(10, 13). Find the lengths of AB, BC and AC. Show that A, B and C lie on a straight line.

6 The vertices of a quadrilateral are A(1, 1), B(4, 7), C(−2, 10) and D(−5, 4). Show that ABCD is a rhombus. Find the lengths of the diagonals AC and BD, and state whether ABCD is a square.

7 A is (2, 1), B is (3, 2) and P is (x, y). Given that PA = PB show that $x + y = 4$.

8 The point A(x, 1) is a distance of 5 units from B(2, 3). Find the two possible values of x.

*9 Positions on Ordnance Survey maps are given by grid references, which are coordinates relative to an origin. A grid reference consists of six digits. The first three digits give the distance east of the origin (in hundreds of metres), the next three digits give the distance north.

a) Calculate the distances between the places with the following grid references.
  (i) 272857 and 237779   (ii) 958845 and 222754

b) A town is 10 km from the origin (which has grid reference 000000). Given that the town is due north of the position that has grid reference 060000, find its grid reference.

Equations of circles

Suppose that a point P(x, y) lies on a circle, with radius r and centre C(a, b). Then we can use the formula for the distance between points to find an equation of the circle.

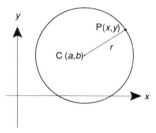

The distance PC is $\sqrt{(x-a)^2 + (y-b)^2}$. This must be equal to r. Squaring the expression gives

$$(x-a)^2 + (y-b)^2 = r^2$$

When this is expanded, we obtain

$$x^2 + y^2 - 2ax - 2by + a^2 + b^2 - r^2 = 0$$

So, given the centre and radius of a circle, we can find its equation. Conversely, if we are given the equation of a circle, we can find its centre and radius.

Fig. 4.2

EXAMPLE 4.2
Find the equation of the circle with centre (3, −2) and radius 4.

Solution
Here $a = 3$, $b = -2$ and $r = 4$. Put these into the formula.

$$(x-3)^2 + (y+2)^2 = 16$$

which gives

$$x^2 + y^2 - 6x + 4y + 13 = 16$$
$$x^2 + y^2 - 6x + 4y - 3 = 0$$

The equation is $x^2 + y^2 - 6x + 4y - 3 = 0$.

EXAMPLE 4.3

A circle has equation $x^2 + y^2 - 2x - 6y - 1 = 0$. Find its centre and radius.

Solution

Group the x terms together and the y terms together.

$$x^2 - 2x + y^2 - 6y - 1 = 0$$

Use the techniques of Chapter 2 to complete the square for the x terms and for the y terms.

$$(x - 1)^2 - 1 + (y - 3)^2 - 9 - 1 = 0$$
$$(x - 1)^2 + (y - 3)^2 = 11$$

The centre is $(1, 3)$ and the radius is $\sqrt{11}$.

EXERCISE 4B

1 Find the equations of the following circles.

a) centre $(0, 2)$ and radius 3   **b)** centre $(2, -1)$ and radius 4

c) centre $(-4, -1)$ and radius 1   **d)** centre $(0.5, -0.5)$ and radius 2

2 For each of the following circles, find the centre and radius.

a) $x^2 + y^2 - 4x + 2y - 3 = 0$ **b)** $x^2 + y^2 + 8x + 8y - 4 = 0$

c) $x^2 + y^2 - 3x + y - 1 = 0$ **d)** $x^2 + y^2 - x - y = 0$

3 Find the centres and radii of the circles with equations $x^2 + y^2 - 9 = 0$ and $x^2 + y^2 - 6x - 8y + 21 = 0$. Find the distance apart of the centres of the circles. Show that the circles touch each other.

4 Let A and B be at $(1, 1)$ and $(3, 3)$ respectively. Let P be at (x, y). Given that P moves so that $PA^2 + PB^2 = 8$, find an equation in x and y. Show that P moves on a circle, and find its centre and radius.

5 O is the origin $(0, 0)$, A is at $(0, 3)$ and P is at (x, y). Given that P moves so that $PA = 2PO$, find an equation in x and y. Show that P moves on a circle, and find its centre and radius.

*6 Let a circle have equation $x^2 + y^2 + ax + by + k = 0$. Show that the origin $(0, 0)$ lies within the circle if and only if k is negative.

*7 Let two circles have equations $x^2 + y^2 - 1 = 0$ and $x^2 + y^2 - 4x - 4y - k = 0$. Find a condition on k for the circles to touch.

4.2 Equations of lines

Gradient of lines

The **gradient** of a line is a measure of its steepness, or angle of slope.

If we have two points, the gradient of the line between them is defined as the vertical change divided by the horizontal change.

So if the points are A at (x_1, y_1) and B at (x_2, y_2), the gradient of the line segment AB is

$$\frac{y_2 - y_1}{x_2 - x_1}$$

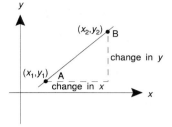

Fig. 4.3

Notes

1 If two different points have the same *x*-coordinate, then this formula is not defined. One point is vertically above the other, and the gradient of the line between them is infinite.

2 If two line segments have the same gradient then they are equally steep, and so are parallel to each other.

3 It does not matter whether we go from A to B or from B to A.

$$\frac{y_2 - y_1}{x_2 - x_1} = \frac{y_1 - y_2}{x_1 - x_2}$$

EXAMPLE 4.4

The points A, B and C have coordinates (1, 1), (2, 3) and (4, 7) respectively. Find the gradients of AB and BC. What can you say about the points?

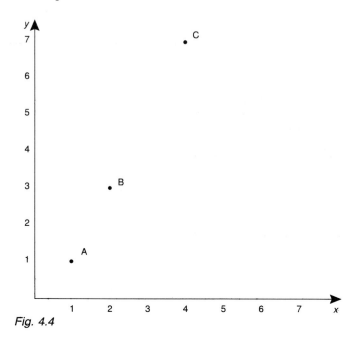

Fig. 4.4

Solution

The points are shown in Fig. 4.4. Use the formula for gradient.

Gradient of AB $= \frac{2}{1} = 2$

Gradient of BC $= \frac{4}{2} = 2$

So the line segment from B to C has the same gradient as the segment from A to B.

A, B and C are in a straight line.

EXERCISE 4C

1 Find the gradients of the line segments joining the following pairs of points.

   a) (1, 1) and (2, 5)   **b)** (2, 3) and (4, 4)   **c)** (10, 3) and (5, 13)

   d) (6, 5) and (3, 2)   **e)** (0, 0) and (2, -1)   **f)** (1, 2) and (-2, -4)

2 Show that the points with coordinates (2, 2), (3, 0) and (5, -4) are in a straight line.

3 The points (3, 4), (5, 5) and (9, y) are in a straight line. Find y.

4 The four corners of a quadrilateral are (1, 1), (2, 2), (4, 3) and (5, 3). Show that the quadrilateral is a trapezium. Is it a parallelogram?

5 The line from (1, 1) to (x, 5) has gradient 2. Find x.

6 The line from (1, 1) to (x, $3x$) has gradient 2. Find x.

*7 Let points A, B and C have coordinates (x_1, y_1), (x_2, y_2) and (x_3, y_3) respectively. If A, B and C are on a straight line show that

$$y_1 x_2 - y_2 x_1 + y_2 x_3 - y_3 x_2 + y_3 x_1 - y_1 x_3 = 0$$

Equations of straight lines

Lines in the plane can be investigated by looking at their equations. We can find the equation of a straight line once we know its gradient and a point that it goes through.

Suppose the gradient is m and the point is (x_1, y_1). Then the equation of the line is

$$y - y_1 = m(x - x_1)$$

Proof

Let (x, y) be a general point on the line. The gradient of the line from (x, y) to (x_1, y_1) is

$$\frac{y - y_1}{x - x_1} = m$$

Multiply through by $(x - x_1)$ and the formula is obtained.

$$y - y_1 = m(x - x_1)$$

The equation can be rewritten in the standard form

$$y = mx + c$$

where c is a constant.

Notes

1 If a line is vertical, then it has infinite gradient. A vertical line will have equation $x + c = 0$, for some c. So a general way of writing the equation of the line is in the form

$$ax + by + c = 0$$

By putting $b = 0$ and $a = 1$, this form includes lines with infinite gradient.

2 The gradient of a straight line is the same everywhere along it. So it does not matter which pair of points we use to find the gradient.

3 If the gradient of a line is a fraction, it sometimes helps to multiply through so that the equation has integer coefficients. For example, convert $y = \frac{3}{4}x + 2$ to $4y = 3x + 8$.

4 A line is uniquely determined if we know its gradient and one point through which it passes.

Figure 4.5 shows lines for which the gradients are positive, negative, zero and infinite.

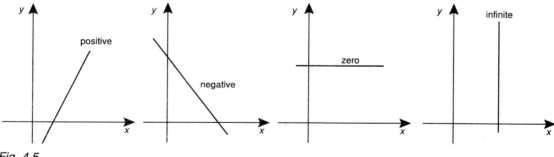

Fig. 4.5

EXAMPLE 4.5

Find the equation of the line through the points $(1, 2)$ and $(4, 11)$. Find where this line meets the line with equation $y = -2x - 16$.

Solution

The gradient of the first line is $\dfrac{11 - 2}{4 - 1} = 3$.

Apply the formula for its equation.

$$y - 2 = 3(x - 1).$$

The equation is $y = 3x - 1$.

Where the lines meet, the two values of y will be equal.

$$3x - 1 = -2x - 16$$

Hence $x = -3$. Substitute for x in either equation to find $y = -10$.

The lines meet at $(-3, -10)$.

EXERCISE 4D

1 Find the equations of the following straight lines.

a) gradient 3, through $(1, 2)$ **b)** gradient -2, through $(1, 3)$

c) gradient $\frac{1}{2}$, through $(5, 6)$ **d)** gradient $-\frac{1}{3}$, through $(4, 7)$

e) through $(1, 1)$ and $(4, 2)$ **f)** through $(3, 2)$ and $(5, 6)$

g) through $(3, 2)$ and $(0, 1)$ **h)** through $(3, -2)$ and $(0, 10)$

i) through $(2, 4)$ and $(5, 4)$ **j)** through $(2, -1)$ and $(2, 5)$

2 Find the gradients of the following lines.

a) $y = 3x - 2$ **b)** $2y = 3x - 4$ **c)** $4y = 3x - 7$

d) $y + 3x + 2 = 0$ **e)** $3y - x - 1 = 0$ **f)** $2y + 3x - 1 = 0$

3 Find the points of intersection of the following pairs of lines.

a) $y = x + 1$ and $y = 2x - 3$ **b)** $y = 2x + 3$ and $y = 5x - 6$

c) $2y + x = 4$ and $3y - x = 7$

4 The three sides of a triangle have equations $y + x = 5$, $y = 2x - 1$ and $2y + x = 3$. Find the coordinates of the vertices.

*5 A straight line goes through the points $(a, 0)$ and $(0, b)$. Show that the equation of the line can be written as

$$\frac{x}{a} + \frac{y}{b} = 1$$

6 ABCD is a parallelogram. Point A is at $(1, 2)$ and point C is at $(3, 7)$. The equation of AB is $2y + x = 5$, and the equation of BC is $y = x + 4$.

a) Find the coordinates of B.   **b)** Find the equation of AD.   **c)** Find the coordinates of D.

*7 The midpoints of the sides BC, CA and AB of triangle ABC are D$(1, -2)$, E$(0, 1)$ and F$(-1, 0)$ respectively. A theorem about triangles says that a line joining the midpoints of two sides is parallel to the third side. Find the coordinates of A, B and C.

4.3 Parallel and perpendicular lines

Suppose two lines have equations $y = mx + c$ and $y = m'x + c'$.

a) If the lines are parallel then $m = m'$.
b) If the lines are perpendicular then $mm' = -1$.

Proof
a) If lines are parallel then they have the same steepness, i.e. their gradients are equal. So if they are parallel $m = m'$.

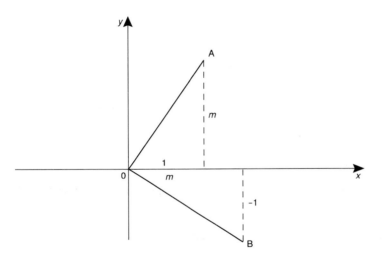

Fig. 4.6

b) Suppose the two lines are perpendicular. In Fig. 4.6, the vertical change between O and A is m. The horizontal change is 1, so the gradient of OA is $\dfrac{m}{1} = m$.

Line OB is obtained by rotating OA through $90°$, so it is perpendicular to OA. The y-change between O and B is -1, and the x-change is m. So OB has gradient $-\dfrac{1}{m}$.

Therefore $m' = -\dfrac{1}{m}$

and $mm' = -1$.

EXAMPLE 4.6
A line L has equation $y = 2x + 3$. Find the equations of the lines

a) parallel to L, through (1, 3)
b) perpendicular to L, through (2, 7).

Solution

a) This line must have gradient 2. Use the formula of Section 4.2.

$$y - 3 = 2(x - 1)$$
$$y - 3 = 2x - 2$$
$$y = 2x + 1$$

The equation is $y = 2x + 1$.

b) This line must have gradient $-\frac{1}{2}$. Use the formula of Section 4.2.

$$y - 7 = -\frac{1}{2}(x - 2)$$
$$y - 7 = -\frac{1}{2}x + 1$$
$$y = -\frac{1}{2}x + 8$$

The equation is $2y + x - 16 = 0$.

EXERCISE 4E

1 Which of the following lines are parallel and which are perpendicular to each other?

a) $y = -2x - 2$ **b)** $2y = x - 3$ **c)** $x - 2y = 7$

2 Find the equations of the following lines.

a) through (1, 2), parallel to $y = 3x + 4$ **b)** through (5, 2), perpendicular to $y = 3x - 7$

c) through (5, 3), parallel to the line through (2, 1) and (0, −1)

d) through (2, 1), perpendicular to the line through (9, 3) and (8, 5)

3 Points A, B and C are at (1, 1), (3, 5) and (−7, 5) respectively. Find the gradients of AB and AC, and show that $\widehat{BAC} = 90°$.

4 A is at (1, 1), B is at (3, 5), P is at (x, y) and PA is perpendicular to PB. Show that

$$x^2 + y^2 - 4x - 6y + 8 = 0$$

Show that P moves on a circle, and find its centre and radius.

*5 A is at $(a, 0)$, B is at $(b, 0)$ and P is at (x, y). If $\widehat{APB} = 90°$, find an equation in x and y.

*6 Four points are at $\left(ap, \dfrac{a}{p}\right)$, $\left(aq, \dfrac{a}{q}\right)$, $\left(ar, \dfrac{a}{r}\right)$ and $\left(as, \dfrac{a}{s}\right)$, where a is not zero and p, q, r and s are all non-zero and different from each other. The line joining any two of these points is perpendicular to the line joining the other two. Show that $pqrs = -1$.

7 ABCD is a rectangle, with A at (1, 1) and C at (5, 7). The equation of BC is $5y = x + 30$. Find the coordinates of B and D.

4.4 Division of a line segment in a ratio

Suppose P lies on the line joining A to B. Then P will divide AB in the ratio AP:PB. If we know the ratio and the coordinates of A and B, then we can find the coordinates of P.

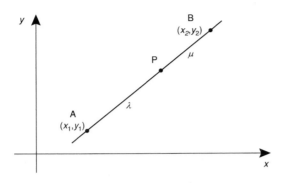

Fig. 4.7

Let P divide the line between $A(x_1, y_1)$ and $B(x_2, y_2)$ in the ratio $\lambda : \mu$. Then the coordinates of P are

$$\left(\frac{\mu x_1 + \lambda x_2}{\lambda + \mu}, \; \frac{\mu y_1 + \lambda y_2}{\lambda + \mu} \right)$$

Proof

AB is divided into $\lambda + \mu$ equal parts, of which AP occupies the first λ. So P can be found by going to A and then travelling $\dfrac{\lambda}{\lambda + \mu}$ of the line segment AB.

The x-coordinate of A is x_1, and the x-component of the segment AB is $x_2 - x_1$. So the x-coordinate of P is given by

$$x_1 + \frac{\lambda(x_2 - x_1)}{\lambda + \mu} = \frac{\lambda x_1 + \mu x_1 + \lambda x_2 - \lambda x_1}{\lambda + \mu} = \frac{\mu x_1 + \lambda x_2}{\lambda + \mu}$$

The y-coordinate is found similarly.

In particular, if P is the midpoint of AB, then it divides AB in the ratio $1:1$. Its coordinates are

$$\left(\frac{x_1 + x_2}{2}, \; \frac{y_1 + y_2}{2} \right)$$

In fact we could think of the midpoint of AB as the **average** of A and B.

EXAMPLE 4.7

A is at $(1, 1)$, and B is at $(3, 8)$. P lies on AB with $AP = \tfrac{1}{5}AB$. Find the coordinates of P.

Solution

If P is a fifth of the way from A to B, then it divides AB in the ratio $\frac{1}{5}$ to $\frac{4}{5}$. This ratio is equal to $1:4$. Apply the formula above.

$$P \text{ is at } \left(\frac{4 \times 1 + 1 \times 3}{1 + 4}, \frac{4 \times 1 + 1 \times 8}{1 + 4} \right)$$

giving

$$P\left(\frac{7}{5}, \frac{12}{5} \right) \quad \text{or} \quad P(1.4, \ 2.4).$$

The coordinates of P are (1.4, 2.4).

EXERCISE 4F

1 Find the coordinates of the points which divide the following points in the ratios shown.

    a) (2, 0) and (4, 6), ratio $2:3$ **b)** (3, 4) and (11, 12), ratio $3:1$

    c) (0, 0) and (1, 1), ratio $1:x$ **d)** (a, 0) and (0, b) ratio $4:5$

2 Find the midpoints of the lines between the following pairs of points.

    a) (1, 1) and (3, 7)   **b)** (3, 6) and (-1, -2)

    c) (0, 0) and (x, y)   **d)** (a, 0) and (0, b)

3 The triangle ABC has vertices A(5, 7), B(1, 1) and C(3, 13). Find the midpoint M of BC. Find the coordinates of the point G which divides AM in the ratio $2:1$.

4 A is at (1, 0), B is at (0, 1). P divides the line from the origin (0, 0) to B in the ratio $1:x$. Q divides AP in the ratio $1:y$. Find the coordinates of Q.

5 A is at (1, 1), B is at (3, 4) and C is at (7, 10). Show that ABC is a straight line. Find the ratio in which B divides AC.

*6 The triangle ABC has vertices A(x_1, y_1), B(x_2, y_2) and C(x_3, y_3). Find the midpoint M of BC. Find the coordinates of the point G which divides AM in the ratio $2:1$. Simplify your expression as far as possible. What can you say about G?

4.5 Angles in terms of coordinates

By the techniques of Section 4.1 we can find the distances between points. We can also find the angles between the lines joining points.

Suppose A is at (x_1, y_1) and B is at (x_2, y_2). The angle P that AB makes with the x-axis is given by

$$\tan P = \frac{y_2 - y_1}{x_2 - x_1}$$

So the tan of the angle is given by the gradient of the line.

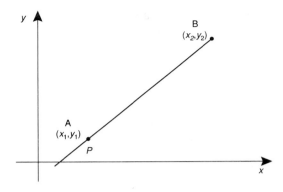

Fig. 4.8

Suppose the line is given in the form $y = mx + c$, where the constant m is the gradient of the line. The angle P the line makes with the x-axis is given directly by

$$P = \tan^{-1} m$$

Suppose we have two lines. The angle between them can be found by finding the angles each makes with the x-axis, and subtracting.

EXAMPLE 4.8
A is at $(1, 1)$, B is at $(3, 4)$ and C is at $(6, 2)$. Find the angle BAC.

Solution
Find the angles using the formula above.

Angle between BA and the x-axis $= \tan^{-1} \dfrac{4 - 1}{3 - 1} = \tan^{-1} \dfrac{3}{2}$.

Angle between CA and the x-axis $= \tan^{-1} \dfrac{2 - 1}{6 - 1} = \tan^{-1} \dfrac{1}{5}$.

$\text{BAC} = \tan^{-1} \dfrac{3}{2} - \tan^{-1} \dfrac{1}{5} = 45°$

EXERCISE 4G

1 Find the angles between the lines joining the following pairs of points and the x-axis.

   a) $(1, 1)$ and $(4, 7)$ **b)** $(0, -3)$ and $(3, 8)$ **c)** $(4, 3)$ and $(3, 5)$

2 Find the angles between the following lines and the x-axis.

   a) $y = 3x + 4$ **b)** $y = 4x - 1$ **c)** $y = 0.5x - 3$

   d) $2y = 3x - 4$ **e)** $y = -3x + 2$ **f)** $2y + 5x + 1 = 0$

3 Find the angles between the following pairs of lines.

a) $y = 2x + 1$ and $y = 3x - 2$ **b)** $y = 2x + 1$ and $y = -3x + 1$

c) $y = 2x + 1$ and $y = 2 - \frac{1}{2}x$ **d)** $y = 4x - 3$ and $2y = x + 5$

e) $3y + x = 4$ and $y - x = 3$ **f)** $2y + x + 1 = 0$ and $3y + x - 1 = 0$

4 A triangle has vertices A(1, 1), B(2, 5) and C(7, 3). Find the angle BAC.

4.6 Geometric problems

The techniques of this chapter can be used to solve a wide variety of problems in geometry. A few examples are presented here.

Perpendicular bisector and circumcentre

The **perpendicular bisector** of the line segment AB is the line perpendicular to AB through its midpoint. If P is on the perpendicular bisector, then PA = PB.

The **circumcentre** X of a triangle ABC is the centre of the circle which goes through A, B and C. So it is equidistant from A, B and C. The circumcentre must lie on the perpendicular bisectors of AB, BC and CA.

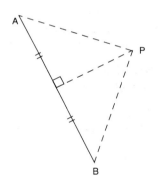

Perpendicular bisectors and circumcentres can be found by coordinate geometry. We shall find the circumcentre of the triangle with vertices A(2, 0), B(6, 2) and C(2, 6).

The midpoint of AB is (4, 1), and its gradient is $\frac{1}{2}$. So the perpendicular bisector of AB will go through (4, 1) and have gradient -2. Its equation is

$$y - 1 = -2(x - 4)$$
$$y + 2x = 9$$

Similarly the perpendicular bisector of BC will have gradient 1 and go through (4, 4). Its equation is

$$y - 4 = (x - 4)$$
$$y = x$$

We do not need to find the third bisector. The lines $y + 2x = 9$ and $y = x$ intersect at (3, 3).

The circumcentre is at (3, 3).

As a check, we can find the distances from the circumcentre to A, B and C. They are all $\sqrt{10}$.

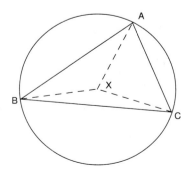

Fig. 4.9

EXERCISE 4H

1 Find the equations of the perpendicular bisectors of the line segments between the following pairs of points.

a) (1, 1) and (4, 4)   **b)** (2, 3) and (4, −9)

c) (3, 4) and (7, 4)   **d)** (3, 2) and (3, −4)

2 Find the circumcentre of the triangle with vertices A(5, −2), B(7, 2) and C(−2, 5). Find the radius of the circle which goes through A, B and C.

*3 Three towns have grid references 430020, 230420 and 160410 respectively. A radio transmitter is to be placed so that it is equidistant from the three towns. Find the grid reference of the position where the transmitter should be sited.

Area of triangle

If we are given the coordinates of the vertices of a triangle, then we can find its area.

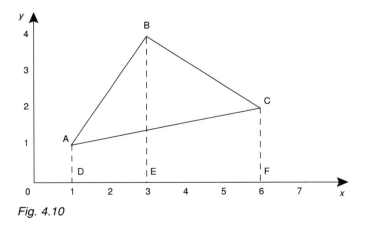

Fig. 4.10

Suppose the triangle has vertices at A(1, 1), B(3, 4) and C(6, 2). Drop perpendiculars from A, B and C, meeting the *x*-axis at D(1, 0), E(3, 0), F(6, 0) respectively.

The area of ABC will be given by

$$\text{area ABC} = \text{area ABED} + \text{area BCFE} - \text{area ACFD}$$

All these shapes are trapezia. Suppose the parallel sides of a trapezium are of length *a* and *b*, and are *h* apart. The area of the trapezium is

$$\text{area} = \tfrac{1}{2}(a + b)h$$

Applying this formula gives

area $\text{ABED} = \frac{1}{2}(1+4)2 = 5$

area $\text{BCFE} = \frac{1}{2}(4+2)3 = 9$

area $\text{ACFD} = \frac{1}{2}(1+2)5 = 7\frac{1}{2}$

So

area $\text{ABC} = 5 + 9 - 7\frac{1}{2} = 6\frac{1}{2}$

The area of ABC is $6\frac{1}{2}$.

EXERCISE 4I

1 Find the areas of the triangles with the following vertices.

a) A(2, 3), B(5, 8), C(7, 1) **b)** A(1, 6), B(3, 2), C(5, 4)

*2 Three points A(x_1, y_1), B(x_2, y_2), C(x_3, y_3) are as shown in Fig. 4.11. Show that the area of triangle ABC is

$\frac{1}{2}(y_1x_2 - y_2x_1 + y_2x_3 - y_3x_2 + y_3x_1 - y_1x_3)$

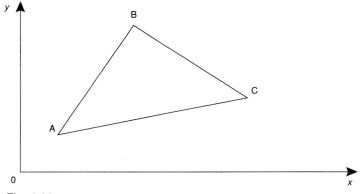

Fig. 4.11

3 Question 7 of Exercise 4C gives a condition for three points A, B and C to be on a straight line. How is this connected with the formula for the area of ABC?

Distance of a point from a line

Suppose we want to find the shortest distance from a point P to a line. This shortest distance will be PD, where D is the point on the line such that PD is perpendicular to the line.

We shall find the shortest distance from P(1, 1) to the line with equation $y = 2x + 9$.

The gradient of the line is 2, so the gradient of PD is $-\frac{1}{2}$. So the equation of PD is

$y - 1 = -\frac{1}{2}(x - 1)$

$2y = -x + 3$

The lines $y = 2x + 9$ and $2y = -x + 3$ meet at $(-3, 3)$. These are the coordinates of D. Now we can use the formula for the distance between two points.

$$\text{Shortest distance} = \sqrt{(-3 - 1)^2 + (3 - 1)^2} = \sqrt{20}$$

EXERCISE 4J

1 In each of the following, find the shortest distance from the point to the line.

   a) $(0, 0)$ to $y + x = 8$ **b)** $(2, 3)$ to $y = 3x - 1$

   c) $(1, 1)$ to $3y + 4x = 12$ **d)** $(2, -1)$ to $2y + 3x = 6$

*2 A straight road connects the points with grid references 950230 and 910190. I am at the point with grid reference 920250. What is the shortest distance from me to the road?

LONGER EXERCISE

Circumcentre, orthocentre, centroid

On page 71 we defined the circumcentre of a triangle. We can now define two other centres.

Orthocentre

An **altitude** of a triangle is a line from a vertex, perpendicular to the opposite side. The **orthocentre** is the intersection of the three altitudes.

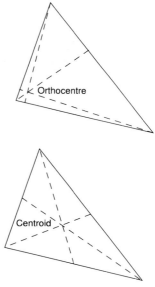

Centroid

A **median** of a triangle is a line from a vertex to the midpoint of the opposite side. The **centroid** is the point of intersection of the three medians. (The centroid is the point of intersection G of Question 6 from Exercise 4F.)

A triangle has vertices at A(4, 5), B(−4, 1) and C(5, −2). Find the circumcentre X, the orthocentre Y and centroid G.

What can you say about the three points X, Y and G? Test any theory on other triangles.

Fig. 4.12

EXAMINATION QUESTIONS

1 The points A, B and C have coordinates $(-2, 3)$, $(5, 4)$ and $(4, 11)$, respectively. Find

   (i)   the equation of the line through A parallel to BC
   (ii) the equation of the perpendicular bisector of AC.
         Verify that this second line passes through B.
         The line through A parallel to BC meets the perpendicular bisector of AC at the point D.
   (iii) Find the coordinates of D.
   (iv) Show that $AD = BC$.   *J AS 1992*

2 The vertices of $\triangle ABC$ are the points A$(-1, 5)$, B$(-5, 2)$ and C$(8, -7)$.

a) Find in the form $px + qy + r = 0$, where p, q and r are integers, an equation of the line passing through B and C.

b) Show, by calculation, that AB and AC are perpendicular.

The point D lies on the line BA produced and is such that $3BA = BD$.

c) Determine the coordinates of D.

L 1992

3 The points A and B have coordinates $(-5, -2)$ and $(7, 4)$ respectively. Find

a) the coordinates of the point C which divides AB internally in the ratio $2 : 1$

b) an equation of the line through C perpendicular to AB.

L AS 1989

Summary and key points

1 The distance between (x_1, y_1) and (x_2, y_2) is

$$\sqrt{(x_1 - x_2)^2 + (y_1 - y_2)^2}$$

Be careful with negative numbers when using the formula. Do not oversimplify the result.

$\sqrt{3^2 + 5^2}$ **is not equal to** $3 + 5$

The equation of a circle with centre (a, b) and radius r is

$$x^2 + y^2 - 2ax - 2by + a^2 + b^2 - r^2 = 0.$$

Given the equation of a circle, its centre and radius can be found by the technique of completing the square.

2 The gradient of the line from (x_1, y_1) to (x_2, y_2) is

$$\frac{y_2 - y_1}{x_2 - x_1}$$

If a line goes through (x_1, y_1) and has gradient m, then its equation is

$$y - y_1 = m(x - x_1)$$

When finding the gradient of a line from its equation, put the equation in the standard form $y = mx + c$. The gradient of the line with equation $2y = 3x + 1$ is 1.5, not 3.

3 Suppose two lines have equations $y = mx + c$ and $y = m'x + c'$.

If the lines are parallel, then $m = m'$.

If the lines are perpendicular, then $mm' = -1$.

4 Suppose P lies on the line segment joining $A(x_1, y_1)$ and $B(x_2, y_2)$, dividing it in the ratio $\lambda : \mu$. The coordinates of P are

$$\left(\frac{\mu x_1 + \lambda x_2}{\lambda + \mu}, \ \frac{\mu y_1 + \lambda y_2}{\lambda + \mu} \right)$$

If the line segment AB is divided in the ratio $\lambda : \mu$, the coordinates of A are multiplied by μ and the coordinates of B by λ.

5 The line $y = mx + c$ makes an angle $\tan^{-1} m$ with the x-axis.

6 The techniques of this chapter can be used to solve a wide variety of geometric problems.

Sequences

A collection of numbers arranged in a definite order is a **sequence**. Each of the numbers of a sequence is called a **term**.

Suppose we know the first term of a sequence and the rule for moving from one term to the next. Then we can write down the subsequent terms of the sequence. Take a sequence which begins with 3, and for which every subsequent term is 2 greater than the one before. The sequence is

$$3, 5, 7, 9, \cdots$$

The sum of the terms of a sequence is a **series**. The series corresponding to the first six terms of the sequence above is

$$3 + 5 + 7 + 9 + 11 + 13 = 48$$

Sometimes a series with infinitely many terms has a finite sum. Consider the recurring decimal $0.111\,11\ldots$. This can be regarded as the following series.

$$0.1 + 0.01 + 0.001 + 0.0001 + \cdots$$

We shall show how to find the sum of such infinite series.

5.1 Notation for sequences and series

Use of dots

The sequence of the first ten square numbers is

$$1, 4, 9, 16, 25, \cdots, 100$$

We have not written down all the terms. The dots indicate the missing terms 36, 49, 64 and 81.

The sequence is **finite**. If a sequence continues indefinitely, we indicate this by writing dots at the end. For example, there are infinitely many odd numbers, and so we write the sequence of odd numbers as

$$1, 3, 5, 7, \cdots$$

The u_n notation

A sequence can be written as

$$u_1, \ u_2, \ u_3, \cdots, \ u_n, \ \ldots$$

So u_1 is the first term of the sequence, u_2 is the second and so on. If this sequence is the sequence of odd numbers, then $u_1 = 1$, $u_2 = 3$ and so on. The **general term** is u_n, the nth odd number.

Suppose a sequence begins with 3, and that each term is 2 greater than the one before. The sequence can be described in the u_n notation by

$$u_1 = 3, \ u_{n+1} = u_n + 2$$

The Σ notation

The series which gives the sum of the first six squares is written

$$1 + 4 + 9 + 16 + 25 + 36 = \sum_{i=1}^{6} i^2$$

Σ is the Greek letter sigma, corresponding to S. In this context it means 'the sum of'.

The $i = 1$ below the Σ sign is the **lower limit**. The 6 at the top is the **upper limit**. The Σ notation means that we work out i^2 for $i = 1$, $i = 2$ and so on all the way up to $i = 6$. We then add the results together.

Using the u_n notation, a series is written as

$$u_1 + u_2 + \cdots + u_n = \sum_{i=1}^{n} u_i$$

Here the letter i takes the values from $i = 1$ up to $i = n$.

If a sequence continues indefinitely, the sum of all the terms may be finite. In this case it is called the **sum to infinity**. It is written as

$$u_1 + u_2 + \cdots + u_n + \cdots = \sum_{i=1}^{\infty} u_i$$

EXERCISE 5A

1 The first term of a sequence is 5. Each term is found by adding 7 to the term before. Write down the first three terms.

2 Write down the first three terms of each of the sequences given by the following rules.

a) Start with 2. Each term is 3 greater than the term before.
b) Start with 3. Each term is twice the term before.
c) Start with 10. Each term is 2 less than the term before.
d) Start with 48. Each term is half the term before.

3 Describe the sequences of Questions 1 and 2 in terms of the u_n notation.

4 Write down the first three terms of the sequences for which the nth term is given by

a) $u_n = 5n + 3$ **b)** $u_n = n^2 + 1$ **c)** $u_n = 2^n + 3^n$

5 Write out the following series in full and evaluate them.

a) $\displaystyle\sum_{i=1}^{6}(2i + 5)$ **b)** $\displaystyle\sum_{i=1}^{5}i^2$ **c)** $\displaystyle\sum_{i=1}^{5}2^i$

d) $\displaystyle\sum_{i=1}^{4}\frac{1}{i}$ **e)** $\displaystyle\sum_{i=1}^{4}\frac{1}{i(i + 1)}$

6 The general term of a sequence is $3n - 1$. Using the Σ notation, write down the sum of the first five terms.

7 Using the Σ notation, write down the sum of the following series.
a) $1 + 2 + 3 + \cdots + 20$
b) $2 + 4 + 6 + \cdots + 22$
c) $1 + 8 + 27 + \cdots + 1000$ (general term n^3)
d) $\dfrac{1}{2} + \dfrac{1}{4} + \dfrac{1}{8} + \cdots$ $\left(\text{general term } \dfrac{1}{2^n}\right)$

*8 $u_1, u_2, \cdots, u_n, \cdots$ is a sequence. Using the Σ notation, write down the sum of all the odd terms, i.e. $u_1 + u_3 + u_5 + \cdots$

*9 The **Fibonacci numbers** is a sequence in which the first two terms are both 1, and then each term is found by adding the previous two. Express the Fibonacci numbers in terms of the u_n notation.

5.2 Arithmetic progressions

An **arithmetic progression** (AP) is a sequence in which the terms increase (or decrease) by a constant amount. In the sequence below, each term is 3 greater than the one before.

1, 4, 7, 10, $\cdots$

Using the u_n notation, this sequence can be described as

$$u_1 = 1, \ u_{n+1} = u_n + 3$$

The constant increase is the **common difference**. So $u_{n+1} - u_n = d$, for every n. Once we know the first term and the common difference we can find any term.

Suppose the first term is a and the common difference is d. So $u_1 = a$, $u_2 = a + d$ and $u_3 = a + 2d$. In general the nth term is obtained by adding d, $n - 1$ times.

$$u_n = a + (n - 1)d$$

EXAMPLE 5.1
An arithmetic progression has third term 5 and fifth term 9. Find the first term and the common difference.

Solution
Use the formula above, putting $n = 3$ and $n = 5$.

$$u_3 = a + 2d, \ 5 = a + 2d$$
$$u_5 = a + 4d, \ 9 = a + 4d$$

Subtract, to find $4 = 2d$. This gives $d = 2$. Substitute into either of the equations to find that $a = 1$.

The first term is 1 and the common difference is 2.

EXERCISE 5B

1 Write down the first three terms for each of the following APs.

   a) first term 5, common difference 3   **b)** first term 3, common difference 7

   c) first term 2, common difference -2 **d)** first term 3, common difference $\frac{1}{2}$

   e) first term 5, common difference -1.5

2 Find expressions for the nth terms of the sequences in Question 1.

3 An AP has fourth term 8 and seventh term 17. Find the first term and the common difference. Find an expression for the nth term.

4 Find the first term, the common difference and an expression for the nth term for each of the following arithmetic progressions.

   a) second term 3, third term 5   **b)** second term 4, fifth term 16

   c) third term 8, fifth term 9   **d)** second term 4, fourth term 2

5 The formulae below give the nth terms of APs. In each case find the first term and the common difference.

  a) $u_n = 5n + 2$ **b)** $u_n = 2n - 3$ **c)** $u_n = 3 - 3n$

  d) $u_n = 2 + 0.1n$ **e)** $u_n = -2n$ **f)** $u_n = 4 - \frac{1}{2}n$

6 A ten-year-old child receives £2 pocket money per week. This amount will increase by 40p after each birthday. How much will she get at age 14? How much will she get at age n (assuming she is still receiving pocket money)?

7 The first three terms of an AP are $x + 3$, $2x + 4$, $4x + 1$. Find x.

*8 The sum of the first three terms of an arithmetic progression is 39. The product of the terms is 2145. Find the first term and the common difference. (Hint: let x be the second term.)

*9 The sum of the first five terms of an AP is 40. The sum of the squares of the terms is 410. Find the first term of the sequence. (Hint: let x be the third term.)

10 The first term of a sequence is 5 and the common difference is 3. How many terms of the sequence are less than 1000?

11 The first term of a sequence is 6.5 and the common difference is -0.3. How many terms of the sequence are positive?

12 The arithmetic mean of two numbers a and b is $m = \frac{1}{2}(a + b)$. Show that a, m and b form an arithmetic progression.

13 Consider the sequence below.

  1, 3, 6, 10, 15, ...

Show that the **differences** between successive terms form an arithmetic progression. Hence continue the sequence for three more terms.

A computer is useful when dealing with sequences like the one of Question 13 above. The spreadsheet investigation on page 169 is about this type of sequence.

Sum of an arithmetic progression

Suppose we want to find the sum of the first n terms of an arithmetic progression. We proceed as follows.

The original progression **increases** at a constant rate. If we write it in reverse order, it will **decrease** at a constant rate. If we add together the two sequences the increases and the decreases will cancel out. We shall be left with a sequence of constant terms.

Formula

Suppose an arithmetic progression has first term a and nth term l. then the sum of the first n terms is

$$S = \tfrac{1}{2}n(a + l)$$

Proof

Let the common difference be d. Write out the original series, and beneath it the series in reverse order.

$$S = a + (a + d) + \cdots + (l - d) + l$$
$$\updownarrow \qquad \updownarrow \qquad \updownarrow \qquad \updownarrow \qquad \updownarrow$$
$$S = l + (l - d) + \cdots + (a + d) + a$$

Now add, combining the linked terms. Notice that the ds cancel out.

$$2S = (a + l) + (a + l) + \cdots + (a + l) + (a + l)$$

Now we have $(a + l)$, repeated n times.

$$2S = n(a + l)$$

Divide by 2, to obtain the formula.

$$S = \tfrac{1}{2}n(a + l)$$

Note that l is the nth term, so we can write it as $l = a + (n - 1)d$. So the sum of the series can also be written as

$$S = \tfrac{1}{2}n[a + a + (n - 1)d] = \tfrac{1}{2}n[2a + (n - 1)d]$$

EXAMPLE 5.2

An arithmetic progression has first term 3 and common difference 2. Find the sum of the first 50 terms.

Solution

Apply the formula, putting $a = 3$, $d = 2$ and $n = 50$.

$$\text{Sum} = \tfrac{1}{2} \times 50[2 \times 3 + (50 - 1)2]$$

The sum is 2600.

EXAMPLE 5.3

The sum of the first n terms of an arithmetic progression is $n + n^2$. Find the first term and the common difference.

Solution

The first term can be found by putting $n = 1$.

$$\text{First term} = 1 + 1^2 = 2$$

The sum of the first two terms can be found by putting $n = 2$.

$$\text{First term} + \text{second term} = 2 + 2^2 = 6$$

By subtracting, we find that the second term is 4.

The first term is 2 and the common difference is 2.

EXERCISE 5C

1 Find the sum of the first n terms for the following arithmetic progressions.

    a) first term 5, common difference 3   **b)** first term 2, common difference 1

    c) first term 9, common difference 3   **d)** $u_1 = 3$, $u_{n+1} = u_n + 5$

    e) first term 5, third term 11   **f)** $u_1 = 6$, $u_5 = 8$

2 Find the sum of the first 50 terms of each of the sequences in Question 1.

3 Find the sum of the first n integers.

4 Find the sum of the first n odd integers, simplifying your answer as far as possible.

5 The sum of the first n terms of an arithmetic progression is $n^2 - n$. Find the first two terms, and hence find the common difference.

6 The formulae below give the sum of the first n terms of arithmetic progressions. In each case find the first term and the common difference.

    a) $3n + 2n^2$ **b)** $\frac{1}{4}(n^2 + 3n)$ **c)** $3n - n^2$

*7 The sum of the first n terms of a sequence is $5n + n^2$. Find an expression for the nth term, and so verify that the sequence is an arithmetic progression.

*8 The sum of the first n terms of a sequence is $f(n)$. What must the function f be, in order for the sequence to be an AP?

*9 Let S_n denote the sum of the first n terms of an AP.

    a) Express the nth term in terms of S_n and S_{n-1}.

    b) Express the common difference in terms of S_n, S_{n-1} and S_{n-2}.

10 The first term of an AP is 2, and the last term is 59. If the sum of the terms is 610, find the common difference.

11 An arithmetic progression has first term 10 and common difference -0.5. How many terms must be taken before the sum is negative?

12 In 1786, in a school in Germany, the teacher wanted to keep his class quiet by setting them the following addition sum.

    $81\ 297 + 81\ 495 + 81\ 693 + \cdots + 100\ 899$ (100 terms in all)

One pupil, aged nine, wrote down the answer as soon as the teacher had finished setting the problem. What was the answer, and how was it obtained?

(The pupil was Karl Friedrich Gauss, who went on to become one of the greatest of mathematicians.)

*13 Figure 5.1 shows a 'House of Cards'. How many cards are there on the top row, without the 'floor'? How many on the row which is second from top? How many on the row which is *n*th from top?

If the house is *n* storeys high, how many cards does it contain? What is the tallest house you can build with 52 cards?

*14 The positive integers are bracketed as follows.

{1} {2,3} {4,5,6} {7,8,9,10} {11,12,13,14,15} ···

There is one number in the first pair of brackets, two in the second, three in the third and so on. What are the numbers in the *n*th pair of brackets?

*15 The first and third terms of an arithmetic progression are *p* and *q* respectively. The sum of the first *n* terms is written S_n. Express S_n in terms of *p* and *q*.

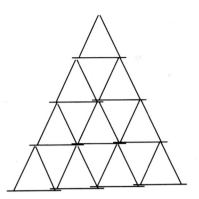

Fig. 5.1

5.3 Geometric progressions

In a **geometric progression** (GP) the ratio between successive terms is constant. An example of a GP is

3, 6, 12, 24, 48, ···

We see that each term is twice the one before, so the ratio between successive terms is 2. This ratio is the **common ratio** of the sequence.

If we know the first term and the common ratio of a sequence, then we can find any term. Suppose the first term is *a* and the common ratio is *r*. In terms of the u_n notation, the sequence can be defined as

$u_1 = a, \quad u_{n+1} = ru_n$

$u_1 = a, \quad u_2 = ar$ and $u_3 = ar^2$. In general the *n*th term is found by multiplying by *r*, $(n-1)$ times.

$u_n = ar^{n-1}$

If the common ratio is negative, then the terms will be alternately positive and negative. If $a = 3$ and $r = -2$, the first six terms of the sequence are

3, −6, 12, −24, 48, −96

The *n*th term is $3 \times (-2)^{n-1}$. Note that it is necessary to put brackets round the −2.

EXAMPLE 5.4

The first term of a GP is 5 and the common ration is 2. Find the 10th term.

Solution

Apply the formula, with $a = 5$, $r = 2$ and $n = 10$.

$$\begin{aligned}
\text{10th term} &= 5 \times 2^{10-1} \\
&= 5 \times 2^9 \\
&= 5 \times 512 \\
&= 2560
\end{aligned}$$

The 10th term is 2560.

EXAMPLE 5.5

The second term of a GP is 2 and the fourth is 18. Find the possible values of the common ratio and the corresponding first terms.

Solution

Put these values into the formula.

$$2 = ar^1 \text{ and } 18 = ar^3$$

Divide to eliminate a.

$$9 = r^2$$

It follows that $r = 3$ or -3. The values of a can be found from either of the equations above.

If $r = 3$,

$$2 = a \times 3$$
$$a = \frac{2}{3}$$

If $r = -3$,

$$2 = a \times (-3)$$
$$a = -\frac{2}{3}$$

The possible values are $r = 3$, $a = \frac{2}{3}$ and $r = -3$, $a = -\frac{2}{3}$.

EXERCISE 5D

1 Find the nth term of the following GPs.

   a) first term 2, common ratio 3   **b)** first term 1, common ratio -2

   c) first term 8, common ratio $\frac{1}{2}$

2 Find the eighth term of each of the sequences in Question 1.

3 The nth term of a sequence is 4×3^n. Find the first term and the common ratio.

4 Find the first term and the common ratio for the sequences with the following nth terms.

    a) 2^n **b)** $\dfrac{3}{2^n}$ **c)** $5 \times (-3)^n$

5 The first term of a geometric progression is 4 and the third term is 1. Find the possible values of the common ratio.

6 Find the possible values of the common ratios of the following geometric progressions.

    a) $u_1 = 3, \ u_3 = 75$ **b)** $u_2 = 8, \ u_5 = 1$ **c)** $u_1 = 2, \ u_4 = -54$ **d)** $u_1 = 1, \ u_3 = 2$

7 The first four terms of a geometric progression are $1, x, y, 125$. Find x and y.

*8 The geometric mean of two positive numbers a and b is $g = \sqrt{ab}$. Show that a, g and b form a geometric progression.

•

Sum of a geometric progression

Suppose we want to find the sum of the first n terms of a geometric progression. The technique is to multiply the original series by the common ratio r. Then when we subtract the original series, all terms will cancel out except the first and the last.

Formula
Let a GP have first term a and common ratio r. The sum of the first n terms is

$$\sum_{i=1}^{n} ar^{i-1} = \frac{a(r^n - 1)}{r - 1}$$

Proof
Write down the original series, and underneath write the series multiplied by r.

$$S = a + ar + ar^2 + \cdots + ar^{n-1}$$
$$\updownarrow \quad \updownarrow$$
$$rS = \quad ar + ar^2 + ar^3 + \cdots + ar^n$$

When we subtract the first row from the second, the ar term will cancel out, as will the ar^2 term, and so on. All the terms will cancel out except the first and the last.

$$rS - S = ar^n - a$$

Now we factorise.

$$S(r - 1) = a(r^n - 1)$$

Now divide by $(r - 1)$ and the formula is obtained.

$$S = \frac{a(r^n - 1)}{r - 1}$$

EXAMPLE 5.6

A GP has first term 3 and common ratio 2. Find the sum of the first ten terms.

Solution

Apply the formula, putting $a = 3$, $r = 2$ and $n = 10$.

$$\text{Sum} = \frac{3(2^{10} - 1)}{2 - 1} = 3(1024 - 1) = 3069$$

The sum of the first ten terms is 3069.

EXERCISE 5E

1 Find the sum of the first n terms for each of the following geometric progressions.

    a) first term 5, common ratio 3   **b)** first term 4, common ratio $\frac{1}{2}$

    c) first term 3, common ratio -2 **d)** first term 10, common ratio $-\frac{1}{3}$

2 Find the sum of the first 10 terms of each of the sequences in Question 1.

3 The sum of the first n terms of a GP is $2^n - 1$. Find the first term and the common ratio.

4 The sum of the first n terms of GPs are given below. In each case find the first term and the common ratio.

    a) $3^n - 1$ **b)** $5 \times 2^n - 5$ **c)** $2^{n+3} - 8$

    d) $8 - (\frac{1}{2})^{n-3}$ **e)** $9 - 9(-\frac{1}{2})^n$

5 There is a legend about the invention of the game of chess: an emperor was so delighted by it that he offered the inventor anything he desired. The inventor asked for one grain of rice on the first square of the board, two grains on the second, four on the third, eight on the fourth and so on for all the 64 squares. How many grains had the inventor asked for?

*6 It is claimed that the sum of the first n terms of a GP is $2^n + 2$. Show that this cannot be true.

7

> *As I was going to St Ives,*
> *I met a man with seven wives,*
> *Every wife had seven sacks,*
> *Every sack had seven cats,*
> *Every cat had seven kits.*
> *Kits, cats, sacks, man and wives,*
> *How many were going to St Ives?*   *(Nursery rhyme)*

*8 A sequence starts with 1, and then every term is found by adding all the previous terms. i.e.

$$u_1 = 1, \quad u_n = \sum_{i=1}^{n-1} u_i$$

Find the first five terms. Find a formula for the nth term.

Compound interest

If money is invested at compound interest, then in successive years the amount invested increases in a geometric progression.

Suppose the rate of interest is 8%. Then after one year each £100 becomes £108. The amount of money will have been multiplied by $\frac{108}{100} = 1.08$.

Hence the common ratio of the progression is 1.08.

EXAMPLE 5.7

£1000 is invested at 9% compound interest. How much will there be after n years?

Solution

At the end of the first year each £100 will have become £109, so the capital will have been multiplied by $\frac{109}{100} = 1.09$.

After n years it will have been multiplied by 1.09, n times.
After n years at 9%, there will be £1000 × 1.09^n.

EXERCISE 5F

1   £1000 is invested at 10% compound interest. What factor is the capital multiplied by each year? How much will there be after n years?

2   A population is increasing at 2% each year. If it is 10 000 000 now, what will it be in n years' time?

3   A radioactive material loses 10% of its mass each year. What proportion will be left after n years?

4   The value of a car decreases by 15% each year. If it cost £8000 when new, what will it cost in n years' time?

5   When a child is born his godmother invests for him £100 at 8% compound interest. She continues to invest £100 on each of the child's birthdays, up to and including his 18th. What will be the total value of the investments on the child's 18th birthday?

6   On his 40th birthday a man starts a pension policy, investing £3000 at 6% compound interest. He invests the same amount on every birthday up to and including his 64th. What will be the total value of the investments on his 65th birthday?

5.4 The sum to infinity of a geometric progression

The decimal form of the fraction $\frac{1}{3}$ is the recurring decimal $0.333\ldots$. This can be written as

$$\frac{3}{10} + \frac{3}{100} + \frac{3}{1000} + \cdots$$

This is a geometric series, with first term $\frac{3}{10}$ and common ratio $\frac{1}{10}$. As we take more terms of the series, their sum approaches $\frac{1}{3}$. So we say that the **sum to infinity** of the series is $\frac{1}{3}$.

Formula

Let a geometric progression have first term a and common ratio r, where $-1 < r < 1$. Then the sum to infinity of the series is

$$\sum_{i=1}^{\infty} ar^{i-1} = \frac{a}{1-r}$$

Proof

The formula for the sum of the first n terms of a GP is

$$\sum_{i=1}^{n} ar^{i-1} = \frac{a(r^n - 1)}{r - 1}$$

As n tends to infinity, $\sum_{i=1}^{n} ar^{i-1}$ tends to $\sum_{i=1}^{\infty} ar^{i-1}$. So the sum to infinity of the series is obtained by letting n tend to infinity in the formula above.

As r is between -1 and 1, r^n will get smaller as n gets larger and so r^n will approach 0 as n increases. The sum will approach the fixed value obtained by putting $r^n = 0$.

$$\text{Sum to infinity} = \frac{-a}{r-1} = \frac{a}{1-r}$$

EXAMPLE 5.8

A geometric progression has first term 3 and common ratio $\frac{1}{2}$. Find the sum to infinity.

Solution

As $-1 < r < 1$, the sum to infinity will exist. Use the formula, putting $a = 3$ and $r = \frac{1}{2}$.

$$\text{Sum to infinity} = \frac{3}{1 - \frac{1}{2}} = \frac{3}{(\frac{1}{2})} = 6$$

The sum to infinity is 6.

EXERCISE 5G

1 State whether or not the sum to infinity of each of the following geometric series exists. Find this sum where it does exist.

    a) first term 4, common ratio $\frac{1}{2}$ **b)** first term 3, common ratio -2

    c) first term 4, common ratio $-\frac{1}{3}$

2 The sum to infinity of a GP is 5. If the first term is 2 find the common ratio.

3 The sum to infinity of a GP is 6. If the common ratio is $-\frac{1}{2}$ find the first term.

4 The sum to infinity of a GP is three times the first term. Find the common ratio.

5 For what values of x does the series $1 + 2x + 4x^2 + 8x^3 + \cdots$ have a sum to infinity? Find this sum to infinity where it does exist.

6 Consider the recurring decimal $0.444\,444\ldots$. Show that it can be written as

$$\frac{4}{10} + \frac{4}{100} + \frac{4}{1000} + \cdots$$

Regarding this as a GP, find its sum to infinity as a fraction. Verify that this is equal to the original decimal.

7 Use the method of Question 6 to express as fractions the following recurring decimals.

    a) $0.777\,77\ldots$ **b)** $0.181\,818\,18\ldots$ **c)** $0.027\,027\,027\ldots$

*8 Achilles and the tortoise run a race. Achilles can run ten times as fast as the tortoise, so the tortoise is given a start of 100 paces.

When Achilles has made up the gap of 100 paces, the tortoise is 10 paces ahead. When Achilles has made up these 10 paces, the tortoise will be 1 pace ahead, and so on. The Ancient Greeks argued that Achilles could never catch up with the tortoise.

Write down the distances Achilles runs to make up the gap between him and the tortoise. By finding the sum to infinity of a geometric progression, find how far Achilles runs to catch up with the tortoise.

*9 A ball is thrown upwards, then 2 seconds later it returns to the ground, where it bounces. After each bounce its speed is reduced, so that the time it takes until the next bounce is multiplied by 0.75. Find the time that elapses until it stops bouncing. How many times will it bounce?

*10 A pendulum is swinging so that the distances of successive swings are in geometric progression. The first swing from left to right is 5 cm, and the second swing from right to left is 4 cm. Find the distance of the nth swing. Find the total distance the pendulum covers before it stops swinging.

*11 The sum to infinity of a geometric progression is 10. The sum of all the odd numbered terms (i.e. the first term, third term, fifth term, etc.) is 8. Find the common ratio and the first term.

LONGER EXERCISE

Mortgages

Most homeowners buy their house or flat with the help of a mortgage. An initial sum is borrowed, which is paid back over a long period such as 25 years. When people take out mortgages they are often surprised by two things.

a) The total repayments are considerably greater than the amount borrowed.
b) After paying back for two or three years they still owe almost as much as when the loan started.

Here we shall analyse a simple mortgage. We shall consider a mortgage of £30 000, to be paid back over a period of 25 years. We shall assume that the interest rate is constant at 8%, and that repayments are made annually at the end of each year. Let the repayments be £x. Say the loan is taken out on 1 January 2000. It will be repaid after the 25th repayment, on 31 December 2024.

1 This question refers to the situation when the loan is repaid, on 31 December 2024, so all money will have accrued interest from the time of investment up to this date.

    a) What will be the value of the £30 000, compounded at 8% interest? (Refer to the compound interest section on page 88.)

    b) In terms of x, what will be the value of the final repayment?

    c) In terms of x, what will be the value of the last but one repayment, paid on 31 December 2023? (This repayment has received one year of 8% interest.)

    d) What will be the total value of all the repayments?

    e) By equating your answers to (a) and (d), find the value of x.

2 What is the total of all the repayments?

3 How much of the loan is outstanding on 1 January 2003?

4 What mortgage can I take out, if I can afford to pay back £3000 per year over 20 years? Assume the rate of interest is always 8%.

EXAMINATION QUESTIONS

1 A 'Yearly Plan' is a National Savings scheme requiring 12 monthly payments of a fixed amount of money on the same date each month. All savings earn interest at a rate of x% per complete calendar month.

A saver decides to invest £20 per month in this scheme and makes no withdrawals during the year. Show that, after 12 complete calendar months, his first payment has increased in value to £$20r^{12}$, where $r = 1 + \dfrac{x}{100}$.

Show that the total value, after 12 complete calendar months, of all 12 payments is

$$£\frac{20r(r^{12} - 1)}{r - 1}$$

Hence calculate the total interest received during the 12 months when the monthly rate of interest is $\frac{1}{2}$ per cent.

J AS 1990

2 In a car with four gears, the maximum speed in bottom gear is 20 km h^{-1} and the maximum speed in top gear is 200 km h^{-1}. Given that the maximum speeds in each gear form a geometric series, calculate, in km h^{-1} to 1 decimal place, the maximum speeds in the two intermediate gears.

L AS 1989

3 a) Find the sum of the arithmetic progression 1, 4, 7, 10, 13, 16, $\cdots$, 1000.

Every third term of the above progression is removed, i.e. 7, 16, etc. Find the sum of the remaining terms.

b) The rth term, u_r, of a series is given by

$$u_r = \left(\frac{1}{3}\right)^{3r-2} + \left(\frac{1}{3}\right)^{3r-1}$$

Express $\sum_{r=1}^{n} u_r$ in the form $A\left(1 - \dfrac{B}{27^n}\right)$, where A and B are constants. Find the sum to infinity of the series.

C 1991

Summary and key points

1 The nth term of a sequence can be written u_n. The sum of the first n terms can be written

$$\sum_{i=1}^{n} u_i$$

Be careful with the notation. Do not confuse u_n (the nth term of a sequence) with u^n (u to the power n) or with

$$\sum_{i=1}^{n} u_i$$

(the sum of the first n terms).

2 In an arithmetic progression there is a constant difference between terms. If the first term is a and the common difference is d

nth term $= a + (n - 1)d$

sum of first n terms $= \frac{1}{2}n[2a + (n - 1)d]$

Be careful with signs. If an AP is **decreasing**, then its common difference is **negative**.

3 In a geometric progression there is a constant ratio between terms. If the first term is a and the common ratio is r

nth term $= ar^{n-1}$

sum of first n terms $= \dfrac{a(r^n - 1)}{r - 1}$

If the common ratio of a GP is negative, then the terms will be alternately positive and negative.

In many problems involving GPs there is more than one possible value of the common ratio. If your information leads to $r^2 = 4$, then r could be $+2$ or -2.

4 Suppose the common ratio of a geometric progression is between -1 and 1. Then as n tends to infinity the sum of n terms approaches the sum to infinity, given by the formula

$\dfrac{a}{1 - r}$

Consolidation section A

Chapter 1

1 Let $P(x) = x^3 + 2x^2 - 3x - 7$ and $Q(x) = 4x^3 - 2x^2 + 7x + 1$. Evaluate the following.

  a) $P(x) + Q(x)$ **b)** $2P(x) - 3Q(x)$ **c)** $P(x)Q(x)$

2 With $P(x)$ and $Q(x)$ as in Question 1, find non-zero a and b such that $aP(x) + bQ(x)$ is a quadratic.

3 Find the quotient and remainder when $x^3 - 2x^2 + 7x + 1$ is divided by $(x - 2)$.

4 Factorise these expressions.

  a) $x^2 - 8x + 7$ **b)** $x^2 + 5x - 84$ **c)** $2x^2 - 5x + 2$

5 Let $R(x) = \dfrac{2x - 3}{x^2 + 1}$ and $S(x) = \dfrac{x - 1}{x^2 + x + 1}$. Express the following as single fractions.

  a) $R(x)S(x)$ **b)** $\dfrac{R(x)}{S(x)}$ **c)** $R(x) + S(x)$

6 Find the remainder when $x^3 + 4x^2 - 3x + 2$ is divided by $(x + 3)$.

7 $x^3 + 2x^2 + ax + b$ has remainder 2 when divided by $(x - 1)$, and remainder -7 when divided by $(x + 2)$. Find a and b.

8 Show that $(x + 1)$ is a factor of $x^3 - 3x^2 + 5x + 9$.

9 X is proportional to the square of Y and $X = 24$ when $Y = 2$. Find an equation giving X in terms of Y.

10 Two variables x and y are inversely proportional to each other. When $x = 3$, $y = 4$. Find an equation giving x in terms of y.

Chapter 2

11 Solve these equations.

  a) $x^2 - 4x - 12 = 0$ **b)** $2x^2 + 5x - 1 = 0$ **c)** $x(2x - 1) = 50$

12 Complete the square for these quadratics.

    a) $x^2 + 4x - 3$ **b)** $x^2 - 3x - 1$ **c)** $3 - 4x - x^2$

13 Find the greatest or least values of the quadratics in Question 12, and the values of x at which they are reached.

14 Solve the following equations.

    a) $x^4 - 8x^2 + 15 = 0$ **b)** $x - 6\sqrt{x} + 8 = 0$ **c)** $x + y = 3$ and $x^2 + y^2 = 5$

Chapter 3

15 In triangle ABC, $AB = 0.96$ cm, $BC = 0.28$ cm and $B = 90°$. Find AC. Evaluate the following ratios.

    a) $\sin A$ **b)** $\cos A$ **c)** $\tan A$ **d)** $\cot A$ **e)** $\sec A$ **f)** $\operatorname{cosec} A$

16 Without the use of a calculator write down the values of the following.

    a) $\cos 60°$ **b)** $\sin 45°$ **c)** $\operatorname{cosec} 30°$ **d)** $\tan 0°$ **e)** $\cot 90°$

17 Express these in terms of the trigonometric ratios of acute angles.

    a) $\sin 145°$ **b)** $\tan 174°$ **c)** $\cos 220°$ **d)** $\operatorname{cosec} 260°$

    e) $\cos 330°$ **f)** $\sec 150°$ **g)** $\cos 500°$ **h)** $\sin -200°$

18 Find the angles between $0°$ and $360°$ for which these are true.

    a) $\sin x = 0.3$ **b)** $\cos x = -0.1$ **c)** $\cot x = 4$ **d)** $\tan x = -0.6$

Chapter 4

19 Find the distances between the following pairs of points.

    a) $(1, 1)$ and $(2, 7)$ **b)** $(2, -3)$ and $(4, 10)$ **c)** $(-3, 2)$ and $(-5, -8)$

20 Find the equation of the circle with centre $(2, 1)$ and radius 5.

21 A circle has equation $x^2 + y^2 - 2x + 6y - 12 = 0$. Find its centre and radius.

22 Find the equations of the following straight lines.

    a) through $(1, 2)$, with gradient 3   **b)** through $(-1, 5)$, with gradient -3

    c) through $(1, 3)$ and $(4, 5)$ **d)** through $(2, 6)$ and $(7, 1)$

    e) through $(1, -2)$, parallel to $y = 3x - 1$ **f)** through $(0, 2)$, perpendicular to $y + 3x = 5$

23 Find the intersection points of the following pairs of lines.

    a) $y + 3x = 2$ and $y = x - 10$ **b)** $2x + y = 3$ and $2y - 4x = 14$

24 Find the equation of the line through (1, 1) which is perpendicular to the line $y = 7 - x$. Find where the two lines intersect.

25 Find the equation of the line joining (3, 4) and (5, −2). Find the midpoint of this line segment.

26 Find the point which divides the line segment between (2, 7) and (12, 2) in the ratio $2:3$.

27 **a)** Find the angle between the line $y = 2x - 5$ and the x-axis.

    b) Find the angle between the lines $3y = 2x + 3$ and $y = 2x - 1$.

Chapter 5

28 An arithmetic progression has first term 2 and common difference 3. Find

    a) the third term   **b)** the nth term   **c)** the sum of the first n terms.

29 The nth term of an arithmetic progression is $6 - 2n$. Find

    a) the first term   **b)** the common difference   **c)** the sum of the first n terms.

30 A geometric progression has first term 81 and common ratio $\frac{2}{3}$. Find

    a) the fourth term   **b)** the nth term   **c)** the sum of the first n terms.

31 Find the sum to infinity of the series in Question 30.

32 The first term of a geometric progression is 8 and the fourth term is 27. Find the common ratio and the sum of the first n terms.

MIXED EXERCISE

1 1, x^2 and x are successive terms of an arithmetic progression. Find two possible values of x.

2 Factorise $(x + 1)(x^2 + x + 1) + (x + 1)(x^2 + 4x + 1)$.

3 The first three terms of a geometric progression are $3x + 3$, $x + 7$ and $2x - 2$. Find the possible values of the sum to infinity.

4 The first term of a geometric progression is 20. The sum to infinity is 25. Find the common ratio.

5 $(x - k)$ is a factor of $x^2 + (k - 3)x + k^2 - k - 4$. Find the possible values of k.

6 $x^2 + bx + c$ has the same remainder when divided by $(x - \alpha)$ or by $(x - \beta)$, where α and β are different. Show that $\alpha + \beta + b = 0$.

7 A point P(x, y) moves so that PA $= 2$PB, where A is at (3, 0) and B is at (0, 3). Show that P moves on the circle with equation

$$x^2 + y^2 + 2x - 8y + 9 = 0$$

and find the centre and radius of this circle.

8 A circle is drawn with centre (f, g) and radius r. Show that it crosses the x-axis at the roots of

$$x^2 - 2fx + f^2 + g^2 - r^2 = 0$$

9 The sum to infinity of a geometric progression is 1. The sum of the first three terms is eight times the sum of the next three terms. Find the first term and the common ratio.

10 A line goes through (a, b) and makes angle θ with the x-axis. Show that the equation of the line can be written as

$$y \cos \theta - x \sin \theta = b \cos \theta - a \sin \theta$$

11 Three points A, B and C are at (1, 1), (2, 5) and (5, 2) respectively. Find the equation of the bisector of angle BAC.

12 The diagram shows a hill of height R which is the shape of a hemisphere. A person whose eyes are h above the ground is standing on the top of the hill, and can just see a point which is x from the base of the hill. If the angle of depression of this point is α, show that

$$x = \frac{h(1 - \sin \alpha)}{\tan \alpha - \sin \alpha}$$

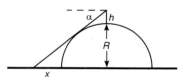

13 The diagram shows a hill in the shape of a cone, with sides that slope at an angle θ to the horizontal. From point A, the angle of elevation of the top of the hill is α, and the angle of elevation of a point halfway up the hill is β. Show that

$$\cot \theta = 2 \cot \alpha - \cot \beta$$

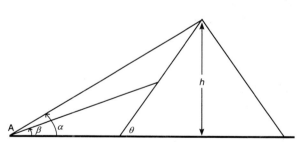

14 The first three terms of an arithmetic progression are $\cos \theta$, $3 \sin \theta$, $4 \cos \theta$. Find θ, given that it is acute.

15 The first three terms of a geometric progression are $\cos \theta$, 1, 3. Find θ, given that it is acute.

16 The common ratio of a geometric progression lies between 0 and 1. The sum to infinity of the series is five times the sum of all the even numbered terms. Find the common ratio.

17 The arithmetic mean of two numbers a and b is $\frac{1}{2}(a + b)$. Their geometric mean is $\sqrt{ab}$. If the numbers are positive, show that the arithmetic mean cannot be less than the geometric mean.

18 f(x) is a quadratic expression for which f(1)=3, and which is divisible by $(x-2)$. It has a remainder of 1 when divided by $(x-3)$. Find the quadratic.

19 A triangle has vertices at A(1, 1), B(2, 4) and C(6, 1). A general point P is at (x, y).

Let $f(x) = \frac{1}{3}(PA^2 + PB^2 + PC^2)$. Show that f(x) can be written in the form

$$f(x) = (x+a)^2 + (y+b)^2 + c$$

Hence find the least possible value of f(x), and the coordinates of P where this is achieved.

20 In the British Museum there is an Egyptian manuscript dating from about 1650BC, called the Rhind papyrus. It contains mathematical problems, one of which is equivalent to the following.

Divide 100 loaves among five men so that the shares shall be in arithmetic progression and one seventh of the sum of the largest three shares shall be equal to the sum of the smallest two shares. Find how many loaves are given to each of the five men.

21 A length of $4k$ metres of wire is bent into a rectangle. If one side of the rectangle is x m, show that the area is $x(2k-x)$ m^2.

Show that the area of the rectangle is greatest when it is a square.

22 x, y and z are positive numbers for which x^2, y^2 and z^2 are in arithmetic progression. Show that $\dfrac{1}{y+z}$, $\dfrac{1}{z+x}$, and $\dfrac{1}{x+y}$ are also in arithmetic progression.

23 What is wrong with the following argument?

Let $f(x) = x + x^2 + x^3 + \cdots$ and $g(x) = 1 + \dfrac{1}{x} + \dfrac{1}{x^2} + \cdots$. By summing geometric series, we have that $f(x) = \dfrac{x}{1-x}$ and that $g(x) = \dfrac{x}{x-1}$. Adding, $f(x) + g(x) = 0$ for all x.

LONGER EXERCISE

Length of day

The length of the day t months after the autumnal equinox (23 September) is approximately $12 - 4.5 \sin 30t$ hours of daylight per day.

1 How many hours of daylight are there on

(i) 3 December   (ii) 28 February   (iii) 4 July?

2 On which days will there be 11 hours of daylight?

3 According to this model, when is the shortest day? When is the longest?

4 The formula is changed to $12 - 4.5 \cos 30t$. Where is time now measured from?

5 If time were measured from the spring equinox, what would the formula be?

EXAMINATION QUESTIONS

1 The points A and B have coordinates (2, 3) and (0, −1) respectively. The angle BAC is a right angle and BC has a length of 5 units. Find the coordinates of each of the two possible positions for the point C.

AEB AS 1991

2 (i) Prove that the polynomial $P(x) = 2x^3 - x^2 + 9$ has $2x + 3$ as one factor.

   (ii) If $P(x)$ is written in the form $(2x + 3)Q(x)$, find the polynomial $Q(x)$.

   (iii) Write $Q(x)$ in the form $(x + b)^2 + c$, where b and c are numbers (positive or negative) to be determined, and hence show that $Q(x)$ is positive for all values of x.

   (iv) Deduce the complete set of values of x for which $P(x)$ is positive.

SMP 1989

3 Points A, B and C are the corners of a triangular field which has angle ABC $= 90°$ and AC $= 60$ m. The point D lies on AC and the point E lies on AB. A straight path joints D to E so that angle ADE $= 90°$ and AD $=$ BE $= 10$ m. Given that angle BAC $= x$,

a) Find, in terms of x, the lengths of AE and AB.

b) Hence prove that

$$6 \cos^2 x - \cos x - 1 = 0$$

c) Determine the values of cos x which satisfy this equation and hence find the values of x, $0° \le x \le 360°$, which also satisfy this equation.

Calculate the length, in m to 1 decimal place,

d) of the path DE

e) of a straight fence which joins B to D.   *L 1989*

4 In an arithmetic progression, the 8th term is twice the 3rd term and the 20th term is 110.

a) Find the common difference.

b) Determine the sum of the first 100 terms.   *AEB 1992*

5 The points A, B and C have coordinates (−1, 5), (3, −3) and (6, 1) respectively. Find the equation of the line L which passes through C and is parallel to AB.

The perpendicular bisector of AB meets L in the point D. Find the coordinates of D.

Show that the area of the trapezium ABCD is 25.   *JMB AS 1991*

6 An enclosure PQRS is to be made as shown in the diagram. PQ and QR are fences of total length 300 m. The other two sides are hedges. The angles Q and R are right angles, and angle S is 135°. The length of QR is x m.

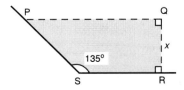

a) Show that the area, A m^2, of the enclosure is given by

$$A = 300x - \tfrac{3}{2}x^2$$

b) It is required that the area shall be at least $11\,250$ m^2. Find the range of values of x.

c) Show that A can be written as $-\tfrac{3}{2}[(x - a)^2 - b]$, where a and b are constants whose values you should discover. Hence show that A cannot exceed $15\,000$.

AEB 1990

Indices and logarithms

The words 'square' and 'cube' come from geometry. If a square has side x units, the square of the number x gives the area of the square. If a cube has side x units, the cube of the number x gives the volume of the cube.

These are written as x^2 and x^3. This notation can be extended to higher powers such as x^4 and x^5, although there is no obvious physical interpretation of these expressions. The **index notation** is used to save us having to write several multiplications.

$$5^6 \text{ means } 5 \times 5 \times 5 \times 5 \times 5 \times 5$$

As defined, the notation is restricted to positive whole numbers, but it can be further extended to negative and fractional powers, so that we can speak of 2^{-3}, or of $3^{\frac{1}{2}}$.

6.1 Index notation

Throughout this chapter we shall consider powers of a number a. In some cases a will be restricted to positive values.

If n is a positive whole number, a^n means n as, all multiplied together.

$$a^n = \underbrace{a \times a \times a \times \cdots \times a}_{n \ as}$$

A **negative power** is the **reciprocal** of the positive power.

$$a^{-n} = \frac{1}{a^n}$$

For this definition a must be non-zero.

The $\frac{1}{n}$th power of a number is its nth root.

$$a^{1/n} = \sqrt[n]{a}$$

If n is an even number, then a must be positive for this to hold.

The 0th power of a number is always 1.

$$a^0 = 1$$

For this definition a must be non-zero. The justification of these definitions comes after the next exercise.

EXERCISE 6A

1 Find the value of each of the following.

   a) 3^4

   b) 9^2

   c) 5^3

   d) $9^{\frac{1}{2}}$

   e) $64^{\frac{1}{3}}$

   f) 8^0

   g) $\dfrac{1}{4^{\frac{1}{2}}}$

   h) $0.001^{\frac{1}{3}}$

   i) 3^{-1}

   j) 2^{-2}

   k) 10^{-3}

   l) $\dfrac{1}{2^{-1}}$

   m) 0.1^{-2}

   n) $4^{-\frac{1}{2}}$

   o) $1000^{-\frac{1}{3}}$

2 Simplify the following.

   a) $\left(\frac{4}{9}\right)^{\frac{1}{2}} \times \left(\frac{81}{16}\right)^{\frac{1}{4}}$

   b) $8^{\frac{1}{3}} + 4^0$

   c) $\left(\frac{1}{2}\right)^{-2} - \left(\frac{1}{27}\right)^{-\frac{1}{3}}$

3 Simplify these

   a) $\dfrac{x - x^{-1}}{x - 1}$

   b) $\dfrac{1 + x^{-1}}{1 + x}$

   c) $\dfrac{x + y}{x^{-1} + y^{-1}}$

4 Find the value of x in the following equations.

   a) $8 = 2^x$

   b) $\frac{1}{4} = 2^x$

   c) $3 = 27^x$

   d) $1 = 4^x$

*5 In the definition of a^0, we specified that a should not be zero. What is 0^0? Try to find a sensible value, by considering a^a for smaller and smaller a.

Laws of indices

1 When powers are multiplied, the indices are added
   $$a^n \times a^m = a^{n+m}$$

2 When powers are divided, the indices are subtracted.
   $$a^n \div a^m = a^{n-m}$$

3 When we take a power of a power, the indices are multiplied.
   $$(a^n)^m = a^{nm}$$

Proofs

1 $\underbrace{(a \times a \times \cdots \times a)}_{n \text{ as}} \times \underbrace{(a \times a \times \cdots \times a)}_{m \text{ as}} = \underbrace{a \times a \times \cdots \times a}_{(n+m) \text{ as}}$

2 $\dfrac{\overbrace{a \times a \times \cdots \times a}^{n \text{ as}}}{\underbrace{a \times a \times \cdots \times a}_{m \text{ as}}} = \overbrace{a \times a \times \cdots \times a}^{(n-m) \text{ as}}$ (after cancelling m as from the top)

3 $\underbrace{\overbrace{(a \times a \times \cdots \times a)}^{n \text{ as}} \times \overbrace{(a \times a \times \cdots \times a)}^{n \text{ as}} \times \cdots \times \overbrace{(a \times a \times \cdots a)}^{n \text{ as}}}_{m \text{ factors}} = \overbrace{(a \times a \times \cdots \times a)}^{nm \text{ as}}$

These rules can provide the justification for the definitions of fractional, zero and negative indices, as follows.

Fractional

$$a^{\frac{1}{2}} \times a^{\frac{1}{2}} = a^{\frac{1}{2}+\frac{1}{2}} = a^1 = a$$

When we square $a^{\frac{1}{2}}$ we get a, so $a^{\frac{1}{2}}$ is the **square root** of a.

Zero

$$a^0 \times a^n = a^{0+n} = a^n$$

So $a^0 = 1$.

Negative

$$\frac{1}{a^n} = a^0 \div a^n = a^{0-n} = a^{-n}$$

To summarise, the rules are

$$a^n \times a^m = a^{n+m} \qquad a^n \div a^m = a^{n-m} \qquad (a^n)^m = a^{nm}$$

$$a^{1/n} = \sqrt[n]{a} \qquad a^0 = 1 \qquad a^{-n} = \frac{1}{a^n}$$

EXAMPLE 6.1
Evaluate $27^{\frac{2}{3}}$.

Solution
Rewrite this as $(27^{\frac{1}{3}})^2$.

$27^{\frac{1}{3}}$ is the cube root of 27, which is 3.

$$27^{\frac{2}{3}} = 3^2 = 9$$

So $27^{\frac{2}{3}} = 9$.

EXAMPLE 6.2
Simplify $x^2 \times x^5 \div x^3$.

Solution
Add and subtract the indices.
$$x^2 \times x^5 \div x^3 = x^{2+5-3} = x^4$$
So $x^2 \times x^5 \div x^3 = x^4$.

EXERCISE 6B

1 Evaluate the following.

a) $8^{\frac{2}{3}}$ **b)** $4^{\frac{3}{2}}$ **c)** $1000^{\frac{2}{3}}$

d) $27^{-\frac{1}{3}}$ **e)** $49^{-\frac{1}{2}}$ **f)** $64^{\frac{5}{6}}$

2 Write the following as single powers.

a) $2^3 \times 2^2 \times 2^4$ **b)** $3^{-1} \times 3^4 \times 3^8$ **c)** $5^2 \times 5^2 \div 5$

d) $2^3 \times 4^5 \times 8^2$ **e)** $3^{-2} \times 9^{-1} \times 27^3$ **f)** $5^3 \times 25^{-2} \div 125^{-4}$

3 Simplify the following.

a) $2^x \times 2^{2x} \times 2^{3x}$ **b)** $2^{4x} \times 4^x \times 16^x$ **c)** $100^y \times 10^x$

d) $x \times x^3 \times x^5$ **e)** $y^3 \times y^{-2} \times y^{-3}$ **f)** $z^a \times z^b \times z^{-c}$

4 For each of the following, write x in terms of the other variables.

a) $x^{\frac{1}{2}} = y$ **b)** $x^n = y$ **c)** $(xy)^{\frac{1}{2}} = z$ **d)** $\left(\dfrac{y}{x}\right)^{-\frac{1}{3}} = z$

5 Simplify $\left[\left(\dfrac{x^4}{9}\right)^{\frac{1}{2}} + \left(\dfrac{x^6}{8}\right)^{\frac{1}{3}}\right]^{\frac{1}{2}}$.

6 Suppose $x = 2^y$. Express

a) $(\frac{1}{8})^y$ in terms of x **b)** $\sqrt{x}$ in terms of y.

Index equations

Suppose powers of a number a are equal. Then provided that a is not 0 or 1, the powers or indices are equal.

       If $a^n = a^m$, then $n = m$.

This gives rise to an equation in the indices.

EXAMPLE 6.3
Solve the equation $2^{3x} = 4^{x-1}$.

Solution
We know that $4 = 2^2$. Rewrite the equation as
$$2^{3x} = (2^2)^{x-1} = 2^{2x-2}$$
By the principle above we can equate the indices.
$$3x = 2x - 2$$
So the solution is $x = -2$.

EXAMPLE 6.4
Solve the equation $4^x - 12 \times 2^x + 32 = 0$.

Solution
Write 4 as 2^2.
$$4^x - 12 \times 2^x + 32 = (2^2)^x - 12 \times 2^x + 32$$
$$= (2^x)^2 - 12 \times 2^x + 32 = 0$$
Writing $y = 2^x$, the equation becomes
$$y^2 - 12y + 32 = 0$$
This is a quadratic equation, with solutions $y = 4$ and $y = 8$.

Now rewrite these equations in terms of x.

If $y = 4$, $2^x = 4 = 2^2$ gives $x = 2$.

If $y = 8$, $2^x = 8 = 2^3$ gives $x = 3$.

So the solution is $x = 2$ or $x = 3$.

EXERCISE 6C

1 Solve the following equations.

a) $3^x = 3^{3x-6}$ **b)** $2^{2x-1} = 2^{3x-1}$ **c)** $5^{4x+3} = 5^{x-6}$

d) $4^x = 2^{x+3}$ **e)** $3^x = 9^x$ **f)** $2^{x+3} = 8^{x-1}$

g) $2^{x-1} = 2 \times 4^{2x}$ **h)** $9 \times 3^{4x} = 27^{x-1}$

2 Solve the following equations.

a) $(3^x)^2 - 12 \times 3^x + 27 = 0$ **b)** $9^x + 3^x - 12 = 0$ **c)** $2 \times 4^x - 5 \times 2^x + 2 = 0$

3 Solve the following pairs of simultaneous equations.

a) $2^x \times 2^y = 8$ and $3^x \div 3^y = 1$ **b)** $x + y = 3$ and $2^x \times 2^{2y} = 16$

4 At the beginning of this section we had the restriction that a was not 0 or 1. Why was this necessary?

6.2 Rational and irrational numbers

Fractions are sometimes called **rational** numbers. A number is rational if it is of the form $\frac{a}{b}$, where a and b are whole numbers and $b \neq 0$.

Not all numbers are rational. Such numbers are called **irrational**. In particular $\sqrt{2}$ is irrational. This was discovered about 500BC, by a member of the Pythagorean community in southern Italy. Nowadays a mathematician who makes an important discovery announces it to the world, but 2500 years ago mathematics was a more secretive and dangerous profession. The Pythagoreans tried to keep the discovery secret, and the person who revealed it to the public was executed.

Proof of irrationality

Suppose that $\sqrt{2}$ is rational, i.e. that $\sqrt{2} = \frac{a}{b}$, where a and b are whole numbers. We may also suppose that we have simplified this fraction as far as possible, so that a and b have no common factor. The following is a **proof by contradiction**: we shall deduce a contradiction from the original assumption that $\sqrt{2} = \frac{a}{b}$.

Square both sides, to obtain $2 = \frac{a^2}{b^2}$

Now multiply both sides by b^2, to obtain $2b^2 = a^2$.

This means that a^2 must be even, and it follows that a itself is even. (If a were odd, its square would also be odd.) We therefore write $a = 2c$, and substitute in the above equation to obtain

$$2b^2 = (2c)^2 = 4c^2$$

Divide by 2, to obtain $b^2 = 2c^2$.

It follows that b^2 is even, and, by the argument above, b itself is also even. The original assumption was that $\sqrt{2}$ could be written as $\frac{a}{b}$, where a and b are whole numbers with no common factor. But we have found that they do have a common factor of 2. The original assumption must be false.

So $\sqrt{2}$ cannot be written as $\frac{a}{b}$, where a and b are whole numbers.

So $\sqrt{2}$ is an example of an irrational number. Many other roots are also irrational. The number π is irrational, though it is very difficult to prove this.

In the Longer Exercise at the end of this chapter we shall find that many other numbers are irrational.

Recurring decimals

A decimal is recurring if it repeats itself. Examples of recurring decimals are

$$0.333\,33\ldots, \ 0.181\,818\,18\ldots, \ 5.343\,543\,543\,5\ldots$$

We already know that $0.3333\ldots$ is equal to the rational number $\frac{1}{3}$. In general, a number is rational if and only if its decimal form either terminates or recurs.

EXAMPLE 6.5
Show that $2 + \sqrt{2}$ is irrational.

Solution
Suppose that $2 + \sqrt{2}$ is rational, i.e. that $2 + \sqrt{2} = \dfrac{a}{b}$, where a and b are integers, $b \neq 0$. Then subtracting 2 from both sides and writing the right-hand side as a single fraction

$$\sqrt{2} = \frac{a}{b} - 2 = \frac{a - 2b}{b}$$

The right-hand side is a ratio of integers, so it is a rational number. But we know from the proof above that $\sqrt{2}$ is not rational. This is a contradiction, so $2 + \sqrt{2}$ must be irrational.

$2 + \sqrt{2}$ is irrational.

EXAMPLE 6.6
Express the recurring decimal $0.181\,818\ldots$ as a rational number.

Solution
Let $x = 0.181\,818\ldots$

Multiply x by 100.

$$100x = 18.181\,818\ldots$$

Subtract x from $100x$, and the recurring parts will cancel.

$$100x - x = 18.181\,818\ldots - 0.181\,818\ldots = 18$$

It follows that $99x = 18$. Divide by 99.

$$x = \frac{18}{99} = \frac{2}{11}$$

So $0.181\,818\ldots = \frac{2}{11}$.

EXERCISE 6D

1 Prove that the following numbers are irrational.

    a) $1 + \sqrt{2}$ **b)** $3\sqrt{2}$ **c)** $\sqrt{8}$

2 Let x and y be non-zero rational numbers, with $x = \frac{a}{b}$ and $y = \frac{c}{d}$. Show that the following are rational.

a) $x + y$ **b)** xy **c)** $\frac{x}{y}$

3 Let x be a non-zero rational number, and α an irrational. Show that the following are irrational.

a) $x + \alpha$ **b)** $x\alpha$ **c)** $\frac{x}{\alpha}$ **d)** $\frac{\alpha}{x}$

4 Let α and β be irrational numbers. Show by example that the following need not be irrational.

a) $\alpha + \beta$ **b)** $\alpha\beta$ **c)** $\frac{\alpha}{\beta}$

5 Express the following as rational numbers.

a) $0.4444\ldots$ **b)** $0.272\,727\ldots$ **c)** $0.123\,123\,123\ldots$

d) $0.233\,333\ldots$ **e)** $1.254\,545\,4\ldots$ **f)** $23.102\,020\,2\ldots$

*6 Given that a and b are non-zero digits, express these as fractions.

a) $0.aaaa$ **b)** $0.ababab\ldots$

Can you generalise?

Surds

A calculator will give $\sqrt{2}$ as $1.414\,213\,562$, to 10 significant figures. This is not the exact value. Because $\sqrt{2}$ is irrational, it can never be represented exactly as a fraction or as a terminating decimal.

A number expressed in terms of root signs is a **surd**. There are many advantages in expressing the number in this way, instead of writing it as a decimal.

Accuracy
The expression $\sqrt{2}$ is accurate. $1.414\,213\,562$ is only an approximation.

Brevity
$\sqrt{2}$ requires two symbols. $1.414\,213\,562$ requires eleven.

Structure
Leaving the square root sign in may reveal how the number was obtained, which would be obscured by writing it as a decimal. For example, if we are told that α is an acute angle for which $\cos\alpha = \dfrac{1}{\sqrt{2}}$, then we know that $\alpha = 45°$ exactly. If we were told that $\cos\alpha = 0.7071$, we would not be so sure.

Recall that when a square root is multiplied by itself, the result is the original number.

$$\sqrt{2} \times \sqrt{2} = 2$$

This can be rewritten as

$$2 \div \sqrt{2} = \sqrt{2}$$

Surds obey the usual rules of arithmetic and algebra. The multiplication of surds may involve expanding brackets.

EXAMPLE 6.7
Simplify $(2 + \sqrt{3})(4 - \sqrt{12})$.

Solution
Multiply out the brackets.

$$8 - 2\sqrt{12} + 4\sqrt{3} - \sqrt{3}\sqrt{12}$$

Now, $\sqrt{12} = \sqrt{4} \times \sqrt{3} = 2\sqrt{3}$. Also $\sqrt{3}\sqrt{12} = \sqrt{36} = 6$.

The expression becomes

$$8 - 2 \times 2\sqrt{3} + 4\sqrt{3} - 6 = 8 - 4\sqrt{3} + 4\sqrt{3} - 6 = 8 - 6$$

$$(2 + \sqrt{3})(4 - \sqrt{12}) = 2$$

Note
This answer is exact. If we had expressed $\sqrt{3}$ and $\sqrt{12}$ as decimals, we might not have obtained exactly 2 as our answer.

EXERCISE 6E

1 Simplify the following expressions.

    a) $(1 + \sqrt{2})(3 - \sqrt{2})$ **b)** $(2 - \sqrt{2})(3 + \sqrt{2})$ **c)** $(1 + \sqrt{3})(2 - \sqrt{3})$

    d) $(3 - \sqrt{2})(1 + \sqrt{8})$ **e)** $(\sqrt{2} - \sqrt{3})(\sqrt{2} + \sqrt{3})$ **f)** $(\sqrt{27} + \sqrt{8})(\sqrt{3} - \sqrt{2})$

    g) $(1 + \sqrt{2})^2$ **h)** $(2 - \sqrt{3})^2$ **i)** $(\sqrt{2} - 2\sqrt{3})^2$

2 Write the following in terms of $\sqrt{2}$.

    a) $\sqrt{18}$ **b)** $\sqrt{8}$ **c)** $\sqrt{24}/\sqrt{12}$ **d)** $\sqrt{14}/\sqrt{28}$

*3 Suppose $a + b\sqrt{2} = c + d\sqrt{2}$, where a, b, c and d are rational numbers. Show that $a = c$ and $b = d$.

Division of surds

When a surd is divided by another, we use a clever technique called 'rationalising the denominator', which sometimes simplifies the answer.

Suppose the expression is

$$\frac{a+b\sqrt{n}}{c+d\sqrt{m}}$$

where a, b, c, d, n and m are all rational numbers.

When $c+d\sqrt{m}$ is multiplied by $c-d\sqrt{m}$, it will become the difference of two squares, $c^2 - d^2m$, which is rational. So when numerator and denominator of the expression are multiplied by $c-d\sqrt{m}$, the denominator becomes rational and we can divide by it.

EXAMPLE 6.8

Simplify $\dfrac{3+\sqrt{5}}{2-\sqrt{3}}$.

Solution
Multiply numerator and denominator of this expression by $2+\sqrt{3}$.

$$\frac{(3+\sqrt{5})(2+\sqrt{3})}{(2-\sqrt{3})(2+\sqrt{3})} = \frac{6+3\sqrt{3}+2\sqrt{5}+\sqrt{5}\sqrt{3}}{4+2\sqrt{3}-2\sqrt{3}-\sqrt{3}\sqrt{3}} = \frac{6+3\sqrt{3}+2\sqrt{5}+\sqrt{15}}{4-3}$$

$$\frac{3+\sqrt{5}}{2-\sqrt{3}} = 6+2\sqrt{5}+3\sqrt{3}+\sqrt{15}$$

EXERCISE 6F

1 Simplify the following by the technique of rationalising the denominator.

  c) $\dfrac{2+\sqrt{5}}{3+\sqrt{8}}$ **b)** $\dfrac{3+\sqrt{5}}{\sqrt{3}-\sqrt{2}}$ **c)** $\dfrac{2+\sqrt{3}}{5-\sqrt{7}}$

2 Show that the following equations are true, (i) by evaluating them on a calculator (ii) by surd operations.

  a) $\sqrt{(17+12\sqrt{2})} = 3+2\sqrt{2}$ **b)** $\sqrt{(22-12\sqrt{2})} = 3\sqrt{2}-2$ **c)** $\sqrt{(9+4\sqrt{5})} = 2+\sqrt{5}$

3 The sides of a rectangle are $\sqrt{2}+1$ and $\sqrt{2}-1$. Find the length of the diagonal, leaving your answer as a surd.

4 An equilateral triangle has sides 3 cm long. Find the area of the triangle, leaving your answer as a surd.

5 Write in terms of surds, simplifying your answer as far as possible:

  a) $\sin 60° \times \sin 45°$ **b)** $(1+\cos 30°)^2$ **c)** $(\cos 45° + \sin 45°)^2$

*6 Simplify $\dfrac{x-1}{x^{\frac{1}{2}}-1}$.

6.3 Logarithms

So far we have been given the index, and have found the resulting power. Sometimes we want to go in the opposite direction, that is to find the index which will give us a particular power. The function which does this is the **logarithm** function.

Suppose $a^x = b$. Then if we raise a to the power x the result is b. We write this as $x = \log_a b$.

When speaking we say, 'The log of b to the base a is x.'

If no base is mentioned then the base is assumed to be 10.

$$\log x = \log_{10} x$$

EXAMPLE 6.9

Find **a)** $\log_{10} 1000$ **b)** $\log_{16} 8$

Solution

a) We want the power of 10 which will give us 1000.

We know that $10^3 = 1000$.

This means that if we raise 10 to the power 3 we get 1000.

$$\log_{10} 1000 = 3$$

b) The situation here is more complicated. Given that $\log_{16} 8 = x$, then we can rewrite the equation as

$$16^x = 8$$

Write both 16 and 8 as powers of 2.

$$(2^4)^x = 2^3$$

Hence $4x = 3$, giving $x = \dfrac{3}{4}$.

$$\log_{16} 8 = \dfrac{3}{4}$$

EXERCISE 6G

1 Find the following.

a) $\log_2 4$	**b)** $\log_{10} 100$	**c)** $\log_3 27$
d) $\log_{25} 625$	**e)** $\log_2 32$	**f)** $\log_2 1024$
g) $\log_{49} 7$	**h)** $\log_8 2$	**i)** $\log_{125} 5$
j) $\log_2 \frac{1}{4}$	**k)** $\log_3 1$	**l)** $\log_9 \frac{1}{3}$
m) $\log_{10} 0.001$	**n)** $\log_8 4$	**o)** $\log_{81} 27$

2 Find the following. Throughout a is a positive number.

   a) $\log_a a^2$ **b)** $\log_a 1$ **c)** $\log_a \sqrt{a}$ **d)** $\log_{a^2} a$

3 Rewrite each of the following in the form $a^b = c$.

   a) $\log_x y = z$ **b)** $\log_2 4 = 2$ **c)** $\log_9 3 = \frac{1}{2}$ **d)** $\log 100 = 2$

4 Find the unknown bases for the following.

   a) $\log_x 8 = 3$ **b)** $\log_x 7 = \frac{1}{2}$ **c)** $\log_x 27 = 1.5$

*5 Suppose that $\log_2 x = y$. What is $\log_4 x$?

*6 Show that $\log_a x = \dfrac{\log x}{\log a}$.

Laws of logarithms

There are laws of logarithms which correspond to those of indices.

a) $\log_a x + \log_a y = \log_a xy$ **b)** $\log_a x - \log_a y = \log_a \frac{x}{y}$

c) $\log_a x^n = n \log_a x$

Proof

Suppose $\log_a x = p$ and $\log_a y = q$.

Then $a^p = x$ and $a^q = y$.

a) Multiply x and y together.

$$xy = a^p a^q = a^{p+q}$$

Hence $\log_a xy = p + q = \log_a x + \log_a y$.

b) Divide x by y.

$$\frac{x}{y} = \frac{a^p}{a^q} = a^{p-q}$$

Hence $\log_a \frac{x}{y} = p - q = \log_a x - \log_a y$.

c) Take the nth power of x.

$$x^n = (a^p)^n = a^{pn}$$

Hence $\log_a x^n = pn = n \log_a x$.

EXAMPLE 6.10

Write the following as a single log.

$$\log x^2 + 5 \log x$$

Solution

Use rule **(c)** above.

$$5 \log x = \log x^5$$

Now use rule **(a)**.

$$\log x^2 + \log x^5 = \log (x^2 \times x^5)$$
$$= \log x^7$$
$$\log x^2 + 5 \log x = \log x^7$$

EXAMPLE 6.11

Solve this equation.

$$\log_4 x - \log_4(x - 2) = 0.5$$

Solution

Use rule **(b)**.

$$\log_4 x - \log_4(x - 2) = \log_4 \left(\frac{x}{x - 2} \right)$$

Rewrite the equation.

$$\frac{x}{x - 2} = 4^{0.5} = 2$$

Multiply both sides by $(x - 2)$, then simplify.

$$x = 2(x - 2) = 2x - 4$$
$$x = 4$$

If $\log_4 x - \log_4(x - 2) = 0.5$ then $x = 4$.

EXERCISE 6H

1 Simplify the following by writing them as single logs.

   a) $\log x + \log 3x$ **b)** $\log 2x + \log 4x$ **c)** $\log x^2 - 2 \log x$

2 Evaluate the following.

   a) $\log 8 - \log 0.8$ **b)** $\log 10 - \log 1000$ **c)** $\log_2 16 - \log_2 8$

3 Express $\log 36$ in terms of $\log 2$ and $\log 3$.

4 Express the following in terms of $\log x$, $\log y$ and $\log z$.

   a) $\log xyz$ **b)** $\log x^2 y$ **c)** $\log \dfrac{x}{z^2}$ **d)** $\log \sqrt{xyz}$

5 a is a positive number for which $\log_a 2 = x$ and $\log_a 3 = y$. Express the following in terms of x and y.

   a) $\log_a 6$ **b)** $\log_a 18$ **c)** $\log_a 0.75$ **d)** $\log_a \sqrt{2}$ **e)** $\log_a \sqrt{6}$

6 Solve the following equations.

   a) $\log_2(x + 1) = 3$ **b)** $\log(x + 5) - \log x = \log 5$ **c)** $\log_3 x + \log_3(3x) = 3$

7 Solve the following pairs of simultaneous equations.

   a) $xy = 1\,000\,000$ and $2 \log x - \log y = 6$

   b) $\log_2 x + \log_2 y = \log_2 36$ and $\log_2 x - \log_2 y = \log_2 9$

   c) $y = 27x$ and $\log_3 x + \log_3 y = 7$

   d) $y = 2 \log_3 x$ and $y = \log_3 9x + 3$

8 Express $\log_{10} 15^{87}$ in terms of $\log_{10} 15$. Hence find the number of digits in 15^{87}.

9 In 1985 the largest known prime was $2^{216091} - 1$. How many digits are there in this number?

*10 Suppose $x = \log_y z$, $y = \log_z x$ and $z = \log_x y$. Show that $xyz = 1$.

*11 Prove that $\log(x + y) = \log x + \frac{1}{2} \log\left(1 + \frac{2y}{x} + \frac{y^2}{x^2}\right)$

The equation $a^x = b$

Suppose we want to solve the equation $a^x = b$, where a and b are known. By the definition of logarithms, this is equivalent to finding the value of $\log_a b$. The technique is to take logs of both sides.

> If $a^x = b$, then $\log a^x = \log b$.

Use rule **(c)** above, to write the left-hand side as $x \log a$. Now we can divide by $\log a$.

$$x = \frac{\log b}{\log a}$$

There are many situations, in particular those involving growth or decay, in which we want to solve an equation of this form.

EXAMPLE 6.12
A population is increasing at 2% each year. At this rate, how long will it be before it doubles?

Solution
Every year, every 100 increases to 102. So every year the population is multiplied by $\frac{102}{100} = 1.02$. After n years, the population will have been multiplied by 1.02 n times over.

After n years, the population is multiplied by 1.02^n.

Suppose the population has been doubled. Then this expression must equal 2.

$$1.02^n = 2$$

This is an equation of the form $a^x = b$. Find the solution by taking logs as indicated.

$$n \log 1.02 = \log 2$$

So

$$n = \frac{\log 2}{\log 1.02}$$

$$= 35$$

The time to double is 35 years.

EXERCISE 6I

1 Solve the following equations.

   a) $2^x = 3$ **b)** $12^x = 20$ **c)** $4^x = 3$

2 Evaluate the following.

   a) $\log_4 5$ **b)** $\log_3 2$ **c)** $\log_4 0.1$

3 Solve the following equations.

   a) $3^{x+1} = 4^x$ **b)** $5^{2x} = 3^{x+1}$ **c)** $2^{2x+1} = 3^{3x-2}$

4 A population is growing at 3% per year. How long will it take to triple?

5 A population is decreasing at 5% each year. How long will it take to halve?

6 The half-life of a radioactive material is defined as the time it takes for half of it to decay into another element. Suppose a material decays at 2% each year. What is its half-life?

7 If you owe money on your credit card, you will have to pay interest every month. The credit card company will inform you of the equivalent **Annual Percentage Rate** (APR).

   a) If the monthly interest is 1.9%, what is the equivalent APR?

   b) What is the monthly interest equivalent to an annual rate of 20%?

LONGER EXERCISE

Irrationals

In Section 6.2 we proved that $\sqrt{2}$ is irrational.

1 Modify the proof to show that $\sqrt{3}$ is irrational.

2 We know that $\sqrt{4} = 2$, which is rational. Why cannot the proof be modified to show that $\sqrt{4}$ is irrational?

3 Modify the proof to show that $\sqrt[3]{2}$ is irrational.

*4 Let p be a prime number. Show that $\sqrt{p}$ is irrational.

*5 Let N be a number which is not a perfect square. Show that $\sqrt{N}$ is irrational.

*6 A wide class of irrationals can be found from the following theorem.

Let α be a solution of the equation $x^n + a_1 x^{n-1} + a_2 x^{n-2} + \cdots + a_n = 0$, where $a_1, a_2, \ldots, a_n$ are whole numbers. Then α is either a whole number or irrational.

Prove this theorem. (If α is a rational root which is not a whole number, then it can be written as $\frac{a}{b}$, where $b > 1$ and a and b have no common factor. Obtain a contradiction from this.)

EXAMINATION QUESTIONS

1 Given that $x = 2^p$ and $y = 4^q$ find, in terms of p and q,

a) $8xy$ as a power of 2

b) $\log_2 x^3 y$ in a form not involving logarithms.

L AS 1990

2 a) Express as a single logarithm in its simplest form

$$\log 2 + 2 \log 18 - \tfrac{3}{2} \log 36$$

b) Solve the equation

$$3^x = 4^{x-2}$$

giving your answer to three significant figures.

O 1990

3 Without using a calculator or mathematical tables, verify that the exact value of

$$(1 - \sqrt{3})^4 - 4(1 - \sqrt{3})^2 - 8(1 - \sqrt{3}) - 4$$

is zero. Hence write down one linear factor of $x^4 - 4x^2 - 8x - 4$ in the form $x + a + \sqrt{b}$, where a and b are integers.

C 1992

Summary and key points

1 Powers involving indices which are negative, fractional or zero are defined by

$$a^{-n} = \frac{1}{a^n} \qquad a^{\frac{1}{n}} = \sqrt[n]{a} \qquad a^0 = 1$$

Indices obey the following laws.

$$a^n \times a^m = a^{n+m} \qquad a^n \div a^m = a^{n-m} \qquad a^{\frac{1}{n}} = \sqrt[n]{a}$$

$$a^0 = 1 \qquad (a^n)^m = a^{nm}$$

Index equations can be solved using the following principle.

If $a^n = a^m$ then $n = m$, provided that a is not 0 or 1.

Notice that $a^n \times a^m$ is **not** equal to a^{nm}.

2 A rational number is one which can be written as $\frac{a}{b}$, where a and b are integers and $b \neq 0$.

A number whose decimal form is either terminating or recurring is rational. All other numbers are irrational.

Surds are expressions involving root signs. If a surd fraction has denominator $a + \sqrt{b}$, then it can be simplified by multiplying numerator and denominator by $a - \sqrt{b}$.

Notice that $\sqrt{a+b}$ is **not** equal to $\sqrt{a} + \sqrt{b}$.

3 If $a^x = b$, then $x = \log_a b$.

Logarithms obey the following rules.

$$\log n + \log m = \log nm \qquad \log n - \log m = \log \frac{n}{m}$$

$$\log n^x = x \log n$$

The solution of the equation $a^x = b$ is $x = \dfrac{\log b}{\log a}$.

To change base, use the rule $\log_a b = \dfrac{\log_{10} b}{\log_{10} a}$.

Note that $\log nm$ is **not** equal to $\log n \times \log m$ and $\log (n + m)$ is **not** equal to $\log n + \log m$.

Radians

In all the geometric and trigonometric applications you have met so far, you have measured angles in degrees. One degree is defined as 1/360 of a complete turn. Fractions of a degree can be written either as decimals, or in terms of minutes and seconds. Each degree is made up of 60 minutes, and each minute is made up of 60 seconds.

These units date from thousands of years ago, when the Babylonian sexagesimal (base 60) number system was in use. During the French Revolution an attempt was made to decimalise the measurement of angles. Each right angle (90 degrees) was divided into 100 grades. Most scientific calculators can work with angles in grades, but they are not used for many purposes.

Degrees and grades are arbitrarily chosen units. They are defined in terms of the number system in use, whether base 60 or base 10. For many scientific and mathematical purposes it is necessary to measure angles in units which do not depend on any particular number system. The units we use are radians.

If there is another advanced species in the Universe, it is very likely that they also measure angles in radians.

7.1 Radians

Consider a sector of a circle of radius r units. Suppose that the arc length is also r units. Then we define 1 **radian** to be the angle of the sector.

If the arc length is a, then the angle of the sector is $\frac{a}{r}$ radians.

$$\text{angle } \alpha = \frac{a}{r}$$

Fig. 7.1   Fig. 7.2

Note
This way of measuring angles does not depend on any number system. It is defined only in terms of the lengths of the radius and the arc.

Conversion between degrees and radians

The circumference of a circle of radius 1 unit is 2π units. So, measured in radians, the angle in a full circle is 2π.

Measured in degrees, the angle in a full circle is $360°$.

$$360 \text{ degrees} = 2\pi \text{ radians}$$

To convert from degrees to radians, multiply by $\dfrac{2\pi}{360}$ or $\dfrac{\pi}{180}$

To convert from radians to degrees, multiply by $\dfrac{180}{\pi}$

When we write an angle in radians, very often we express it as a multiple of π.

EXAMPLE 7.1

Express $30°$ in radians, leaving your answer as a multiple of π.

Solution

Use the formula above.

$$30° = \frac{30 \times \pi}{180} \text{ radians}$$

$30°$ is $\dfrac{\pi}{6}$ radians.

EXERCISE 7A

1 Convert the following to radians.

   a) $90°$ **b)** $270°$ **c)** $180°$ **d)** $45°$ **e)** $60°$ **f)** $120°$

2 Convert the following radian measurements to degrees.

   a) π **b)** $\frac{\pi}{2}$ **c)** $\frac{5\pi}{6}$ **d)** $\frac{2\pi}{3}$ **e)** $\frac{\pi}{4}$

3 Convert $\frac{\pi}{6}$ radians to degrees. Hence find $\sin \frac{\pi}{6}$.

4 Without the use of a calculator, find expressions for the following. (All angles are measured in radians.)

   a) $\sin \frac{\pi}{2}$ **b)** $\cos \frac{\pi}{6}$ **c)** $\tan \frac{\pi}{3}$

5 A circle has radius 5 cm. An arc has length 2 cm. What angle does the arc subtend at the centre? Express your answer in radians and in degrees.

6 A sector of a circle of radius 10 cm has angle $40°$. Convert $40°$ to radians, and hence find the arc length of a sector.

7 What is $1'$ in radians? ($1'$ is a minute, one sixtieth of a degree.) What is $1''$ in radians? ($1''$ is a second, one sixtieth of a minute.)

8 Grades were defined in the introduction to this chapter. Find rules to convert between radians and grades.

Use of a calculator

Scientific calculators can work in radians as well as in degrees (and also in grades). When we change to radian mode, a small 'rad' or 'R' appears on the display, and all the trigonometric functions will now be calculated in terms of radians.

EXERCISE 7B

For these exercises, make sure your calculator is set to radian mode.

1 Find the value of each of the following:

    a) $\sin 1$ **b)** $\cos 0.6$ **c)** $\tan 2$

    d) $\sin^{-1} 0.7$ **e)** $\cos^{-1} 0.3$ **f)** $\tan^{-1} 2$

2 Find the value of each of the following.

    a) $\cot 0.1$ **b)** $\operatorname{cosec} 1.1$ **c)** $\sec 1.5$

    d) $\sec^{-1} 2$ **e)** $\cot^{-1} 0.8$ **f)** $\operatorname{cosec}^{-1} 1.8$

3 a) Use the π button on your calculator to enter $\frac{\pi}{2}$. Find $\sin \frac{\pi}{2}$.

    b) Find $\cos^{-1} 0$. Multiply by 2. What number do you have now?

    c) Find $\sin^{-1} 0.5$. What number is this?

4 Find the value of each of the following.

    a) $\sin \frac{\pi}{6}$ **b)** $\cos \frac{\pi}{2}$ **c)** $\tan \frac{\pi}{4}$

5 Enter a very large number on your calculator. Find its $\tan^{-1}$. What number do you have now?

6 Enter a small number on your calculator. Find its $\tan$. What do you notice? What happens as the number gets smaller?

7.2 Arcs, sectors and segments

Suppose we have a sector of angle α radians from a circle of radius r.

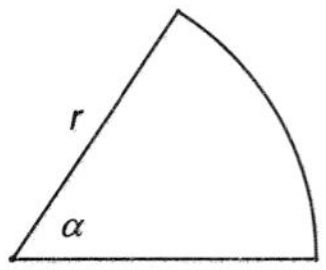

Arcs
By the definition of radians

      arc length $= r\alpha$

Fig. 7.3

Sectors
The area of the whole circle is πr^2. If this is reduced in the ratio $\alpha{:}2\pi$ we obtain the area of the sector.

$$\text{area of sector} = \pi r^2 \times \frac{\alpha}{2\pi} = \frac{1}{2}r^2\alpha$$

Note

These formulae are simpler than the corresponding formulae in terms of degrees. This confirms that the radian is a more natural unit to use for angles than the degree.

Segments

The segment shown shaded is the difference between a sector and a triangle. Find the two areas and subtract.

$$\text{area of sector} = \frac{1}{2}r^2\alpha$$

$$\text{area of triangle} = \frac{1}{2}r \times r \sin \alpha$$

$$\text{area of segment} = \frac{1}{2}r^2(\alpha - \sin \alpha)$$

Fig. 7.4

EXAMPLE 7.2

A chord of length 6 cm is drawn in a circle of radius 6 cm. Find the area of the smaller segment cut off by the chord.

Solution

The triangle formed by the radii and the chord is equilateral. Hence the angle subtended by the arc is $\frac{\pi}{3}$.

Apply the formula.

$$\text{area of segment} = \frac{1}{2} \times 6^2\left(\frac{\pi}{3} - \sin\frac{\pi}{3}\right)$$

$$= 3.26$$

Area of segment $= 3.26\,\text{cm}^2$

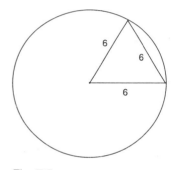

Fig. 7.5

EXERCISE 7C

Throughout this exercise, angles are measured in radians.

1 Find the lengths of the arcs in the sectors below.

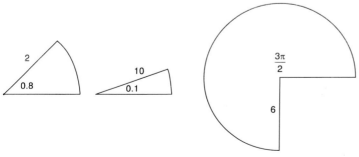

Fig. 7.6

2 Find the areas of the sectors below.

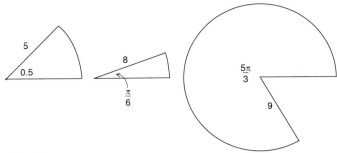

Fig. 7.7

3 Find the areas of the segments shown shaded below.

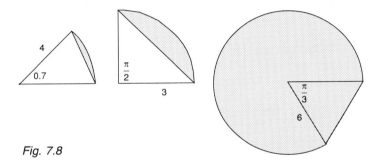

Fig. 7.8

4 An arc of length 5 cm subtends an angle of 1.2 at the centre of a circle. Find the radius of the circle.

5 A sector of area 4 cm² is taken from a circle of radius 7 cm. Find the angle of the sector.

6 A sector of angle 0.6 has area 10 square inches. Find the radius of the circle from which it is taken.

7 A segment of area 6 cm² is cut from a sector of angle 0.5. Find the radius of the circle from which it is taken.

8 A sector of a circle of radius 5 cm has area 50 cm² less than that of the circle. Find the angle of the sector.

9 The area of a segment is a third the area of the circle from which it is taken. Show that the angle, α radians which the segment subtends obeys the following equation:

$$3\alpha = 3 \sin \alpha + 2\pi$$

*10 Two circles each have radius 5 cm, and their centres are 8 cm apart. Find the area which is common to both circles.

*11 Find the total area covered by the shape formed by the two circles in Question 10.

*12 Find the perimeter of the shape formed by the two circles in Question 10.

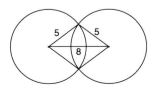

Fig. 7.9

*13 Three discs of radius 2 cm are placed on a table, so that each disc touches the other two. Find the area surrounded by the discs.

*14 A string passes round the outside of the three discs in Question 13. Find the length of the string.

15 A wire is wrapped tightly round the equator. (Suppose this to be possible.) By how much would the wire have to be lengthened in order to raise it 1 m above the surface all the way round? (Take the radius of the Earth as 6 400 000 m.)

16 A **nautical mile** is the length, along a circle of longitude on the surface of the Earth which subtends 1′ of latitude. Find the radius of the Earth in nautical miles.

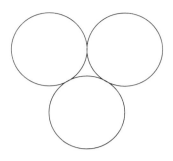

Fig. 7.10

Angular speed

The rate at which a wheel is turning can be measured by the number of radians it turns through per second. This is the **angular speed** of the wheel.

Suppose the angular speed of a wheel is ω radians per second. Suppose the radius of the wheel is r cm. Then in every second a point on the rim will move through an arc of length $r\omega$. So the actual speed of the point is $r\omega \, \text{cm s}^{-1}$.

So there is a direct connection between angular speed measured in radians per second and actual speed measured in cm per second. This is another justification for the use of radians.

Fig. 7.11

EXERCISE 7D

1 A wheel turns through five radians in two seconds. Find its angular speed.

2 A wheel is turning at 23 revolutions per second. Find its angular speed in radians per second.

3 A wheel turns at 53 radians per second. Find its speed of rotation in revolutions per second.

4 A wheel turns at ω radians per second. Find its speed in r.p.m. (revolutions per minute).

5 A wheel turns at r r.p.m. Find its angular speed in radians per second.

6 Find the angular speeds of the following in radians per second.

    a) the hour hand of a clock   **b)** the minute hand of a clock

    c) the Earth about the Sun   **d)** the Earth about its axis

7 A wheel of radius 3 cm rotates at 3.9 radians per second. Find the speed of a point on its rim.

8 A wheel has radius 10 cm. A point on its rim is moving at $27\,\text{cm s}^{-1}$. Find the angular speed of the wheel.

9 The wheels of a bicycle have radius 60 cm. Find the angular speed of the wheels when the bicycle is travelling at $10\,\text{m s}^{-1}$.

7.3 Small angles

If an angle α is small, then the approximate values of its sine, cosine and tangent can be found quickly.

1 $\sin \alpha \approx \alpha$ **2** $\tan \alpha \approx \alpha$ **3** $\cos \alpha \approx 1 - \frac{1}{2}\alpha^2$

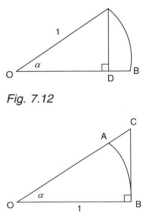

Fig. 7.12

Proof
1 Figure 7.12 shows a sector of angle α, taken from a circle of radius 1 unit. The arc AB has length α. The perpendicular AD has length $\sin \alpha$. If α is small, then these two lengths are approximately equal.

    $\sin \alpha \approx \alpha$

2 In Fig. 7.13 the perpendicular CB has length $\tan \alpha$. Again, if α is small CB is approximately equal in length to the arc length AB.

    $\tan \alpha \approx \alpha$

Fig. 7.13

3 In Fig. 7.14, triangle OAB is isosceles, hence the angle OBA is $\frac{\pi}{2} - \frac{1}{2}\alpha$. Therefore angle BAD is $\frac{1}{2}\alpha$.

Using trigonometry in triangle BAD

    $DB = AD \tan \frac{1}{2}\alpha = \sin \alpha \tan \frac{1}{2}\alpha$

Now we use the approximations **1** and **2** above.

    $DB \approx \alpha \times \frac{1}{2}\alpha = \frac{1}{2}\alpha^2$

Subtract DB from OB, to obtain OD.

    $OD \approx 1 - \frac{1}{2}\alpha^2$

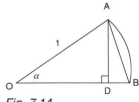

Fig. 7.14

The length OD is equal to $\cos \alpha$. This gives the required approximation:

    $\cos \alpha \approx 1 - \frac{1}{2}\alpha^2$

EXAMPLE 7.3
If x is a small angle, find the approximate value of $\dfrac{\sin x \tan x}{1 - \cos x}$.

Solution

Apply the approximations above.

$$\frac{\sin x \tan x}{1 - \cos x} \approx \frac{x \times x}{1 - (1 - \frac{1}{2}x^2)} = \frac{x^2}{\frac{1}{2}x^2} = \frac{1}{\frac{1}{2}} = 2$$

The value of $\dfrac{\sin x \tan x}{1 - \cos x}$ is approximately 2.

EXERCISE 7E

Throughout this exercise all angles are measured in radians.

1 Use your calculator to complete the following table:

Angle x	$\sin x$	$\tan x$	$\dfrac{\sin x}{x}$	$\dfrac{\tan x}{x}$
0.1				
0.01				
0.001				

Does this confirm the approximations at the beginning of this section?

2 Find the error in the approximation $\cos x \simeq 1 - \frac{1}{2}x^2$ in the following cases.

a) $x = 1$ **b)** $x = 0.1$

3 If x is a small angle, find the approximate values of the following.

a) $\sin 2x$ **b)** $\tan \frac{1}{2}x$ **c)** $\cos 4x$

d) $\dfrac{\sin 2x}{\tan 4x}$ **e)** $\dfrac{1 - \cos x}{x}$ **f)** $\dfrac{1 - \cos 2x}{1 - \cos 3x}$

4 If n is large, find the approximate values of the following.

a) $n \sin \dfrac{2}{n}$ **b)** $\dfrac{\tan \frac{2}{n}}{\sin \frac{3}{n}}$ **c)** $\dfrac{\tan \frac{1}{n} \sin \frac{1}{n}}{1 - \cos \frac{3}{n}}$

*5 Use the approximations of this section to find a solution when x is small to this equation.

$$\sin x + \cos x = 1.1$$

6 Find small values of x which are approximate solutions of the following equations.

a) $\tan 2x = \sin x + 0.1$ **b)** $\cos x + x \sin x = 1.045$

c) $\cos x - 4 \sin x = x^2$ **d)** $\cos x = 11x$

*7 a) By trial and improvement, find the value of x such that the approximation $\sin x \approx x$ is accurate to within a 1% degree of accuracy. Give your answer correct to two decimal places.

b) By trial and improvement, find the value of x such that the approximation $\cos x \approx 1 - \frac{1}{2}x^2$ is accurate to within a 1% degree of accuracy. Give your answer correct to two decimal places.

c) Which of the approximations above is more accurate?

8 On the same graph paper, draw the graphs of **a)** $y = x$ **b)** $y = \sin x$ **c)** $y = \tan x$, all for $0 \le x \le 0.5$. How does this connect with the approximations of this section?

LONGER EXERCISE

Series for sin x, tan x and cos x

The approximations for $\sin x$ and $\cos x$ can be improved as follows:

$$\sin x \approx x - \tfrac{1}{6}x^3$$
$$\cos x \approx 1 - \tfrac{1}{2}x^2 + \tfrac{1}{24}x^4$$

1 Use these new expressions to find the approximate values of the following.

a) $\sin 0.1$ **b)** $\cos 0.1$ **c)** $\sin 0.5$ **d)** $\cos 0.5$

2 What are the errors in your answers to Question 1?

3 Use your answers to Question 1 to find the approximate values of

a) $\tan 0.1$ **b)** $\tan 0.5$

4 Suppose that the approximation $\sin \alpha \approx \alpha$ is accurate to within 1%. By putting $\alpha = 1.01 \sin \alpha$ into the improved approximation above, find the value of α.

*5 The approximation $\tan x \approx x$ can be improved to

$$\tan x \approx x + kx^3$$

where k is a constant.

Use your calculator to find the value of k.

EXAMINATION QUESTIONS

1 A sector OAB of a circle, of radius a and centre O, has $\angle AOB = \theta$ radians. Given that the area of the sector OAB is twice the square of the length of the arc AB, find θ.   *C 1991*

2 Figure 7.15 shows a triangle ABC in which AB = 5 cm, AC = BC = 3 cm. The circle, centre A, radius 3 cm, cuts AB at X; the circle, centre B, radius 3 cm, cuts AB at Y.

a) Determine the size of the angle CAB, giving your answer in radians to four decimal places.

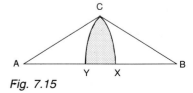

Fig. 7.15

b) The region R, shaded in the diagram, is bounded by the arcs CX, CY and the straight line XY.

Calculate

(i)   the length of the perimeter of R
(ii) the area of the sector ACX
(iii) the area of the region R.

O 1991

3 a) The angle in a sector of a given circle is 20° and the area of the sector is 2 m². Calculate the arc-length of the sector.

b) Given that θ is so small that powers of θ above the second may be neglected, find an approximate value for the acute angle θ (measured in radians) which satisifies the equation

$$\cos 3\theta - \cos 5\theta = \frac{2}{625}$$

NI 1991

Summary and key points

1 An arc of length a units from a circle of radius r units subtends $\frac{a}{r}$ radians at the centre.

To convert from degrees to radians multiply by $\frac{\pi}{180}$.

To convert from radians to degrees multiply by $\frac{180}{\pi}$.

Often we express angles as multiples of π. A right angle, for example, is $\frac{\pi}{2}$ radians.

When using a calculator, check whether it is in degree or radian mode.

2 Take an angle α enclosed between two radii of a circle of radius r.

The arc length is $r\alpha$.

The sector area is $\frac{1}{2}\alpha r^2$.

The segment between the chord and the arc has area $\frac{1}{2}r^2(\alpha - \sin \alpha)$.

Suppose a wheel of radius r metres is rotating at ω radians per second. Then a point on the rim is moving at $r\omega$ metres per second.

3 If α is a small angle, the following approximations hold:

$$\sin \alpha \approx \alpha \qquad \tan \alpha \approx \alpha \qquad \cos \alpha \approx 1 - \frac{1}{2}\alpha^2$$

These approximations only hold if α is measured in radians.

Differentiation

With this chapter you begin your study of calculus, which is the most important invention in the whole of science. It provides the language within which the behaviour of very many phenomena can be described and predicted. With calculus it is possible to analyse the behaviour of electricity, magnetism, planetary motion, gases and many other physical phenomena. Calculus has also become increasingly important in social sciences such as economics.

Calculus deals with quantities that are changing. For example, when a car is being driven its position on the road is changing. Suppose the speedometer of the car registers 30 m.p.h. What does this mean?

In 1 hour, the car travels 30 miles.

But it is very unlikely that the car will keep travelling at exactly the same speed for one hour. It could be rewritten.

In $\frac{1}{2}$ hour, the car travels 15 miles.

Again, it is unlikely that the car will keep up the same speed for half an hour.

In 1 minute, the car travels 0.5 mile.

A car is unlikely to travel at a constant speed, even for 1 minute. If the car is accelerating, it will not maintain the same speed for any period of time, however small. The 30 m.p.h. registered by the speedometer tells us the **instantaneous** speed. It is speed registered over an infinitesimally small period of time.

To investigate instantaneous speed we need **calculus**. Calculus deals with the infinitely small and the infinitely large.

8.1 The gradient of $y = x^2$

Figure 8.1 shows the graph of $y = x^2$. Suppose we want to find the gradient of the curve at the point $(1, 1)$. We could do it by drawing the tangent to the curve as shown. The tangent goes through $(2, 3)$ and $(0, -1)$, and so its gradient is $\frac{4}{2} = 2$.

Finding a gradient by drawing a tangent is time-consuming and inaccurate. It is better to find it by calculation. The initial stages are complicated, but by the end of this section we shall be able to find the gradient at any point on the curve immediately.

We shall find the gradient at the point $(1, 1)$ by considering chords which go from $(1, 1)$ to points higher up the curve. These points will get closer to $(1, 1)$.

The chord from $(1, 1)$ to $(2, 4)$ has gradient $\dfrac{4 - 1}{2 - 1} = 3$

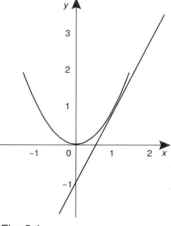

Fig. 8.1

The chord from $(1, 1)$ to $(1.5, 2.25)$ has gradient $\dfrac{2.25 - 1}{1.5 - 1} = 2.5$

The chord from $(1, 1)$ to $(1.1, 1.21)$ has gradient $\dfrac{1.21 - 1}{1.1 - 1} = 2.1$

The chord from $(1, 1)$ to $(1.01, 1.0201)$ has gradient

$$\frac{1.0201 - 1}{1.01 - 1} = 2.01$$

Notice two things.

- The chords are getting closer and closer to the tangent.
- The gradients are getting closer and closer to 2.

It seems very likely that the gradient of the tangent is 2, i.e. that the gradient of the curve at $(1, 1)$ is 2.

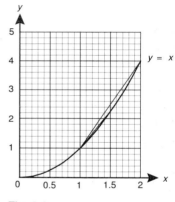

Fig. 8.2

EXERCISE 8A

1 Find the gradient of $y = x^2$ at $(2, 4)$ by drawing a tangent at that point and finding its gradient. Then follow the procedure below.

   a) Show that $(3, 9)$, $(2.5, 6.25)$, $(2.1, 4.41)$ and $(2.01, 4.0401)$ all lie on the curve $y = x^2$.

   b) For each point of part **(a)**, find the gradient of the chord joining it to $(2, 4)$.

   c) Are these gradients getting closer to a particular value? What is this value?

2 Find the gradient of $y = x^2$ at $(3, 9)$.

Proof that gradient at (1, 1) is 2

Numerical evidence is not enough to prove a result. A proof is needed.

We need a notation to describe the small increases in x and y. We use the expressions δx and δy. Note that δx does not mean the product of δ and x. The δ is not a number, it means 'a small increase in'. It applies to other variables: δy means 'a small increase in y'.

The chord starts at (1, 1). Suppose we increase x by a small amount δx. Then x has become $1 + \delta x$ and y has become $(1 + \delta x)^2$.

The chord now goes from (1, 1) to $(1 + \delta x, \ (1 + \delta x)^2)$. The gradient of this chord is given by

$$\text{gradient} = \frac{(1 + \delta x)^2 - 1}{1 + \delta x - 1} = \frac{1 + 2\delta x + (\delta x)^2 - 1}{\delta x} = \frac{2\delta x + (\delta x)^2}{\delta x} = 2 + \delta x$$

Now we let δx get smaller and smaller. The chord gets closer and closer to the tangent. The gradient of this chord is $2 + \delta x$, and as δx becomes smaller this gets closer to 2.

EXERCISE 8B

1 Amend the proof above, to show that the gradient of $y = x^2$ at (2, 4) is 4.

2 Show that the gradient of $y = x^2$ at (3, 9) is 6.

3 Can you guess what the gradient is, at the general point $(x, \ x^2)$?

The general gradient

Let us take a general point on the curve $y = x^2$. For a general value of x, the corresponding value of y is x^2. So the general point is $(x, \ x^2)$.

Increase x by a small amount δx, so that it becomes $x + \delta x$. The new value of y is $(x + \delta x)^2$. The increase in y is written δy.

The chord now goes from $(x, \ x^2)$ to $(x + \delta x, \ (x + \delta x)^2)$. The gradient of this chord is given by

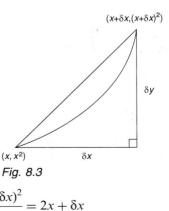

Fig. 8.3

$$\text{gradient} = \frac{\delta y}{\delta x} = \frac{(x + \delta x)^2 - x^2}{\delta x} = \frac{x^2 + 2x\delta x + (\delta x)^2 - x^2}{\delta x} = \frac{2x\delta x + (\delta x)^2}{\delta x} = 2x + \delta x$$

Now we let δx get smaller and smaller. The chord gets closer and closer to the tangent. The gradient of this chord is $2x + \delta x$, and as

δx becomes very small we can ignore it. The gradient gets closer to $2x$.

The gradient of $y = x^2$ at $(x, \ x^2)$ is $2x$.

EXAMPLE 8.1
Find the gradient of the curve $y = x^2 - 3x + 1$.

Solution
Apply the procedure above. Let x increase by a small amount δx, and let the corresponding increase in y be δy.

$$\delta y = ((x + \delta x)^2 - 3(x + \delta x) + 1) - (x^2 - 3x + 1)$$

Expand and simplify.

$$\delta y = x^2 + 2x\delta x + (\delta x)^2 - 3x - 3\delta x + 1 - x^2 + 3x - 1$$
$$= 2x\delta x - 3\delta x + (\delta x)^2$$

Now divide by δx to obtain the gradient of the chord.

$$\frac{\delta y}{\delta x} = 2x - 3 + \delta x$$

Now let δx get smaller and smaller. The chord gets closer to the tangent. The gradient of the chord approaches the gradient of the tangent. In the expression above we can ignore δx.

The gradient of $y = x^2 - 3x + 1$ at $(x, \ y)$ is $2x - 3$.

EXERCISE 8C

1 Find the gradients of the following functions.

   a) $y = 2x^2$ **b)** $y = x^2 + 2x - 1$ **c)** $y = 2x^2 + x + 1$

   d) $y = x^2 - 2x + 1$ **e)** $y = 2x^2 - 3x + 4$ **f)** $y = 7x - x^2$

*2 Let a, b and c be constants. Find the gradients of the following.

   a) $y = ax^2$ **b)** $y = x^2 + bx$ **c)** $y = ax^2 + bx + c$

8.2 Differentiation of $y = x^n$

The process of Section 8.1 is called **differentiation**. We have differentiated x^2 and obtained $2x$. A notation to describe the differentiation is as follows.

The gradient of the chord is $\dfrac{\delta y}{\delta x}$. We then let δx approach zero, so the gradient of the chord approaches the gradient of the tangent. The value that

$$\frac{\delta y}{\delta x}$$

approaches, as δx approaches zero, is written

$$\frac{dy}{dx}$$

Note that this is not an ordinary algebraic fraction. The ds cannot be cancelled to give

$$\frac{y}{x}$$

Another way of saying this is that

the **limit** of $\dfrac{\delta y}{\delta x}$ is $\dfrac{dy}{dx}$.

This phrase occurs so often that there is a special notation for it

$$\lim_{\delta x \to 0} \frac{\delta y}{\delta x} = \frac{dy}{dx}$$

The main result of the previous section can be written as

if $y = x^2$, then $\dfrac{dy}{dx} = 2x$

EXAMPLE 8.2

Let $y = 2x^2 + 3$. Find $\dfrac{dy}{dx}$.

Solution
Follow the procedure of the previous section.

$$\delta y = (2(x + \delta x)^2 + 3) - (2x^2 + 3)$$
$$\delta y = 2x^2 + 4x\delta x + 2(\delta x)^2 + 3 - 2x^2 - 3 = 4x\delta x + 2(\delta x)^2$$

Divide by δx to obtain the gradient of the chord.

$$\frac{\delta y}{\delta x} = 4x + \delta x$$

Let δx tend to zero. The gradient of the chord becomes $\dfrac{dy}{dx}$

$$\frac{dy}{dx} = 4x$$

EXERCISE 8D

1 Find $\dfrac{dy}{dx}$ for each of the following functions.

a) $y = x^2 + 5x$ **b)** $y = 2 - x^2$ **c)** $y = 3x^2 - 2x - 1$

2 Let $y = x^3$. Use the procedure of the previous section to find $\dfrac{dy}{dx}$.

You will need the expansion $(x + h)^3 = x^3 + 3x^2h + 3xh^2 + h^3$.

3 The expansion of $(x + h)^4$ is $x^4 + 4x^3h + 6x^2h^2 + 4xh^3 + h^4$. Use this expansion to differentiate $y = x^4$.

4 Let $y = x^n$, where n is a constant. Make a guess at the expression for $\dfrac{dy}{dx}$.

Differentiation of $\dfrac{1}{x}$

Let $y = \dfrac{1}{x}$. We shall follow the procedure above to find $\dfrac{dy}{dx}$.

Increase x by a small amount δx. So x has increased from x to $x + \delta x$.

Now y will be $\dfrac{1}{x + \delta x}$. So the increase δy of y is

$$\frac{1}{x + \delta x} - \frac{1}{x}$$

Write down the gradient of the chord joining $\left(x, \dfrac{1}{x} \right)$ and $\left(x + \delta x, \dfrac{1}{x + \delta x} \right)$.

$$\frac{\delta y}{\delta x} = \frac{1/(x + \delta x) - 1/x}{\delta x} = \frac{x - (x + \delta x)}{(x + \delta x)x\delta x} = \frac{-\delta x}{(x + \delta x)x\delta x} = \frac{-1}{(x + \delta x)x}$$

Now let δx get smaller and smaller, so that $x + \delta x$ tends to x.
Then the gradient tends to $\dfrac{-1}{x^2}$.

If $y = \dfrac{1}{x}$, $\dfrac{dy}{dx} = \dfrac{-1}{x^2}$

EXERCISE 8E

1 Let $y = \dfrac{1}{x^2}$. Find $\dfrac{dy}{dx}$.

*2 Let $y = \sqrt{x}$. Find $\dfrac{dy}{dx}$.

You will need the identity $(\sqrt{x + h} - \sqrt{x})(\sqrt{x + h} + \sqrt{x}) = x + h - x$.

Rules for differentiating

This process of finding the gradient is called **differentiating from first principles**. It is very time consuming, and it would be better to have a set of general rules.

In every case we have looked at so far, whenever we have differentiated a power of x, we have multiplied by the power and reduced the power by 1. The following rule has been found to hold.

If $y = x^n$, then $\dfrac{dy}{dx} = nx^{n-1}$

This rule is always true, even when n is negative or fractional.

There are two special cases which should be mentioned.

n = 0
If $y = x^0$, then $y = 1$, which is constant. The graph of $y = 1$ is horizontal, with zero gradient. This fits the rule above.

n = 1
If $y = x^1$, then $y = x$. The graph of $y = x$ is a straight line with gradient 1. Again, this fits the rule above.

EXAMPLE 8.3
Differentiate $y = x^4 \times \sqrt{x}$.

Solution
Our rule works for powers of x. The expression above can be written in that form.

$$y = x^4 \times \sqrt{x} = x^4 \times x^{\frac{1}{2}} = x^{4\frac{1}{2}}$$

Now the rule for differentiation can be applied.

$$\frac{dy}{dx} = 4\tfrac{1}{2}x^{3\frac{1}{2}}$$

EXERCISE 8F

1 Differentiate the following.

a) $y = x^5$

b) $y = x^{11}$

c) $y = x^{-2}$

d) $y = \dfrac{1}{x^3}$

e) $y = \dfrac{1}{x^5}$

f) $y = x^{\frac{1}{4}}$

g) $y = \sqrt[3]{x}$

h) $y = \sqrt[4]{x}$

i) $y = \dfrac{1}{\sqrt{x}}$

2 Differentiate the following, by first writing them as single powers of x.

a) $y = x^2 \times x^3$ **b)** $y = x^4 \times x^5$ **c)** $y = (x^3)^4$

d) $y = x^3 \div x^6$ **e)** $y = x \div x^4$ **f)** $y = \left(\dfrac{1}{x^4}\right)^3$

g) $y = x\sqrt{x}$ **h)** $y = x^2 \times \sqrt[3]{x}$ **i)** $y = x^2 \div \sqrt{x}$

3 Let $y = x^2$. Find the point where $\dfrac{dy}{dx} = 6$.

4 Let $y = x^3$. Find the points where $\dfrac{dy}{dx} = 12$.

5 Let $y = x^{-1}$. Find the points where $\dfrac{dy}{dx} = -4$.

8.3 Differentiation of functions of the form $ax^n + bx^m$

When we differentiate, we find the function $\dfrac{dy}{dx}$.

This function is called the **derivative** of y with respect to x. The result of the previous section can be reworded.

The derivative of x^n is nx^{n-1}.

Multiplying by a constant

If x^n is multiplied by a constant a, its derivative is also multiplied by a. So the derivative of ax^n is nax^{n-1}.

Justification
Suppose that $y = az$, where a is a constant and z is a function of x. Then if z increases by a small amount δz, the increase in y will be $\delta y = a\delta z$. Dividing by δx gives

$$\frac{\delta y}{\delta x} = a\frac{\delta z}{\delta x}$$

Now letting δx tend to zero, we obtain the derivatives.

$$\frac{dy}{dx} = a\frac{dz}{dx}$$

Adding functions

If functions are added, their derivatives are also added. So the derivative of $x^n + x^m$ is $nx^{n-1} + mx^{m-1}$.

Justification

Suppose that $y = z + w$, where z and w are functions of x. If z and w increase by small amounts δz and δw respectively, the increase in y will be the sum of these increases.

$$\delta y = \delta z + \delta w$$

Dividing by δx gives

$$\frac{\delta y}{\delta x} = \frac{\delta z}{\delta x} + \frac{\delta w}{\delta x}$$

Letting δx tend to zero, we obtain the derivatives.

$$\frac{dy}{dx} = \frac{dz}{dx} + \frac{dw}{dx}$$

Now we can put these together.

The derivative of $ax^n + bx^m$ is $nax^{n-1} + mbx^{m-1}$.

Note

The equation of a straight line is $y = ax + b$, where a is the gradient and b is the intercept on the y-axis. Differentiating, we confirm that its gradient is

$$\frac{dy}{dx} = a.$$

EXAMPLE 8.4

Let $y = 3x^4 + 7x^2$. Find $\dfrac{dy}{dx}$.

Solution

The derivatives of x^4 and x^2 are $4x^3$ and $2x$ respectively. Multiply these by 3 and 7 respectively and add.

$$\frac{dy}{dx} = 12x^3 + 14x$$

EXERCISE 8G

1 Differentiate the following.

a) $y = 5x^2$

b) $y = 12x^4$

c) $y = 5\sqrt{x}$

d) $y = 7x^{-3}$

e) $y = \frac{8}{x}$

f) $y = \frac{3}{\sqrt{x}}$

2 Differentiate the following.

a) $y = x^3 + x^5$

b) $y = x^4 + \dfrac{1}{x^2}$

c) $y = \sqrt{x} + \dfrac{1}{\sqrt{x}}$

d) $y = x^3 + x^2 + x$

e) $y = x^2 + x + 1$

3 Differentiate the following.

a) $y = 3x^6 + 7x^3$ **b)** $y = 8x^3 - 5x^4$ **c)** $y = 5x^3 - 2\sqrt{x}$

d) $y = \dfrac{3}{x} + \dfrac{2}{x^2}$ **e)** $y = 6\sqrt[3]{x} + \dfrac{2}{\sqrt{x}}$ **f)** $y = 4x^2 - 3x - 2$

4 Find the value of $\dfrac{dy}{dx}$ for the following functions at the points indicated.

a) $y = x^2 + 3x + 1$ at $(1, 5)$ **b)** $y = x^3 - x - 2$ at $(2, 4)$ **c)** $y = 2x - x^2$ at $(-1, -3)$

d) $y = \sqrt{x}$ at $(4, 2)$ **e)** $y = 2x + \dfrac{1}{x}$ at $(\frac{1}{2}, 3)$ **f)** $y = \sqrt[3]{x} - \dfrac{8}{x^2}$ at $(-1, -9)$

5 Let $y = 2x^2 + 3x + 1$. Find the value of x for which $\dfrac{dy}{dx} = -9$.

6 Let $y = 2x + \dfrac{50}{x}$. Find the values of x for which $\dfrac{dy}{dx} = 0$.

7 Let $y = x^3 + x$. Show that $\dfrac{dy}{dx}$ is never zero.

8 Find where $y = x^2 - 3x + 2$ crosses the x-axis. Find the gradients of the curve at these points.

9 Find where the curve $y = x^2 - 3x + 8$ is parallel to the line $y = 5x - 1$.

*10 Let $y = zw$, where z and w are functions of x. Is it true that $\dfrac{dy}{dx} = \dfrac{dz}{dx} \times \dfrac{dw}{dx}$?

When we work with more complicated expressions, we often have to do some algebra before we can differentiate.

EXAMPLE 8.5
Differentiate these functions.

a) $y = (x + 1)(x - 3)$ **b)** $y = \dfrac{2x + 1}{x}$

Solution
a) Before this can be differentiated, the brackets must be expanded out.

$$y = (x + 1)(x - 3)$$
$$= x^2 - 2x - 3$$

Now differentiate.

$$\frac{dy}{dx} = 2x - 2$$

b) Before this can be differentiated, we need to divide through by x.

$$y = \frac{2x + 1}{x}$$

$$= 2 + \frac{1}{x}$$

Now differentiate.

$$\frac{dy}{dx} = -\frac{1}{x^2}$$

EXERCISE 8H

1 Differentiate the following.

a) $y = (x - 3)(x - 4)$ b) $y = (2x + 1)(x - 3)$ c) $y = x(x - 7)$

d) $y = (x + 3)^2$ e) $y = 3x^2 \times 4x^6$ f) $y = (3x + 1)^2$

g) $y = \dfrac{1 + 3x^2}{x}$ h) $y = \dfrac{2x - 4}{\sqrt{x}}$ i) $y = \dfrac{\sqrt{x} + 1}{x^2}$

2 Let $y = (x + 3)^2$. Show that $(x + 3)$ is also a factor of $\dfrac{dy}{dx}$.

3 Let $y = (ax + b)^2$, where a and b are constants. Show that $(ax + b)$ is also a factor of $\dfrac{dy}{dx}$.

8.4 The equations of tangents and normals to curves

The derivative $\dfrac{dy}{dx}$ gives the gradient of a tangent to a graph, so

the process of differentiation can be used to find the equation of the tangent.

EXAMPLE 8.6

Find the equation of the tangent to the curve $y = x^2 + 2x + 2$ at the point $(1, 5)$.

Solution

Differentiating, we get the gradient of the graph.

$$\frac{dy}{dx} = 2x + 2$$

When $x = 1$, $\dfrac{dy}{dx} = 2 \times 1 + 2 = 4$.

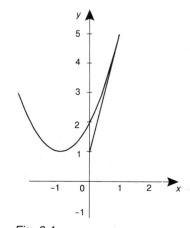

Fig. 8.4

This is the gradient of the tangent, so the equation of the tangent must be

$$y = 4x + c$$

where c is a constant.

The tangent must go through $(1, 5)$. Put this into the tangent equation.

$$5 = 4 \times 1 + c$$

So $c = 1$.

The tangent is $y = 4x + 1$.

EXERCISE 8I

1 Find the equations of the tangents at the points indicated.

a) $y = x^2$, at $(1, 1)$ **b)** $y = 2x^2 - 3x + 4$, at $(2, 6)$

c) $y = 2 - 3x - x^3$, at $(-2, 16)$ **d)** $y = (x - 2)(x + 3)$, at $(4, 14)$

e) $y = \dfrac{x + 3}{x^2}$, at $(1, 4)$ **f)** $y = \sqrt{x}$, at $(4, 2)$

2 Find the equation of the tangent to $y = x^3 - x$ at the point $(2, 6)$. Find where this tangent cuts the curve again.

3 Find where on the curve $y = x^2 + 3x - 2$ the tangent is parallel to $y = 3x - 2$.

4 Find the value of k for which $y = 2x + k$ is a tangent to $y = x^2 + 4x + 3$.

*5 The equation of the tangent to $y = ax^2 + bx$ at $x = 1$ is $y = 2x + 3$. Find a and b.

*6 Find the equation of the tangent to $y = ax^2 + bx + c$ at the point (x_0, y_0).

7 Let $y = x^2 - 8x + 7$. Find the equations of the tangents at the points where the curve crosses the x-axis. Find the coordinates of the point of intersection of these tangents.

*8 Find the two values of k such that $y = 9x + k$ is a tangent to $y = x^3 - 3x + 1$.

Normals

Suppose a tangent to a curve touches it at **P**. The line through **P** perpendicular to the tangent is the **normal**.

If the gradient of the tangent is m, the gradient of the normal is $-\dfrac{1}{m}$. So the equation of the normal to a curve can also be found by differentiation.

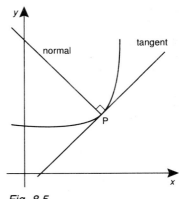

EXAMPLE 8.7

Find the equation of the normal to the curve $y = x^2$ at the point $(2, 4)$.

Fig. 8.5

Solution

The gradient of the curve is given by $\dfrac{\mathrm{d}y}{\mathrm{d}x} = 2x$.

At $x = 2$, the gradient is $2 \times 2 = 4$. So the gradient of the normal is $-\frac{1}{4}$.

The normal has equation $y = -\frac{1}{4}x + c$.

The normal goes through $(2, 4)$, so put in $x = 2$ and $y = 4$.

$$4 = -\tfrac{1}{4} \times 2 + c$$

So $c = 4\frac{1}{2}$.

The normal has equation $y = -\frac{1}{4}x + 4\frac{1}{2}$.

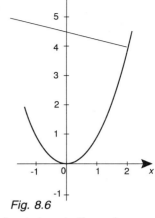

Fig. 8.6

EXERCISE 8J

1 Find the equations of the normals to the following curves, at the points indicated.

   a) $y = x^3$, at $(2, 8)$ **b)** $y = x^2 - 3x + 4$, at $(1, 2)$

   c) $y = 4 - x^2 - x^3$, at $(-1, 4)$ **d)** $y = (x - 2)(x - 5)$, at $(0, 10)$

   e) $y = \dfrac{6}{x} + \dfrac{4}{x^2}$, at $(2, 4)$ **f)** $y = \dfrac{1}{\sqrt{x}}$, at $(9, \frac{1}{3})$

2 Find the equation of the normal to the curve $y = x^2 - x$ at the point $(2, 2)$. Find where this normal cuts the curve again.

3 A normal to the curve $y = x^2$ is also parallel to the line $y = 3x - 2$. Find the equation of the normal.

4 At the point where $x = 1$, the normal to $y = ax^2 + c$ has equation $y = 2 - 3x$. Find a and c.

*5 Find the value of k for which $y = 2x + k$ is a normal to $y = x^2 - 3x$.

Angles between lines

Suppose a line has gradient m, and makes an angle θ with the horizontal or positive x-axis. Then $m = \tan\theta$. From this we can find the angle that a straight line makes with the x-axis. We did this in Chapter 4.

Using differentiation we can find the gradients of curves, so we can find the angles that these curves make with the horizontal axis.

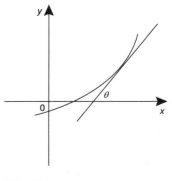

EXAMPLE 8.8

Find the angles the curve $y = x^2 - 2x - 3$ makes with the horizontal at the points where it crosses the x-axis.

Fig. 8.7

Solution

The two points where the curve cross the x-axis can be found by solving the equation $y = 0$.

$$x^2 - 2x - 3 = 0$$

$$(x + 1)(x - 3) = 0$$

$$x = -1 \text{ or } x = 3$$

Differentiating gives

$$\frac{dy}{dx} = 2x - 2$$

For $x = -1$, $\dfrac{dy}{dx} = -4$.

For $x = 3$, $\dfrac{dy}{dx} = 4$.

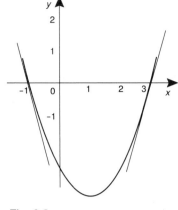

Fig. 8.8

Find the $\tan^{-1}$ for each of these gradients.

$$\tan^{-1}(-4) = -76°$$

$$\tan^{-1}4 = 76°$$

The angles are $-76°$ and $76°$.

EXERCISE 8K

1 Find the angles made by the following curves with the x-axis, at the points indicated.

a) $y = x^2$, at $(1, 1)$ **b)** $y = 2x^2 - 3x - 1$, at $(-1, 4)$ **c)** $y = \dfrac{1}{x^2}$, at $(\frac{1}{2}, 4)$

*2 Find where the line $y = 3x - 3$ crosses the curve $y = x^2 + x - 3$. Find the angles between the line and the curve at these points.

*3 Find where the line $y = x$ crosses the curve $y = x^3 - 3x$. Find the angles between the line and the curve at these points.

LONGER EXERCISE

Parabolic telescopes and searchlights

A large telescope such as that at Jodrell Bank consists essentially of a mirror which collects light from very far away and concentrates it on a single point, the **focus**. A searchlight works in the other direction: light is sent from a point, and is reflected so that it emerges in a parallel beam to light up an aircraft.

Here you will show that a mirror in the shape of the curve $y = x^2$ (a parabola) will have these properties. All you need to know

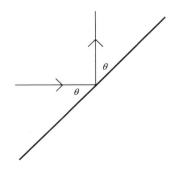

Fig. 8.9

about light is that when a ray is reflected by a mirror, the angles it makes with the tangent to the mirror before and after reflection are equal.

We shall consider the searchlight case. Let the mirror have a cross-section in the shape of the curve $y = x^2$. Place the light source at F$(0, \frac{1}{4})$. Consider a ray of light leaving F and striking the mirror at A$(1, 1)$.

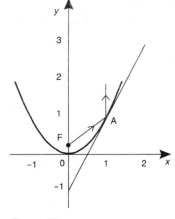

1   Find the gradient of the tangent at A.

2   Find the angle this tangent makes with the vertical.

3   Find the gradient of the chord joining F to A.

4   Find the angle this chord makes with the vertical.

Fig. 8.10

5   What is the connection between the angles found in steps 2 and 4? Use the property of reflection above to show that the reflected ray will be vertical.

6   Repeat steps 1 to 5 for other points on the curve, for example $(2, 4)$ or $(\frac{1}{2}, \frac{1}{4})$. In every case you should find that the reflected ray is vertical.

EXAMINATION QUESTIONS

1   Find the equation of the straight line which passes through the point $(3, 1)$ and has gradient -4.

Given that this line is parallel to the tangent to the curve $y = \dfrac{k}{x^2}$ at the point where $x = 2$, calculate the value of k.

C AS 1992

2   a) Verify that the point P with coordinates $(1, 3)$ lies on the curve with equation

$y = x^3 - x + 3$

b) Find an equation of the tangent to the curve at P.

c) Find an equation of the normal to the curve at P.

The tangent to the curve at P meets the x-axis at A and the normal to the curve at P meets the y-axis at B.

d) Calculate the area of the triangle AOB, where O is the origin.

L AS 1990

Summary and key points

1 The gradient of $y = x^2$ is $\dfrac{\mathrm{d}y}{\mathrm{d}x} = 2x$.

2 The gradient of $y = x^n$ is nx^{n-1}.

3 If a function is multiplied by a constant a, its derivative is also multiplied by a. If two functions are added, their derivatives are also added.

   When differentiating a product of brackets or a quotient, expand or divide through before differentiating.

4 When we know the gradient of a curve at a point, we can find the equations of the tangent and normal at that point.

Chapter 9

Sine rule and cosine rule

In Chapter 3 trigonometric functions for acute angles were defined in terms of right-angled triangles. They can also be used to find the sides and angles of triangles which are not necessarily right-angled.

Suppose you know that the three sides of a triangle are 7 cm, 8 cm and 9 cm. Then the triangle can be constructed using compasses and ruler. Having constructed the triangle, you can measure the angles with a protractor. You should find, for example, that the largest angle is 73°.

The angles of the triangle are determined by the sides of the triangle. There is a rule to calculate the angles from the sides. There are many advantages to finding the angles by calculation rather than measurement.

Accuracy
With a protractor, the largest angle is found to be 73° to the nearest degree. Using a calculator, the value is 73.398 450 40°.

Speed
To construct a triangle and measure an angle takes several minutes. Once you have learned the method, using a calculator to find an angle takes less than half a minute.

Equipment and expertise
Accurate drawing requires skill in draughtsmanship, as well as good equipment such as a sharp pencil, non-wobbly compasses, and an unkinked ruler. Calculation requires only a scientific calculator.

These trigonometric rules illustrate the success of mathematics in solving problems. If the three sides of a triangle are known, the angles of the triangle must be determined by the sides. Mathematics provides a way to calculate the angles from the sides.

Labelling

Throughout this chapter, we shall be using the following convention to label the sides and angles of a triangle.

In the triangle in Fig. 9.1 the angles are A, B and C. The sides are of length a, b and c units. Notice that side a is opposite angle A, side b is opposite angle B, side c is opposite angle C.

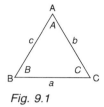
Fig. 9.1

The sine rule (sin rule)

$$\frac{\sin A}{a} = \frac{\sin B}{b} = \frac{\sin C}{c}$$

The rule can also be put the other way up.

$$\frac{a}{\sin A} = \frac{b}{\sin B} = \frac{c}{\sin C}$$

The proof of this result is in Section 9.4 at the end of the chapter. In this section we show how to use the rule.

Finding unknown sides

Suppose we know two angles of a triangle and one of the sides. Then we can use the sine rule to find the other sides.

Suppose that we know the angles A and B and side a. Then side b is given by the first part of the sine rule.

$$\frac{b}{\sin B} = \frac{a}{\sin A}$$

$$b = \frac{a}{\sin A} \times \sin B$$

By subtraction from $180°$ we can find the third angle C. The sin rule can be used again to find the side c.

$$\frac{c}{\sin C} = \frac{a}{\sin A}$$

$$c = \frac{a}{\sin A} \times \sin C$$

EXAMPLE 9.1

In triangle ABC, $a = 7$ cm, $B = 40°$ and $C = 55°$. Find b.

Solution

The diagram shows the triangle. Find A by subtraction.

$$A = 180° - 40° - 55° = 85°$$

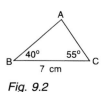
Fig. 9.2

The sine rule can now be used to find b.

$$\frac{b}{\sin 40°} = \frac{7}{\sin 85°}$$

$$b = \frac{7}{\sin 85°} \times \sin 40°$$

$$= 4.52$$

$$b = 4.52 \text{ cm}$$

EXERCISE 9A

1 Find the unknown sides in the diagram below, all lengths are in cm.

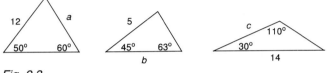

Fig. 9.3

2 In triangle ABC, BC = 8 cm, $B = 63°$, $C = 72°$. Find AB and AC.

3 Find the unknown sides for the following triangles.

    a) $a = 3$ cm, $B = 40°$, $C = 60°$. Find b.   **b)** $b = 23$ cm, $B = 45°$, $A = 67°$. Find a.

    c) $b = 12$ cm, $B = 63°$, $C = 39°$. Find a.   **d)** $a = 4.3$ m, $B = 123°$, $A = 29°$. Find c.

4 A surveyor wants to find the width of a river. From two positions, 200 m apart, the direction of the position of a tree on the opposite bank makes 83° and 73° with the bank as shown. Find the width of the river.

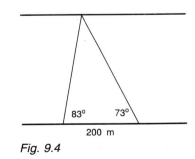

Fig. 9.4

5 From a gun site, the bearing of its target is 350°. From an observation post 50 m west of the gun, the bearing is 005°. Find the distance of the target from the gun.

6 Two men stand 100 m apart on flat ground. There is a tree on the line directly between them, and they measure the angle of elevation of its top as 12° and 15° respectively. How high is the tree?

*7 Two men are due south of a tower, D metres apart. They measure the angle of elevation of the top of the tower as α and β respectively, where $\alpha > \beta$. Show that the height of the tower is

$$\frac{D \sin \alpha \sin \beta}{\sin (\alpha - \beta)}$$

*8 The angle of elevation of the top of a tower is measured from two places A and B. A is on the ground and B is h vertically above A. The angle from A is α, and from B it is β. Show that the height of the tower is

$$h\left(1 + \frac{\sin \beta \cos \alpha}{\sin(\alpha - \beta)}\right)$$

9 Let ABC be a right-angled triangle. Check that the sine rule gives the same results as ordinary trigonometry.

Finding unknown angles

Suppose we know two sides of a triangle and an angle that is not enclosed by them. The sine rule can be used to find the other angles.

Suppose we know a, b and A. The sin of angle B can be found from the sine rule.

$$\frac{\sin B}{b} = \frac{\sin A}{a}$$

$$\sin B = \frac{\sin A}{a} \times b$$

Once B is found, C can be found by subtraction.

EXAMPLE 9.2

In triangle PQR, PQ = 8 cm, QR = 11 cm and $P = 55°$. Find R.

Solution

Use the formula above.

$$\sin R = \frac{\sin 55°}{11} \times 8 = 0.595\,75$$

Find the $\sin^{-1}$ of this value.

$$R = \sin^{-1} 0.595\,75 = 36.6°$$

$R = 36.6°$.

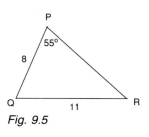

Fig. 9.5

EXERCISE 9B

In this exercise, assume all lengths are in cm unless otherwise stated.

1 Find the unknown angles in the diagram on the right.

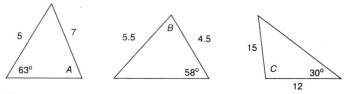

Fig. 9.6

2 In triangle ABC, $AB = 6$, $BC = 7.3$, $A = 68°$. Find C and B.

3 Find the unknown angles for the following triangles.

a) $a = 3$, $b = 4$, $B = 48°$. Find A. **b)** $b = 6$, $c = 5$, $B = 64°$. Find C.

c) $a = 12$, $c = 18$, $C = 78°$. Find B. **d)** $a = 32$, $b = 41$, $B = 63°$. Find C.

4 Two squares of wood, of sides 13 inches and 15 inches, are hinged together and placed on the ground. The shorter square makes 45° with the horizontal. What angle does the longer square make with the horizontal?

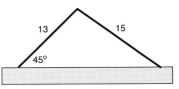

Fig. 9.7

5 In triangle, RST, $RS = 23$ cm, $ST = 32$ cm, $R = 58°$. Find T. Use the sine rule again to find the length of RT.

6 In triangle PQR, $P = 39°$, $PQ = 7.3$, $QR = 8.9$. Find PR.

The ambiguous case

As we have seen in Chapter 3, the sine function takes the same value twice in the range $0°$ to $180°$. So when we take the $\sin^{-1}$ of a ratio, there are two possible values for the angle.

Suppose we know that $AB = 9$, $BC = 8$ and $A = 30°$. The point C could be in two possible places as shown. There are two possible values for the angle C, P_1 and P_2. Notice that they are related by $P_2 = 180° - P_1$.

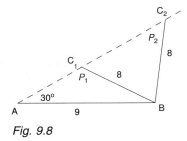

Fig. 9.8

EXAMPLE 9.3
Find the possible values of C in the triangle in Fig. 9.8.

Solution
Use the formula for finding angles.

$$\frac{\sin C}{9} = \frac{\sin 30°}{8}$$

$$\sin C = \frac{\sin 30°}{8} \times 9 = 0.5625$$

A calculator gives the $\sin^{-1}$ of 0.5625 as 34.2°. The other possible value of C is found by subtraction from $180°$.

$$180° - 34.2° = 145.8°$$

$$C = 34.2° \text{ or } C = 145.8°.$$

EXERCISE 9C

In this exercise, assume all lengths are in cm unless otherwise stated.

1 Find the possible values of the unknown angles in Fig. 9.9.

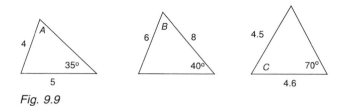

Fig. 9.9

2 In triangle ABC, AB $= 7$, BC $= 6$ and $A = 35°$. Find the possible values of C and of B.

3 The sine rule is not always ambiguous. Suppose AB $= 7$, BC $= 8$ and $A = 58°$. Show that there is only one possible value for C.

4 Suppose that AB $= 7$, BC $= 5$ and $A = 58°$. Try to find C. Can you show by a diagram what is wrong?

*5 All the problems in Exercise 9B were arranged so that ambiguity would not arise. Can you see how this was done?

9.2 The cosine rule

Suppose, as before, that the three sides of triangle ABC are a, b and c. Side a is opposite angle A and so on.

The cosine rule (cos rule)

$$a^2 = b^2 + c^2 - 2bc \cos A$$

This formula gives a^2 in terms of b, c and A. It can equally well be stated in either of the following forms.

$$b^2 = c^2 + a^2 - 2ca \cos B$$
$$c^2 = a^2 + b^2 - 2ab \cos C$$

The proof of the cosine rule is in Section 9.4 at the end of the chapter. Here we show how to use the rule.

Finding unknown sides

Suppose we know two sides of a triangle and the angle between them. The cosine rule can be used to find the third side.

EXAMPLE 9.4

The two sides of an uneven stepladder are 6 ft and 7 ft long. The
sides can be opened to an angle of 35°. How far apart are the feet
of the sides?

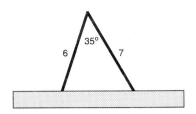

Fig. 9.10

Solution

Here we know two sides of a triangle and the angle between them. The
cosine rule can be used.

Letting d be the distance apart of the feet.

$$d^2 = 6^2 + 7^2 - 2 \times 6 \times 7 \cos 35° = 16.19$$

The distance apart is 4.02 feet.

EXERCISE 9D

In this exercise assume all lengths are in cm unless otherwise stated.

1 Find the unknown sides in Fig. 9.11.

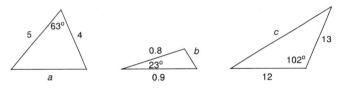

Fig. 9.11

2 In triangle ABC, AB $= 8$ cm, AC $= 12$ cm, $A = 36°$. Find BC.

3 Find the unknown sides in the following triangles.

a) $a = 5$, $b = 6$, $C = 48°$. Find c.   **b)** $b = 5$, $c = 3$, $A = 83°$. Find a.

c) $a = 12$, $b = 18$, $C = 132°$. Find c.   **d)** $a = 6.7$, $c = 5.1$, $B = 73°$. Find b.

4 The hour hand of a clock is 5 inches long, and the minute hand 7 inches. How far apart are the
tips of the hands at 2 o'clock?

5 In triangle ABC, $A = 90°$. Write down the cosine rule giving BC in terms of AB and AC. What
result is this?

6 In triangle ABC, $a = 5$, $b = 6$ and $C = 60°$. Show that $c = \sqrt{31}$.

*7 From a point on the seashore, Ship A is m miles away on a bearing of α. Ship B is $2m$ miles away
on a bearing of β. Show that the distance apart of the ships is

$$m\sqrt{5 - 4 \cos (\alpha - \beta)}$$

Finding unknown angles

The cosine rule can be rewritten to make cos A the subject of the formula.

$$\cos A = \frac{b^2 + c^2 - a^2}{2bc}$$

This version enables us to find A. The other two forms are

$$\cos B = \frac{c^2 + a^2 - b^2}{2ca}$$

$$\cos C = \frac{a^2 + b^2 - c^2}{2ab}$$

EXAMPLE 9.5

The three sides of a triangle are 7 cm, 8 cm and 9 cm long. Find the angle opposite the longest side.

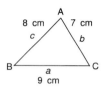

Fig. 9.12

Solution

Label the triangle as shown. The longest side is 9 cm long, which we have labelled a. We want to find A. The cosine rule gives

$$\cos A = \frac{7^2 + 8^2 - 9^2}{2 \times 7 \times 8} = 0.2857$$

Find the $\cos^{-1}$ of this value.

$$\cos^{-1} 0.2857 = 73.4°$$

The angle opposite the longest side is 73.4°.

EXERCISE 9E

In this exercise, assume all lengths are in cm unless otherwise stated.

1 Find the unknown angles in Fig. 9.13.

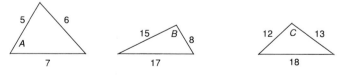

Fig. 9.13

2 In triangle ABC, AB = 12, BC = 13.5 and CA = 12.9. Find B.

3 Find the unknown angles in the following triangles.

    a) $a = 4$, $b = 3$, $c = 2$. Find A.   **b)** $a = 8$, $b = 9.5$, $c = 12$. Find B.

    c) $a = 34$, $b = 43$, $c = 51$. Find C.   **d)** $a = 0.3$, $b = 0.35$, $c = 0.41$. Find A.

4 In triangle PQR, PQ = 13.7, PR = 6.3 and QR = 7.2. Try to find R. Can you show by a diagram what is wrong?

5 The sides of a triangle are in the ratio $3:5:7$. Without use of a calculator show that the largest angle is $120°$.

*6 In triangle ABC, $c^2 = a^2 + b^2 - ab$. Without use of a calculator, find C.

*7 A and B are on the same horizontal plane as the foot F of a tower. A is due north of the tower, and B is on a bearing of $060°$. The angles of elevation of the top of the tower from A and B are α and β respectively. If h is the height of the tower, show that the distance AB is given by

$$AB^2 = h^2(\cot^2\alpha + \cot^2\beta - \cot\alpha\cot\beta)$$

9.3 Solving triangles

Sometimes it can be difficult to decide whether to use the sine or the cosine rule. Below is a guide for deciding which rule to apply.

SSS
If we know the three sides of a triangle, then use the cosine rule to find the angles.

SAS
If we know two sides of a triangle and the angle between them, then use the cosine rule to find the third side.

ASA
If we know two angles of a triangle and one of the sides, then use the sine rule to find the other sides.

SSA
If we know two sides of a triangle and an angle **not** between them, then use the sine rule to find another angle. There may be two possible answers.

Sometimes the rules have to be applied twice. If there is a choice of which rule to use to find an angle, then it is safer to use the cosine rule as there is no risk of ambiguity.

EXAMPLE 9.6
In the triangle shown, AB = 5 cm, AC = 8 cm and $A = 38°$. Find B.

Solution
In this triangle, we have two sides and the angle between them. This is the SAS condition, so we use the cosine rule to find the side BC.

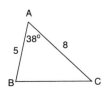

Fig. 9.14

$$BC^2 = 5^2 + 8^2 - 2 \times 5 \times 8 \cos 38°$$

$$= 25.959\ 14$$

$$BC = 5.095\ \text{cm}$$

We now have all the sides of the triangle and one angle. We could use either the sine rule or the cosine rule. Use the cosine rule to avoid the risk of ambiguity.

$$\cos B = \frac{5^2 + 5.095^2 - 8^2}{2 \times 5 \times 5.095}$$

$$= -0.25596$$

$$B = \cos^{-1}(-0.25596) = 104.8°$$

$B = 104.8°$.

EXERCISE 9F

In this exercise, assume all lengths are in cm unless otherwise stated.

1 Find the unknown sides and angles in Fig. 9.15.

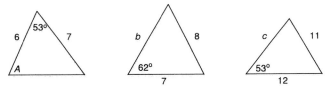

Fig. 9.15

2 In triangle ABC, $AB = 7$, $BC = 6.3$ and $B = 43°$. Find A.

3 In triangle PQR, $PQ = 8$, $QR = 9.5$ and $P = 68°$. Find PR.

4 In triangle LMN, $m = 5$, $n = 6$ and $L = 53°$. Find M.

5 In triangle XYZ, $x = 12$, $y = 13$ and $X = 47°$. Find z.

6 Can you show why there is no ambiguity in using the cosine rule to find an angle?

*7 In triangle ABC, $B = 60°$. Show that $c = \frac{1}{2}a \pm \sqrt{b^2 - \frac{3}{4}a^2}$.

If a and b are given, find the condition on a and b for the triangle to exist. Find the condition that there are two possible values of c.

9.4 Proofs of the rules

Here we provide the proofs of the sine and cosine rules.

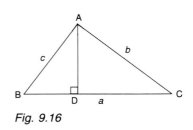

Fig. 9.16

Proof of the sine rule

Drop a perpendicular from A to BC. The foot of this perpendicular is D.

From ordinary trigonometry in the left-hand triangle

$$AD = c \sin B$$

From ordinary trigonometry in the right-hand triangle

$$AD = b \sin C$$

We now equate these two expressions for AD, to obtain

$$c \sin B = b \sin C$$

Now divide through by b and c.

$$\frac{\sin B}{b} = \frac{\sin C}{c}$$

This is half the sine rule. The other half can be proved in exactly the same manner, by dropping a perpendicular from C to AB.

Proof of the cosine rule

Use the same diagram as for the sine rule. By ordinary trigonometry

$$BD = c \cos B$$

Subtract this from a to find CD.

$$CD = a - c \cos B$$

Use Pythagoras's theorem in the left-hand triangle.

$$AD^2 = c^2 - (c \cos B)^2$$

Use Pythagoras's theorem in the right-hand triangle.

$$AD^2 = b^2 - (a - c \cos B)^2$$

Equate these two.

$$c^2 - (c \cos B)^2 = b^2 - (a - c \cos B)^2$$

Expand the right-hand side of this equation.

$$c^2 - (c \cos B)^2 = b^2 - a^2 - (c \cos B)^2 + 2ac \cos B$$

The $(c \cos B)^2$ term cancels. The other terms can be collected.

$$b^2 = a^2 + c^2 - 2ac \cos B$$

This is the cosine rule.

EXERCISE 9G

*1 The cosine rule can be found from coordinates. Put A at the origin (0, 0), C on the x-axis at $(c, 0)$, and B at (p, q), where $p = b \cos A$ and $q = b \sin A$. Find an expression for BC^2, and show that it reduces to the cosine rule.

*2 Let the triangle ABC be inscribed within a circle of diameter D. Show that $D = \dfrac{a}{\sin A}$.

*3 The proofs given for the sine rule and cosine rule were illustrated using acute-angled triangles. Rewrite the proofs using obtuse-angled triangles.

LONGER EXERCISE

Rowing across rivers

In this exercise, a river is flowing between straight banks at 2 m s^{-1}. The banks are 1000 m apart.

1 A man can row at 3 m s^{-1}. He starts at A and wants to reach a point B directly opposite. In which direction should he row and how long will it take him?

2 Suppose instead that he wants to reach C, which is 1000 m upstream from B. In which direction should he row and how long will it take him?

3 A woman can row at 1.8 m s^{-1}. She wishes to reach point D, 1000 m downstream from B. Show that there are two possible directions in which she could row, and find them.

*4 Suppose the river flows at $v \text{ m s}^{-1}$ and that a man can row at $u \text{ ms}^{-1}$. Show that, in order to reach D from A, there is a choice of two possible directions provided that

$$u < v < u\sqrt{2}$$

EXAMINATION QUESTIONS

1 On a circular clock face, with centre O, the minute hand OA is of length 10 cm and the hour hand OB is of length 6 cm. Prove that the difference between the distances that A and B travel during the period of 1 hour from 12 o'clock to 1 o'clock is 19π cm. Calculate, correct to the nearest millimetre, how far A is from B at 1 o'clock.

J 1992

2 Three towns denoted by A, B and C are as shown in Fig. 9.17, which is not drawn to scale. BC is 19 miles, and AC is 28 miles.

The bearings of A and B from C are 228° and 260° respectively. Calculate the distance and bearing of B from A.

AEB AS 90   *Fig. 9.17*

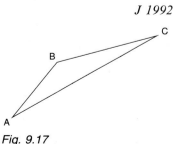

3 The lengths of the sides of a triangle are 4 cm, 5 cm and 6 cm. The size of the largest angle of the triangle is θ.

a) Calculate the value of $\cos \theta$.

b) Hence, or otherwise, show that $\sin \theta = a\sqrt{7}/b$, where a and b are integers.

L 1992

Summary and key points

1 A triangle is lettered so that side a is opposite angle A and so on. The sine rule states that

$$\frac{\sin A}{a} = \frac{\sin B}{b} = \frac{\sin C}{c}$$

This rule can be used to find sides or to find angles.

When the sine rule is used to find an angle, there may be two possible answers. If one value is x, another value with the same sine is $180° - x$. This will be a possible value provided that the sum of this angle and the known angle is not greater than $180°$.

2 The cosine rule states that

$$a^2 = b^2 + c^2 - 2bc \cos A$$

In this form it can be used to find the third side of a triangle, given two sides and the angle between them. If we know the three sides of a triangle, we can rewrite the rule so that we can find the angles.

$$\cos A = \frac{b^2 + c^2 - a^2}{2bc}$$

When using a calculator with this formula be careful to divide the **whole** of the top line by $2bc$, and be careful to **divide** by b and c.

3 Which rule is used depends on which sides and angles are known. Make sure that the triangle is labelled correctly. Sometimes the rules have to be used more than once.

4 The proofs of the sine and cosine rules involve ordinary trigonometry and Pythagoras's theorem.

Probability

The results of mathematics are, for the most part, absolutely true. We can say with certainty that two plus two equals four, but we cannot be so sure about statements in ordinary life. We cannot say with certainty that it will rain tomorrow, or that we shall win a tennis match, or that a tossed coin will land heads uppermost.

Though we cannot be absolutely sure about these events, we can give a number to them, to show how likely we think they are. This number is the **probability** of the event.

Probability is measured on a scale between 0 and 1. If an event is certain to happen, its probability is 1. If an event is impossible, its probability is 0. If an event is just as likely to happen as not to happen, its probability is $\frac{1}{2}$.

The probability of an event A is written $P(A)$.

10.1 Probability of equally likely events

Suppose we are about to perform an experiment, for which there are n equally likely outcomes. Then the probability of each outcome is $\frac{1}{n}$. For example, suppose an unbiased die is rolled. Each of the six faces is equally likely to be uppermost, and so the probability that the face labelled 5 will be uppermost is $\frac{1}{6}$.

EXAMPLE 10.1
Two dice are rolled and the total score is found. What is the most likely score, and what is its probability?

Solution
Draw up a table, showing all the possible scores on the dice.

			First die				
	+	**1**	**2**	**3**	**4**	**5**	**6**
	1	2	3	4	5	6	7
	2	3	4	5	6	7	8
Second	**3**	4	5	6	7	8	9
die	**4**	5	6	7	8	9	10
	5	6	7	8	9	10	11
	6	7	8	9	10	11	12

The most frequent total is 7.

The most likely score is 7.

There are 36 boxes, and 7 occurs in 6 of them.

This gives a probability of scoring 7 as $\dfrac{6}{36}$ or $\dfrac{1}{6}$.

The probability of a total of 7 is $\dfrac{1}{6}$.

EXERCISE 10A

1 Two fair dice are rolled. Refer to the table in the example above to find the probabilities that

    a) the total score is 10   **b)** the total score is even

    c) the dice show the same number   **d)** the total score is less than 11.

2 Two tetrahedral (four-sided) dice are rolled. By means of a table or otherwise, find the most likely total score and its probability.

3 A six-sided die and a four-sided die are rolled simultaneously. Find the probability that

    a) the total score is 8   **b)** the total score is odd.

4 Two coins are spun. List the possible arrangements of heads and tails. Find the probability that there is one head and one tail.

5 Three coins are spun. List the possible arrangements of heads and tails. Find the probability that there are exactly two heads.

6 A bag contains four balls, which are white, black, brown and grey. Two balls are selected at random. Write down the possible selections of two balls. Find the probability that

    a) the white and the black ball are selected   **b)** the grey ball is selected.

7 Three people A, B and C play a game which is purely determined by chance. Write down the possible orders in which they could finish. Find the probability that they finish in the order A, B, C.

8 Two dice are rolled, and the **product** of the scores is found. Find the probability that this product is

    a) 1   **b)** odd   **c)** prime.

9 A secretary has three letters to put into three envelopes. Being in a rush, she puts them in at random. Find the probabilities that

   a) each letter is in its correct envelope   **b)** no letter is in its correct envelope.

   c) exactly two letters are in their correct envelopes.

*10 Four passengers on a flight have suitcases which are externally identical. After the flight, the suitcases appear on the carousel at the terminal. Two of the passengers each pick up one of the four identical suitcases at random. Find the probabilities that

   a) both pick the correct suitcase   **b)** neither picks the correct suitcase

   c) exactly one of them picks the correct suitcase.

11 A bag contains two black marbles, two white marbles and two grey marbles. Two marbles are selected at random. What is the probability they are both black?

12 Blood transfusions were given before the different blood groups were known. The results were often fatal, if a patient received blood of the wrong group. Suppose in a group of six people three have group O, two have group A and one has group B blood. If two are picked at random what is the probability that they have the same group?

10.2 Combined probabilities

Often we are interested in combinations of events. Given events A and B, we might be concerned with A **not** happening, or with A **or** B happening, or with A **and** B happening.

It often helps to illustrate combinations of probabilities on a **Venn diagram**. Events are shown by regions within the diagram. The oval region shown corresponds to event A.

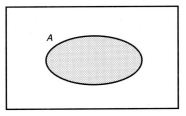

Fig. 10.1

Not

The event of A **not** happening is sometimes written A'. Its probability is given by

   $P(A') = 1 - P(A)$

In the Venn diagram, the event A' is shown by the region **outside** the A-region.

Or

The event of A **or** B happening is sometimes written as $A \cup B$. Note that we use the **inclusive or**: A or B means A or B or both.

Two events are **exclusive** if they cannot occur together. In this

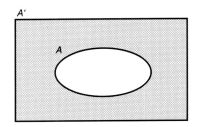

Fig. 10.2

case, the probability that either one of them occurs is the sum of their individual probabilities.

If events *A* and *B* are exclusive, then P(*A* or *B*) = P(*A*) + P(*B*).

In the Venn diagram, the *A* and *B* regions do not overlap, hence they cannot both occur. The region included in either *A* or *B* is obtained by combining the two regions.

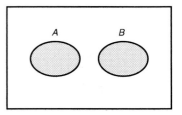

Fig. 10.3

And

The event of both *A* and *B* happening is sometimes written as *A* ∩ *B*, or as *A* & *B*.

In the Venn diagram, the region corresponding to *A* & *B* is the overlap of the two regions.

Two events are **independent** if knowledge of the outcome of one of them does not alter the probability of the other. In this case, the probability of them both occurring is the product of their individual probabilities.

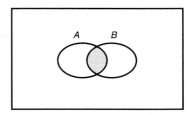

Fig. 10.4

If events *A* and *B* are independent, then P(*A* & *B*) = P(*A*)P(*B*)

EXAMPLE 10.2
Suppose P(*A*) = 0.5, P(*B*) = 0.6 and P(*A* & *B*) = 0.4. Are *A* and *B* exclusive? Are they independent? Find P(*A* or *B*).

Solution
Note that P(*A* & *B*) is not zero, so they are not exclusive. Note also that P(*A* & *B*) ≠ P(*A*)P(*B*).

A and B are neither exclusive nor independent.

In the diagram, the overlap of *A* and *B* has probability 0.4. If we add the probabilities of *A* and *B*, the overlap will have been counted twice, so the total probability covered by *A* and *B* is obtained by adding the probabilities of *A* and *B* and then subtracting the probability of *A* & *B*.

P(*A* or *B*) = 0.5 + 0.6 − 0.4 = 0.7

P(*A* or *B*) = 0.7

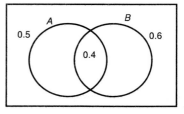

Fig. 10.5

EXERCISE 10B

1 Events *A* and *B* are such that P(*A*) = 0.4, P(*B*) = 0.6 and P(*A* & *B*) = 0.2. Are *A* and *B* independent? Are they exclusive? Find P(*A* or *B*).

2 Events *A* and *B* are such that P(*A*) = 0.5 and P(*A* or *B*) = 0.8. Find P(*B*) in the cases that

   a) *A* and *B* are exclusive　　　　***b)** *A* and *B* are independent.

3 Events *A* and *B* are such that P(*A*) = 0.4, P(*B*) = 0.3 and P(*A* or *B*) = 0.6. Are *A* and *B* independent? Find P(*A'* & *B'*).

4 Events A and B are such that $P(A) = x$ and $P(B) = y$.

 a) If A and B are independent, find $P(A \,\&\, B)$ and $P(A \text{ or } B)$.

 b) If A and B are exclusive, find $P(A \,\&\, B)$ and $P(A \text{ or } B)$.

5 Two dice are rolled. Events are defined as follows.

 A: the total score is 7. B: the total score is 8.

 C: the first die shows a 3. D: the first die shows a 1.

 a) Which pairs of these events are exclusive?

 b) Show that A and C are independent, but not B and C. Can you explain why?

6 A card is drawn from the standard pack. Events are defined as follows.

 A: the card is a king. B: the card is a spade.

 C: the card is an 8. D: the card is red.

 a) Which pairs of events are exclusive?   **b)** Which pairs of events are independent?

7 A person is picked at random from the Electoral Register. Which pairs of the following events do you think are independent?

 A: the person is a homeowner. B: the person votes Labour.

 C: the person's surname begins with M. D: the person's birthday falls in October.

More than two events

The principles of multiplying and adding probabilities apply to more than two events.

Suppose we have a succession of independent events, A_1, $A_2, \cdots, A_n$. Then the probability of them all occurring is the product of their individual probabilities.

$$P(A_1 \,\&\, A_2 \,\&\, \cdots \,\&\, A_n) = P(A_1)P(A_2) \cdots P(A_n)$$

In particular, if these events all have the same probability p, the result is the nth power of p.

$$P(A_1 \,\&\, A_2 \,\&\, \cdots \,\&\, A_n) = p^n$$

Often we want to know the probability that at least one of the A_is occurs. It is often best to turn this problem round, and find the probability that **none** of them occurs. This probability can then be subtracted from 1.

If $P(A_i) = p$, $P(A_i') = 1 - p$. So $P(\text{no } A_i \text{ occurs}) = (1 - p)^n$.

$$P(\text{at least one } A_i \text{ occurs}) = 1 - P(\text{no } A_i \text{ occurs}) = 1 - (1 - p)^n$$

EXAMPLE 10.3

I throw 12 darts at the dartboard. With each dart, the probability that I hit the bullseye is 0.1, and each throw is independent of the others. What is the probability that I hit the bullseye at least once?

Solution

For each dart, the probability that I **don't** hit the bullseye is $1 - 0.1 = 0.9$. The probability that I don't hit the bullseye with any of the 12 darts is 0.9^{12}.

To find the probability that I hit with at least one dart, subtract from 1 the probability that I miss with all the darts.

$P(\text{no hits}) = 0.9^{12} = 0.282$

$P(\text{at least one hit}) = 1 - 0.282 = 0.718$

The probability of at least one hit is $1 - 0.9^{12} = 0.718$

EXERCISE 10C

1 A multiple choice test contains ten questions, each of which has five possible answers. A candidate is totally ignorant of the subject of the test, and answers at random. What is the probability that the candidate gets

a) all right   **b)** none right   **c)** at least one right.

2 An electrical device has five fuses. Each fuse has probability 0.2 of blowing by the end of a day's operation. What is the probability that the device will still be working after a day if

a) the device requires all the fuses to be working

b) the device requires at least one of the fuses to be working.

3 For each mile that I drive, the probability that I have an accident is 0.001. Each mile is independent of the others. What is the probability that I complete a journey of 100 miles safely?

4 A prisoner is up before a firing squad of ten soldiers. Each soldier has probability 0.1 of firing a fatal shot. What is the probability the prisoner will survive?

5 A certain sort of tulip bulb is such that one in three of them will flower. If five are bought, what is the probability that at least one will flower?

*6 Assuming that birthdays are spread at random throughout the year, what is the probability that a class of 27 children will contain at least two children with the same birthday?

10.3 Tree diagrams

When we want to combine probabilities, it is often best to draw a **tree diagram**.

Suppose we have two or more experiments. Lay the diagram out as shown, and indicate the possible outcomes of each experiment by branches. Label each branch with its probability. Combined probabilities can then be found by multiplying the individual probabilities along each branch.

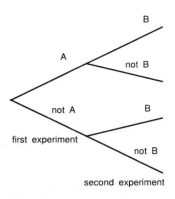

Fig. 10.6

EXAMPLE 10.4

In my cupboard I have seven black socks and eight white socks. In the morning I pick out two socks at random. What is the probability I have a matching pair?

Solution

For the first draw, the probability of a black sock is $\frac{7}{15}$ and of a white sock is $\frac{8}{15}$. Write these probabilities on the tree.

If the first sock is black, then there are six black socks left and eight white. So the probability that the second will also be black is $\frac{6}{14}$, and the probability that the second is white is $\frac{8}{14}$. Write these probabilities on the tree.

If the first sock is white, the probabilities of a black or a white second sock are both $\frac{7}{14}$. The complete tree is as shown.

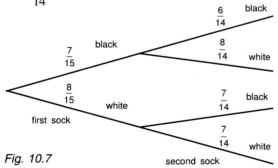

Fig. 10.7

The probability that both socks are black is the probability of the top branch, $\frac{7}{15} \times \frac{6}{14}$. The probability that both socks are white is $\frac{8}{15} \times \frac{7}{14}$. Add these to find the probability that I have a matching pair.

$$P(\text{black pair}) = \frac{7}{15} \times \frac{6}{14} = \frac{3}{15}$$

$$P(\text{white pair}) = \frac{8}{15} \times \frac{7}{14} = \frac{4}{15}$$

$$P(\text{pair}) = \frac{3}{15} + \frac{4}{15} = \frac{7}{15}$$

The probability that I have a pair is $\frac{7}{15}$.

EXERCISE 10D

1 A bag contains three black balls and four white balls. Two are drawn at random. What is the probability they are both black?

2 A box of chocolates contains six hard-centred sweets and four soft-centred sweets. If I pick out two, what is the probability that one is hard-centred and the other is soft-centred?

3 One in ten of a large batch of flash-cubes is defective. If I buy two, what is the probability that exactly one will be defective?

4 A and B are playing chess. If A has the white pieces, he will win with probability 0.4 and draw with probability 0.3. If B has the white pieces, he will win with probability 0.5 and draw with probability 0.4. Before the game, they spin a coin to see who will have the white pieces. What is the probabililty that A will win the game?

5 Barchester City and Coketown United are due to play a football match. If it is dry, Barchester's probability of winning is 0.25, and of drawing is 0.3. If it is wet, Barchester's probabilities of winning and drawing are 0.4 and 0.1 respectively. If the probability of a dry day is $\frac{2}{3}$, what is the probability that Barchester will win the match?

6 I go to work by bus, car or on foot with probabilities $\frac{1}{6}$, $\frac{1}{3}$ and $\frac{1}{2}$ respectively. For each mode of transport, the probability that I shall be late for work is $\frac{1}{20}$, $\frac{1}{30}$ and $\frac{1}{100}$ respectively. Find the probability that I shall be late for work.

7 In a psychology experiment, a mouse runs through a maze, part of which is shown in Fig. 10.8. At the first junction A it is equally likely to turn left or right. If it turns left, the probability that the next turn will also be left is 0.4. If it turns right, the probability that the next turn will also be right is 0.3. It never retraces its path. Find the probability that it reaches point B.

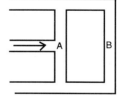

Fig. 10.8

*8 A foreign tourist has large numbers of 50p and 20p coins. When asked to pay a bill of 70p, the tourist presents coins, which are equally likely to be 50p or 20p, until the total is at least 70p. What is the probability that the bill is overpaid?

*9 Two girls Anne and Jane play tennis. The winner will be the first to gain two sets. If Anne wins a set, the probability she will win the next is 0.7. If she loses a set, the probability she will win the next is 0.2. If Anne has probability 0.6 of winning the first set, what is the probability she will win the match?

10.4 Conditional probability

Suppose we have two events *A* and *B*, which may or may not be connected. The **conditional probability** of '*A* given *B*' is the probability that *A* occurs, given that we know *B* has occurred.

P(*A*|*B*) = P(*A*, given that *B* is true)

Conditional probability is given by the formula

$$P(A|B) = \frac{P(A \ \& \ B)}{P(B)}$$

EXAMPLE 10.5

Two dice are rolled. What is the probability that the dice show the same number, given that the total score is 8?

Solution

Here A is the event that the dice show the same number, and B that the total score is 8.

A & B will only occur if both dice show 4. This has probability $\frac{1}{36}$.

Refer to the table in Section 10.1, to find that the probability of B is $\frac{5}{36}$.

$$P(A|B) = \frac{P(A \ \& \ B)}{P(B)} = \frac{\frac{1}{36}}{\frac{5}{36}} = \frac{1}{5}$$

The probability that both dice will show the same number, given a total of 8 is $\frac{1}{5}$.

EXAMPLE 10.6

A disease afflicts 2% of the population. There is a test for the disease, which is positive for 90% for those with the disease, and for 5% of those without. If you are tested and the result is positive, what is the probability that you have the disease?

Solution

Let D be the event of having the disease, and T the event of the test proving positive. Note that T could occur in two ways: a person with the disease could be correctly tested, or a person without the disease could be incorrectly tested.

By a tree diagram or otherwise, note that

$$P(T) = 0.02 \times 0.9 + 0.98 \times 0.05$$

Now apply the formula for conditional probability.

$$P(D|T) = \frac{P(D \ \& \ T)}{P(T)} = \frac{0.02 \times 0.9}{0.02 \times 0.9 + 0.98 \times 0.05} = 0.269$$

The probability you have the disease is 0.269.

EXERCISE 10E

1 Two dice are rolled. Find the following probabilities.

   a) That the total was 6, given that a double was thrown.

   b) That the total was 12, given that it was even.

   c) That a double was thrown, given that the total was 7.

2 Suppose that in Example 10.6 above, the results of the test are negative. What is the probability that you do not have the disease?

3 Suppose A and B are events for which $P(A) = 0.3$, $P(B) = 0.4$ and $P(A|B) = 0.2$. Find $P(A \text{ or } B)$.

4 Suppose A and B are independent. Show that $P(A|B) = P(A)$.

5 A school contains girls and boys in the ratio $6:5$. It is known that 15% of the girls and 10% of the boys are left-handed. A piece of offensive graffiti is found, and it is shown that the person who wrote it is left-handed. Ignoring all other considerations, what is the probability that the culprit is a girl?

6 You have two coins. One of them is fair. The other has a probability of $\frac{1}{3}$ that it lands heads uppermost. A coin is selected and spun, and it lands heads uppermost. What is the probability it is the fair coin?

7 A woman goes to work by bus, car or bicycle with probabilities 0.2, 0.3 and 0.5 respectively. For each mode of transport, the probability that she has an accident are $\frac{1}{10000}$, $\frac{1}{1000}$ and $\frac{1}{500}$ respectively. If she had an accident, what is the probability that she went by car?

*8 Parliament is divided into two parties, the Reactionaries and the Progressives, in the ratio $2:3$. Reactionary MPs tell the truth with probability 0.8, and Progressives with probability 0.7. You meet an MP who says, 'I am a Reactionary.' What is the probability he is telling the truth?

*9 If a person drives with a level of alcohol above the limit, the probability that he or she will be involved in an accident is 0.01. The corresponding probability for drivers not above the limit is 0.001. The police find that one third of the drivers involved in accidents have alcohol levels above the limit. What proportion of drivers have alcohol levels above the limit?

LONGER EXERCISE

Genes

People's bodily characteristics, such as hair colour or eye colour, are determined by their **genes**. For each characteristic, a person has two genes, one from each parent. If the genes are the same, then they will determine the characteristic. If the genes are different, then one will be **dominant** and the other **recessive**. The dominant gene will determine the characteristic.

For simplicity, let us suppose there are only two possible eye colours, brown and blue. Brown is dominant over blue, so a person will only have blue eyes if **both** his eye genes are blue.

1 A mother and father each have a blue-eye gene and a brown-eye gene (so both are brown-eyed). What is the probability that their child has blue eyes?

2 Suppose that in the whole population, the proportion of blue-eyed genes is p. What proportion of the population will have blue eyes?

3 A man has brown eyes. Find, in terms of p, the probability he carries a blue-eye gene.

4 A blue-eyed woman marries a brown-eyed man. What is the probability, in terms of p, that their child will have blue eyes?

*5 The rare condition *Brod's Syndrome* is carried on a recessive gene. It is known that 0.01% of the population has Brod's Syndrome.

 a) What proportion of the gene pool of the whole population carry Brod's Syndrome?

 b) The mother of a child has Brod's Syndrome, but the father doesn't. What is the probability that the child has the syndrome?

EXAMINATION QUESTIONS

1 In a child's game there should be 7 triangles, 3 of which are blue and 4 of which are red, and 11 squares, 5 of which are blue and 6 of which are red. However, two pieces are lost. Assuming the pieces are lost at random, find the probability that they are

 (i) the same shape (ii) the same colour (iii) the same shape and colour

 (iv) the same shape given that they are the same colour.

MEI 1991

2 A golfer observes that, when playing a particular hole at his local course, he hits a straight drive on 80 per cent of the occasions when the weather is not windy but only on 30 per cent of the occasions when the weather is windy. Local records suggest that the weather is windy on 55 per cent of all days.

 (i) Show that the probability that, on a randomly chosen day, the golfer will hit a straight drive at the hole is 0.525.

 (ii) Given that he fails to hit a straight drive at the hole, calculate the probability that the weather is windy.

JMB 1991

3 The probability that a door-to-door salesman convinces a customer to buy is 0.7. Assuming sales are independent, find the probability that the salesman makes a sale before reaching the fourth house.

L 1992

Summary and key points

1 The probability of an event measures our belief that it occurs. If there are n equally likely outcomes to an experiment, the probability of each of them is $\frac{1}{n}$.

Do not ignore the 'equally likely' condition.

2 Events A and B are exclusive if they cannot happen together. In this case, $P(A \text{ or } B) = P(A) + P(B)$.

Events A and B are independent if knowledge of one of them does not alter the probability of the other. In this case, $P(A \& B) = P(A)P(B)$.

Do not apply these rules unless you are sure that the events are exclusive or independent.

If events are dependent, that does not mean that one event causes the other.

3 Combinations of probabilities can be calculated with the help of a tree diagram.

4 The conditional probability of A given B is

$$P(A|B) = \frac{P(A \& B)}{P(B)}$$

Do not write the top of this fraction as $P(A)P(B)$, unless you know that A and B are independent.

Computer investigations I

1 The method of differences

In an arithmetic progression, the differences between successive terms are constant. In a more complicated sequence, the differences of the differences (called **second differences**) may be constant, or the third differences may be constant, and so on. This principle can be used to check the terms of a sequence.

A spreadsheet is ideal for doing the calculations connected with this method of differences. For these investigations it is assumed that you have access to a spreadsheet and are familiar with the basic processes of entering words, numbers and formulae, absolute and relative addresses, and of copying from one group of cells to another.

In A1 to A10 enter the numbers 1 to 10.

In B1 to B10 enter the terms of the sequence. If the formula is $n^2 + n + 41$, enter in B1 the formula

$$+A1\char`^2 + A1 + 41$$

Copy this formula down to B10.

Enter the first differences in the C column, the second differences in the D column and so on up to sixth differences. So in C1 enter

$$+B2 - B1$$

Copy this to C1 to C9, D1 to D8, E1 to E7, F1 to F6, G1 to G5, H1 to H4. (Note that it is easiest to copy the formula to the rectangular block with corners C1, H1, H9, C9, and to erase the ones not needed.) The spreadsheet should now look like this.

	A	B	C	D	E	F	G	H
1	1	43	4	2	0	0	0	0
2	2	47	6	2	0	0	0	0
3	3	53	8	2	0	0	0	0
4	4	61	10	2	0	0	0	0
5	5	71	12	2	0	0	0	
6	6	83	14	2	0	0		
7	7	97	16	2	0			
8	8	113	18	2				
9	9	131	20					
10	10	151						

Note that the second differences are all 2. The next term in the C column is $20 + 2 = 22$. The next term in the B column is $151 + 22 = 173$. Check that this agrees with the formula.

Find the 12th term in the sequence by this method, and check that it is correct.

Try some other functions in the B column. Suggestions are

$$n^3 - n, \ n^4 + n^3 - 3, \ 0.5n^6 - 0.5n^2$$

What are the connections between the polynomial and the column of constant differences?

Correcting errors

The method of differences can also be used to find errors in a list of numbers. In the B column enter the following.

3   7   12   18   25   33   42   53   63   75

One of these numbers is wrong. Notice that the D column begins with five 1s. In fact, all the entries should be 1. By making this change and working through the C and B columns find the incorrect entry.

In each of the following sequences there is an incorrect term. Find it.

a) 1   1   3   4   7   11   16   22   29   37

b) 0   4   18   48   100   180   298   448   648   900

c) −1   4   27   80   175   324   539   830   1215   1700

d) 0   14   78   252   620   1290   2394   4088   6554   9990

The use of a computer is particularly appropriate for this problem. One of the forerunners of the computer was the 'Difference machine', designed by Charles Babbage in the 1830s. One model is shown in the photograph.

Babbage built a machine which could evaluate formulae such as $n^2 + n + 41$, by using constant second differences. He planned a machine which would use constant sixth differences to evaluate formulae involving n^6. This machine would have been able to calculate tables for navigation, and it would have been very quick and accurate. Unfortunately the Government cut off its subsidy for the machine and it was never built.

2 Quadratic iteration

In the Longer Exercise of Chapter 2, you were asked to solve the quadratic equation $x^2 + 3x - 2 = 0$ by means of iteration. The starting point was 0.5, and the iterations were

$$x_{n+1} = -3 + \frac{2}{x_n} \text{ or } x_{n+1} = -\frac{1}{3}x_n^2 + \frac{2}{3}$$

A spreadsheet can be used to do all the calculation involved in iteration. Enter 0.5 in both B1 and C1. In B2 and C2 enter the formulae

$$-3 + 2/\text{B1}$$

and

$$-C1\hat{}2/3 + 2/3$$

respectively.

Copy the formulae down to B20 and C20. Do the numbers converge to the solutions of the equation? Vary the starting points to see when the numbers converge to the solutions. What happens if the starting point is

a) between the solutions

b) to the left of the solutions

c) to the right of the solutions?

So far this spreadsheet works for only one quadratic. It can be adjusted to work for any quadratic.

The general quadratic equation is $ax^2 + bx + c = 0$. On division by a this can be written as $x^2 + dx + e = 0$. The iterations can be written as

$$x_{n+1} = -d - \frac{e}{x_n} \quad \text{and} \quad x_{n+1} = -\frac{x_n^2}{d} - \frac{e}{d}$$

Enter the starting value 0.5 in A1. Change the entries in B1 and C1 to +A1. Then the starting point can be changed automatically.

Enter the values of d and e in A2 and A3 respectively. Change the formulae in B2 and C2 to

$$-A\$2 - A\$3/B1$$

and

$$-C1\hat{}2/A\$2 - A\$3/A\$2$$

respectively.

Copy the formulae down to B20 and C20. The first five rows should be as shown.

	A	B	C
1	0.5	0.5	0.5
2	3	1	0.583333
3	−2	−1	0.553240
4		−5	0.564641
5		−3.4	0.560393

By changing the entries in A2 and A3 you can investigate other quadratics. How does this affect the numbers in the B and C columns? Things to look out for are

What happens if d and e are **a)** large **b)** small?

What hapens if d is much smaller than e, or much larger?

How do the signs of d and e affect the results?

3 Trigonometric graphs

It takes a long time to draw graphs by hand. There are many computer packages which will draw graphs automatically. On a spreadsheet, also, you can plot graphs of functions.

Adding sin and cos graphs

Take the x-range of the graphs to be $0°$ to $360°$. In the following plot the graphs on the same picture. Plot the graph of $y = \sin x$ on the screen.

Plot the graph of $y = \sin x + \cos x$ on top of the original graph. How do they compare? How much higher is the new graph?

In fact, $\sin x + \cos x = A \sin (x + k)$, for some constants A and k. Try plotting the right-hand side of this equation, for different A and k, until the graphs overlap. What are A and k?

It is also true that $3 \sin x + 4 \cos x = A \sin (x + k)$, for some A and k. Find these values.

In general, $a \sin x + b \cos x = A \sin (x + k)$. How are A and k related to a and b?

Beats

The vibration caused by a musical note can be modelled by $y = \sin kx$, where k can be found from the frequency of the note. Suppose the note is played by two instruments, of which one is slightly out of tune. The vibration will be modelled by

$$y = \sin kx + \sin k'x$$

Plot the graph of this function, for $k = 40$ and $k' = 41$. What does the graph look like? What does the music sound like?

Consolidation section B

Chapter 6

1 Find the values of the following without using a calculator.

   a) 7^3 **b)** $1000^{\frac{1}{3}}$ **c)** 5^{-2} **d)** $\frac{1}{3}^{-2}$ **e)** $32^{\frac{3}{5}}$

2 Simplify the following.

   a) $3^x \times 9^{1-x}$ **b)** $x^{\frac{1}{2}} \times (x^2)^{\frac{3}{4}}$ **c)** $10^{1+x} \times 1000^{x-3}$

3 Solve the following equations.

   a) $10^{3x} = 100^{x-1}$ **b)** $4 \times 2^x = 8^{1-x}$ **c)** $25^x - 10 \times 5^x + 25 = 0$

4 Express $0.363\,636\,363\,6\ldots$ as a fraction.

5 Simplify the following surd expressions.

   a) $\sqrt{2}(\sqrt{2} + \sqrt{8})$ **b)** $(1 + \sqrt{3})^2$ **c)** $\dfrac{3 + \sqrt{2}}{\sqrt{2} - 1}$

6 Evaluate the following.

   a) $\log_3 27$ **b)** $\log 100$ **c)** $\log_{16} 2$ **d)** $\log_4 0.5$

7 Solve the following equations.

   a) $\log_3(x - 1) = 2$ **b)** $\log_2(x + 1) - \log_2 x = 3$ **c)** $4^x = 6$

Chapter 7

8 A sector of angle 0.77 radians has area 5 cm^2. What is the radius of the circle from which it is taken?

9 A chord subtends 0.8 radians at the centre of a circle of radius 3 cm. What is the area of the smaller segment cut off by the chord?

10 If x is a small angle, find approximations for

   a) $3 \sin x - \tan x$ **b)** $1 - \cos 2x$

11 Find a small angle which is an approximate solution to the equation $\cos x = 1.02 - x$.

Chapter 8

12 Differentiate the following.

a) x^6 **b)** $\sqrt{x}$ **c)** $\dfrac{1}{x^4}$ **d)** $3x^2 - 4x^3$

e) $\sqrt[3]{x} - \dfrac{2}{x^3}$ **f)** $(x+2)(x-9)$ **g)** $\dfrac{x^2 + 3}{x}$

13 Find the gradient of the curve $y = 2x^3 - 4x - 1$ at the point (2, 7).

14 Find the equation of the tangent to $y = x^2 + 3x - 1$ at the point where $x = 2$.

15 Find the equation of the normal to $y = \dfrac{3}{x}$ at the point where $x = 3$.

Chapter 9

16 In triangle ABC, $A = 45°$, AB = 6 cm and BC = 7 cm. Find B and C.

17 In triangle PQR, $P = 37°$, RP = 13 cm and RQ = 11 cm. Find the possible values of Q and R.

18 In triangle LMN, LM = 5 cm, MN = 6.3 cm and NL = 4.9 cm. Find L.

19 In triangle XYZ, $X = 43°$, XY = 63 cm and XZ = 71 cm. Find YZ.

20 Find the unknown lengths in the triangles below.

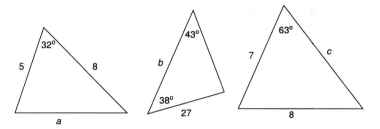

21 Find the unknown angles in the triangles below.

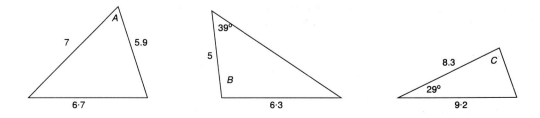

Chapter 10

22 An octahedral (eight-sided) die has the numbers 1 to 8 on its faces. It is rolled twice and the numbers are added to give a total score. What is the probability that

a) the score is 16   **b)** the score is 15   **c)** the score is 4, given that it is a double.

What is the most likely score, and what is its probability?

23 A woman has probability $\frac{1}{3}$ of passing her driving test, each test being independent of the others. What is the probability that she passes before the fifth attempt?

24 A house insurance firm covers houses in three areas A, B and C. The number of houses insured in the three areas are 100 000, 200 000 and 300 000 respectively. In each year, for each house, the probability that there will be a claim is 0.05 for houses in area A, 0.1 for houses in area B, and 0.01 for houses in area C.

a) If a house insured by the company is picked at random, what is the probability there will be a claim for it in a given year?

b) If the insurance company receives a claim, what is the probability it comes from a house in area C?

25 A and B are events for which $P(A) = 0.1$ and $P(A\ \&\ B) = 0.05$.

a) If A and B are independent, find $P(B)$.   **b)** If $P(B) = 0.2$, find $P(A\ \text{or}\ B)$.

MIXED EXERCISE

1 Find the gradient of the tangent to the curve $y = x^2 + 3$ at the point (2, 7). Find the angle, in radians, that this tangent makes with the x-axis.

2 Find the angle, in radians, between the lines $y = 2x + 1$ and $y = 3x - 2$.

3 The sides of a triangle have equations $y = x + 1$, $y = 3 - 2x$ and $y = \frac{1}{2}x - 2$. Find, in radians, the largest angle of the triangle.

4 In triangle ABC, $AB = 8$ cm, $BC = 5$ cm and $B = \pi/3$. Without the use of a calculator, find AC.

5 A river is 200 m across, and flows at 2 km h^{-1}. A man can row at 4 km h^{-1}. Where should he steer to reach a place on the far bank 40 m upstream from the point immediately opposite him? Give your answer in radians.

6 A strut AC is at $\dfrac{\pi}{4}$ to the horizontal, with A on the ground.

AC is supported by a second strut BC of length 25 ft, with B 35 ft distant from A. Find the possible lengths of AC.

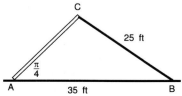

7 A sector of angle θ radians is taken from a circle of radius x cm. If the perimeter of the sector is constant at 20 cm, find an expression for the area A in terms of x only. Hence find the maximum area of the sector.

8 A wheel of radius a cm is rotated, so that after t seconds a spoke makes an angle of $\frac{1}{4}t^2$ radians with the horizontal. Find the speed of a point on the rim after 10 seconds.

9 If $\log_x 125 = 3$, find x. If $\log_y 64 = 3$, find y.

10 $\log_a 10 = x$ and $\log_a 4 = y$. Find in terms of x or y or both

    a) $\log_a 2$ **b)** $\log_a 5$ **c)** $\log_a \frac{1}{20}$

11 $\log_a xy = p$ and $\log_a \dfrac{x}{y} = q$. Find $\log_a x$ and $\log_a y$ in terms of p and q.

12 Solve the equation $\log_2 x + \dfrac{1}{\log_2 x} = 2$

13 In the diagram shown, the arc AB subtends an angle of θ at the centre of the circle. The tangents at A and B meet at T. If AT is equal to the arc length AB, show that $\theta = \tan \frac{1}{2}\theta$.

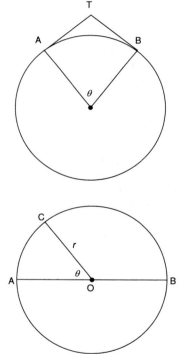

14 The figure shows a circle with centre O and radius r. AB is a diameter and OC is a radius making angle θ with the diameter. Find the area of the sector OCA. If the arc BC exceeds the arc AC by the radius, find θ.

15 a) a and b are rational numbers, with $a < b$. Show that between a and b there is a rational number and an irrational number.

    b) α and β are irrational numbers, with $\alpha < \beta$. Show that between α and β there is a rational number and an irrational number.

16 Use the identity $a^3 - b^3 \equiv (a - b)(a^2 + ab + b^2)$ to write $\dfrac{1}{\sqrt[3]{2} - 1}$ as an expression with a rational denominator.

17 If k is an integer greater than 1, show that $\sqrt{k^2 - 1}$ is irrational. Hence show that $\sqrt{k + 1} + \sqrt{k - 1}$ is also irrational.

18 Two wheels, of radii 10 cm and 15 cm, are in the same plane with their centres 25 cm apart. The wheels are connected by a belt tied tightly round them. Find the length of the belt.

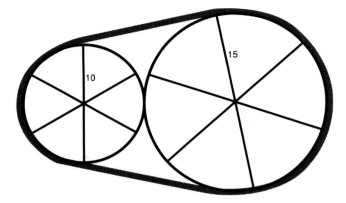

19 The quadratic $y = ax^2 + bx + c$ passes through the point (1, 4) and has gradient -1 at the point $(-1, -2)$. Find a, b and c.

20 A biased coin is such that the probability of 'heads' is p. It is found that when it is tossed twice, the probability of exactly one 'head' is 0.42. Find p.

21 In Russian Roulette, a live bullet is loaded into one of the six chambers of a revolver. The other five chambers are empty. The chambers are spun. The first player points the revolver at his head and pulls the trigger.

If the gun fires the game is over. Otherwise the chambers are spun again and the second player has a turn. This is repeated until the gun fires.

Boris and Ivan play Russian Roulette, Boris starting. Find the probabilities that the revolver fires

a) on the first go   **b)** on the third go (Boris's second turn)

c) on the fifth go (Boris's third turn).

What is the probability that Boris loses the game?

22 An aunt has three nephews, Alan, Brian and Charles. At Christmas she will give an expensive present to just one of the three. On Christmas Eve Alan goes to his aunt and says, 'One of Brian and Charles won't be getting the present, you won't be telling me anything if you let me know which one.' The aunt tells him that Brian won't be receiving the present. Alan now goes to bed happier, as he thinks his chance of receiving the present have improved from $\frac{1}{3}$ to $\frac{1}{2}$. How can this be, if the aunt hasn't given away any real information?

LONGER EXERCISE
Equal temperament

Take two musical notes, with frequencies f and $f*$. (The frequency of middle C, for example, is about 260 cycles per second.) The notes will be in harmony with each other if $f*/f$ is a simple rational number. In particular, if $f*/f = 2$, then the second note is an **octave** above the first.

The note an octave above C is C′, and has frequency 520 cycles per second.

Two notes for which $f*/f = \frac{3}{2}$ form a **fifth**.

1 Find the frequency of the note which is a fifth above middle C. (This is G.) Find the frequency of the note a fifth above G.

2 This note is higher than C′, so divide its frequency by 2, to obtain D.

3 Repeat this process, of going up by a fifth and then halving if the note is higher than C′, until you get a note with frequency almost equal to that of C′.

4 The note you obtained in step 3 is not exactly C′. An **equal temperament** scale provides notes equally spaced between C and C′. The notes are C, C#, D, D#, E, F, F#, G, G#, A, A#, B, C′. Suppose the ratio between the frequencies of successive notes is k. Show that k obeys the equation $k^{12} = 2$.

5 Solve the equation above. Is the solution rational?

6 With the value of k you have found, find some of the ratios between the frequencies of notes on the equal temperament scale. How close are they to simple rational numbers?

EXAMINATION QUESTIONS

1 In the triangle PQR, PQ $= 2$ cm, QR $= 3$ cm and the angle PQR is θ radians. The length of PR is a cm and the area of the triangle is A cm^2.

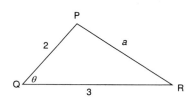

For small values of θ, show that

i) $a^2 \approx 1 + 6\theta^2$ **ii)** $A \approx k\theta$, where k is a constant to be determined.

JMB 1990

2 The points P, Q and R have coordinates (2, 4), (8, −2) and (6, 2) respectively.

a) Find the equation of the straight line l which is perpendicular to the line PQ and which passes through the midpoint of PR.

b) The line l cuts PQ at S. Find the ratio PS : SQ.

c) The circle passing through P, Q and R has centre C. Find the coordinates of C and the radius of the circle.

d) Given the angle PCQ $= \theta$ radians, show that $\tan \theta = \frac{24}{7}$.

Prove that the smaller segment of the circle cut off by the chord PQ has area $25\theta - 24$.

AEB 1991

3 In the diagram, A and B are two points on the circumference of a circle, centre O and radius r. Angle AOB is θ radians, and C is the mid-point of OA. The length of BC is x, and the length of the arc BA is s. Express x^2 in terms of r and θ.

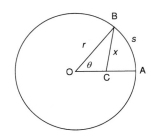

Hence show that, if θ is small, then $x^2 \approx \frac{1}{4}r^2 + \frac{1}{2}s^2$.

C 1992

4 Given that $y = x^3 - 4x^2 + 5x - 2$, find $\dfrac{dy}{dx}$.

P is the point on the curve where $x = 3$.

(i)   Calculate the y coordinate of P.   (ii) Calculate the gradient at P.

(iii) Find the equation of the tangent at P.   (iv) Find the equation of the normal at P.

Find the values of x for which the curve has a gradient of 5.

MEI 1992

5 The equation of the curve C is $y = \dfrac{x^2}{4}$.

(i)   Show that the equation of the normal to C at the point P(-4, 4) is

$$x - 2y + 12 = 0.$$

(ii) Find the coordinates of the point Q where this normal meets the curve C again.

(iii) Given that the tangents to C at P and Q intersect at the point R, find the coordinates of R.

(iv) Show that the length of the perpendicular from R to the line PQ is equal to the length of PQ.

W AS 1992

6 The figure shows a sketch of part of the curve C with equation

$$2y = 3x^3 - 7x^2 + 4x$$

which meets the x-axis at the origin O, the point A(1, 0) and the point B.

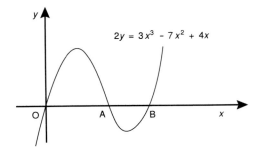

a) Find the coordinates of B.

The normals to the curve C at the points O and A meet at the point N.

b) Find the coordinates of N.   c) Calculate the area of $\triangle$OAN.

At the points P and Q, on the curve C, $\dfrac{dy}{dx} = 0$.

Given that the x-coordinates of P and Q are x_1 and x_2

d) find the values of (i) $x_1 + x_2$ (ii) $x_1 x_2$.

L 1993

7 Show that for a an integer, $a > 1$ and p, $q > 0$

$\log_a p^2$, $2 \log_a pq$ and $2 \log_a pq^2$

are three terms in an arithmetic sequence whose common difference is $2 \log_a q$.

Given that $pq^2 = a$, show that the sum of the first 5 terms of the arithmetic sequence with first term $\log_a p^2$ and common difference $2 \log_a q$ is 10.

L 1988

8 The probability that a particular man will survive the next twenty-five years is 0.6, and independently, the probability that the man's wife will survive the next twenty-five years is 0.7. Calculate the probability that in twenty-five years' time

(i) only the man will be alive   (ii) at least one of the two will be alive.

W AS 1990

CHAPTER 11

The binomial theorem

In Chapter 1 we handled expansions such as $(x + y)^2$ and $(x + y)^3$. The expansion of a higher power such as $(x + y)^{10}$ could be done by direct methods, but the working would be very tedious and we would be very liable to make errors. This chapter deals with methods of finding such expansions quickly and accurately.

We shall also find expansions for the roots and inverses of algebraic expressions.

11.1 Pascal's triangle

The expansion of $(x + y)^2$ is

$$(x + y)^2 = x^2 + 2xy + y^2$$

To find the expansion of $(x + y)^3$, we multiply this expansion by a further factor of $(x + y)$.

$$(x + y)^3 = (x + y)(x + y)^2 = (x + y)(x^2 + 2xy + y^2)$$

When this is multiplied out, we will obtain terms involving x^3, x^2y, xy^2 and y^3.

$$(x + y)(x^2 + 2xy + y^2) = x(x^2 + 2xy + y^2) + y(x^2 + 2xy + y^2)$$
$$= x^3 + 2x^2y + xy^2 + x^2y + 2xy^2 + y^3$$
$$= x^3 + 3x^2y + 3xy^2 + y^3$$

Note that the x^2y term came from multiplying the $2xy$ term by x, and the x^2 term by y. So this term is obtained by combining previous terms.

Instead of working through the expansion of $(x+y)^4$, we could apply this principle. Each term can be obtained by combining together two terms from the expansion of $(x+y)^3$.

The x^3y term will come from the $3x^2y$ term and the x^3 term, so it will be $4x^3y$. Each coefficient of the expansion of $(x+y)^4$ is obtained by adding two coefficients of $(x+y)^3$.

A figure which gives the coefficients of these expansions is the arrangement of numbers known as **Pascal's triangle**.

$$
\begin{array}{cccccccccccccc}
n=0 & & & & & & & 1 & & & & & & \\
n=1 & & & & & & 1 & & 1 & & & & & \\
n=2 & & & & & 1 & & 2 & & 1 & & & & \\
n=3 & & & & 1 & & 3 & & 3 & & 1 & & & \\
n=4 & & & 1 & & 4 & & 6 & & 4 & & 1 & & \\
n=5 & & 1 & & 5 & & 10 & & 10 & & 5 & & 1 & \\
n=6 & 1 & & 6 & & 15 & & 20 & & 15 & & 6 & & 1
\end{array}
$$

Note that each term is obtained by adding the two immediately to the left and right above it. For example, the 20 in the bottom row is obtained by adding the 10s in the row immediately above it.

The triangle can be used to find the expansion of $(x+y)^n$. This expansion will contain terms x^n, $x^{n-1}y$, $x^{n-2}y^2$ and so on to y^n. The n-row will give the coefficients of each of these terms.

EXAMPLE 11.1
Use Pascal's triangle to write out the expansion of $(x+y)^4$.

Solution
Look at the row labelled $n=4$. This gives the coefficients of x^4, x^3y, x^2y^2, xy^3, y^4, as 1, 4, 6, 4 and 1.
$$(x+y)^4 = x^4 + 4x^3y + 6x^2y^2 + 4xy^3 + y^4$$

EXERCISE 11A

1 Continue Pascal's triangle for two more rows.

2 Use Pascal's triangle to write out the expansions of the following.
   a) $(x+y)^5$ b) $(x+y)^6$ c) $(x+y)^7$

3 Does the expansion of $(x+y)^2$ fit the expansion from the $n=2$ row of Pascal's triangle?

4 The top two rows of Pascal's triangle correspond to the expansions of $(x+y)^0$ and $(x+y)^1$. Convince yourself that they are correct.

*5   Pascal's triangle was known to Chinese mathematicians many centuries before Pascal.

The picture on the right is from a book dated 1302.

There is a mistake in one of the entries. Can you find it? No knowledge of Chinese or of the Chinese system of numbers is necessary.

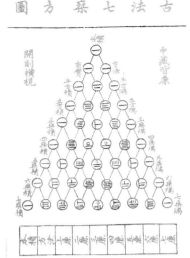

More expansions

The expansion of $(x + y)^n$ can become more complicated if x or y is multiplied by a constant, or if one term is negative. Care must be taken with the algebra.

Fig. 11.1

EXAMPLE 11.2
Find the expansion of $(2x - 3y)^4$.

Solution
The expansion of $(x + y)^4$ is

$$(x + y)^4 = x^4 + 4x^3y + 6x^2y^2 + 4xy^3 + y^4$$

To find the expansion of $(2x - 3y)^4$, replace x by $2x$, and y by $-3y$. Be sure to use brackets.

$$(2x - 3y)^4 = (2x)^4 + 4(2x)^3(-3y) + 6(2x)^2(-3y)^2$$
$$+4(2x)(-3y)^3 + (-3y)^4$$
$$(2x - 3y)^4 = 16x^4 - 96x^3y + 216x^2y^2 - 216xy^3 + 81y^4$$

EXAMPLE 11.3
Find the expansion of $(x + y)(2x - 3y)^4$.

Solution
Using the previous result

$$(x + y)(2x - 3y)^4 = (x + y)(16x^4 - 96x^3y$$
$$+216x^2y^2 - 216xy^3 + 81y^4)$$

Multiplying first by x and then by y, we obtain

$$16x^5 - 96x^4y + 216x^3y^2 - 216x^2y^3 + 81xy^4$$
$$+16x^4y - 96x^3y^2 + 216x^2y^3 - 216xy^4 + 81y^5$$
$$(x + y)(2x - 3y)^4 = 16x^5 - 80x^4y + 120x^3y^2 - 135xy^4 + 81y^5$$

EXERCISE 11B

1 Find the expansions of the following.

a) $(x + 2y)^4$
b) $(a - 2b)^5$
c) $(2p + 3q)^4$

d) $(3m - 5n)^6$
e) $\left(x + \dfrac{1}{x}\right)^6$
f) $\left(y - \dfrac{1}{y}\right)^7$

g) $(x + y)(x + 2y)^4$
h) $(a - b)(a - 2b)^5$
i) $(p + q)(2p + 3q)^4$

2 Without the aid of a calculator expand and simplify $(\sqrt{3} + 1)^4 + (\sqrt{3} - 1)^4$.

3 Without the aid of a calculator expand and simplify $(2 + \sqrt{2})^5 - (2 - \sqrt{2})^5$.

4 Without the aid of a calculator evaluate 1.1^6.

5 Find the following terms.

a) the $x^3 y^2$ term in $(x + 2y)^5$
b) the x^2 term in $(x - 3)^5$

c) the $x^2 y^2$ term in $(2x - 5y)^4$
d) the term without x in $\left(x + \dfrac{1}{x}\right)^8$

6 The a^4 and a^5 terms in the expansion of $(2 + a)^7$ are equal. Find a.

7 The x^4 term in the expansion of $(x + 3)^6$ is twice the x^5 term. Find x.

11.2 Factorials, permutations and combinations

n factorial

The entries in Pascal's triangle involve a function known as the **factorial** function.

$$n! = n \times (n - 1) \times (n - 2) \times \cdots 3 \times 2 \times 1$$

So $n!$ is the product of all the positive integers less than or equal to n. A button on your calculator will give you $n!$. The function is called 'n factorial' or 'n shriek' or 'n bang'.

A special case occurs when $n = 0$. By convention, $0! = 1$.

Ordering

Suppose there are five objects, labelled A, B, C, D and E.

There are five choices for which object comes first. Say D is chosen.

There are now four choices for which object comes second, as D has been already chosen. The second must be one of A, B, C or E.

There are now three choices for the third object.

There are two choices for the fourth object.

The fifth object must be the one remaining object.

So the total number of ways these choices could be made is $5 \times 4 \times 3 \times 2 \times 1 = 5! = 120$.

In general, $n!$ is the number of ways that n objects can be arranged in any order.

EXAMPLE 11.4

How many arrangements are there of the letters of these words?
a) HOUSE   **b)** BUILDING

Solution

a) In HOUSE we have five different letters. The number of orders these letters could be put in is 5!

The number of arrangements is 120.

b) In BUILDING we have eight letters. If they were all different, they could be arranged in 8! ways. We have two identical Is, which themselves can be arranged in 2! ways. So divide 8! by 2!.

$$\frac{8!}{2!} = \frac{40\,320}{2} = 20\,160$$

The number of arrangements is 20 160.

EXERCISE 11C

1 Evaluate the following without using a calculator.

a) 4!   **b)** 6!   **c)** 7!

2 Write down all the possible arrangements of the letters A, B and C. Make sure you have all 3! of them.

3 Write down all the possible arrangements of the letters A, B, C and D. Make sure you have all 4! of them.

*4 Describe a systematic way of writing down all the arrangements of the letters A, B, C, D and E.

5 What is the largest factorial your calculator can handle?

6 Simplify the following.

a) $50! \times 51$ **b)** $n! \times (n+1)$ **c)** $10! \div 10$

d) $n! \div n$ **e)** $\dfrac{1}{n!} + \dfrac{1}{(n+1)!}$

7 A competition lists ten desirable qualities of a mathematics textbook, and asks you to arrange them in order of importance. How many entries must you submit to be sure that one of them will give the correct order?

8 How many arrangements are there of the letters of these words?

    a) FEAR   **b)** DREAD   **c)** TERROR   **d)** HORROR

Permutations

We have established that $n!$ is the number of ways of ordering n objects. Suppose we want to select and order only r of these objects. This is called **permuting** r from n.

The number of ways of selecting r objects from n objects in an order is denoted by ${}^n\mathrm{P}_r$. It is given by the formula

$$^n\mathrm{P}_r = n(n-1)(n-2)\cdots(n-r+1) = \frac{n!}{(n-r)!}$$

Justification
The first object can be picked in n ways. The second object can be picked in $n-1$ ways. We will be selecting r objects and not selecting $n-r$ objects. So we will include all the factors of $n!$ down to $(n-r+1)$.

This product can also be obtained by dividing $n!$ by $(n-r)!$, to get rid of all the factors we don't want.

EXAMPLE 11.5
Ten horses run in a race. How many ways are there of predicting the first three places?

Solution
Here we have ten objects, and we want to select three of them in an order. The number of ways of doing this is

$$^{10}\mathrm{P}_3 = \frac{10!}{7!}$$

The number of ways is 720.

EXERCISE 11D

1 Evaluate the following without a calculator.

    a) ${}^5\mathrm{P}_2$ **b)** ${}^6\mathrm{P}_3$ **c)** ${}^7\mathrm{P}_2$

2 Show that ${}^n\mathrm{P}_n = n!$.

3 A form of 25 pupils will be awarded first, second, third and fourth prizes. In how many ways can this be done?

4 The governing body of a college has 20 members. In how many ways can the Principal, Dean and Bursar be selected?

5 The key-word of a cipher consists of five different letters from the alphabet. How many possible key-words are there?

6 The car number plates of a country consist of three different digits followed by three different letters. How many possible number plates are there?

Combinations

The number of ways we can select r objects out of n in an order is denoted by ^nP_r. Often we are not concerned with the order in which the objects are selected.

The number of ways we can select r objects out of n, without regard to order is written ^nC_r or $\binom{n}{r}$.

It is given by

$$^nC_r = \frac{^nP_r}{r!} = \frac{n!}{(n-r)!r!}$$

Justification
The number of ways we can select the object **with** regard to order is ^nP_r.

If we are not bothered with the order, these r objects can be arranged in $r!$ ways, without altering the selection. Hence

$$^nC_r = \frac{^nP_r}{r!}$$

EXAMPLE 11.6
Four equal prizes are to be awarded to children from a class of 25. In how many ways can this be done?

Solution
We need to select four children out of 25. As the prizes are equal, it does not matter in which order the children are selected. Hence the number of ways is $^{25}C_4$.

The number of ways is 12 650.

EXAMPLE 11.7
The chorus of an opera will contain ten men and eight women, who are to be selected from 30 men and 20 women. In how many ways can this be done?

Solution

There are $^{30}C_{10}$ ways of selecting the men. For each of these selections, there are $^{20}C_8$ ways of selecting the women. Multiply these together.

The number of ways is $^{30}C_{10} \times {}^{20}C_8 = 3.78 \times 10^{12}$

The number of ways is 3.78×10^{12}.

EXERCISE 11E

1 Evaluate the following.

 a) $^5C_3$ **b)** $^{10}C_4$ **c)** $^{10}C_6$

2 Show that $^9C_2 = {}^9C_7$. Can you show why this is true?

3 Show that $^nC_0 = 1$. Can you give a reason why this is true?

4 Show that $^nC_r = {}^nC_{n-r}$. Can you give a reason why this must be true?

5 Show that $^nC_r + {}^nC_{r-1} = {}^{n+1}C_r$. Can you give a reason why this must be true?

6 List the selections of two letters from A, B, C, D and E. Check that you have $^5C_2$ of them.

7 In how many ways can a football team of 11 players be selected from a class of 20?

8 In how many ways can you select four library books from a shelf of 12 books?

9 There are ten points in a plane, no three of which are in a straight line. How many triangles can be drawn using the points as vertices?

10 There are 12 points on the circumference of a circle. How many chords can be drawn joining these points?

11 A football tournament is arranged between teams from Europe and America. There are to be four European teams and four American teams. If the European teams are to be chosen from ten national sides, and the American teams from eight national sides, find how many possible selections there are.

12 Three people eat at a restaurant. They will each pick a starter from a choice of six items, and a main course from a choice of ten items. They will be sharing the dishes, so it doesn't matter who orders what, as long as they are different. How many possible choices are there?

Probability

Permutations and combinations involve different ways of selecting objects. If these ways are **equally likely**, then the ideas of this section can be used to solve probability problems.

EXAMPLE 11.8

There are six runners in a race, in which skill plays no part. A punter places a bet by selecting the runners which will come first and second. What is the probability he wins the bet?

Solution

The number of selections for the first two places is $^6P_2 = 30$. As skill plays no part, all the selections are equally likely to be the winning one.

The probability he wins is $\frac{1}{30}$.

EXAMPLE 11.9

In bridge, each player is dealt 13 out of the 52 cards. It does not matter which order the cards are dealt. A **Yarborough** hand is one containing no ace, king, queen, jack or 10. What is the probability that a bridge hand is a Yarborough?

Solution

There are $^{52}C_{13}$ possible bridge hands. The number of cards which are not aces, kings, queens, jacks or 10s is 32. Hence there are $^{32}C_{13}$ possible Yarborough hands. If the deal is fair, all the possible hands are equally likely.

The probability of a Yarborough hand is $\dfrac{^{32}C_{13}}{^{52}C_{13}} = 0.000\,547$.

EXERCISE 11F

1 The combination of a lock consists of three different digits. If I guess the digits, what is the probability I get the correct combination?

2 A 'Lucky Dip' at a bazaar contains 50 boxes, of which 20 are empty and 30 contain prizes. If I pick out three boxes, what is the probability that

   a) all three are empty   **b)** all three contain prizes.

3 A class contains twelve girls and eight boys. Four are chosen at random to go on a trip. What is the probability that the four chosen are all girls?

4 Three letters are picked at random from A, B, C, D, E and F and laid out in order. What is the probability they will spell the word BAD?

*5 In each packet of a certain breakfast cereal there is a plastic model of a footballer. There are 20 different models available. If you buy ten packets, what is the probability that all the models will be different?

*6 What is the probability that a bridge hand will contain seven cards of the same suit?

11.3 The binomial theorem for positive integer powers

Now we can link together the previous two sections. The entries in Pascal's triangle are all of the form ^nC_r. In the n-row, the entries are $^nC_0$, $^nC_1$, $\ldots$, ^nC_n. This gives rise to the **binomial theorem**. In this context ^nC_r is more usually written as $\binom{n}{r}$.

$$(x+y)^n = x^n + \binom{n}{1}x^{n-1}y + \binom{n}{2}x^{n-2}y^2 + \cdots$$

$$+\binom{n}{r}x^{n-r}y^r + \cdots + y^n = \sum_{r=1}^{n} \binom{n}{r}x^{n-r}y^r$$

Justification
Regard $(x+y)^n$ as having n factors all equal to $(x+y)$.

$$(x+y)(x+y)\cdots(x+y) \qquad (n \text{ factors})$$

When these are multiplied together, the x^n term occurs only once. So the coefficient of x^n is 1.

The $x^{n-1}y$ terms can be obtained by picking y from one of the brackets, and x from the remaining $n-1$ brackets. The number of ways this can be done is $\binom{n}{1}$. So the coefficient of $x^{n-1}y$ is $\binom{n}{1}$.

The $x^{n-2}y^2$ terms can be obtained by picking y from two of the brackets, and x from the remaining $n-2$ brackets. The number of ways this can be done is $\binom{n}{2}$.

The general $x^{n-r}y^r$ term can be obtained by picking y from r brackets, and x from the remaining $n-r$ brackets. The number of ways this can be done is $\binom{n}{r}$.

So the coefficients of the expansion are $\binom{n}{0}$, $\binom{n}{1}$, $\cdots$, $\binom{n}{n}$. This gives the theorem.

EXAMPLE 11.10
Use the binomial theorem to find the first three terms in the expansion of $(a-2b)^{20}$.

Solution
Note that with the binomial theorem we can go directly to the expansion. With Pascal's triangle we would have to write out the first 19 rows before we reached the 20th.

$$(x+y)^{20} = x^{20} + \binom{20}{1}x^{19}y + \binom{20}{2}x^{18}y^2 + \cdots$$

By calculator or by ordinary calculation, we find that $\binom{20}{1} = 20$ and $\binom{20}{2} = 190$. Substitute a for x and $-2b$ for y.

$$a^{20} + 20a^{19}(-2b) + 190a^{18}(-2b)^2$$

The first three terms are $a^{20} - 40a^{19}b + 760a^{18}b^2$.

EXERCISE 11G

1 Expand the following using the binomial theorem.

a) $(a + b)^6$ 　　　　**b)** $(p - q)^7$ 　　　　**c)** $(m + 2n)^5$

d) $(2a + b)^5$ 　　　　**e)** $(2z + 3w)^6$ 　　　　**f)** $(2x - 5y)^4$

2 Use the binomial theorem to find the first three terms in the expansions of these expressions.

a) $(x + y)^{18}$ 　　　**b)** $(a + b)^{15}$ 　　　**c)** $(p - q)^{20}$

d) $(x + 2y)^{17}$ 　　**e)** $(2c + 3d)^{12}$ 　　**f)** $(p - 3q)^{14}$

3 In the following, find the given term in the expansion indicated.

a) x^5y^7 term, in $(x + y)^{12}$ 　　**b)** x^3y^5 term, in $(x - y)^8$ 　　**c)** z^4w^7 term, in $(2z + w)^{11}$

d) p^7q^2 term, in $(2p + 3q)^9$ 　　**e)** x^4y^4 term, in $(2x - 5y)^8$ 　　**f)** w^5z^3 term, in $(w - 2z)^8$

g) x^4y^3 term, in $(x^2 + y)^5$ 　　**h)** z^6w^6 term, in $(z^2 - w^3)^5$ 　　**i)** p^{-6} term, in $\left(2 + \dfrac{1}{p}\right)^8$

4 Find the constant term in the expansion of $\left(x + \dfrac{1}{x}\right)^{14}$.

5 Find the constant term in the expansion of $\left(y - \dfrac{2}{y}\right)^8$.

6 Find the constant term in the expansion of $\left(x^2 - \dfrac{1}{x}\right)^9$.

7 Write down the first three terms in the expansion of $(1 + x)^{10}$. Hence expand $(1 + y + y^2)^{10}$, up to the y^2 term.

8 Expand $(1 - x + x^2)^8$, up to the x^2 term.

9 Expand $(1 + x)(1 + 2x)^{10}$ up to the x^3 term.

10 Expand $(1 - 2x)(2 + x)^8$ up to the x^2 term.

11.4 The binomial expansion for other powers

In the previous section we used the binomial theorem to expand expressions of the form $(a + b)^n$. In every case n was a positive integer.

The expansion involved $\binom{n}{r}$. If this is defined as 'the number of ways of picking r objects from n' then the definition cannot be extended to cases when n is fractional or negative.

But if we use the formula

$$\binom{n}{r} = \frac{n(n-1)(n-2)\cdots(n-r+1)}{r!}$$

then it can be evaluated for any n.

Binomial theorem for all powers

If $-1 < x < 1$, then

$$(1+x)^n = 1 + nx + \frac{n(n-1)}{2!}x^2 + \frac{n(n-1)(n-2)}{3!}x^3 + \cdots + \frac{n(n-1)(n-2)\cdots(n-r+1)}{r!}x^r + \cdots$$

If n is not a positive integer, none of the coefficients will be zero. The series will continue for ever. The restriction on x is necessary in order for the infinite series to converge.

EXAMPLE 11.11
Find the expansion of $\sqrt{1+x}$, up to the term in x^3.

Solution
We use the formula above, with $n = \frac{1}{2}$.

$$(1+x)^{\frac{1}{2}} \approx 1 + \frac{1}{2}x + \frac{\frac{1}{2}(\frac{1}{2}-1)x^2}{2} + \frac{\frac{1}{2}(\frac{1}{2}-1)(\frac{1}{2}-2)x^3}{6}$$

$$\sqrt{1+x} \approx 1 + \frac{1}{2}x - \frac{1}{8}x^2 + \frac{1}{16}x^3$$

EXAMPLE 11.12
Find the expansion of $\dfrac{1}{1-2x}$, up to the x^3 term.

For what values of x is the expansion valid?

Solution
Regard the expression as $(1-2x)^{-1}$. Apply the formula, with $n = -1$.

$$(1-2x)^{-1} = 1 + (-1)(-2x) + \frac{(-1)(-2)}{2!}(-2x)^2 + \frac{(-1)(-2)(-3)}{3!}(-2x)^3 + \cdots$$

Simplify the coefficients. Note that all the fractions are equal to 1 or -1.

$$\frac{1}{1-2x} \approx 1 + 2x + 4x^2 + 8x^3$$

Note
This expansion is the geometric series with common ratio $2x$.

The series will converge provide that $-1 < 2x < 1$.

The expansion is valid for $-\frac{1}{2} < x < \frac{1}{2}$.

EXERCISE 11H

1 Find the expansions of the following, up to the x^3 terms.

    a) $(1 + x)^{\frac{1}{3}}$ **b)** $(1 + 2x)^{\frac{1}{2}}$ **c)** $(1 - 3x)^{\frac{1}{2}}$

    d) $(1 + 4x)^{\frac{3}{2}}$ **e)** $(1 - \frac{1}{2}x)^{\frac{1}{3}}$ **f)** $(1 - x)^{-2}$

    g) $(1 + x)^{-\frac{1}{2}}$ **h)** $(1 + 2x)^{-2}$ **i)** $(1 - 4x)^{-3}$

    j) $\dfrac{1}{(1 + x)^2}$ **k)** $\dfrac{1}{(1 + 3x)^3}$ **l)** $\dfrac{1}{\sqrt[3]{1 + x}}$

2 For each of the expansions of Question 1, find the range of values of x for which it is valid.

3 Find the first three terms in the expansion of each of the following.

    a) $(1 + x^2)^{\frac{1}{2}}$
    b) $(1 - 2x^2)^{-3}$
    c) $(1 + x^3)^{-\frac{1}{2}}$

4 Show that $\sqrt{\dfrac{1 + x}{1 - x}} = \dfrac{1 + x}{\sqrt{1 - x^2}}$. Find the expansion of this expression up to the term in x^6.

5 Find the expansion of $\sqrt{\dfrac{1 + x}{1 - 2x}}$ up to the term in x^5.

6 The first four terms in the expansion of $(1 + kx)^{-1.5}$ are $1 + ax + 30x^2 + cx^3$. Find k, a and c.

7 The first three terms in the expansion of $(1 + ax)^n$ are $1 + 2x - 2x^2$. Find a and n.

8 In the expansion of $(1 + x)^n$, the coefficient of x^2 is three times that of x. Find n.

9 Show that in the expansion of $(1 + 2x)^{\frac{3}{2}} + (1 - 9x)^{\frac{1}{3}}$ there is no x term. Find the x^2 term. For what range of values of x is the expansion valid?

10 If $\sqrt[3]{1 + ax}$ and $\dfrac{1 + 2bx}{1 + bx}$ are expanded, the constant terms are equal and the x terms are equal. Show that the x^2 terms are also equal. Find a relationship between a and b.

11 Find k such that $\dfrac{1 + kx}{\sqrt{1 + x}}$ has no x term. Find the x^2 term.

11.5 Approximations using the binomial expansion

The binomial expansion can be used to find approximations for powers.

The binomial expansion for n a positive integer

Suppose we find the expansion of $(x + y)^n$, where n is a positive whole number. The expansion involves terms with successively greater powers of y. If y is small in comparison with x, then an approximation can be obtained by ignoring the later terms with higher powers of y.

The binomial expansion for other values of n

Suppose we find the expansion of $(1 + x)^n$, where n is not necessarily a positive whole number. If x is small, we can find an approximation by using the first few terms of the expansion.

Sometimes it is necessary to adjust the expression to get it in the form $(1 + x)^n$.

EXAMPLE 11.13
Use the first three terms in the expansion of $(x + y)^{10}$ to find the approximate value of 1.01^{10}.

Solution
We can write 1.01 as $1 + 0.01$. If we let $x = 1$ and $y = 0.01$, then y is small in comparison with x. Write out the first three terms of the expansion of $(x + y)^{10}$.

$$x^{10} + 10x^9y + 45x^8y^2$$

Now put $x = 1$ and $y = 0.01$ into this expression.

$$1^{10} + 10 \times 1^9 \times 0.01 + 45 \times 1^8 \times 0.01^2$$

1.01^{10} is approximately 1.1045.

EXAMPLE 11.14
By using the binomial expansion up to the x^3 term, find an approximation for $\sqrt{5}$.

Solution
Note that we cannot write $\sqrt{5}$ as $\sqrt{1 + 4}$ and use the binomial expansion with $x = 4$. This would not obey the requirement that $-1 < x < 1$. Instead we write it as $\sqrt{4 + 1}$ and take out a factor of 2.

$$\sqrt{5} = \sqrt{4 + 1} = \sqrt{4(1 + \tfrac{1}{4})} = \sqrt{4}\sqrt{1 + \tfrac{1}{4}} = 2\sqrt{1 + \tfrac{1}{4}}$$

Now put $x = \tfrac{1}{4}$. This does obey the requirement that $-1 < x < 1$. We can now use the expansion of $(1 + x)^{\frac{1}{2}}$ found in the example of the previous section.

$$\sqrt{1 + \tfrac{1}{4}} \approx 1 + \tfrac{1}{2}(\tfrac{1}{4}) - \tfrac{1}{8}(\tfrac{1}{4})^2 + \tfrac{1}{16}(\tfrac{1}{4})^3 = 1.118\,164\,063$$

$\sqrt{5}$ is approximately 2.236 328 125.

EXERCISE 11I

1 Write out the first three terms in the expansion of $(x + y)^{12}$. Hence find approximations for the following.

a) 1.01^{12} b) 1.1^{12} c) 0.98^{12}

2 Write out the first three terms in the expansion of $(x + y)^{8}$. Hence find approximations for the following.

a) 2.1^{8} b) 1.95^{8}

3 Find the expansion of $(1 + x)(1 - x)^{10}$ up to the x^2 term. Hence find approximations for the following.

a) 1.01×0.99^{10} b) 0.98×1.02^{10}

4 Find the first term in the expansion of $(1 + 0.02)^{10}$ which is less than 0.005. Hence, without using a calculator, find 1.02^{10} to an accuracy of two decimal places.

5 Without using a calculator, find the following to an accuracy of two decimal places.

a) 1.1^{12} b) 0.9^{12}

6 By using the binomial expansion up to the x^3 term, find approximations for the following.

a) $\sqrt{1.02}$ b) $\sqrt{0.98}$ c) $\sqrt[3]{1.02}$

d) $\sqrt{17}$ e) $\sqrt{99}$ f) $\sqrt[3]{9}$

7 By expanding $\left(1 - \dfrac{1}{8}\right)^{\frac{1}{3}}$, find an approximation for $\sqrt[3]{7}$.

8 Obtain the expansion of $\sqrt{\dfrac{1 + x}{1 - x}}$ up to the x^3 term. By putting $x = \frac{1}{10}$, find a rational approximation for $\sqrt{11}$.

*9 Find constants a and b such that the expansions of $(1 + x)^{\frac{1}{4}}$ and $\dfrac{4 + ax}{4 + bx}$ agree for the first three terms. Show that $\sqrt[4]{\dfrac{6}{5}}$ is approximately $\dfrac{45}{43}$.

*10 Find constants a and b such that the expansions of $\sqrt{1 - 2x}$ and $\dfrac{1 + ax}{1 + bx}$ agree for the first three terms. By putting $x = 0.01$ find a rational approximation for $\sqrt{2}$.

LONGER EXERCISE

More about $^{n}C_{r}$

1 Show that $^{n+1}C_{r} = {}^{n}C_{r} + {}^{n}C_{r-1}$.

2 Show that $^{n+2}C_{r} = {}^{n}C_{r} + 2\,{}^{n}C_{r-1} + {}^{n}C_{r-2}$.

3 Can you find a similar expression for $^{n+3}C_{r-2}$?

4 Can you guess what the similar expression is for $^{n+9}C_r$?

5 Show that $^nC_0 + {^nC_1} + {^nC_2} + \cdots + {^nC_n} = 2^n$.

EXAMINATION QUESTIONS

1 Given that $(1 + x)^{10} = 1 + px + qx^2 + \cdots + x^{10}$, find the values of the constants p and q.

L 1993

2 Write down the first three terms in the binomial expansion, in ascending powers of x, of $(1 + ax)^n$, where $a \neq 0$ and $n \geq 2$.

Given that the coefficient of x in this expansion is twice the coefficient of x^2

a) show that $n = \dfrac{1 + a}{a}$

b) find the value of the coefficient of x^2 when $a = \frac{1}{4}$.

L 1992

3 a) Find the term independent of x in the expansion of

$$\left(x^2 - \frac{1}{2x}\right)^9$$

b) If $-1 < x < 1$, find the term independent of x in the binomial expansion of $(x - x^4)^{-9}$ in ascending powers of x.

NI 1991

4 Write down and simplify the binomial expansion of $(1 + 2x)^{-\frac{1}{2}}$ up to and including the term in x^3.

By putting $x = \frac{1}{8}$, use your expansion to obtain an approximation to $\sqrt{5}$, giving your answer as a ratio of two integers.

W 1990

Summary and key points

1 Each term of Pascal's triangle is found by adding the terms immediately above.

The *n*-row of Pascal's triangle can be used to find the expansion of $(x + y)^n$.

Be careful when expanding something like $(2x - 3y)^6$. Take powers of $2x$, not just of x, and take powers of $(-3y)$, not of $3y$.

2 The number of ways in which n objects can be arranged is

$$n! = n \times (n-1) \times (n-2) \times \cdots \times 3 \times 2 \times 1$$

The number of ways r objects can be chosen in order from n objects is

$$^n\text{P}_r = n(n-1)(n-2)\cdots(n-r+1) = \frac{n!}{(n-r)!}$$

The number of ways r objects can be chosen from n objects in any order is

$$^n\text{C}_r = \binom{n}{r} = \frac{n!}{r!(n-r)!}$$

Be careful to distinguish between $^n\text{P}_r$ and $^n\text{C}_r$.

3 The binomial theorem states that

$$(x+y)^n = x^n + \binom{n}{1}x^{n-1}y + \binom{n}{2}x^{n-2}y^2 + \cdots$$
$$+ \binom{n}{r}x^{n-r}y^r + \cdots + y^n$$

As with Pascal's triangle, be careful if x or y is negative or multiplied by a constant.

4 The binomial expansion can be extended to cases when n is fractional or negative.

If $-1 < x < 1$, then

$$(1+x)^n = 1 + nx + \frac{n(n-1)}{2!}x^2 + \frac{n(n-1)(n-2)}{3!}x^3 + \cdots$$

$$+ \frac{n(n-1)(n-2)\cdots(n-r+1)}{r!}x^r + \cdots$$

Do not ignore the restriction on x.

5 If y is small in comparison to x, an approximation to $(x+y)^n$ can be found by taking the first few terms in the expansion.

Provided x is small, approximations can be found using the expansion of $(1+x)^n$.

Differentiation and graphs

In Chapter 8 we defined differentiation in terms of gradients of tangents to graphs. The derivative $\dfrac{dy}{dx}$ gives the gradient of a graph, even when that graph is curved.

Now we shall continue with applications of differentiation to graphs. We shall see how differentiation can be used to solve many practical problems.

12.1 The second derivative of a function

If we differentiate a function twice, the result is the **second derivative**. This is the derivative of the first derivative, or the rate of change of the rate of change.

If y represents distance and x represents time, then $\dfrac{dy}{dx}$ represents velocity. The second derivative represents the rate of change of velocity, which is **acceleration**.

Notation

The **first derivative** (usually called just 'the derivative') is $\dfrac{dy}{dx}$.

Differentiate the derivative.

The **second derivative** is

$$\frac{d}{dx}\left(\frac{dy}{dx}\right) = \frac{d^2y}{dx^2}.$$

For example, let $y = x^3$. Then $\dfrac{dy}{dx} = 3x^2$.

The second derivative is $\dfrac{d}{dx}(3x^2) = 6x$.

The process can be repeated.

The **third derivative** is $\dfrac{d^3y}{dx^3}$, and so on.

The f′ notation

The original function is y. If we write y as $f(x)$, then the derivative is written $f'(x)$.

$$\text{If } y = f(x), \quad \text{then} \quad \frac{dy}{dx} = f'(x)$$

The second derivative is written $f''(x)$, the third derivative $f'''(x)$ and so on.

Sign of $\dfrac{d^2y}{dx^2}$

If $\dfrac{d^2y}{dx^2}$ is positive, then $\dfrac{dy}{dx}$ is increasing. If the function itself is increasing, the curve will be getting steeper. If the function itself is decreasing, the curve will be getting shallower. Both cases are shown in Fig. 12.1. Notice that the shape of the curve is **concave**.

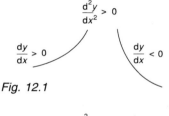

If $\dfrac{d^2y}{dx^2}$ is negative, then $\dfrac{dy}{dx}$ is decreasing. If the function itself is increasing, the curve will be getting shallower. If the function itself is decreasing, the curve will be getting steeper. Both cases are shown in Fig. 12.2. Notice that the shape of the curve is **convex**.

Fig. 12.1

There may be a point at which the curve changes from being convex to concave or vice versa. This is a **point of inflection**. At a point of inflection $\dfrac{d^2y}{dx^2} = 0$.

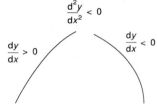

Fig. 12.2

EXAMPLE 12.1

Let $y = 4x^3 + 3x^2 - x - 2$. Find $\dfrac{d^2y}{dx^2}$. Where is the graph of the function concave?

Solution

Differentiate once to find the first derivative.

$$\frac{dy}{dx} = 12x^2 + 6x - 1$$

Differentiate again for the second derivative.

$$\frac{d^2y}{dx^2} = 24x + 6$$

The graph will be concave if $\dfrac{d^2y}{dx^2}$ is positive.

$\dfrac{d^2y}{dx^2} > 0$ if $24x + 6 > 0$

which simplifies to $4x + 1 > 0$ or $x > -\frac{1}{4}$.

The graph is concave if $x > -\frac{1}{4}$.

EXERCISE 12A

1 Find the second derivatives of the following functions.

a) $y = x^3$ **b)** $y = 2x^2 - 3x + 1$ **c)** $y = 4x - x^2$

d) $y = (3 - x)(4 + x)$ **e)** $y = \dfrac{1}{x}$ **f)** $y = \sqrt{x}$

g) $y = \dfrac{2}{x^3}$ **h)** $y = \dfrac{x + 1}{x}$ **i)** $y = \sqrt{x}(x + 1)$

2 Where is the graph of $y = x^3 - 12x$ concave?

3 Let $y = 4 - x^4$. Show that the graph of this function is always convex.

4 Show that the second derivative of the quadratic function $y = ax^2 + bx + c$ is constant.

5 Let $y = x^3 - 6x^2 + 3$. Find the point at which $\dfrac{d^2y}{dx^2} = 0$.

12.2 Stationary points

The graph in Fig. 12.3 has a high point at A. This point is called a **maximum**. At this point the graph is flat, as it has reached a high point.

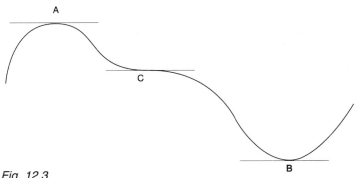

Fig. 12.3

At a maximum $\dfrac{dy}{dx} = 0$.

The same applies to a low point like B, which is called a **minimum**.

There is also a third possibility, a point like C which is neither a maximum nor a minimum. Notice that at C the graph changes from concave to convex, so C is a point of inflection.

All these points are **stationary** points.

At a stationary point, $\dfrac{dy}{dx} = 0$

To find the stationary points, we differentiate and put $\dfrac{dy}{dx} = 0$. This will give an equation in x. The solutions of the equation will give the x-values of the stationary points.

There are various ways to determine whether a stationary point is a maximum, a minimum or a point of inflection. Below are three methods.

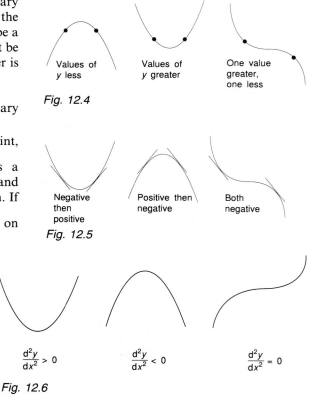

1 Evaluate y on both sides of the stationary point. If the values of y are both less than the value of y at the stationary point, it must be a maximum. If they are both greater, it must be a minimum. If one is greater and the other is less then it is a point of inflection.

Values of
y less

Values of
y greater

One value
greater,
one less

Fig. 12.4

2 Evaluate $\dfrac{dy}{dx}$ on both sides of the stationary

point. If $\dfrac{dy}{dx}$ is negative on the left of the point,

and positive on the right, then it is a minimum. If it is positive on the left and negative on the right then it is a maximum. If $\dfrac{dy}{dx}$ is positive on both sides or negative on both sides then it is a point of inflection.

Negative
then
positive

Positive then
negative

Both
negative

Fig. 12.5

3 Evaluate $\dfrac{d^2y}{dx^2}$ at the stationary point.

If $\dfrac{d^2y}{dx^2}$ is positive, then the curve is

concave and we have a minimum.

If $\dfrac{d^2y}{dx^2}$ is negative, then the curve is

convex and we have a maximum.

$\dfrac{d^2y}{dx^2} > 0$

$\dfrac{d^2y}{dx^2} < 0$

$\dfrac{d^2y}{dx^2} = 0$

Fig. 12.6

If we have a point of inflection, then $\dfrac{d^2y}{dx^2} = 0$.

Notes

If $\dfrac{d^2y}{dx^2} = 0$ at a stationary point, we do not know whether the point is a maximum, a minimum or a point of inflection. We have to use **1** or **2** to see which of the three it is. Note also that at a point of inflection, $\dfrac{dy}{dx}$ is not necessarily 0.

The maxima and minima that we find may be only **local** maxima and minima. The function may achieve greater or smaller values elsewhere, as shown.

Fig. 12.7

EXAMPLE 12.2

Find the stationary point of $y = x^2 - 4x + 5$ and determine whether it is a maximum or a minimum.

Solution

Differentiating, we get $\dfrac{dy}{dx} = 2x - 4$.

Put this equal to zero.

$$2x - 4 = 0$$
So $x = 2$.

If $x = 2$, then $y = 2^2 - 4 \times 2 + 5 = 1$

$(2, 1)$ is a stationary point.

We use the second method for determining whether $(2, 1)$ is a maximum or minimum. Evaluate $\dfrac{dy}{dx}$ on either side of $x = 2$.

For $x = 1.9$ $\quad 2x - 4 = -0.2$

For $x = 2.1$ $\quad 2x - 4 = +0.2$

So $\dfrac{dy}{dx}$ is negative to the left of the stationary point, and positive to the right. The point is a minimum.

$(2, 1)$ is a minimum.

Note

The same result could have been obtained by the 'completing the square' method of Chapter 2. Differentiation is a much more powerful method, it works for many other functions besides quadratics.

EXERCISE 12B

1 Find the stationary points of the following functions, determining whether they are maxima, minima or points of inflection.

a) $y = x^2 + 4x - 3$ $\qquad$ b) $y = x^3 - x$ $\qquad$ c) $y = x + \dfrac{1}{x}$

d) $y = x^3 - 3x^2 + 3x - 2$ $\quad$ e) $y = 9x^4 + 4x^3$ $\qquad$ f) $y = (x + 3)(x - 1)$

2 Find the points of inflection of the following functions.

    a) $y = x^3 - 6x^2 + 4x - 1$ **b)** $y = x^2 - \dfrac{1}{x}$

3 Show that a quadratic function $y = ax^2 + bx + c$ has no points of inflection.

4 Show that the following functions have no points of inflection.

    a) $y = x + \dfrac{1}{x}$ **b)** $y = x^4$ **c)** $y = x^4 + 60x^2$

5 Consider the three functions

    a) $y = x^3$ **b)** $y = x^4$ **c)** $y = 1 - x^4$

In each case find $\dfrac{d^2y}{dx^2}$ and find the points where $\dfrac{d^2y}{dx^2} = 0$. Show that one of the points is a maximum, one a minimum and one a point of inflection.

*6 Let $y = ax^3 + bx^2 + cx + d$. Show that y will have no maximum or minimum if $b^2 < 3ac$.

Sketching graphs

Once we have found the maxima, minima and points of inflection of a function, we can make a rough sketch of its graph, showing these features.

EXAMPLE 12.3
Find the stationary points of the curve $y = x^3 - 3x^2 - 9x + 2$. For each point, state whether it is a maximum, a minimum or a point of inflection. Find any point of inflection of the curve. Sketch the curve.

Solution
We start by differentiating.

$$\frac{dy}{dx} = 3x^2 - 6x - 9$$

To find the stationary points, we put $\dfrac{dy}{dx} = 0$.

$$0 = 3x^2 - 6x - 9 = 3(x^2 - 2x - 3) = 3(x + 1)(x - 3)$$

So the stationary points are at $x = -1$ and $x = 3$. At these points, $y = 7$ and $y = -25$ respectively.

Differentiating again gives

$$\frac{d^2y}{dx^2} = 6x - 6$$

This is negative for $x = -1$ and positive for $x = 3$.

There is a maximum at $(-1, 7)$, a minimum at $(3, -25)$.

When $x = 1$, $\dfrac{d^2y}{dx^2} = 0$. For this value, $y = -9$.

For $x < 1$, $\dfrac{d^2y}{dx^2}$ is negative.

For $x > 1$, $\dfrac{d^2y}{dx^2}$ is positive.

So the graph is convex up to $x = 1$, and concave after $x = 1$.

There is a point of inflection at $(1, -9)$.

Now we can sketch the graph by putting in the stationary points and the point of inflection.

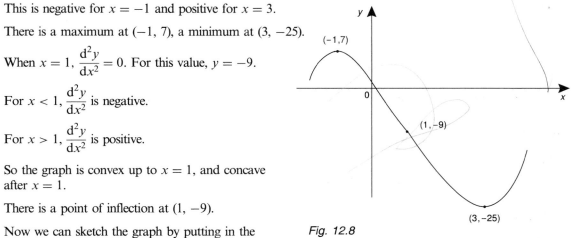

Fig. 12.8

EXERCISE 12C

1 Find the maxima, minima and points of inflection of the following functions. Hence make rough sketches of their graphs.

a) $y = x^3 - 3x$

b) $y = x^3 - 6x^2 + 9x - 5$

c) $y = 7 + 6x + 3x^2 - 4x^3$

d) $y = x^4 - 8x^2$

d) $y = x^3 + 3x^2 + 3x + 2$

f) $y = 3x^4 - 4x^3 + 1$

*2 Find the stationary points of $y = x + \dfrac{1}{x}$. Why is the minimum greater than the maximum? Sketch the graph of the function.

*3 Let $y = x^n(x - 1)$. Find the stationary points for a) $n = 1$ b) $n = 2$ c) $n = 3$.

Sketch the graph of the function in these three cases. Can you generalise?

4 Find the stationary points of $y = x^3 - 12x$. Sketch a graph of the function. For what values of k are there three solutions of the equation $x^3 - 12x = k$?

12.3 Applications of stationary points

There are many situations in which we want to maximise something – for example a business will want to make its profits as large as possible. Similarly there are many situations in which a quantity should be minimised, for example a car designer will want to ensure that the wind resistance is as small as possible.

The technique of finding stationary points can be used in many of these situations.

EXAMPLE 12.4

The cost of producing a book is £2 per copy. If the book is priced at £x, the sales manager of the company reckons that $10\,000 - 1000x$ copies will be sold. Find an expression for the profit made by the book, and find the price for which this profit is a maximum.

Solution

The profit on each copy is $£(x - 2)$. The total profit in £ will be

$$P = (x - 2)(10\,000 - 1000x) = -20\,000 + 12\,000x - 1000x^2$$

Differentiate this expression, and put the derivative equal to zero.

$$\frac{\mathrm{d}P}{\mathrm{d}x} = 12\,000 - 2000x$$

If $\dfrac{\mathrm{d}P}{\mathrm{d}x} = 0$ then $12\,000 - 2000x = 0$

$$x = 6$$

Differentiating the derivative gives

$$\frac{\mathrm{d}^2 P}{\mathrm{d}x^2} = -2000$$

which is negative.

So we have a maximum.

The book should be priced at £6.

EXAMPLE 12.5

An open box is to be made from thin metal, in the shape of a cuboid with a square base. The volume of the box is 4 m³. What shape should the box be, in order to use as little metal as possible for the sides and base?

Solution

The diagram shows the box. Let the side of the base be x m and the height h m. If the volume is V m³ then

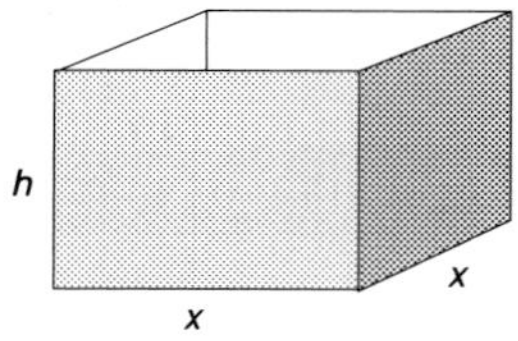

$$V = hx^2 = 4$$

The total area A m² of sides and base is given by

$$A = 4xh + x^2$$

This gives A as a function of two variables, x and h. We can write h in terms of x by using the formula above for the volume.

$$h = \frac{4}{x^2}$$

Fig. 12.9

$$A = 4 \times x \times \frac{4}{x^2} + x^2 = \frac{16}{x} + x^2 = 16x^{-1} + x^2$$

We want to make A as small as possible. Differentiate to find the stationary points of A.

$$\frac{\mathrm{d}A}{\mathrm{d}x} = -16x^{-2} + 2x$$

At a minimum, $\frac{\mathrm{d}A}{\mathrm{d}x} = 0$.

$$-16x^{-2} + 2x = 0$$
$$2x = 16x^{-2}$$

So $x^3 = 8$.

Hence $x = 2$ m. We must now show that this value of x does give a minimum.

Differentiating again gives

$$\frac{\mathrm{d}^2 A}{\mathrm{d}x^2} = 32x^{-3} + 2$$

If x is positive, $32x^{-3}$ is positive, 2 is positive, so $\frac{\mathrm{d}^2 A}{\mathrm{d}x^2} > 0$, ensuring

that we do have a minimum. Now find the height.

$$h = \frac{4}{2^2} = 1$$

The box should have a base of side 2 m and height 1 m.

EXERCISE 12D

1 The sales manager of a firm calculates that if the price of an item is £p, the number of items sold will be $5000 - 40p$. Find the value of p which will maximise the income from sales of the item.

2 When a vessel is travelling at v miles per hour, its fuel consumption is $(1 + 0.1v^2)$ tonnes per hour. If a journey of 1000 miles is to be undertaken, find expressions in terms of v for the time taken and the fuel consumed. Find the value of v for which this is a minimum.

3 Find the equation of the straight line through $(2, 3)$ with gradient m. Find where this line crosses the axes. Find the minimum area of the triangle enclosed between the line and the axes.

4 Metal trays are to be made from sheets of metal which are 20 cm square. Square corners of side x cm are cut out, and the metal along the dotted lines is folded upwards. Show that the volume of the tray is V cm^3 where

$$V = 4x^3 - 80x^2 + 400x$$

Find the greatest possible volume of the tray.

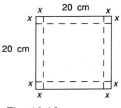

Fig. 12.10

5 Cylindrical cans are to be made from thin metal. The cans are to contain 1000 cm^3. If the base radius is x cm and the height is h cm, show that the surface area is A cm^2, where

$$A = 2\pi x^2 + \frac{2000}{x}$$

Find the dimensions of the can which minimises the amount of metal required to make it.

6 The skeleton of a cuboid is to be made from 100 cm of wire. The length of the base is to be twice the width. Find the maximum volume of the cuboid.

7 A closed box in the shape of a cuboid is to be made from thin metal. The length of the base is to be three times the width. If the volume of the cuboid is to be 750 cm^3 find the minimum area of metal that can be used.

*8 A cylinder is inserted inside a sphere of radius a. Find the maximum volume of the cylinder.

*9 A cone is inserted inside a sphere of radius a. Find the maximum volume of the cone.

*10 A 100 cm length of thin wire is cut into two pieces. One part is bent into a square, the other into an equilateral triangle. Find the minimum total area that could be enclosed.

LONGER EXERCISE

Radius of curvature

How curvy is a curve? This is of importance to motorists, who must take a sharp bend more slowly than a gradual bend. One way of measuring the curviness of a bend is by giving the radius of the circle which best fits that bend. This is the **radius of curvature**.

The radius of curvature is given by $\rho = \dfrac{\left[1 + \left(\dfrac{dy}{dx}\right)^2\right]^{\frac{3}{2}}}{\dfrac{d^2y}{dx^2}}$

1 Find the radius of curvature of the following curves at the points indicated.

  a) $y = x^2$ at $(0, 0)$ **b)** $y = x^2$ at $(1, 1)$ **c)** $y = x^3$ at $(0, 0)$

2 The **centre of curvature** of a curve at a point is the centre of the circle which fits the curve at that point. Find the centre of curvature of $y = x^2$ at $(0, 0)$.

*3 Let P be a point on the y-axis at $(0, p)$. There is always at least one normal from P to the curve $y = x^2$. Find the condition on p for there to be two more normals. How is this connected to your answer to part 2?

*4 Let $y = ax^2 + bx + c$, where a, b and c are constants. Show that the radius of curvature is least at the stationary point of the curve.

Cubics

Every quadratic graph has essentially the same shape. By contrast, there is more than one possible shape for a cubic graph. To investigate the possible shapes it will help if you have access to a computer running a spreadsheet or graph-drawing program.

1 Sketch the graphs of the following functions.

   a) $y = x^3 - 3x^2 + 3x + 4$ **b)** $y = x^3 - 3x + 4$ **c)** $y = x^3 + 3x - 4$

   How do you explain the difference between these graphs in terms of maxima, minima and points of inflection?

2 Sketch more graphs, to confirm or refute any conjecture you may have made in step 1.

3 Can you find a condition on $y = ax^3 + bx^2 + cx + d$ to determine which shape it has?

EXAMINATION QUESTIONS

1 The equation of a curve C is

   $$y = x^3 - 3x^2 - 9x$$

   Find the coordinates of the points on C at which $\dfrac{dy}{dx} = 0$ and, for each such point, determine whether it is a maximum or a minimum.

   Sketch the curve.

   Find the equation of the tangent to C at the point where $x = 0$.

   Find the coordinates of the point where this tangent meets the curve C again.   *J AS 1990*

2 For a particular journey of a ship, the running cost, C, in hundreds of pounds, is given in terms of its average speed for the journey, v km h^{-1}, by the equation

   $$C = \frac{16\,000}{v} + v^2$$

   Use differentiation to calculate the value of v for which C is a minimum, and show that C is a minimum and not a maximum for this value of v.   *C AS 1993*

3 A square piece of cardboard has side 1 metre. A smaller square of side x metres is cut from each of the four corners and the sides are folded up to form an open box (see the diagram).

   (i)   Between which two values must x lie?

   (ii) Express the volume of the box in terms of x.

   (iii) What is the greatest possible volume of the box?

   (iv) Given that the four squares of cardboard which have been cut out can be glued together to form a cover for the box, what is now the greatest possible volume of the box?

   NI 1992

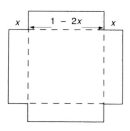

Fig. 12.11

Summary and key points

1 The second derivative of a function is written as $\dfrac{d^2y}{dx^2}$ or $f''(x)$. The sign of $\dfrac{d^2y}{dx^2}$ indicates whether the curve is concave or convex. At a point of inflection the curve changes from convex to concave or vice versa.

2 At a stationary point, $\dfrac{dy}{dx} = 0$. A stationary point is a maximum, a minimum or a point of inflection. The nature of the stationary point can be found by testing either y or $\dfrac{dy}{dx}$ on either side of the point, or by examining $\dfrac{d^2y}{dx^2}$.

If $\dfrac{d^2y}{dx^2} = 0$, this does not guarantee that the stationary point is a point of inflection.

Once we have found the stationary points and the points of inflection, we can make a rough sketch of the curve.

3 The technique of finding stationary points can be used to obtain the greatest or least values of quantities.

If the quantity to be maximised or minimised involves two variables, express one of them in terms of the other before differentiating.

Integration

In Chapters 8 and 12 we discussed differentiation, the process of finding $\dfrac{dy}{dx}$ from y. Often we need to go in the other direction, to find y from $\dfrac{dy}{dx}$. This process is called **integration**.

Integration is the inverse of differentiation. There are many uses for integration, other than just going from $\dfrac{dy}{dx}$ to y. In Section 13.3 we shall use integration to find areas.

13.1 Finding y from $\dfrac{dy}{dx}$

Integration is a 'backwards' procedure. We are given $\dfrac{dy}{dx}$ and have to find y. For example, suppose we know that $\dfrac{dy}{dx} = 2x$. Then we try to think of a function with derivative $2x$. This function could be x^2.

EXAMPLE 13.1

Find a possible equation for y, if $\dfrac{dy}{dx} = 2x + x^4 - 3$.

Solution

We know that

$$\frac{d(x^2)}{dx} = 2x$$

so the first part of the solution could be x^2.

The derivative of x^5 is $5x^4$. We want a function with derivative x^4, so divide x^5 by 5, to obtain $\frac{1}{5}x^5$.

The derivative of kx is k. So the final part of the solution is $-3x$.

A possible equation could be $y = x^2 + \frac{1}{5}x^5 - 3x$.

EXERCISE 13A

1 In each of the following you are given $\frac{dy}{dx}$ and you are asked to find a possible expression for y.

    a) $3x^2$ **b)** $\dfrac{2}{x^2}$ **c)** $10x^4$

    d) $2x^{-\frac{1}{2}}$ **e)** $2x + 2$ **f)** $x^2 + x + 1$

    g) $3x^{-2} + 2x^2$ **h)** $\dfrac{5}{x^2} + \dfrac{2}{x^3}$ **i)** $4x^{-\frac{1}{2}} - 6x^{-\frac{1}{3}}$

Constant of integration

You may have noticed that in the previous example we were asked for a possible solution rather than an exact answer. The reason is as follows.

Suppose that the derivative of y is $2x$. We know that

$$\frac{d(x^2)}{dx} = 2x$$

so y could be x^2.

But it is also true that

$$\frac{d(x^2 + 17)}{dx} = 2x$$

so y could be $x^2 + 17$.

In general, y could be $x^2 + c$, where c is any constant.

So when we go from $\frac{dy}{dx}$ to y, a constant should always be added on. From $\frac{dy}{dx} = 2x$ we can say that $y = x^2 + c$, where c is a constant.

This c is called the **constant of integration**.

Integration of x^n

Suppose $\frac{dy}{dx} = x^n$, where n is constant.

When we differentiate x^{n+1}, we obtain $(n+1)x^n$. So divide by $n+1$, and the formula will be correct.

The integral of x^n is $\dfrac{x^{n+1}}{n+1} + c$

This formula works for every value of n, with one exception. We cannot divide by 0, so n cannot be -1. The integral of x^{-1} turns out to be a function which is completely different from a power of x. It will be found in Chapter 18.

EXAMPLE 13.2

Suppose $\dfrac{dy}{dx} = (x+3)(x-2)$. Find y.

Solution

First expand the brackets.

$$\frac{dy}{dx} = x^2 + x - 6$$

Now integrate, being sure to put in the constant of integration.

$$y = \frac{x^3}{3} + \frac{x^2}{2} - 6x + c$$

EXAMPLE 13.3

Suppose the derivative of y is $\dfrac{2x^2 + 6x^3}{x}$, and that $y = 8$ when $x = 1$. Find y.

Solution

First divide through by x.

$$\frac{dy}{dx} = 2x + 6x^2$$

$$y = x^2 + 2x^3 + c$$

Substitute $y = 8$ and $x = 1$ into this equation.

$$8 = 1 + 2 + c$$

Hence $c = 5$.

$$y = x^2 + 2x^3 + 5$$

EXERCISE 13B

1 Integrate the following expressions.

a) $4x^3 + 8x$

b) $2x^2 - 3x^4 + 2$

c) $4 - 3x^3$

d) $x^{\frac{1}{2}} + 4x$

e) $x^{-2} + 3x^{-3} + 8$

f) $\dfrac{3}{x^2} - \dfrac{2}{x^{\frac{1}{2}}}$

2 Integrate the following expressions.

a) $(x + 2)(x + 3)$ b) $(2x - 3)(x - 1)$ c) $(1 - x)(2 + x)$

d) $x(x^2 + 1)$ e) $(x^2 - 1)(x^2 + 1)$ f) $2x^3(x^2 + x + 1)$

g) $\dfrac{x + 1}{x^3}$ h) $\dfrac{3 + x^2}{x^2}$ i) $\dfrac{x^2 - 1}{x + 1}$

j) $x^{\frac{1}{2}}(x - 1)$ k) $\sqrt{x}(2x - 3)$ l) $\sqrt[3]{x}\left(x - \dfrac{1}{x}\right)$

m) $\dfrac{1 + x}{\sqrt{x}}$ n) $\dfrac{x + x^2}{x^{\frac{1}{2}}}$ o) $\dfrac{x + x^{\frac{1}{2}}}{\sqrt{x}}$

p) $(x + 3)^2$ q) $(2x^2 - 1)^2$ r) $(x + \sqrt{x})^2$

*3 Find the integrals of these expressions.

a) $\dfrac{(x - 1)(x + 1) + 1}{x}$ b) $\dfrac{x - 1}{x^{\frac{1}{2}} - 1}$

4 Find the integrals of the following, using the extra condition given to find the constant of integration.

a) $2x - 6x^2$, $y = 3$ when $x = 0$ b) $8x^3 + 2x + 1$, $y = 2$ when $x = 1$

c) $6\sqrt{x}$, $y = 3$ when $x = 9$ d) $\dfrac{2}{x^2}$, $y = 5$ when $x = \frac{1}{2}$

e) $(x - 1)(x + 3)$, $y = 2\frac{1}{3}$ when $x = 1$ f) $\dfrac{1 + x^3}{x^2}$, $y = 3$ when $x = 1$

13.2 The integral notation

It is a nuisance to have to say, 'given $\dfrac{dy}{dx}$, find y'. We need a notation which describes the process of going from $\dfrac{dy}{dx}$ to y.

Suppose $\dfrac{dy}{dx} = f(x)$. Then we say that

$$y = \int f(x)\, dx$$

As we shall see in the next section, integration is a process of adding up, or **summation**. At the time of the invention of calculus the letter s was written as $\int$, so we use this symbol to describe the process of $\int$umming f(x).

The rule above for the integration of x^n is written as

$$\int x^n \, dx = \frac{x^{n+1}}{n+1} + c$$

Recall two rules for differentiating.

If $y = z + w$, $\dfrac{dy}{dx} = \dfrac{dz}{dx} + \dfrac{dw}{dx}$

If $y = az$, $\dfrac{dy}{dx} = a\dfrac{dz}{dx}$ (*a* is constant)

These provide two related rules for integrating.

$\int (z + w) \, dx = \int z \, dx + \int w \, dx$

$\int az \, dx = a \int z \, dx$ (*a* is constant)

EXAMPLE 13.4
Evaluate $\int (8x^3 + 2x^2 + 3) \, dx$.

Solution
Use the formula for integration.

$\int (8x^3 + 2x^2 + 3) \, dx = 8 \times \frac{1}{4}x^4 + 2 \times \frac{1}{3}x^3 + 3x + c$

$\int (8x^3 + 2x^2 + 3) \, dx = 2x^4 + \frac{2}{3}x^3 + 3x + c$

EXERCISE 13C

1 Evaluate the following.

   a) $\int 4x \, dx$ **b)** $\int (6x^2 - 4x^3) \, dx$ **c)** $\int 5x^{1.5} \, dx$

2 Rewrite the following using the $\int$ symbol.

   a) If $\dfrac{dy}{dx} = 3x^2$, then $y = x^3 + c$.

   b) If $\dfrac{dy}{dx} = -\dfrac{4}{x^2}$, then $y = \dfrac{4}{x} + c$.

Definite and indefinite integrals

In the examples above, we integrated $\dfrac{dy}{dx}$ and found y. The solution involved a constant of integration c. This is an **indefinite** integral.

Sometimes we want to know the change of y between two values of x. Then we evaluate y at these two values and subtract the results. This is a **definite** integral.

Suppose $\dfrac{dy}{dx} = 6x$. Then the indefinite integral is

$$y = \int 6x \, dx = 3x^2 + c$$

To find the change of y between $x = 1$ and $x = 3$, we can put these values into the expression for y and subtract.

$$\text{Change} = (3 \times 3^2 + c) - (3 \times 1^2 + c) = 24$$

This is the definite integral. Notice that the c has cancelled out.

We show that the integral is to be evaluated between 1 and 3 by putting these numbers at the top and bottom of the integral sign, and at top and bottom of the square brackets, like this.

$$\text{Total change in } y = \int_1^3 6x \, dx = \left[3x^2\right]_1^3 = 27 - 3 = 24$$

So to distinguish between definite and indefinite integrals

$\int 6x \, dx$ is an **indefinite** integral. It is a function of x.

$\int_1^3 6x \, dx$ is a **definite** integral. It is an actual number.

Note: For definite integrals we need not include the constant of integration.

EXAMPLE 13.5

Evaluate $\displaystyle\int_4^9 6x^{\frac{1}{2}} \, dx$.

Solution
Evaluate the indefinite integral and put in the values.

$$\int_4^9 6x^{\frac{1}{2}} \, dx = \left[4x^{1\frac{1}{2}}\right]_4^9 = 4(9^{1\frac{1}{2}}) - 4(4^{1\frac{1}{2}}) = 4 \times 27 - 4 \times 8$$

$$\int_4^9 6x^{\frac{1}{2}} \, dx = 76$$

EXERCISE 13D

1 Evaluate the following definite integrals.

a) $\displaystyle\int_0^1 10x^4 \, dx$ **b)** $\displaystyle\int_0^2 (2x + 3x^2) \, dx$ **c)** $\displaystyle\int_3^6 (x+1)(x+5) \, dx$

d) $\displaystyle\int_1^2 \frac{1}{x^2}\,dx$
 e) $\displaystyle\int_1^4 2x^{-\frac{1}{2}}\,dx$
 f) $\displaystyle\int_1^4 \frac{x+1}{x^{\frac{1}{2}}}\,dx$

g) $\displaystyle\int_{-1}^2 3x^2\,dx$
 h) $\displaystyle\int_{-1}^3 4x^3\,dx$
 i) $\displaystyle\int_{-1}^1 (2x+4)\,dx$

j) $\displaystyle\int_1^{10} \frac{2}{x^2}\,dx$
 k) $\displaystyle\int_1^8 8x^{\frac{1}{3}}\,dx$
 l) $\displaystyle\int_1^{100} 6x^{-3}\,dx$

2 In the following a and b are constants. Evaluate the definite integrals.

a) $\displaystyle\int_a^b 4x\,dx$
 b) $\displaystyle\int_a^b (6x+3)\,dx$
 c) $\displaystyle\int_{-a}^a (x^3 + x^5 + x^7)\,dx$

13.3 Finding area by integration

Integration is a process of adding up. So it can be used to add up area.

Suppose we have an area under the curve $y = f(x)$, as shown in the diagram. We have the following result.

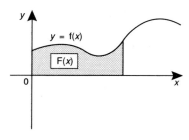

Fig. 13.1

Areas and integration

Let the area under the curve $y = f(x)$, from 0 to x, be $F(x)$. Then $F'(x) = f(x)$.

Justification

Suppose we increase x by a small amount δx. Then the extra bit of area is the thin strip shown. This small increase in area is approximately $f(x)\delta x$.

$$\delta F \approx f(x)\delta x$$

Divide by δx.

$$\frac{\delta F}{\delta x} \approx f(x)$$

Now take the limit as δx tends to 0. The left-hand side tends to $\dfrac{dF}{dx}$.

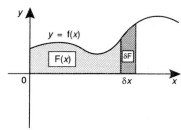

Fig. 13.2

$$\frac{dF}{dx} = f(x)$$

Hence the derivative of $F(x)$ is $f(x)$. Conversely, the area under the curve $y = f(x)$ is the integral of $f(x)$.

The area between $x = a$ and $x = b$ is the change in area between these values. Hence it is the definite integral of f(x) between these values.

$$\text{Area between } x = a \text{ and } x = b \text{ is } \int_a^b \text{f}(x)\,\text{d}x$$

Note: This result, linking integration and area, is a simplified version of a very important theorem called the **Fundamental Theorem of Calculus**.

EXAMPLE 13.6
Find the area underneath the curve $y = 9x^2$, between $x = 2$ and $x = 3$.

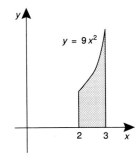

Solution
Find the definite integral of this function between these values.

$$\text{Area} = \int_2^3 9x^2\,\text{d}x = \left[3x^3\right]_2^3 = 3 \times 3^3 - 3 \times 2^3 = 81 - 24$$

$$= 57$$

The area is 57 square units.

Fig. 13.3

EXERCISE 13E

1 Find the following areas.

a) under $y = 4x$, between $x = 0$ and $x = 3$ **b)** under $y = 6x^3$, between $x = 1$ and $x = 3$

c) under $y = x^2 + 1$, between $x = -1$ and $x = 2$ **d)** under $y = \dfrac{1}{x^2}$, between $x = 1$ and $x = 10$

e) under $y = 3x^{\frac{1}{2}}$, between $x = 4$ and $x = 9$

f) under $y = x^2 + x + 1$, between $x = -1$ and $x = 1$

2 Draw the line $y = x$. Find the area under this curve between $x = 0$ and $x = 1$

a) by integration   **b)** from the formula for the area of a triangle.

Do your answers agree?

3 Express the areas shown shaded in Fig. 13.4, in terms of definite integrals, and evaluate them.

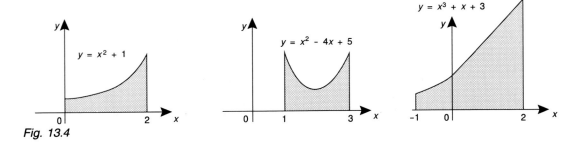

Fig. 13.4

*4 Figure 13.5 shows a semicircle of radius 1 unit. The equation of the semicircle is $y = \sqrt{1 - x^2}$. Write down but do not evaluate the area in terms of a definite integral. Hence find an expression for π in terms of an integral.

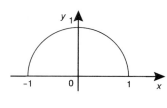

*5 Let $f(x) = 1 + x^2 + x^4 + x^6$. Without evaluating the integrals show that

$$\int_{-a}^{a} f(x) \, dx = 2 \int_{0}^{a} f(x) \, dx$$

Fig. 13.5

Area under the x-axis

If the curve of $y = f(x)$ lies below the x-axis, we find that the integral of $f(x)$ is negative. So the area below the x-axis is counted as negative. If we want to find the actual area, we must take account of this.

EXAMPLE 13.7
Sketch the curve $y = x^2 - 1$. Find the actual area between the curve and the x-axis, from $x = 0$ to $x = 2$.

Solution
The graph is shown in Fig. 13.6. The area we wish to find is shaded. Note that the function is negative between $x = 0$ and $x = 1$, so we shall have to do the integration in two separate parts.

$$\int_{0}^{1} (x^2 - 1) \, dx = \left[\tfrac{1}{3}x^3 - x \right]_{0}^{1} = -\tfrac{2}{3}$$

So area between $x = 0$ and $x = 1$ is $\tfrac{2}{3}$.

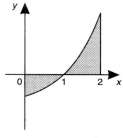

Fig. 13.6

$$\int_{1}^{2} (x^2 - 1) \, dx = \left[\tfrac{1}{3}x^3 - x \right]_{1}^{2} = \tfrac{4}{3}$$

So area between $x = 1$ and $x = 2$ is $\tfrac{4}{3}$.

Total area is $\tfrac{2}{3} + \tfrac{4}{3} = 2$.

The total area is 2 square units.

EXERCISE 13F

1 For each of the following, make a sketch of the curve. Evaluate the actual area between the curve and the x-axis, between the values shown.

a) $y = x - 1$ between $x = 0$ and $x = 2$

b) $y = x^3$ between $x = -1$ and $x = 2$

c) $y = \dfrac{4}{x^2} - 1$ between $x = 1$ and $x = 3$

d) $y = 3x^2 - 3$ between $x = -1$ and $x = 2$

*2 Let $f(x) = x + x^3 + x^5$. Without evaluating the integral show that

$$\int_{-a}^{a} f(x)\,dx = 0$$

Area between curves

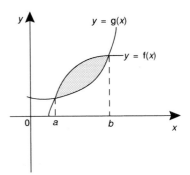

Suppose an area is enclosed between two curves $y = f(x)$ and $y = g(x)$ as shown. Then we can find the areas under $y = f(x)$ and $y = g(x)$ by finding the integrals. The shaded area is the **difference** between these two integrals.

$$\text{Area between curves} = \int_{a}^{b} f(x)\,dx - \int_{a}^{b} g(x)\,dx$$

$$= \int_{a}^{b} (f(x) - g(x))\,dx$$

Fig. 13.7

EXAMPLE 13.8
Find where the curves $y = x^2$ and $y = x^3$ cross. Evaluate the area enclosed between the two curves.

Solution
Putting $x^2 = x^3$, we obtain $x = 0$ or $x = 1$. A sketch of the two curves is shown.

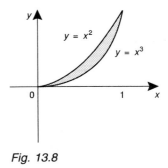

The area can be found by integrating between 0 and 1.

$$\text{Area} = \int_{0}^{1} (x^2 - x^3)\,dx = \left[\tfrac{1}{3}x^3 - \tfrac{1}{4}x^4 \right]_{0}^{1} = \tfrac{1}{3} - \tfrac{1}{4}$$

$$= \tfrac{1}{12}$$

The area enclosed is $\tfrac{1}{12}$ square units.

Fig. 13.8

EXERCISE 13G

1 Find where the curve $y = x^2$ meets the line $y = x$. Find the area enclosed between the line and the curve.

2 Show that the line $y = 7 - 3x$ meets the curve $y = \dfrac{4}{x^2}$ at $x = 1$ and $x = 2$. Find the area enclosed by the line and the curve between these points.

3 Find where the line $y = 2x - 1$ meets the curve $y = x^2 + 2x - 2$. Find the area enclosed between the line and the curve.

4 Show that the curves $y = \dfrac{9}{x^2}$ and $y = 10 - x^2$ meet at $x = 1$ and $x = 3$. Find the area between these curves.

5 Find where the curves $y = 1 - x^2$ and $y = x^2 - 7$ meet. Find the area enclosed between the curves.

6 Find the area enclosed between the curves $y = x^n$ and $y = x^m$ in the positive quadrant. (Assume n and m are positive constants, with $n > m$.)

7 Find the area enclosed between $y^2 = 4x$ and $y = 4x$.

8 Find the points where the line $y = x$ meets the curve $y = 2x - x^2$. The line divides the area under the curve into two regions A and B as shown. Find the ratio of the areas of these regions.

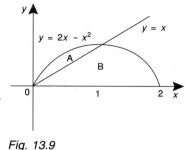

Fig. 13.9

9 Find the equations of the tangents to $y = 4 + 4x - x^2$ at $x = 1$ and $x = 3$. Find where these tangents meet. Find the area enclosed by the tangents and the curve.

*10 Find the equation of the normal to $y = x^{\frac{1}{2}}$ at $(1, 1)$. Find where this normal crosses the x-axis. Find the area of the region enclosed by the curve, the normal and the x-axis.

*11 Show that the line joining $(-1, 0)$ to $(1, 1)$ is a tangent to the curve $y = \sqrt{x}$. Find the area enclosed by the curve, the line and the x-axis.

LONGER EXERCISE

Segments of parabolas

Calculus was invented in the 17th century, independently, by Newton and Leibnitz. For many years before that, however, areas had been found by methods similar to integration. In particular, the Greek mathematician Archimedes invented a form of integration. In 1906 a forgotten manuscript by Archimedes was found in Constantinople. It contained a proof by integration of the following.

> The area of a segment of a parabola is $\frac{4}{3}$ the area of the triangle with the same base and height.

Let the equation for the parabola be $y = x^2$. Consider segments cut off by the line $y = k$ as shown.

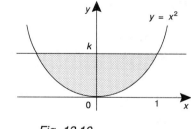

1 Find where the line crosses the curve.

Fig. 13.10

2 By integration, find the area of the segment cut off by the line.

3 Find the area of the triangle with the same base and height as the segment. Do the areas obey the relationship above?

**4 We have considered the special case in which the line is perpendicular to the y-axis. The proof is much more difficult if the line is slanting. See if you can prove the result for the segment cut off by the line with equation $y = 2 - x$.

EXAMINATION QUESTIONS

1 The figure shows a sketch of the curve with equation

$$y = 12x^{\frac{1}{2}} - x^{\frac{3}{2}} \text{ for } 0 \le x \le 12$$

a) Show that $\dfrac{dy}{dx} = \dfrac{3}{2}x^{-\frac{1}{2}}(4 - x)$.

At the point B, on the curve, the tangent to the curve is parallel to the *x*-axis.

b) Find the coordinates of the point B.

c) Find the area of the finite region bounded by the curve and the *x*-axis. *L 1992*

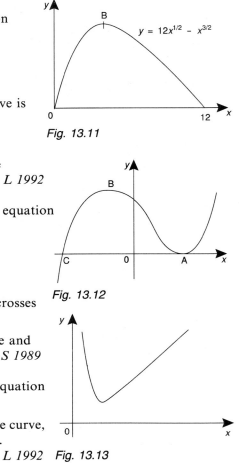

Fig. 13.11

2 The figure shows a sketch of part of the curve with equation

$$y = x^3 - 12x + 16$$

Calculate

a) the coordinates of the turning points A and B

b) the coordinates of C, the point where the curve crosses the *x*-axis

Fig. 13.12

c) the area of the finite region enclosed by the curve and the *x*-axis. *L AS 1989*

3 The figure shows a sketch of the curve with equation $y = 3x^2 + 4x^{-2}$, for $x > 0$.

Calculate the area of the finite region bounded by the curve, the lines with equations $x = 1$, $x = 4$ and the *x*-axis.

L 1992 *Fig. 13.13*

Summary and key points

1 The process of going from $\dfrac{dy}{dx}$ to y is called integration. The integral of x^n is $\dfrac{x^{n+1}}{n+1} + c$.

Do not forget the constant c.

2 The integral of $f(x)$ is written $\int f(x)\,dx$. This is the indefinite integral.

Suppose $\int f(x)\,dx = F(x)$. The definite integral is $\displaystyle\int_a^b f(x)\,dx = F(b) - F(a)$.

3 The area under the curve $y = f(x)$ between $x = a$ and $x = b$ is $\displaystyle\int_a^b f(x)\,dx$.

Area below the *x*-axis is negative.

To find the area between two curves, integrate the difference between the curves.

Trigonometric identities and equations

In Chapter 3 we defined the six trigonometric functions sin, cos, tan, sec, cosec and cot in terms of the sides of a right-angled triangle. There are many relationships between these functions.

In this chapter we must be careful to distinguish between **identities** and **equations**. An identity is something that is true for **all** values of the unknowns. Sometimes, but not always, we write an identity with the sign $\equiv$. An algebraic example of an identity, true for all values of x, is

$$(x - 3)(x + 3) \equiv x^2 - 9$$

An equation is true for some value or values of the unknown. It can be solved to find these values. An example of an equation, for which the solutions are $x = 1$ and $x = 3$, is

$$x^2 - 4x + 3 = 0$$

The same distinction applies with trigonometric expressions. An example of an identity, true for all angles, is

$$\sin(180° - x) \equiv \sin x$$

An example of an equation, true for $x = 30°$, is

$$\sin x = 0.5$$

When proving an identity, make sure you do not assume the answer. Write down one side of the identity and reduce it to the other side, or reduce both sides to the same expression.

A trigonometric identity is true for all values of the angle. It also doesn't matter whether the angle is measured in degrees or radians. Throughout this chapter we shall use both radians and degrees, and it will be clear from the context which is being used.

14.1 Relations between functions

Fig. 14.1

If P is an angle in a right-angled triangle as shown, the six trigonometric functions are defined in terms of the sides as follows.

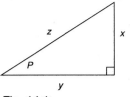

$$\sin P = \frac{x}{z} \qquad \cos P = \frac{y}{z} \qquad \tan P = \frac{x}{y}$$

$$\sec P = \frac{z}{y} \qquad \operatorname{cosec} P = \frac{z}{x} \qquad \cot P = \frac{y}{x}$$

In Chapter 3 we pointed out that cosec, sec and cot could be found from sin, cos and tan respectively, by the relationships

$$\operatorname{cosec} P \equiv \frac{1}{\sin P} \qquad \sec P \equiv \frac{1}{\cos P} \qquad \cot P \equiv \frac{1}{\tan P}$$

These are identities. Other identities follow from the definitions of the functions given above.

EXAMPLE 14.1

Express tan P in terms of sin P and cos P.

Solution

Using the triangle above, $\tan P = \frac{x}{y}$, $\sin P = \frac{x}{z}$, $\cos P = \frac{y}{z}$. Dividing sin P by cos P will eliminate the z term.

$$\sin P \div \cos P = \frac{x}{z} \div \frac{y}{z} = \frac{x}{y}$$

Notice that this is the definition of tan P.

$$\tan P = \frac{\sin P}{\cos P}$$

EXAMPLE 14.2

Prove the identity $\operatorname{cosec} P \div \sec P = \cot P$.

Solution

Write the left-hand side in terms of the x, y and z of the triangle above.

$$\operatorname{cosec} P \div \sec P = \frac{z}{x} \div \frac{z}{y} = \frac{y}{x}$$

This is the definition of cot P.

$$\operatorname{cosec} P \div \sec P = \cot P$$

EXERCISE 14A

1 Which of the following are identities and which equations? Find a solution for those which are equations.

a) $x^2 - 4x + 4 = (x - 2)^2$ **b)** $x^2 - 6x + 9 = x^2$ **c)** $\sin x = 1$

d) $\sin x = \cos(90° - x)$ **e)** $\cos x = \cos(-x)$ **f)** $\sin x = \sin(-x)$

2 Show that the following are identities.

a) $\sin P \cot P = \cos P$ **b)** $\operatorname{cosec} P \tan P = \sec P$ **c)** $\sec P \div \operatorname{cosec} P = \tan P$

d) $\dfrac{1 + \sin x}{1 + \cos x} \times \dfrac{1 + \sec x}{1 + \operatorname{cosec} x} = \tan x$ **e)** $\dfrac{1 + \cos x}{1 - \cos x} = \dfrac{1 + \sec x}{\sec x - 1}$ **f)** $\dfrac{\cot x + \tan y}{\cot y + \tan x} = \cot x \tan y$

3 Express $\cot P$ in terms of $\sin P$ and $\cos P$.

4 Express $\sec P$ in terms of $\cot P$ and $\operatorname{cosec} P$.

5 Let $\sin P = x$ and $\cos P = y$. Express in terms of x and y the following.

a) $\tan P$ **b)** $\cot P$ **c)** $\operatorname{cosec} P$ **d)** $\sec P$

14.2 Pythagorean relationships

Suppose we know the value of $\sin P$ to be $\dfrac{x}{z}$. In the triangle shown, we can let the opposite side be x and the hypotenuse z. Then by use of Pythagoras's theorem, we can find the third side of the triangle. We will then be able to find the other trigonometric functions.

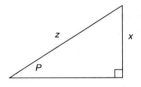

Fig. 14.2

Pythagoras's theorem can also be used to find relationships between the functions, as follows.

a) $\sin^2 P + \cos^2 P = 1$ **b)** $\sec^2 P - \tan^2 P = 1$

c) $\operatorname{cosec}^2 P - \cot^2 P = 1$

Note
$\sin^2 P$ means $(\sin P)^2$ and so on.

Proof
a) Take a right-angled triangle with hypotenuse 1 unit. Then the opposite side is $\sin P$, and the adjacent side is $\cos P$. Pythagoras's theorem gives

$$(\sin P)^2 + (\cos P)^2 \equiv 1$$

Now we can leave out the brackets, and write it in the form above.

$$\sin^2 P + \cos^2 P \equiv 1$$

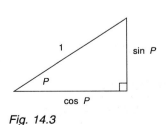

Fig. 14.3

b) Divide both sides of the first identity by $\cos^2 P$.

$$\frac{\sin^2 P}{\cos^2 P} + \frac{\cos^2 P}{\cos^2 P} \equiv \frac{1}{\cos^2 P}$$

Recall that $\dfrac{\sin P}{\cos P} = \tan P$ and that $\dfrac{1}{\cos P} = \sec P$.

Put these into the identity.

$$\tan^2 P + 1 \equiv \sec^2 P$$

Rearrange to give the identity

$$\sec^2 P - \tan^2 P \equiv 1$$

c) Divide the first identity by $\sin^2 P$.

$$\frac{\sin^2 P}{\sin^2 P} + \frac{\cos^2 P}{\sin^2 P} \equiv \frac{1}{\sin^2 P}$$

$\dfrac{\cos P}{\sin P} = \cot P$ and $\dfrac{1}{\sin P} = \operatorname{cosec} P$. This gives

$$1 + \cot^2 P \equiv \operatorname{cosec}^2 P$$

Rearrange to give the identity

$$\operatorname{cosec}^2 P - \cot^2 P \equiv 1$$

EXAMPLE 14.3
P is an acute angle for which $\sin P = \frac{5}{13}$. Find $\cos P$.

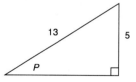

Fig. 14.4

Solution
In the triangle shown, the opposite side is 5 units and the hypotenuse is 13 units. Use of Pythagoras's theorem gives

$$\text{adjacent}^2 = 13^2 - 5^2 = 144$$

Hence the adjacent side is 12 units.

$$\cos P = \frac{12}{13}$$

EXAMPLE 14.4
Prove that $\cos^4 x - \sin^4 x \equiv \cos^2 x - \sin^2 x$.

Solution
The left-hand side is the difference of two squares. It can be written as

$$\cos^4 x - \sin^4 x \equiv (\cos^2 x + \sin^2 x)(\cos^2 x - \sin^2 x)$$

We know that the expression in the first pair of brackets is identically equal to 1. Hence

$$\cos^4 x - \sin^4 x \equiv \cos^2 x - \sin^2 x$$

EXERCISE 14B

1 P is an acute angle for which $\sin P = \frac{4}{5}$. Find $\cos P$ and $\tan P$.

2 P is an acute angle for which $\cos P = \frac{8}{17}$. Find $\sin P$ and $\cot P$.

3 P is an acute angle for which $\tan P = 2$. Find $\cos P$ and $\sin P$, leaving your answers in terms of surds.

4 Use an identity to show that $\cos x = \sqrt{1 - \sin^2 x}$.

Use the identities to express

a) $\sin x$ in terms of $\cos x$ **b)** $\tan x$ in terms of $\sec x$

c) $\sec x$ in terms of $\tan x$ **d)** $\operatorname{cosec} x$ in terms of $\cot x$.

5 Use the identities to write each of the following in terms of a single trigonometric function.

a) $\cos^2 x + 3 \sin x$ **b)** $\sin^2 x + 2 \cos^2 x - \cos x + 3$

c) $\sec x - 3 \tan^2 x$ **d)** $\cot^2 x + 5 \operatorname{cosec} x$

6 In the triangle in Fig. 14.5 the adjacent side is 1 unit in length. Express the other sides in terms of trigonometric functions. Use Pythagoras's theorem to derive one of the identities above.

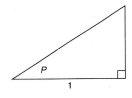

Fig. 14.5

7 In the triangle in Fig. 14.6 the opposite side is 1 unit long. Express the other sides in terms of trigonometric functions. Use Pythagoras's theorem to derive one of the identities above.

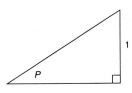

Fig. 14.6

8 Prove the following identities.

a) $\dfrac{1}{\sin^2 x} + \dfrac{1}{\cos^2 x} \equiv \sec^2 x \operatorname{cosec}^2 x$ **b)** $\tan x + \cot x \equiv \sec x \operatorname{cosec} x$

c) $\sin^3 x + \cos^3 x \equiv (\sin x + \cos x)(1 - \sin x \cos x)$

d) $\operatorname{cosec} x - \sin x \equiv \cot x \cos x$ **e)** $(\sin x + \cos x)^2 + (\sin x - \cos x)^2 \equiv 2$

f) $\cot^4 x + \cot^2 x \equiv \operatorname{cosec}^4 x - \operatorname{cosec}^2 x$

*9 Let $X = x \sin \theta + y \cos \theta$ and $Y = x \cos \theta - y \sin \theta$. Show that $X^2 + Y^2 = x^2 + y^2$. Find an expression for $\tan \theta$ in terms of x, y, X and Y.

*10 If $a \cos^2 x + b \sin^2 x = k$, show that $\tan^2 x = \dfrac{k - a}{b - k}$.

14.3 Values taken by trigonometric functions

A trigonometric function takes the same value infinitely many times, so there may be infinitely many solutions to a trigonometric equation. The situation depends on which function we are using. We shall need the following rules, from Section 3.4 of Chapter 3.

$$\sin x = \sin(180° - x)$$

$$\cos x = \cos(360° - x)$$

$$\tan x = \tan(x - 180°)$$

Sine and cosecant

Suppose we want to solve $\sin x = 0.5$. A calculator will give the solution $x = 30°$. From the first of the three rules above, $180° - 30° = 150°$ is also a solution.

Sine repeats itself every 360°. So extra solutions can be found by adding or subtracting multiples of 360° to the solutions already found. Adding or subtracting multiples of 360° to 30° we obtain

$$x = 390°, \ 750°, \ -330°, \ -690° \text{ etc.}$$

Adding or subtracting multiples of 360° to 150° we obtain

$$x = 510°, \ 870°, \ -210°, \ -570° \text{ etc.}$$

Cosine and secant

Suppose we want to solve $\cos x = 0.3$. A calculator gives $x = 72.5°$. From the second of the equations above, another solution is $360° - 72.5° = 287.5°$. Other solutions are

$$x = 432.5°, \ 792.5°, \ -287.5°, \ -647.5° \text{ etc.}$$

$$x = 647.5°, \ 1007.5°, \ -72.5°, \ -432.5° \text{ etc.}$$

Tangent and cotangent

Suppose we want to solve $\tan x = 0.7$. A calculator gives $x = 35°$. From the third of the equations above, tangent repeats itself every 180° as well as every 360°. Other solutions are

$$x = 215°, \ 395°, \ -145°, \ -325° \text{ etc.}$$

Notes

1 All these results have been given in terms of degrees. They can be translated to radians, by changing 180° to π and 360° to 2π.

2 These results have been given for sine, cosine and tangent. Similar calculations hold for cosecant, secant and cotangent respectively.

3 Recall that $\cos(-P) = \cos(+P)$. In the results for cosine, $-72.5°$ is a solution as well as $+72.5°$ etc.

EXAMPLE 14.5

Find four solutions for the equation $\operatorname{cosec} x = 4$.

Solution

Cosecant is the reciprocal of sine. So the equation is equivalent to

$$\sin x = \tfrac{1}{4}$$

A calculator gives the solution as $14.5°$. Subtract this from $180°$ to give a second solution, $165.5°$. Two more solutions can be found by adding $360°$ to the ones already found.

Four solutions are $x = 14.5°$, $x = 165.5°$, $x = 374.5°$, $x = 525.5°$.

EXERCISE 14C

1 Find four solutions (in degrees) for each of the following equations.

a) $\sin x = 0.4$ **b)** $\cos x = 0.3$ **c)** $\tan x = 3$

d) $\tan x = -2$ **e)** $\sin x = -0.5$ **f)** $\cos x = -0.8$

g) $\sec x = 1.6$ **h)** $\cot x = 0.3$ **i)** $\operatorname{cosec} x = -4$

j) $\sin x = \dfrac{1}{\sqrt{2}}$ **k)** $\cot x = \sqrt{3}$ **l)** $\sec x = \dfrac{-2}{\sqrt{3}}$

2 Find four solutions in radians for the equations of Question 1.

*3 Suppose $\sin(x + 30°) = \sin(x - 30°)$. Find a possible value of x.

4 Suppose $\sin x = 0.6$ and $\cos x = -0.8$. Find two possible values of x.

5 $\sin x = k$ has only one solution in the range $0 \leq x \leq 2\pi$. What is k?

14.4 Trigonometric equations

Trigonometric functions repeat themselves. When solving an equation involving trigonometric functions there are often several solutions within a given range.

Sometimes a trigonometric equation can be solved directly.

Sometimes we have to use one of the Pythagorean identities of Section 14.2.

If the range is given in terms of degrees, as $0° \leq x \leq 360°$ for example, then give the answers in degrees. If the range is, for example, $0 \leq x \leq 2\pi$, then give the answers in radians.

EXAMPLE 14.6

Find the solutions of sec $x = 2$, for $0° \leq x \leq 360°$.

Solution

The range is given in degrees, so the solutions will also be in degrees.

If sec $x = 2$, then cos $x = \frac{1}{2}$. Hence $x = 60°$.

Looking at the **cos and sec** paragraph of the previous section, we see that another solution is $360° - 60° = 300°$. If we add or subtract $360°$, we will go outside the range $0°$ to $360°$.

The solutions are $60°$ and $300°$.

EXAMPLE 14.7

Solve the equation $\cos\left(x - \frac{\pi}{2}\right) = 0.3$, for $0 \leq x \leq 2\pi$.

Solution

Here the answers must be given in radians. Find the cos $^{-1}$ of 0.3, to obtain 1.266.

$$x - \frac{\pi}{2} = 1.266$$

We can subtract this answer from 2π. We can also add or subtract 2π as often as we like. So the solution could also be

$$x - \frac{\pi}{2} = 5.017,\ 7.549,\ 11.300,\ -5.017,\ -1.266 \text{ etc.}$$

Add $\frac{\pi}{2}$ to both sides.

$$x = 2.837,\ 6.588,\ 9.120,\ 12.87,\ -3.446,\ 0.305$$

Of these, only the first and the last lie in the range 0 to 2π.

The solution is $x = 2.837$ or $x = 0.305$.

EXAMPLE 14.8

Solve the equation tan $3x = 1$, for $0° \leq x \leq 180°$.

Solution

Here the solution is to be given in degrees. The tan^{-1} of 1 is $45°$.

$$3x = 45°$$

Referring to Section 14.2, we can add or subtract $180°$ as often as we like. So solutions could also be

$$3x = 225°,\ 405°,\ 585°,\ -135° \text{ etc.}$$

Divide by 3, to obtain

$$x = 15°, \ 75°, \ 135°, \ 195°, \ -45° \text{ etc.}$$

Of these, the first three lie within the range $0°$ to $180°$.

The solution is $x = 15°$, $x = 75°$, or $x = 135°$.

EXAMPLE 14.9

Solve the equation $2\cos^2 x + \sin x - 1 = 0$ for $0 \le x \le 2\pi$, giving your answers as multiples of π.

Solution

Two trigonometric functions are involved here, but we can write $\cos^2 x$ in terms of $\sin x$, by using the identity $\cos^2 x + \sin^2 x = 1$.

$$2(1 - \sin^2 x) + \sin x - 1 = 0$$

Expanding and rearranging gives

$$2\sin^2 x - \sin x - 1 = 0$$

This can be regarded as a quadratic in $\sin x$. Factorise to obtain

$$(2\sin x + 1)(\sin x - 1) = 0$$

So either $\sin x = 1$ or $\sin x = -\frac{1}{2}$. Solve these equations in the range 0 to 2π.

From $\sin x = 1$ we obtain $x = \dfrac{\pi}{2}$.

From $\sin x = -\frac{1}{2}$ we obtain $x = \pi + \dfrac{\pi}{6} = \dfrac{7\pi}{6}$ and

$x = 2\pi - \dfrac{\pi}{6} = \dfrac{11\pi}{6}$.

The solution is $x = \dfrac{\pi}{2}$ or $x = \dfrac{7\pi}{6}$ or $x = \dfrac{11\pi}{6}$.

Strategy

We have shown examples of several types of equation. It is often difficult to decide what method to use to solve them. Below is a guide for how to proceed in some cases.

An equation of the form $\sin x = k$ can be solved directly. Usually there will be two solutions in the range $0°$ to $360°$, or 0 to 2π.

For an equation of the form $\cot(x - c) = k$, first convert it to the tan equation $\tan(x - c) = \dfrac{1}{k}$. Take the $\tan^{-1}$ of $\dfrac{1}{k}$, to obtain $x - c$. Then add on c to obtain x.

Suppose you are asked for solutions in the range $0°$ to $360°$. You may have to consider values of $\tan^{-1}\dfrac{1}{k}$ **outside** this range, because when c is added the result may fall **within** the range.

For an equation of the form $\sec 2x = k$, convert it to the cos equation $\cos 2x = \dfrac{1}{k}$. Take the $\cos^{-1}$ of $\dfrac{1}{k}$, to obtain $2x$. Divide by 2 to obtain x.

Suppose you are asked for solutions in the range 0 to 2π. You may have to consider values of $\cos^{-1} \dfrac{1}{k}$ outside this range, because when you divide by 2 the result may fall within the range.

For an equation which involves $\sin^2 x$ as well as $\cos x$, convert the $\sin^2 x$ to $1 - \cos^2 x$. You will now have a quadratic in $\cos x$. Solve this quadratic, by factorisation or by the formula, and then find the values of x.

EXERCISE 14D

1 Solve the following equations in the range $0° \leq x \leq 360°$.

    a) $\sin 2x = 0.7$ **b)** $\cos(x + 10°) = 0.5$ **c)** $\cot 2x = 2$

    d) $\sec(x - 20°) = 4$ **e)** $\tan 3x = \sqrt{3}$ **f)** $\operatorname{cosec}(x - 40°) = 1.5$

2 Find the solutions of the following equations in the range $0 \leq x \leq 2\pi$. Where relevant, give your answers as multiples of π.

    a) $\cos^2 x = 0.5$ **b)** $\sin^2 x = 0.25$ **c)** $\tan x = 2 \cot x$

    d) $\sin x = 3 \cos x$ **e)** $2 \cos x - 3 \sin x = 0$ **f)** $\tan x = 3 \sin x$

    g) $4 \sin x \cos x = 3 \cos x$ **h)** $\sin x = 3 \sin^2 x$ **i)** $\sec x = 2 \cos x$

3 Solve the following equations in the range $0 \leq x \leq \pi$.

    a) $6 \sin^2 x - \sin x - 1 = 0$ **b)** $3 \cos^2 x + \cos x - 2 = 0$ **c)** $2 \sin^2 x - \cos x - 1 = 0$

    d) $8 \cos^2 x + 10 \sin x - 11 = 0$ **e)** $3 \tan x = 2 \cos x$ **f)** $\sec^2 x - 5 \sec x + 6 = 0$

    g) $\tan^2 x + 2 \sec x = 7$ **h)** $2 \cot^2 x - \cot x = 3$ **i)** $\cot^2 x + \operatorname{cosec} x = 11$

4 Find the solutions of the following equations, in the range $0° \leq x \leq 360°$.

    a) $\operatorname{cosec}^2 x + \cot^2 x = 3$ **b)** $\cot^2 x - \sin^2 x = 1$ **c)** $\tan(x - 25°) = 43°$

    d) $\cos(x - 27°) = \sin 53°$ **e)** $2 \sin x \tan x = 3$ **f)** $\cot x \cos x = 1 + \sin x$

LONGER EXERCISE

Latitude and longitude

Assume that the Earth is a perfect sphere. Let its radius be R. We can define the position of any point on its surface by means of two angles, the **latitude** and **longitude** of the point.

In the diagram N and S are the north and south poles.

The equator is the circle perpendicular to NS, going through the centre of the Earth at O.

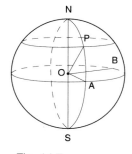

Fig. 14.7

A semicircle on the surface of the Earth connecting N and S is a line of **longitude**, or a **meridian**. The longitudes are measured by the angle they make with the **meridian** through Greenwich, called the **Greenwich Meridian**. The Greenwich meridian meets the equator at A. B is a point on the equator, such that OB is perpendicular to OA.

A plane parallel to the equator meets the Earth in a circle of **latitude**. For a point P, its latitude is measured by the angle between OP and the plane through the equator. Suppose a point P has longitude θ and latitude ϕ .

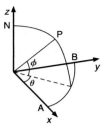

Fig. 14.8

1 Suppose we take three coordinate axes, the x-axis along OA, the y-axis along OB, and the z-axis along ON. Show that the coordinates of P are

$$x = R \cos \phi \cos \theta, \quad y = R \cos \phi \sin \theta, \quad z = R \sin \phi$$

2 With x, y and z as above, show that $x^2 + y^2 + z^2 = R^2$.

3 Find an expression for the direct distance (through the Earth) from P to the point A.

4 Find an expression for the shortest distance along the surface of the Earth from P to A.

EXAMINATION QUESTIONS

1 Find, correct to the nearest degree, all the values of θ between $0°$ and $360°$ satisfying the equation

$$8 \cos^2\theta + 2 \sin \theta = 7$$

W AS 92

2 Sketch, on a single diagram, the graphs of $y = \sin 6x$ and $y = \cos 2x$, for values of x such that $0 \le x \le \pi$.

Given that the smallest positive value of x satisfying the equation $\sin 6x = \cos 2x$ is α, write down, in terms of α

(i) one other value of x, such that $0 < x < \pi$, which satisfies the same equation

(ii) one value of x, such that $\pi < x < 2\pi$, which satisfies the same equation.

By writing the equation $\sin 6x = \cos 2x$ in the form

$$\cos \left(\frac{\pi}{2} - 6x \right) = \cos 2x$$

or otherwise, find, in terms of π, the three smallest positive values of x which satisfy the equation.

C 1991

3 (i) The curve with equation

$$y = 2 + k \sin x$$

passes through the point with coordinates $\left(\dfrac{\pi}{2}, -2\right)$.

Find the value of k and the greatest value of y.

(ii) Find, to the nearest tenth of a degree, the angles θ between $0°$ and $360°$ for which

$$2 \sec^2\theta = 8 - \tan \theta$$

L 1992

Summary and key points

1 Many trigonometric identities follow directly from their definitions.

2 The three Pythagorean identities are

$$\sin^2 P + \cos^2 P = 1 \qquad \sec^2 P - \tan^2 P = 1$$
$$\operatorname{cosec}^2 P - \cot^2 P = 1$$

3 If we have found one solution of a trigonometric function, other solutions can be found as follows.

sin and cosec   Subtract from $180°$, then add or subtract multiples of $360°$.

cos and sec   Subtract from $360°$, then add or subtract multiples of $360°$.

tan and cot   Add or subtract multiples of $180°$.

4 Equations involving trigonometric functions often involve the identities of Sections 14.1 and 14.2.

Suppose you are solving something like $\sin 2x = 0.5$ or $\cot\left(x + \dfrac{\pi}{3}\right) = 2$, where the answers are to be given within a certain range. You may have to consider values outside the range before dividing by 2 or subtracting $\dfrac{\pi}{3}$.

Consolidation section C

Chapter 11

1 Find the expansions of the following expressions.

    a) $(p+q)^5$ **b)** $(3x-4y)^4$ **c)** $\left(\dfrac{3}{x}+2x^2\right)^3$

2 Find the values of the following.

    a) $3! \times 2!$ **b)** $^7C_4$ **c)** $^8P_4$

3 Three awards are to be given to a staff of 50 employees. In how many ways can this be done if

    a) the awards are all of £200   **b)** the awards are of £100, £200 and £300?

4 In a raffle, 1000 tickets are sold to be drawn for 20 prizes. If I buy ten tickets, what is the probability I won't win any prizes?

5 Find the terms indicated in the following expansions.

    a) $(x+y)^{23}$ $x^{10}y^{13}$ term   **b)** $(2x-3y)^{12}$ x^8y^4 term   **c)** $\left(2x-\dfrac{3}{x}\right)^{14}$ constant term

6 Expand each of the following up to the term in x^2.

    a) $(1+x)^{2.5}$ **b)** $(1-3x)^{-3}$ **c)** $\sqrt{1+0.5x}$

7 For each part of Question 6, state the range of values of x for which the expansion is valid.

8 Use the binomial expansion to find approximations for the following.

    a) $\sqrt{1.03}$ **b)** $0.96^{\frac{1}{3}}$ **c)** $\sqrt[3]{1001}$

Chapter 12

9 Find the second derivatives of the following.

    a) $y=x^4$ **b)** $y=1-3x-x^2$ **c)** $y=\dfrac{3}{x^3}$

10 Find the stationary points of the following functions, identifying them as maxima, minima or points of inflection.

    a) $y=x^2-6x-1$ **b)** $y=x^3+2x^2+x+1$ **c)** $y=3x+\dfrac{1}{x^3}$

11 Sketch the graphs of the functions in Question 10.

12 A standard size of parcel is such that the length is twice the width. The volume of the parcel is $10\,000$ cm³. Letting the width be x cm, show that the surface area A cm² of the parcel is given by

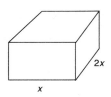

$$A = 4x^2 + \frac{30\,000}{x}$$

Find the least possible surface area of the parcel.

Chapter 13

13 Integrate the following functions.

a) $4x^3 - 4x$ b) $10x^{1.5}$ c) $\dfrac{4}{x^2}$

d) $(2 + x)(3 - x)$ e) $(4 + x)^2$ f) $\dfrac{2 + x}{\sqrt{x}}$

14 Find the following integrals.

a) $\displaystyle\int_0^3 (2x + 3)\, dx$ b) $\displaystyle\int_1^9 \sqrt{x}\, dx$ c) $\displaystyle\int_1^3 \left(x + \frac{1}{x}\right)^2 dx$

15 Find the following areas.

a) under $y = 3x^2$, from $x = 1$ to $x = 2$ b) under $y = 3\sqrt{x}$, from $x = 0$ to $x = 4$

c) between the curves $y = 1 + 3x - x^2$ and $y = 4 - x$, from $x = 1$ to $x = 3$

Chapter 14

16 Prove the following identities.

a) $\sec x \cot x = \operatorname{cosec} x$ b) $(1 + \sin x + \cos x)^2 = 2(1 + \sin x)(1 + \cos x)$

c) $\dfrac{\sin x}{1 + \cos x} = \dfrac{1 - \cos x}{\sin x}$

17 Find the solutions of the following equations, in the range $0 \leq x \leq \pi$. Give your answers as multiples of π.

a) $\sin x = 0.5$ b) $\cos 2x = -0.5$ d) $\tan 3x = 1$

18 Solve the following equations in the range indicated.

a) $\tan^2 x = 2$, for $0° \leq x \leq 180°$ b) $\sin(x + 70°) = 0.5$, for $0° \leq x \leq 360°$

c) $\sin x = 2 \cos x$, for $0 < x < \pi$ d) $3\sin^2 x + 4\cos x - 4 = 0$, for $0 \leq x \leq \pi$

MIXED EXERCISE

1 The diagram shows a rectangle with a semicircle along one side. If the perimeter is fixed, show that the total area is a maximum when the height of the rectangle is zero.

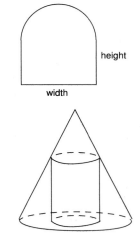

height

width

2 The sides of a triangle are a, b and c, and the angle between the sides a and b is θ. Show that

$$b = a \cos \theta \pm \sqrt{c^2 - a^2 \sin^2 \theta}$$

For what values of a, c and θ does a solution exist? For what values does a unique solution exist?

3 A cylinder is placed inside a fixed cone. If the volume of the cylinder is a maximum, find the ratio of its volume to that of the cone.

4 The diagonal of a cylinder goes from a point on the rim of the top to the point on the rim of the bottom which is diametrically opposite. If the diagonal is held constant at D, show that the volume of the cylinder is a maximum when the ratio of height h to radius r is $h : r = \sqrt{2} : 1$.

5 Find the first three terms in the expansion of $(1 + x^2)^{10}$. Hence find the approximate value of

$$\int_0^{0.25} (1 + x^2)^{10} \, dx$$

6 Solve the equation $\cos^2 \theta + 5 \sin \theta = 2$, giving answers in the range $0 \le \theta \le \pi$.

7 Show that the positive root of $x^2 - 2x - k = 0$ can be written as

$$x = 1 + \sqrt{1 + k}$$

Use the binomial theorem to express x as a series in k up to the term in k^3. Hence find an approximation for the positive root of $x^2 - 2x - 0.1 = 0$.

8 Let $f(x) = \dfrac{\sqrt{1 - x} + \sqrt{1 - 2x}}{\sqrt{1 - 3x + 2x^2}}$. Show that $f(x)$ can be written as $f(x) = (1 - x)^{-\frac{1}{2}} + (1 - 2x)^{-\frac{1}{2}}$.

Hence find the expansion of $f(x)$ up to the term in x^2. For what range of values of x is the expansion correct?

9 An approximation given in Chapter 7 was $\sin x \approx x$ for x small. Use the identity $\cos x = \sqrt{1 - \sin^2 x}$ and the binomial expansion of $\sqrt{1 - x}$ to find an approximation for $\cos x$. Does it agree with the approximation for $\cos x$ given in Chapter 7?

10 The quadratic $y = ax^2 + bx + c$ passes through the point $(1, -3)$ and has a minimum at the point $(2, -5)$. Find a, b and c.

11 Let P be the point at $(0, 1)$. Find an expression for the distance from P to the point (x, x^2) on the curve $y = x^2$. What is the least value of this distance?

12 Show that the positive solutions of $\tan x = 0.4$ form an arithmetic progression, and find its first term and common ratio.

13 The point (x, y) lies on the line $ax + by = c$. Find the minimum value of $x^2 + y^2$.

14 Let $y = x^n$, where n is an integer greater than 1. What is the nature of the stationary point at $x = 0$?

15 a) The area of a circle is $A = \pi r^2$, and the circumference is $C = 2\pi r$. Why is C equal to $\dfrac{dA}{dr}$?

b) The volume of a sphere is $V = \frac{4}{3}\pi r^3$, and its surface area is $S = 4\pi r^2$. Why is S equal to $\dfrac{dV}{dr}$?

16 Write out the first three terms in the expansion of $\sqrt{1 + x}$. Hence find an approximation for the integral

$$\int_0^{\frac{1}{2}} \sqrt{1 + x}\, dx$$

17 Find the points of intersection of the curve $y = 4 - x^2$ and the line $y = x - 1$. Find the area of the region enclosed between the line and the curve, leaving your answer in surd form.

18 In a national lottery, there is a very large number of tickets. The proportion of prize-winning tickets is p, where p is very small. If I buy ten tickets, show that the probability that I shall win at least one prize is approximately $10p$.

19 How many arrangements are there of the letters of MURRAY?

How many three-letter arrangements can be made from the letters of MURRAY?

LONGER EXERCISE
Finding π

This is a method for finding π, which in principle could work for any degree of accuracy.

1 The point (x, y) lies on the circle with centre $(0, 0)$ and radius 1. Show that $y = \sqrt{1 - x^2}$.

2 Show that the area under the curve $y = \sqrt{1 - x^2}$ between $x = 0$ and $x = 1$ is $\frac{\pi}{4}$.

3 Expand $\sqrt{1 - x^2}$ up to the term in x^6.

4 Integrate your expansion of part 3 and find the value between $x = 0$ and $x = 1$. What value does this give for π? How could you make the result more accurate?

Poker hands

In the card game of poker, you are dealt five cards from the pack of 52. It does not matter which order the cards are dealt in. Find the number of possible poker hands.

1 A **flush** consists of five cards all of the same suit, such as five spades or five clubs. Find the number of possible flushes, and hence find the probability that you will be dealt a flush.

2 A **four of a kind** consists of five cards including four with the same number, such as four kings, or four 10's. Find the probability you will be dealt a four of a kind.

3 A **full house** or **tight** consists of three of a kind and a pair, such as three kings and two 10's. Find the probability of being dealt this hand.

4 A **straight** consists of five cards in sequence, regardless of suit. Find the probability of being dealt this hand.

5 Find the probability of being dealt **three of a kind**: a hand containing three cards with the same number, but not a four of a kind or a full house.

*6 Find the probability of being dealt **two pairs**: a hand containing two pairs of cards with the same number, but not a four of a kind or a full house.

*7 Find the probability of being dealt **one pair**: a hand containing a pair of cards with the same number, but not two pairs or three of a kind.

EXAMINATION QUESTIONS

1 In $\triangle ABC$, $AB = 3$ cm, $AC = 4$ cm, $BC = d$ cm and $\cos A = x$.

a) Show that $d = 5\sqrt{(1 - 0.96x)}$

b) Given that x^3 and higher powers of x may be neglected, use the binomial expansion to express d in the form $p + qx + rx^2$, where p, q and r are constants whose values are to be found.

c) Show that, when $x = 0.05$, the approximation found in part (b) is correct to 3 decimal places.

L 1992

2 You are given that $f(x) = 4 + 2x^3 - 3x^4$.

(i) Find $f'(x)$ and the values of x for which $f'(x) = 0$. Hence find the coordinates of the stationary points.

(ii) Find $f''(x)$ and its value at each stationary point.

(iii) Determine the nature of the stationary point at which $f'(x)$ and $f''(x)$ are both zero, explaining your method and showing your working.

(iv) State the nature of the other turning point and, without further calculation, sketch the graph of $y = f(x)$ for values of x between $x = -1$ and $x = 1$.

MEI 1993

3 a) If x is so small that x^3 and higher powers of x may be neglected, show that

$$(8 - 4x + 9x^2)\sqrt{1 + 2x} = 8 + 4x + x^2$$

b) Show that the coefficient of x^{-24} in the binomial expansion of $(2x^4 - x^{-3})^{15}$ is 3640.

NI 1992

4 A curve passes through the point (2, 3). The gradient of the curve is given by

$$\frac{dy}{dx} = 3x^2 - 2x - 1$$

(i) Find y in terms of x.

(ii) Find the coordinates of any stationary points of the graph of y.

(iii) **Sketch** the graph of y against x, marking the coordinates of any stationary points and the point where the curve cuts the y-axis.

MEI 1992

5 Given the function $y = 3x^4 + 4x^3$

(i) find $\dfrac{dy}{dx}$,

(ii) show that the graph of the function y has stationary points at $x = 0$ and $x = -1$ and find their coordinates,

(iii) determine whether each of the stationary points is a maximum, minimum or point of inflection, giving reasons for your answer,

(iv) sketch the graph of the function y, giving the coordinates of the stationary points and the points where the curve cuts the axes.

MEI 1992

6 Solve the equation

$$9 \cos^2 x - 6 \cos x - 0.21 = 0, \quad 0° \leq x \leq 360°$$

giving each answer in degrees to 1 decimal place.

L 1993

Vectors

Chapter outline

15.1 The definition of vectors   **15.4** The product of vectors
15.2 Vectors in terms of   **15.5** Lines and vectors
     coordinates
15.3 The modulus or
     magnitude of a vector

A great deal of applied mathematics is expressed in terms of
vectors. A vector describes the direction of a quantity as well as its
magnitude. For example, when dealing with a force, we want to
know the force's direction as well as its magnitude. When dealing
with velocity, we want to know the direction a body is moving in,
as well as its speed. Both force and velocity are described in terms
of vectors. By contrast, mass, temperature and so on have no
direction associated with them, and they are called **scalars**.

The use of vectors is not limited to mechanics. Economics, games
theory and many other subjects involve vectors. And in pure
mathematics itself, vectors are a topic of study in their own right.
One single vector equation can summarise several ordinary
equations.

15.1 The definition of vectors

A **vector** is a quantity with direction as well as magnitude.
Physical examples of vectors are force, velocity, acceleration and
electric current.

A **scalar** is a quantity which has only magnitude. Physical
examples of scalars are mass, temperature and electrical charge.

We write vector variables in **bold** type, as **v**, or underlined, as v̲, or
with a bar on top, as v̄. When writing by hand it is easier to use v̄,
when printing it is easier to use **v**.

Scalar variables are written in ordinary type, as k, λ, μ etc.

Displacement and position vectors

A vector which represents the **displacement** from point A to point B is written as $\overrightarrow{AB}$.

If O represents the origin, then the vector $\overrightarrow{OP}$ is the **position vector** of P.

Fig. 15.1   *Fig. 15.2*

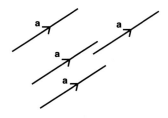

Fig. 15.3

Equal vectors

Two vectors are equal if they have the same direction and the same magnitude. In Fig. 15.3, all the lines labelled **a** are parallel to each other and have the same length.

Note that vectors do not have a **position** in space. It does not matter where they start from. In Fig. 15.4 $\overrightarrow{AB} = \overrightarrow{CD}$.

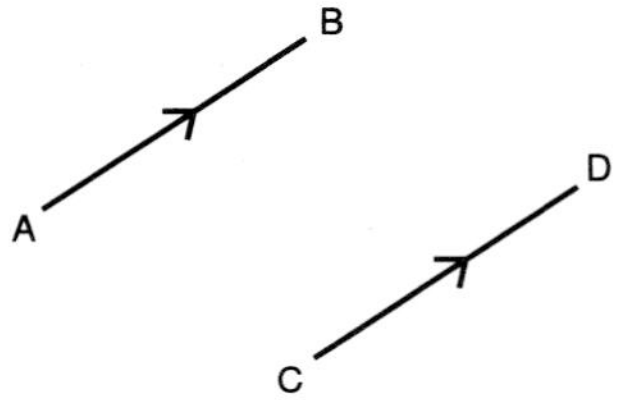

Fig. 15.4

Adding vectors

Vectors are added by forming a triangle as shown in Fig. 15.5.

If the vectors represent displacement, then the sum of the vectors represents the combination of the displacements.

$$\overrightarrow{AB} + \overrightarrow{BC} = \overrightarrow{AC}$$

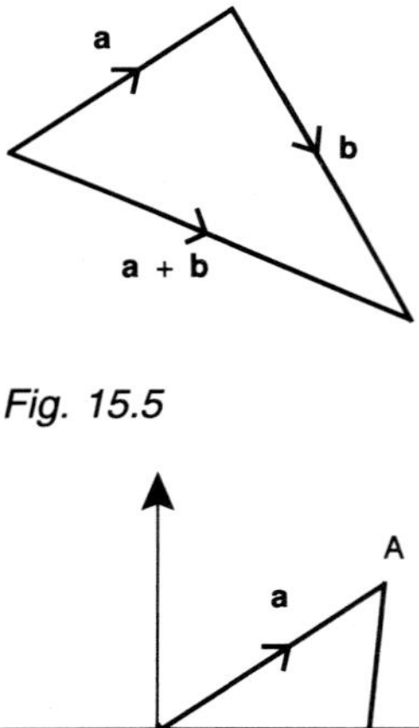

If vectors represent forces, their sum represents the single force which is equivalent to them, called the **resultant** of the forces.

Fig. 15.5

Negative vectors

The vector $-\mathbf{a}$ has the same magnitude as **a** but is opposite in direction. For example, if **a** is the displacement vector $\overrightarrow{AB}$, then $-\overrightarrow{AB} = \overrightarrow{BA}$, which represents the displacement from B to A.

Subtracting vectors

Suppose the position vectors of A and B are **a** and **b**. To go from A to B we can go via the origin.

$$\overrightarrow{AB} = \overrightarrow{AO} + \overrightarrow{OB} = -\overrightarrow{OA} + \overrightarrow{OB} = -\mathbf{a} + \mathbf{b} = \mathbf{b} - \mathbf{a}$$

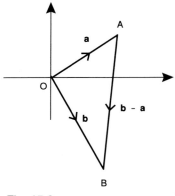

Fig. 15.6

Multiplying vectors by scalars

If a vector is multiplied by a positive scalar, then its length is multiplied by the scalar but its direction is unaltered. If the vector is multiplied by a negative scalar, then its direction is reversed.

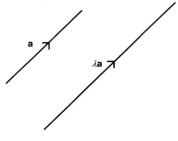

Zero vector

If we multiply **a** by 0, then we obtain the **zero vector**, **0**, which has zero magnitude. The zero vector represents a displacement which goes nowhere at all.

Fig. 15.7

Parallel vectors

If two non-zero vectors are parallel to each other, then they are scalar multiples of each other.

If **a** and **b** are parallel, then $\mathbf{a} = k\mathbf{b}$ for some scalar k.

If A, B and C are three points, then ABC is a straight line if $\overrightarrow{AB}$ is parallel to $\overrightarrow{BC}$.

Vector equations

Suppose **a** and **b** are non-parallel non-zero vectors, and $\lambda\mathbf{a} + \mu\mathbf{b} = k\mathbf{a} + m\mathbf{b}$.

Then $\lambda = k$ and $\mu = m$.

Proof

Rearrange the equation, to give

$$(\lambda - k)\mathbf{a} = (m - \mu)\mathbf{b}$$

As **a** and **b** are not parallel, they cannot be scalar multiples of each other. Hence the multiplying scalars must be zero.

$$\lambda = k \text{ and } \mu = m$$

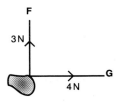

Fig. 15.8

EXAMPLE 15.1

The diagram shows two forces **F** and **G** acting on a body. **F** is 3 newtons acting due north, **G** is 4 newtons acting due east. Copy the diagram, and on it draw the resultant force **F** + **G**. What is the magnitude of this force?

Solution

Draw **F** + **G** by completing the triangle as shown.

The length of this vector is $\sqrt{3^2 + 4^2} = 5$.

The magnitude of **F** + **G** is 5 newtons.

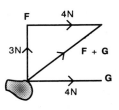

Fig. 15.9

EXERCISE 15A

1 A, B, C and D are points in space. Simplify the following.

a) $\overrightarrow{AB} + \overrightarrow{BC}$ **b)** $\overrightarrow{DC} + \overrightarrow{CB} + \overrightarrow{BA}$ **c)** $\overrightarrow{AB} + 2\overrightarrow{AC} + \overrightarrow{BC}$ **d)** $\overrightarrow{AB} - \overrightarrow{DB}$

2 The vectors **a** and **b** are shown in Fig. 15.10.
Copy the diagram, and on it draw the following.

a) a + b   **b) 2a**   **c) 2a + b**

d) a − b   **e) −a − b**

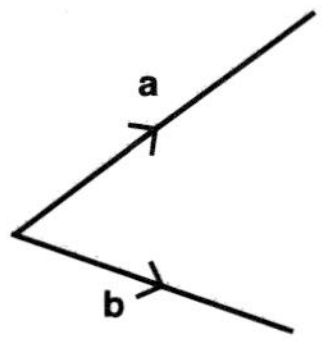

Fig. 15.10

3 **a** and **b** are non-parallel vectors. Which of the following vectors are parallel to each other?

a) a − 2b   **b) 2a − b**   **c) 2a − 4b**   **d) b − 2a**   **e) 4b − 2a**

4 **a** and **b** are non-parallel vectors. Solve the following equations.

a) $2\mathbf{a} + x\mathbf{b} = y\mathbf{a} - 3\mathbf{b}$ **b)** $(x + y)\mathbf{a} + (x - y)\mathbf{b} = \mathbf{a} + 3\mathbf{b}$ **c)** $2x\mathbf{a} + 3y\mathbf{b} = y\mathbf{a} - 2x\mathbf{b} + \mathbf{a}$

5 **F** and **G** are forces of 5 newtons acting south and 13 newtons acting west, respectively. Illustrate **F**, **G** and **F + G** on a diagram. What is the magnitude and direction of **F + G**?

6 **F** is a force of 8 newtons acting due north. **G** is a force of 7 newtons acting north-east. What is the magnitude and direction of **F + G**?

7 **F** is a force of 20 newtons acting south. The resultant of **F** and another force **G** is 25 newtons acting south-east. What is the magnitude and direction of **G**?

Geometry by vectors

Vectors can be used in geometry, to find results about the lengths and directions of lines.

EXAMPLE 15.2

ABC is a triangle, for which $\overrightarrow{AB} = \mathbf{b}$ and $\overrightarrow{AC} = \mathbf{c}$. L and M are the midpoints of AB and AC respectively. Express the following vectors in terms of **b** and **c**.

a) $\overrightarrow{AL}$ **b)** $\overrightarrow{AM}$ **c)** $\overrightarrow{BC}$ **d)** $\overrightarrow{LM}$

What is the relationship between LM and BC?

Solution

a) $\overrightarrow{AL}$ is in the same direction as $\overrightarrow{AB}$ and half its length. Hence

$$\overrightarrow{AL} = \tfrac{1}{2}\mathbf{b}$$

b) $\overrightarrow{AM}$ is in the same direction as $\overrightarrow{AC}$ and half its length. Hence

$$\overrightarrow{AM} = \tfrac{1}{2}\mathbf{c}$$

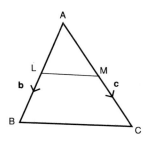

Fig. 15.11

c) To go from B to C, go through A.

$$\overrightarrow{BC} = \overrightarrow{BA} + \overrightarrow{AC} = -\mathbf{b} + \mathbf{c}$$

$$\overrightarrow{BC} = \mathbf{c} - \mathbf{b}$$

d) To go from L to M, go through A.

$$\overrightarrow{LM} = \overrightarrow{LA} + \overrightarrow{AM} = -\overrightarrow{AL} + \overrightarrow{AM} = -\tfrac{1}{2}\mathbf{b} + \tfrac{1}{2}\mathbf{c}$$

$$\overrightarrow{LM} = \tfrac{1}{2}\mathbf{c} - \tfrac{1}{2}\mathbf{b}$$

Notice that $\overrightarrow{LM} = \tfrac{1}{2}\overrightarrow{BC}$.

LM is parallel to BC and half its length.

EXAMPLE 15.3

OABC is a parallelogram, with $\overrightarrow{OA} = \mathbf{a}$ and $\overrightarrow{OC} = \mathbf{c}$. P is the midpoint of AB. Express $\overrightarrow{OB}$, $\overrightarrow{AC}$ and $\overrightarrow{OP}$ in terms of $\mathbf{a}$ and $\mathbf{c}$.

Let OP and AC meet at X. Show that $\overrightarrow{OX}$ can be expressed as $\lambda\overrightarrow{OP}$ and as $\mathbf{a} + \mu\overrightarrow{AC}$. Hence find $\overrightarrow{OX}$.

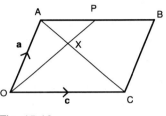

Fig. 15.12

Solution

Remembering that $\overrightarrow{AB} = \overrightarrow{OC}$,

$$\overrightarrow{OB} = \mathbf{a} + \mathbf{c}$$

Go from A to C through O.

$$\overrightarrow{AC} = -\mathbf{a} + \mathbf{c} = \mathbf{c} - \mathbf{a}$$

$\overrightarrow{AP}$ is half $\overrightarrow{AB}$. Hence

$$\overrightarrow{OP} = \mathbf{a} + \tfrac{1}{2}\mathbf{c}$$

Now let $\dfrac{OX}{OP} = \lambda$, and $\dfrac{AX}{AC} = \mu$.

Then

$$\overrightarrow{OX} = \lambda\overrightarrow{OP}$$

and

$$\overrightarrow{OX} = \overrightarrow{OA} + \overrightarrow{AX} = \mathbf{a} + \mu\overrightarrow{AC}$$

Putting these in terms of $\mathbf{a}$ and $\mathbf{c}$

$$\overrightarrow{OX} = \lambda\overrightarrow{OP} = \lambda(\mathbf{a} + \tfrac{1}{2}\mathbf{c})$$

and

$$\overrightarrow{OX} = \mathbf{a} + \mu(\mathbf{c} - \mathbf{a})$$

so

$$\lambda(\mathbf{a} + \tfrac{1}{2}\mathbf{c}) = \mathbf{a} + \mu(\mathbf{c} - \mathbf{a})$$

$$\lambda\mathbf{a} + \tfrac{1}{2}\lambda\mathbf{c} = \mathbf{a}(1 - \mu) + \mu\mathbf{c}$$

Equating the **a** terms and the **c** terms gives

$\lambda = 1 - \mu$ and $\frac{1}{2}\lambda = \mu$

$\lambda = \frac{2}{3}$

$\overrightarrow{OX} = \lambda\overrightarrow{OP} = \frac{2}{3}(\mathbf{a} + \frac{1}{2}\mathbf{c}) = \frac{2}{3}\mathbf{a} + \frac{1}{3}\mathbf{c}$

$\overrightarrow{OX} = \frac{2}{3}\mathbf{a} + \frac{1}{3}\mathbf{c}$

EXERCISE 15B

1 OAB is a triangle, with $\overrightarrow{OA} = \mathbf{a}$ and $\overrightarrow{OB} = \mathbf{b}$. P and Q are the midpoints of OA and OB respectively. Express $\overrightarrow{OP}$, $\overrightarrow{OQ}$, $\overrightarrow{PB}$ and $\overrightarrow{QA}$ in terms of **a** and **b**.

   PB and QA meet at R. Express $\overrightarrow{OR}$ in the forms $\overrightarrow{OP} + k\overrightarrow{PB}$ and $\overrightarrow{OQ} + m\overrightarrow{QA}$. Hence find $\overrightarrow{OR}$.

*2 OAB is a triangle, with $\overrightarrow{OA} = \mathbf{a}$ and $\overrightarrow{OB} = \mathbf{b}$. P and Q are points on OA and OB respectively, with $OP = \frac{1}{3}OA$ and $OQ = \frac{2}{3}OB$. Let PB and QA meet at R. Find $\overrightarrow{OR}$ in terms of **a** and **b**.

3 OABC is a parallelogram, with $\overrightarrow{OA} = \mathbf{a}$ and $\overrightarrow{OC} = \mathbf{c}$. P lies on AB, with $AP = \frac{1}{3}AB$. Find $\overrightarrow{OB}$, $\overrightarrow{OP}$ and $\overrightarrow{AC}$ in terms of **a** and **c**.

   Let OP and AC meet at R. Find $\overrightarrow{OR}$ in terms of **a** and **c**.

4 OABC is a parallelogram, with $\overrightarrow{OA} = \mathbf{a}$ and $\overrightarrow{OC} = \mathbf{c}$. The diagonals OB and AC meet at D. Write $\overrightarrow{OB}$ and $\overrightarrow{OD}$ in terms of **a** and **c**.

   ODEC is also a parallelogram. Express $\overrightarrow{OE}$ in terms of **a** and **c**.

   BDEF is also a parallelogram. Express $\overrightarrow{OF}$ in terms of **a** and **c**.

5 OABCDE is a regular hexagon, with $\overrightarrow{OA} = \mathbf{a}$ and $\overrightarrow{OE} = \mathbf{e}$. Express the following in terms of **a** and **e**.

   a) $\overrightarrow{EA}$ b) $\overrightarrow{AB}$ c) $\overrightarrow{AD}$ d) $\overrightarrow{OC}$

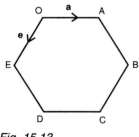

Fig. 15.13

6 **a** and **b** are non-parallel vectors and A, B, C, D are points with position vectors $\mathbf{a} + \mathbf{b}$, $2\mathbf{a} + \mathbf{b}$, $\mathbf{a} - \mathbf{b}$ and $\mathbf{b} - \mathbf{a}$ respectively. Find which three of the points lie on a straight line.

7 OAB is a triangle, with $\overrightarrow{OA} = \mathbf{a}$ and $\overrightarrow{OB} = \mathbf{b}$. P lies on OA, Q on AB and R on OB. $OP = \frac{1}{2}OA$, $AQ = \frac{2}{3}AB$, $OR = 2OB$. Find the position vectors of P, Q and R. Show that P, Q and R lie on a straight line.

15.2 Vectors in terms of coordinates

Vectors are represented as directed line segments as shown. If they are on a coordinate system, they can be given in terms of their components.

$$\mathbf{v} = \begin{pmatrix} x \\ y \end{pmatrix}$$

The vector illustrated is $\begin{pmatrix} 2 \\ 3 \end{pmatrix}$.

Vectors can be added by adding their components.

$$\begin{pmatrix} x \\ y \end{pmatrix} + \begin{pmatrix} w \\ z \end{pmatrix} = \begin{pmatrix} x + w \\ y + z \end{pmatrix}$$

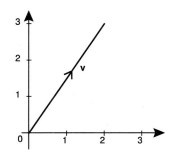

Fig. 15.14

Vectors can be multiplied by scalars. Each component of the vector is multiplied by the scalar.

$$t\begin{pmatrix} x \\ y \end{pmatrix} = \begin{pmatrix} tx \\ ty \end{pmatrix}$$

So far we have dealt with vectors with two components. Vectors representing quantities in three dimensions have three components. Similar rules apply.

$$\begin{pmatrix} x \\ y \\ z \end{pmatrix} + \begin{pmatrix} a \\ b \\ c \end{pmatrix} = \begin{pmatrix} x + a \\ y + b \\ z + c \end{pmatrix} \qquad t\begin{pmatrix} x \\ y \\ z \end{pmatrix} = \begin{pmatrix} tx \\ ty \\ tz \end{pmatrix}$$

Position vector

Recall that the position vector of P is $\overrightarrow{OP}$. This can be written in terms of coordinates.

The position vector of the point (x, y) is $\begin{pmatrix} x \\ y \end{pmatrix}$.

EXAMPLE 15.4

Let $\mathbf{a} = \begin{pmatrix} 2 \\ 4 \end{pmatrix}$ and $\mathbf{b} = \begin{pmatrix} 3 \\ -3 \end{pmatrix}$. Find

a) $\mathbf{a} + \mathbf{b}$　　　**b)** $3\mathbf{a}$

Solution

a) Add the components together.

$$\mathbf{a} + \mathbf{b} = \begin{pmatrix} 2 \\ 4 \end{pmatrix} + \begin{pmatrix} 3 \\ -3 \end{pmatrix} = \begin{pmatrix} 2 + 3 \\ 4 - 3 \end{pmatrix} = \begin{pmatrix} 5 \\ 1 \end{pmatrix}$$

$$\mathbf{a} + \mathbf{b} = \begin{pmatrix} 5 \\ 1 \end{pmatrix}$$

b) Multiply each component of **a** by 3.

$$3\mathbf{a} = 3\binom{2}{4} = \binom{6}{12}$$

$$3\mathbf{a} = \binom{6}{12}$$

EXAMPLE 15.5

With **a** and **b** as in Example 15.4, solve this equation.

$$x\mathbf{a} + y\mathbf{b} = \binom{1}{11}$$

Solution

Rewrite the equation.

$$x\binom{2}{4} + y\binom{3}{-3} = \binom{1}{11}$$

This is equivalent to two ordinary equations.

$$2x + 3y = 1$$

$$4x - 3y = 11$$

These simultaneous equations can be solved.

The solution is $x = 2$, $y = -1$.

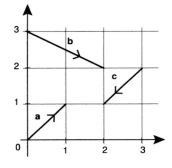

EXERCISE 15C

1 Express the vectors in Fig. 15.15 in terms of coordinates. *Fig. 15.15*

2 Let $\mathbf{u} = \binom{2}{3}$ and $\mathbf{v} = \binom{5}{-1}$. Find

   a) $\mathbf{u} + \mathbf{v}$ **b)** $\mathbf{u} - \mathbf{v}$ **c)** $2\mathbf{u}$ **d)** $2\mathbf{u} + 3\mathbf{v}$ **e)** $3\mathbf{u} - 2\mathbf{v}$

3 Let $\mathbf{a} = \begin{pmatrix} 1 \\ -2 \\ 2 \end{pmatrix}$ and $\mathbf{b} = \begin{pmatrix} 2 \\ 0 \\ -1 \end{pmatrix}$. Express the following in terms of coordinates.

   a) $\mathbf{a} + \mathbf{b}$ **b)** $\mathbf{a} - \mathbf{b}$ **c)** $3\mathbf{a}$ **d)** $2\mathbf{a} - 3\mathbf{b}$ **e)** $-\mathbf{a} + 4\mathbf{b}$

4 With **u** and **v** as in Question 2, solve the following equations.

   a) $x\mathbf{u} + y\mathbf{v} = \binom{9}{5}$ **b)** $\lambda\mathbf{u} + \mu\mathbf{v} = \binom{1}{-7}$ **c)** $p\mathbf{u} + q\mathbf{v} = \binom{1}{10}$

5 ABC is a triangle with vertices at A(1, 1), B(2, 3) and C(4, 1). Find these vectors.

   a) $\overrightarrow{AB}$ **b)** $\overrightarrow{BC}$ **c)** $\overrightarrow{CA}$

6 Which of the following vectors are parallel to each other?

a) $\begin{pmatrix} 2 \\ 1 \end{pmatrix}$ b) $\begin{pmatrix} 1 \\ 2 \end{pmatrix}$ c) $\begin{pmatrix} -1 \\ -2 \end{pmatrix}$ d) $\begin{pmatrix} -2 \\ 4 \end{pmatrix}$ e) $\begin{pmatrix} \frac{1}{2} \\ \frac{1}{4} \end{pmatrix}$

7 The vertices of a quadrilateral are at A(1, 1), B(3, 5), C(6, 4) and D(4, 0). Find the vectors $\overrightarrow{AB}$, $\overrightarrow{BC}$, $\overrightarrow{CD}$ and $\overrightarrow{DA}$. Show that ABCD is a parallelogram.

8 A is at (1, 0), B at (3, 4) and C at (0, 2). Find the coordinates of D, given that ABDC is a parallelogram.

9 **F** is a force of 5 newtons along the *x*-axis, and **G** is a force of 7 newtons along the *y*-axis. Express **F**, **G** and **F** + **G** in terms of coordinates.

15.3 The modulus or magnitude of a vector

The **modulus** or **magnitude** of a vector gives its size. Suppose the vector **v** is $\begin{pmatrix} x \\ y \end{pmatrix}$. Then by Pythagoras's theorem the length of the line representing the vector is

$$|\mathbf{v}| = \sqrt{x^2 + y^2}$$

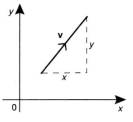

Fig. 15.16

If **v** is a vector in three dimensions, $\mathbf{v} = \begin{pmatrix} x \\ y \\ z \end{pmatrix}$, then its modulus is

$$|\mathbf{v}| = \sqrt{x^2 + y^2 + z^2}$$

If **F** is a force, then |**F**| is the size of the force.

EXAMPLE 15.6
Let **a** and **b** be as stated in Example 15.4. Find the modulus of each of the following vectors.

a) **a**   b) **b**   c) **a** + **b**

Solution

Apply the formula to $\mathbf{a} = \begin{pmatrix} 2 \\ 4 \end{pmatrix}$ and $\mathbf{b} = \begin{pmatrix} 3 \\ -3 \end{pmatrix}$.

a) $|\mathbf{a}| = \sqrt{2^2 + 4^2} = \sqrt{20} = 4.47$

b) $|\mathbf{b}| = \sqrt{3^2 + (-3)^2} = \sqrt{18} = 4.24$

c) Add the vectors and then find the modulus.

$$\mathbf{a} + \mathbf{b} = \begin{pmatrix} 2 \\ 4 \end{pmatrix} + \begin{pmatrix} 3 \\ -3 \end{pmatrix} = \begin{pmatrix} 5 \\ 1 \end{pmatrix}$$

$$|\mathbf{a} + \mathbf{b}| = \left| \begin{pmatrix} 5 \\ 1 \end{pmatrix} \right| = \sqrt{5^2 + 1^2} = \sqrt{26} = 5.10$$

EXERCISE 15D

1 Find the modulus of each of the following vectors.

a) $\begin{pmatrix} 3 \\ 4 \end{pmatrix}$ **b)** $\begin{pmatrix} 2 \\ 3 \end{pmatrix}$ **c)** $\begin{pmatrix} -1 \\ 5 \end{pmatrix}$ **d)** $\begin{pmatrix} 1 \\ 2 \\ 1 \end{pmatrix}$ **e)** $\begin{pmatrix} 3 \\ -1 \\ 2 \end{pmatrix}$ **f)** $\begin{pmatrix} 5 \\ 8 \\ -3 \end{pmatrix}$

2 ABCD is a quadrilateral with vertices at A(1, 1), B(4, 5), C(9, 5) and D(6, 1). Find the vectors $\overrightarrow{AB}$, $\overrightarrow{BC}$, $\overrightarrow{CD}$ and $\overrightarrow{DA}$. Show that ABCD is a rhombus.

3 $\mathbf{r} = \begin{pmatrix} 1 \\ 1 \end{pmatrix} + \lambda \begin{pmatrix} 1 \\ 2 \end{pmatrix}$ where λ is a scalar. Find $|\mathbf{r}|$ when **a)** $\lambda = 1$ **b)** $\lambda = -2$.

Find λ for which $|\mathbf{r}| = 1$.

*4 a) Show by means of a diagram that for any vectors **a** and **b**, $|\mathbf{a} + \mathbf{b}| \leq |\mathbf{a}| + |\mathbf{b}|$.

b) If $|\mathbf{a} + \mathbf{b}| = |\mathbf{a}| + |\mathbf{b}|$, what can you say about **a** and **b**?

Unit vectors

A **unit vector** has modulus 1. If we have a non-zero vector, then a unit vector in the same direction is found by dividing the vector by its modulus.

$\dfrac{\mathbf{v}}{|\mathbf{v}|}$ is a unit vector parallel to **v**

Special unit vectors of particular importance are:

$\mathbf{i} = \begin{pmatrix} 1 \\ 0 \end{pmatrix}$ (the unit vector parallel to the *x*-axis)

$\mathbf{j} = \begin{pmatrix} 0 \\ 1 \end{pmatrix}$ (the unit vector parallel to the *y*-axis)

If we are dealing with three dimensions, the three special unit vectors are

$$\mathbf{i} = \begin{pmatrix} 1 \\ 0 \\ 0 \end{pmatrix} \quad \mathbf{j} = \begin{pmatrix} 0 \\ 1 \\ 0 \end{pmatrix} \quad \mathbf{k} = \begin{pmatrix} 0 \\ 0 \\ 1 \end{pmatrix}$$

where **k** is the unit vector parallel to the z-axis.

We can write a vector in terms of these basic vectors.

If $\mathbf{v} = \begin{pmatrix} x \\ y \end{pmatrix}$, then $\mathbf{v} = x\mathbf{i} + y\mathbf{j}$.

In this case, x and y are the **components** of **v**, in the **i** and **j** directions respectively.

EXAMPLE 15.7

Find a unit vector parallel to $\begin{pmatrix} 1 \\ 2 \\ 2 \end{pmatrix}$.

Solution

First find the modulus of this vector.

$$\left| \begin{pmatrix} 1 \\ 2 \\ 2 \end{pmatrix} \right| = \sqrt{1^2 + 2^2 + 2^2} = \sqrt{9} = 3$$

Now divide by vector by 3.

$$\frac{1}{3} \begin{pmatrix} 1 \\ 2 \\ 2 \end{pmatrix} = \begin{pmatrix} \frac{1}{3} \\ \frac{2}{3} \\ \frac{2}{3} \end{pmatrix}$$

The unit vector is $\begin{pmatrix} \frac{1}{3} \\ \frac{2}{3} \\ \frac{2}{3} \end{pmatrix}$.

EXERCISE 15E

1 Write the following using **i, j**.

a) $\begin{pmatrix} 2 \\ 1 \end{pmatrix}$　　b) $\begin{pmatrix} 1 \\ -2 \end{pmatrix}$　　c) $\begin{pmatrix} 3 \\ 0 \end{pmatrix}$

2 Write the following using **i, j** and **k**.

a) $\begin{pmatrix} 1 \\ 2 \\ 3 \end{pmatrix}$　　b) $\begin{pmatrix} 4 \\ -1 \\ -1 \end{pmatrix}$　　c) $\begin{pmatrix} 2 \\ 0 \\ 1 \end{pmatrix}$

3 Write these as vectors in the form $\begin{pmatrix} x \\ y \end{pmatrix}$.

a) $\mathbf{i} + 3\mathbf{j}$　　b) $2\mathbf{i} - 3\mathbf{j}$　　c) $4\mathbf{j}$

4 Write these vectors in the form $\begin{pmatrix} x \\ y \\ z \end{pmatrix}$.

   a) $\mathbf{i} - 2\mathbf{j} + 3\mathbf{k}$ **b)** $2\mathbf{i} - 3\mathbf{k}$

5 Find unit vectors parallel to the following.

   a) $\begin{pmatrix} 3 \\ 4 \end{pmatrix}$ **b)** $\begin{pmatrix} 2 \\ 5 \end{pmatrix}$ **c)** $\begin{pmatrix} 1 \\ 2 \\ -1 \end{pmatrix}$ **d)** $2\mathbf{i} - \mathbf{j}$ **e)** $\mathbf{i} + 2\mathbf{j} - 2\mathbf{k}$

6 Which of the following vectors are parallel to each other?

   a) $3\mathbf{i} - 6\mathbf{j}$ **b)** $2\mathbf{i} - \mathbf{j}$ **c)** $2\mathbf{j} - \mathbf{i}$ **d)** $4\mathbf{j} - 2\mathbf{i}$

15.4 The product of vectors

Scalar product

So we have added vectors and multiplied them by scalars. There are various ways we could define the product of two vectors. A definition that proves very useful is of the **scalar product**, also called the **dot product**, written $\mathbf{a} \cdot \mathbf{b}$. It can be defined either geometrically or algebraically.

Geometric definition
$\mathbf{a} \cdot \mathbf{b} = |\mathbf{a}| \, |\mathbf{b}| \cos \theta$, where θ is the angle between $\mathbf{a}$ and $\mathbf{b}$.

Algebraic definition

$$\begin{pmatrix} x \\ y \end{pmatrix} \cdot \begin{pmatrix} a \\ b \end{pmatrix} = xa + yb$$

This definition is for vectors in two dimensions. For three dimensions it is defined similarly.

$$\begin{pmatrix} x \\ y \\ z \end{pmatrix} \cdot \begin{pmatrix} a \\ b \\ c \end{pmatrix} = xa + yb + zc$$

Note that by either definition the result of the scalar product is a scalar, not a vector.

Proof of equivalence
The vectors $\mathbf{a}$ and $\mathbf{b}$ are shown in Fig. 15.18. By the cosine rule, the length of the third side of the triangle is given by

$$c^2 = |\mathbf{a}|^2 + |\mathbf{b}|^2 - 2|\mathbf{a}| \, |\mathbf{b}| \cos \theta$$

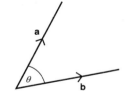

Fig. 15.17

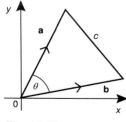

Fig. 15.18

If $\mathbf{a} = \begin{pmatrix} x \\ y \end{pmatrix}$ and $\mathbf{b} = \begin{pmatrix} a \\ b \end{pmatrix}$, then the length of the third side can also be found from Pythagoras's theorem.

$$c^2 = (x - a)^2 + (y - b)^2 = x^2 + a^2 - 2ax + y^2 + b^2 - 2by$$

We know from the definition of modulus that $|\mathbf{a}|^2 = x^2 + y^2$ and $|\mathbf{b}|^2 = a^2 + b^2$. Substituting these in and equating the two expressions for c^2 gives

$$x^2 + y^2 + a^2 + b^2 - 2|\mathbf{a}|\,|\mathbf{b}|\cos\theta$$
$$= x^2 + a^2 - 2ax + y^2 + b^2 - 2by$$
$$-2|\mathbf{a}|\,|\mathbf{b}|\cos\theta = -2ax - 2by$$
$$|\mathbf{a}|\,|\mathbf{b}|\cos\theta = xa + yb$$

Hence the geometric definition is equivalent to the algebraic definition.

Notes

1 Let $\mathbf{v} = \begin{pmatrix} x \\ y \end{pmatrix}$. When we form the scalar product of $\mathbf{v}$ with itself, we obtain the square of the modulus of $\mathbf{v}$.

$$\mathbf{v} \cdot \mathbf{v} = x^2 + y^2 = |\mathbf{v}|^2$$

2 $\mathbf{i}$, $\mathbf{j}$ and $\mathbf{k}$ are unit vectors at right angles to each other. Hence the following hold.

$$\mathbf{i} \cdot \mathbf{i} = \mathbf{j} \cdot \mathbf{j} = \mathbf{k} \cdot \mathbf{k} = 1 \qquad \mathbf{i} \cdot \mathbf{j} = \mathbf{j} \cdot \mathbf{k} = \mathbf{k} \cdot \mathbf{i} = 0$$

3 An interpretation in terms of force is as follows. If the point of application of a force $\mathbf{F}$ moves a distance $\mathbf{d}$, then the work done is $\mathbf{F} \cdot \mathbf{d}$.

EXAMPLE 15.8

Find the angles between the following pairs of vectors:

a) $\begin{pmatrix} 2 \\ 3 \end{pmatrix}$ and $\begin{pmatrix} 4 \\ 1 \end{pmatrix}$ b) $\mathbf{i} + \mathbf{k}$ and $\mathbf{i} - \mathbf{j}$

Solution

a) Write down the scalar product formula for these vectors.

$$\left|\begin{pmatrix} 2 \\ 3 \end{pmatrix}\right|\left|\begin{pmatrix} 4 \\ 1 \end{pmatrix}\right| \cos\theta = 2 \times 4 + 3 \times 1 = 11$$

We know that $\left|\begin{pmatrix} 2 \\ 3 \end{pmatrix}\right| = \sqrt{2^2 + 3^2} = \sqrt{13}$ and that

$$\left|\begin{pmatrix} 4 \\ 1 \end{pmatrix}\right| = \sqrt{4^2 + 1^2} = \sqrt{17}$$

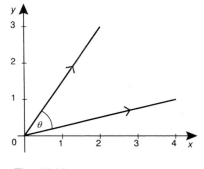

Fig. 15.19

$$\cos \theta = \frac{11}{\sqrt{13}\sqrt{17}} = 0.739\,94$$

$$\theta = \cos^{-1} 0.739\,94$$

The angle between them is 42.3°.

b) The scalar product of these vectors is

$$(\mathbf{i}+\mathbf{k}) \cdot (\mathbf{i}-\mathbf{j}) = \mathbf{i} \cdot \mathbf{i} - \mathbf{i} \cdot \mathbf{j} + \mathbf{k} \cdot \mathbf{i} - \mathbf{k} \cdot \mathbf{j}$$

$\mathbf{i} \cdot \mathbf{i} = 1$. All the other terms are zero. Hence the scalar product is 1.

The moduli of $\mathbf{i} + \mathbf{k}$ and $\mathbf{i} - \mathbf{j}$ are both $\sqrt{2}$.

Hence the angle θ is given by

$$\cos \theta = \frac{1}{\sqrt{2}\sqrt{2}} = \frac{1}{2}$$

The angle between the two vectors is 60°.

EXERCISE 15F

1 Evaluate the following scalar products.

a) $\begin{pmatrix} 2 \\ 3 \end{pmatrix} \cdot \begin{pmatrix} 1 \\ 4 \end{pmatrix}$ **b)** $\begin{pmatrix} 2 \\ -3 \end{pmatrix} \cdot \begin{pmatrix} 1 \\ -5 \end{pmatrix}$ **c)** $\begin{pmatrix} 1 \\ 2 \\ 4 \end{pmatrix} \cdot \begin{pmatrix} 3 \\ 1 \\ -2 \end{pmatrix}$ **d)** $\begin{pmatrix} -1 \\ 0 \\ 2 \end{pmatrix} \cdot \begin{pmatrix} 4 \\ 2 \\ 2 \end{pmatrix}$

e) $(\mathbf{i} + \mathbf{j}) \cdot (2\mathbf{i} - \mathbf{j})$ **f)** $(\mathbf{i} - 2\mathbf{j}) \cdot (3\mathbf{i} - 2\mathbf{j})$ **g)** $(\mathbf{i} + 2\mathbf{j} - \mathbf{k}) \cdot (-\mathbf{i} + 2\mathbf{j} - \mathbf{k})$

2 Find the angles between the following pairs of vectors.

a) $\begin{pmatrix} 2 \\ 1 \end{pmatrix}$ and $\begin{pmatrix} 3 \\ 1 \end{pmatrix}$ **b)** $\begin{pmatrix} 3 \\ -2 \end{pmatrix}$ and $\begin{pmatrix} 4 \\ -1 \end{pmatrix}$ **c)** $\begin{pmatrix} 2 \\ -1 \end{pmatrix}$ and $\begin{pmatrix} 1 \\ 3 \end{pmatrix}$

d) $\begin{pmatrix} 1 \\ 2 \\ 4 \end{pmatrix}$ and $\begin{pmatrix} 2 \\ 1 \\ 5 \end{pmatrix}$ **e)** $\begin{pmatrix} -2 \\ 1 \\ 4 \end{pmatrix}$ and $\begin{pmatrix} 1 \\ -1 \\ -2 \end{pmatrix}$ **f)** $\begin{pmatrix} 0.1 \\ 0.2 \\ 0.2 \end{pmatrix}$ and $\begin{pmatrix} 1 \\ 1 \\ 1 \end{pmatrix}$

g) $\mathbf{i} + \mathbf{j}$ and $\mathbf{i} + 2\mathbf{j}$ **h)** $2\mathbf{i} + 3\mathbf{j}$ and $2\mathbf{i} - 3\mathbf{j}$ **i)** $\mathbf{i} + 2\mathbf{j} - \mathbf{k}$ and $2\mathbf{i} + \mathbf{j} + 2\mathbf{k}$

3 Express $(\mathbf{a} + 2\mathbf{b}) \cdot (\mathbf{a} - 2\mathbf{b})$ in terms of $|\mathbf{a}|$ and $|\mathbf{b}|$.

*4 The angle between $\mathbf{i} + \mathbf{j}$ and $x\mathbf{i} + \mathbf{j}$ is 45°. Find the value of x.

*5 $\mathbf{u}$ and $\mathbf{v}$ are vectors with equal modulus, inclined at 60° to each other. Find the angle between $3\mathbf{u} + 2\mathbf{v}$ and $\mathbf{u} - \mathbf{v}$.

*6 Suppose you are going shopping. Write the numbers of the items you want to buy as the components of a vector $\mathbf{s}$. Write the prices of these items as the components of another vector $\mathbf{p}$. What does $\mathbf{s} \cdot \mathbf{p}$ represent?

Components

Suppose we want to find the component of **u** which lies along the direction of **v**, as shown in Fig. 15.20. We want $|\mathbf{u}| \cos \theta$. This the **component** of **u** in the direction of **v**.

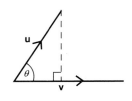

Fig. 15.20

We know that $\mathbf{u} \cdot \mathbf{v} = |\mathbf{u}| \, |\mathbf{v}| \cos \theta$, so we divide by $|\mathbf{v}|$.

The component of **u** in the direction of **v** is $\dfrac{\mathbf{u} \cdot \mathbf{v}}{|\mathbf{v}|}$.

In particular, the component of a vector **u** along the x-axis is $\mathbf{u} \cdot \mathbf{i}$.

$$\mathbf{u} \cdot \mathbf{i} = \begin{pmatrix} x \\ y \end{pmatrix} \cdot \begin{pmatrix} 1 \\ 0 \end{pmatrix} = x + 0 = x$$

Similar results hold for the other coordinate axes.

EXAMPLE 15.9

A force is given by $\mathbf{F} = \begin{pmatrix} 4 \\ 6 \end{pmatrix}$. Find the component of **F** in the direction of $\begin{pmatrix} 3 \\ 4 \end{pmatrix}$.

Solution

Apply the formula.

$$\frac{\begin{pmatrix} 4 \\ 6 \end{pmatrix} \cdot \begin{pmatrix} 3 \\ 4 \end{pmatrix}}{\left| \begin{pmatrix} 3 \\ 4 \end{pmatrix} \right|} = \frac{4 \times 3 + 6 \times 4}{\sqrt{3^2 + 4^2}} = \frac{36}{5} = 7.2$$

The component of **F** in the direction of $\begin{pmatrix} 3 \\ 4 \end{pmatrix}$ is 7.2.

EXERCISE 15G

1 Find the components of the following vectors in the directions indicated.

a) $\begin{pmatrix} 5 \\ 3 \end{pmatrix}$ in the direction $\begin{pmatrix} 5 \\ 12 \end{pmatrix}$

b) $\begin{pmatrix} 1 \\ -2 \end{pmatrix}$ in the direction $\begin{pmatrix} 3 \\ -4 \end{pmatrix}$

c) $\begin{pmatrix} 1 \\ 2 \\ 1 \end{pmatrix}$ in the direction $\begin{pmatrix} 2 \\ 1 \\ 2 \end{pmatrix}$

d) $\mathbf{i} + \mathbf{j}$ in the direction $\mathbf{i} + 2\mathbf{j}$

2 Let **u** be a unit vector in three dimensions. Show that **u** can be written as

$$\cos \alpha \, \mathbf{i} + \cos \beta \, \mathbf{j} + \cos \gamma \, \mathbf{k}$$

where α, β and γ are the angles between **u** and the three axes.

3 A force is given by $\mathbf{F} = \begin{pmatrix} 3 \\ 7 \end{pmatrix}$. Find the components of **F**

a) along the x-axis   **b)** along the y-axis   **c)** along the line $y = x$.

*4 A force **F** has component 3 along the line $y = x$, and 4 along the line $y = 2x$, both components being up and to the right. Find **F** in terms of components along the x and y-axes

Perpendicular vectors

If two vectors are perpendicular to each other, then the angle θ between them is 90°. The cosine of 90° is 0. Hence

If $\mathbf{a}$ and $\mathbf{b}$ are perpendicular, then $\mathbf{a} \cdot \mathbf{b} = 0$.

EXAMPLE 15.10

Find the value of x which will make $\begin{pmatrix} x \\ 2 \\ 3 \end{pmatrix}$ perpendicular to

$\begin{pmatrix} 2 \\ -1 \\ -4 \end{pmatrix}$.

Solution

Apply the formula for scalar product, and put it equal to 0.

$$\begin{pmatrix} x \\ 2 \\ 3 \end{pmatrix} \cdot \begin{pmatrix} 2 \\ -1 \\ -4 \end{pmatrix} = 2x - 2 - 12$$

$$2x - 2 - 12 = 0$$
$$x = 7$$

For $\begin{pmatrix} x \\ 2 \\ 3 \end{pmatrix}$ to be perpendicular to $\begin{pmatrix} 2 \\ -1 \\ -4 \end{pmatrix}$, x must be 7.

EXERCISE 15H

1 Which of the following vectors are perpendicular to each other?

   a) $\begin{pmatrix} 1 \\ 2 \end{pmatrix}$ **b)** $\begin{pmatrix} 2 \\ 1 \end{pmatrix}$ **c)** $\begin{pmatrix} 1 \\ -2 \end{pmatrix}$ **d)** $\begin{pmatrix} -2 \\ 1 \end{pmatrix}$ **e)** $4\mathbf{i} - 2\mathbf{j}$ **f)** $3\mathbf{i} - 6\mathbf{j}$

2 In each of the following, find x to ensure that the two vectors given are perpendicular.

   a) $\begin{pmatrix} 1 \\ 3 \end{pmatrix}$ and $\begin{pmatrix} x \\ 6 \end{pmatrix}$ **b)** $\begin{pmatrix} 4 \\ 8 \end{pmatrix}$ and $\begin{pmatrix} -12 \\ x \end{pmatrix}$ **c)** $\mathbf{i} + 2\mathbf{j}$ and $-2\mathbf{i} + x\mathbf{j}$

   d) $\begin{pmatrix} 1 \\ 2 \\ 1 \end{pmatrix}$ and $\begin{pmatrix} 2 \\ 1 \\ x \end{pmatrix}$ **e)** $\begin{pmatrix} 3 \\ 1 \\ 1 \end{pmatrix}$ and $\begin{pmatrix} 6 \\ -2 \\ x \end{pmatrix}$ **f)** $\mathbf{i} + \mathbf{j} + \mathbf{k}$ and $x\mathbf{i} + 2\mathbf{j} - \mathbf{k}$

3 Find x and y to ensure that $\begin{pmatrix} 1 \\ x \\ y \end{pmatrix}$ is perpendicular to both $\begin{pmatrix} 2 \\ -2 \\ 1 \end{pmatrix}$ and $\begin{pmatrix} 0 \\ 1 \\ -1 \end{pmatrix}$.

4 Find

   a) a unit vector perpendicular to $\begin{pmatrix} 3 \\ 4 \end{pmatrix}$ **b)** a vector of length 26, perpendicular to $\begin{pmatrix} 5 \\ 12 \end{pmatrix}$

c) a unit vector perpendicular to both $\begin{pmatrix} 2 \\ 0 \\ 2 \end{pmatrix}$ and $\begin{pmatrix} 1 \\ 1 \\ 0 \end{pmatrix}$

d) a unit vector perpendicular to both $\begin{pmatrix} 1 \\ 0 \\ 0 \end{pmatrix}$ and $\begin{pmatrix} 0 \\ 1 \\ 1 \end{pmatrix}$.

5 If $\mathbf{a} - \mathbf{b}$ is perpendicular to $\mathbf{a} + \mathbf{b}$, show that $|\mathbf{a}| = |\mathbf{b}|$.

6 If $|\mathbf{a} - \mathbf{b}| = |\mathbf{a} + \mathbf{b}|$, show that $\mathbf{a}$ is perpendicular to $\mathbf{b}$.

7 OABC is a rhombus (all sides equal). Letting $\overrightarrow{OA} = \mathbf{a}$ and $\overrightarrow{OC} = \mathbf{c}$, find expressions for $\overrightarrow{OB}$ and $\overrightarrow{AC}$ in terms of $\mathbf{a}$ and $\mathbf{c}$. Show that the diagonals of the rhombus are perpendicular.

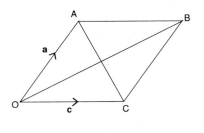

Fig. 15.21

*8 If $(\mathbf{a} \cdot \mathbf{b})\mathbf{c} = \mathbf{a}(\mathbf{b} \cdot \mathbf{c})$, what is the relationship between $\mathbf{a}$ and $\mathbf{c}$?

*9 The angle between $\mathbf{a}$ and $\mathbf{b}$ is $60°$, and $\mathbf{b}$ is perpendicular to $\mathbf{a} - \mathbf{b}$. Show that $|\mathbf{a}| = 2|\mathbf{b}|$.

15.5 Lines and vectors

Vectors can be used to describe lines in the plane and in space.

The section theorem

Suppose the position vectors of A and B are $\mathbf{a}$ and $\mathbf{b}$ respectively. Then the point P which divides AB in the ratio $\lambda : \mu$ has position vector

$$\frac{\mu \mathbf{a} + \lambda \mathbf{b}}{\mu + \lambda}$$

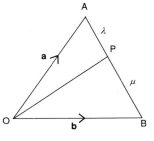

Fig. 15.22

Proof
The line segment AB is divided into $\mu + \lambda$ parts, of which AP occupies the first λ.

Hence $\overrightarrow{AP} = \dfrac{\lambda}{\lambda + \mu}(\mathbf{b} - \mathbf{a})$

The position vector of P is $\overrightarrow{OP}$, which is $\overrightarrow{OA} + \overrightarrow{AP}$.

$$\overrightarrow{OP} = \mathbf{a} + \frac{\lambda}{\mu + \lambda}(\mathbf{b} - \mathbf{a}) = \frac{\mu \mathbf{a} + \lambda \mathbf{a} + \lambda \mathbf{b} - \lambda \mathbf{a}}{\mu + \lambda} = \frac{\mu \mathbf{a} + \lambda \mathbf{b}}{\mu + \lambda}$$

Notes

1 Notice that the order of λ and μ has been reversed.

2 If P is the midpoint of AB, then its position vector is $\frac{1}{2}(\mathbf{a} + \mathbf{b})$.

3 This theorem is essentially the same as the one in Chapter 4, Section 4.4.

EXAMPLE 15.11

A, B and C are three points with position vectors **a**, **b** and **c** respectively. P and Q lie on AB and AC respectively, dividing them in the ratio $m : n$. Show that PQ is parallel to BC, and find the ratio of their lengths.

Solution

Use the theorem to find the position vectors of P and Q to be

$$\frac{n\mathbf{a} + m\mathbf{b}}{n + m} \quad \text{and} \quad \frac{n\mathbf{a} + m\mathbf{c}}{n + m}$$

Now find the vectors $\overrightarrow{BC}$ and $\overrightarrow{PQ}$.

$$\overrightarrow{BC} = \mathbf{c} - \mathbf{b}$$

$$\overrightarrow{PQ} = \frac{n\mathbf{a} + m\mathbf{c}}{n + m} - \frac{n\mathbf{a} + m\mathbf{b}}{n + m} = \frac{m\mathbf{c} - m\mathbf{b}}{n + m} = \frac{m}{n + m}(\mathbf{c} - \mathbf{b})$$

So $\overrightarrow{PQ} = \dfrac{m}{n + m} \overrightarrow{BC}$

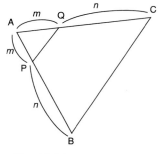

Fig. 15.23

PQ is parallel to BC, and the ratio of their lengths is $m : m + n$.

EXERCISE 15I

1 In each of the following, find the position vector of the point which

a) divides **a** and **b** in the ratio $3 : 5$ **b)** divides $\mathbf{i} + \mathbf{j}$ and $2\mathbf{j}$ in the ratio $1 : 2$

c) divides $\mathbf{i} + 2\mathbf{j} - \mathbf{k}$ and $-3\mathbf{i} + 6\mathbf{j} + 3\mathbf{k}$ in the ratio $1 : 3$.

2 Find the midpoints of the lines joining the points with these position vectors. Write down their position vectors.

a) $\mathbf{i} + \mathbf{j}$ and $3\mathbf{j}$ **b)** $3\mathbf{i} - 2\mathbf{j}$ and $\mathbf{i} + 4\mathbf{j}$

c) $\mathbf{i} + \mathbf{j} + \mathbf{k}$ and $3\mathbf{i} - \mathbf{j} + 3\mathbf{k}$ **d)** $4\mathbf{i} + 3\mathbf{j}$ and $\mathbf{j} - 2\mathbf{k}$

3 A, B, C and D are points with position vectors **a**, **b**, **c** and **d**. P, Q, R and S are the midpoints of AB, BC, CD and DA respectively. Find the position vectors of P, Q, R and S. Show that PQRS is a parallelogram.

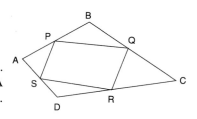

Fig. 15.24

*4 A, B and C are points with position vectors **a**, **b** and **c**. P and Q divide AB and AC respectively in the ratio $m : n$. Find the position vectors of P and Q.

PC and QB cross at R. Find the position vector of R, and the ratio in which it divides PC.

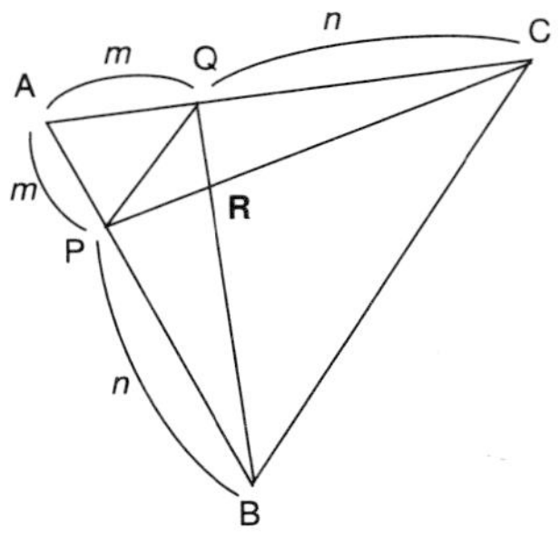

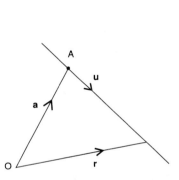

Fig. 15.25

Vector equations of lines

Suppose we have a line which goes through a point A, and is parallel to the vector **u**. Let the position vector of A be **a**. Then we can reach any point on the line by going to A and then moving parallel to **u**. So the position vector of a general point on the line is

$$\mathbf{r} = \mathbf{a} + t\mathbf{u}$$

Here t is a scalar. By making t negative, we can move in the direction opposite to **u**.

This is the **vector equation** of the line.

Suppose two lines are given by $\mathbf{r} = \mathbf{a} + t\mathbf{u}$ and $\mathbf{r} = \mathbf{b} + s\mathbf{v}$. We can see whether the lines intersect, by trying to solve the equation $\mathbf{a} + t\mathbf{u} = \mathbf{b} + s\mathbf{v}$. The angle between the lines is the angle between their directions, so we can find the angle between the lines by finding the angle between **u** and **v**.

Fig. 15.26

Cartesian equations of lines

The vector equation of a line in three dimensions can be converted to Cartesian equations.

Suppose the line has vector equation $\mathbf{r} = \mathbf{a} + t\mathbf{u}$, where $\mathbf{u} = \begin{pmatrix} u \\ v \\ w \end{pmatrix}$

and $\mathbf{a} = \begin{pmatrix} a \\ b \\ c \end{pmatrix}$.

Letting $\mathbf{r} = \begin{pmatrix} x \\ y \\ z \end{pmatrix}$, we have $\begin{pmatrix} x \\ y \\ z \end{pmatrix} = \begin{pmatrix} a \\ b \\ c \end{pmatrix} + t \begin{pmatrix} u \\ v \\ w \end{pmatrix}$. This gives

three ordinary equations.

$$x = a + tu \qquad y = b + tv \qquad z = c + tw$$

For each of these equations, we can make t the subject.

$$t = \frac{x - a}{u} \qquad t = \frac{y - b}{v} \qquad t = \frac{z - c}{w}$$

So the Cartesian equations of the line are

$$\frac{x - a}{u} = \frac{y - b}{v} = \frac{z - c}{w}$$

Notes

1 Note that the word 'equations' is plural. There are two equality signs, as it requires two equations to describe a line in three dimensions.

2 It is possible that u, for example, is zero. In this case we will get the expression

$$\frac{x - a}{0}$$

Normally it is not permissible to have a zero divisor. Here the expression implies that the value of x is constant at a.

EXAMPLE 15.12

A line goes through the point (2, 3) and is parallel to the vector $\mathbf{i} + 2\mathbf{j}$. Find the vector equation of this line. Find where this line meets the line with vector equation

$$\mathbf{r} = 4\mathbf{i} + \mathbf{j} + s(2\mathbf{i} - \mathbf{j})$$

Solution

The position vector of (2, 3) is $2\mathbf{i} + 3\mathbf{j}$, and $\mathbf{u} = \mathbf{i} + 2\mathbf{j}$. Using the method above, the vector equation of the line is

$$\mathbf{r} = 2\mathbf{i} + 3\mathbf{j} + t(\mathbf{i} + 2\mathbf{j})$$

When the two lines cross, the values of $\mathbf{r}$ are equal.

$$\mathbf{r} = 4\mathbf{i} + \mathbf{j} + s(2\mathbf{i} - \mathbf{j}) = 2\mathbf{i} + 3\mathbf{j} + t(\mathbf{i} + 2\mathbf{j})$$

Equating the coefficients of $\mathbf{i}$ and $\mathbf{j}$ gives us two equations.

$$4 + 2s = 2 + t$$
$$1 - s = 3 + 2t$$

We can solve these equations to find that $t = -\frac{2}{5}$. Put this value of t into the vector equation of the line.

$$\mathbf{r} = 2\mathbf{i} + 3\mathbf{j} - \tfrac{2}{5}(\mathbf{i} + 2\mathbf{j}) = \tfrac{8}{5}\mathbf{i} + \tfrac{11}{5}\mathbf{j}$$
$$= 1.6\mathbf{i} + 2.2\mathbf{j}$$

The intersection is at (1.6, 2.2).

EXAMPLE 15.13

Find the shortest distance from the origin to the line with vector equation

$$\mathbf{r} = 2\mathbf{i} + 3\mathbf{j} + t(2\mathbf{i} - \mathbf{j})$$

Solution

Let P lie on the line. P is as close as possible to the origin O when OP is perpendicular to the line, as shown in Fig. 15.27. Hence the vector OP is perpendicular to $2\mathbf{i} - \mathbf{j}$.

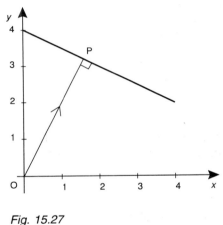

$$(2\mathbf{i} + 3\mathbf{j} + t(2\mathbf{i} - \mathbf{j})) \cdot (2\mathbf{i} - \mathbf{j}) = 0$$
$$4 - 3 + t(4 + 1) = 0$$
$$1 + 5t = 0$$
$$t = -\tfrac{1}{5}$$

Put this value of t into the equation of the line.

$$\mathbf{r} = 2\mathbf{i} + 3\mathbf{j} - \tfrac{1}{5}(2\mathbf{i} - \mathbf{j}) = \tfrac{8}{5}\mathbf{i} + \tfrac{16}{5}\mathbf{j} = 1.6\mathbf{i} + 3.2\mathbf{j}$$

P is at (1.6, 3.2).

Now find the modulus of OP.

$$|\overrightarrow{OP}| = \sqrt{1.6^2 + 3.2^2} = \sqrt{12.8} = 3.58$$

The shortest distance OP is 3.58.

Fig. 15.27

EXAMPLE 15.14

The vector equation of a line in space is $\mathbf{r} = \mathbf{i} + 2\mathbf{j} + t(\mathbf{i} + \mathbf{k})$. The Cartesian equation of a second line is

$$\frac{x - 0}{0} = \frac{y - 1}{1} = \frac{z - 0}{-1}$$

Convert the equation of the first line to Cartesian form, and the equations of the second line to vector form. Do the lines intersect?

Solution

Writing the first line as $\mathbf{r} = \mathbf{a} + t\mathbf{u}$, we have $\mathbf{a} = \mathbf{i} + 2\mathbf{j}$ and $\mathbf{u} = \mathbf{i} + \mathbf{k}$. Convert to Cartesian form.

The Cartesian equations are $\dfrac{x - 1}{1} = \dfrac{y - 2}{0} = \dfrac{z - 0}{1}$.

For the second line, we see that

$$\mathbf{a} = \begin{pmatrix} 0 \\ 1 \\ 0 \end{pmatrix} \text{ and } \mathbf{u} = \begin{pmatrix} 0 \\ 1 \\ -1 \end{pmatrix}$$

Convert to vector form.

The vector equation is $\mathbf{r} = \mathbf{j} + s(\mathbf{j} - \mathbf{k})$

To find whether the lines intersect, we shall use the vector forms.
Equate the two expressions for $\mathbf{r}$.

$$\mathbf{i} + 2\mathbf{j} + t(\mathbf{i} + \mathbf{k}) = \mathbf{j} + s(\mathbf{j} - \mathbf{k})$$

This is equivalent to three ordinary equations:

$$1 + t = 0 \qquad \text{(equating } \mathbf{i} \text{ terms)}$$
$$2 = 1 + s \qquad \text{(equating } \mathbf{j} \text{ terms)}$$
$$t = -s \qquad \text{(equating } \mathbf{k} \text{ terms)}$$

From the first equation, $t = -1$. From the second equation, $s = 1$.
These values satisfy the third equation.

The lines do intersect.

EXERCISE 15J

1 Find the vector equations of the following lines. Find the Cartesian equations of the lines in three dimensions.

a) through $(3, 1)$, parallel to $\mathbf{i} + 2\mathbf{j}$

b) through $(-1, 1)$, parallel to $2\mathbf{i} - \mathbf{j}$

c) through $(1, 2, 1)$, parallel to $\mathbf{i} + 2\mathbf{j} - \mathbf{k}$

d) through $(1, 1)$, perpendicular to $\mathbf{i} + \mathbf{j}$

e) through $(1, 2)$, perpendicular to $3\mathbf{i} - \mathbf{j}$

f) through $(1, 2)$ and $(3, 1)$

g) through $(-1, -2)$ and $(3, 4)$

h) through $(1, 2, 0)$ and $(3, 1, 2)$

i) through $(1, 3)$, parallel to $\mathbf{r} = 3\mathbf{i} + 2\mathbf{j} + t(\mathbf{i} - \mathbf{j})$

j) through $(1, 0)$, perpendicular to $\mathbf{r} = 3\mathbf{i} - 4\mathbf{j} + t(\mathbf{i} + \mathbf{j})$

k) through $(1, 2, 1)$, parallel to $\mathbf{r} = \mathbf{i} + \mathbf{j} + \mathbf{k} + t(\mathbf{i} + 2\mathbf{j} - 3\mathbf{k})$

2 Find the intersection points of the following pairs of lines.

a) $\mathbf{r} = \mathbf{i} + \mathbf{j} + t(\mathbf{i} - \mathbf{j})$ and $\mathbf{r} = \mathbf{i} - \mathbf{j} + s(\mathbf{i} + \mathbf{j})$ **b)** $\mathbf{r} = 2\mathbf{i} + 3\mathbf{j} + t(\mathbf{i} - 2\mathbf{j})$ and $\mathbf{r} = 2\mathbf{j} + s(2\mathbf{i} + \mathbf{j})$

3 Show that the pairs of lines below intersect, and find their intersection points.

a) $\mathbf{r} = \mathbf{i} + 3\mathbf{j} + 2\mathbf{k} + t(2\mathbf{i} - \mathbf{j} + \mathbf{k})$ and $\mathbf{r} = \mathbf{i} + \mathbf{k} + s(\mathbf{i} + \mathbf{j} + \mathbf{k})$

b) $\dfrac{x - 3}{0} = \dfrac{y - 0}{1} = \dfrac{z + 1}{1}$ and $\dfrac{x + 1}{1} = \dfrac{y - 1}{0} = \dfrac{z - 0}{0}$

4 Find the value of x which will ensure that the following pair of lines meet.

$$\mathbf{r} = \mathbf{i} + \mathbf{j} + \mathbf{k} + t(\mathbf{i} - \mathbf{j} + \mathbf{k}) \text{ and } \mathbf{r} = 2\mathbf{i} + 2\mathbf{j} + x\mathbf{k} + s(\mathbf{i} + \mathbf{j} + \mathbf{k})$$

5 For each of the following, find the shortest distance between the point and the line.

a) $(0, 0)$ and $\mathbf{r} = \mathbf{i} + \mathbf{j} + t(\mathbf{i} - \mathbf{j})$ **b)** $(0, 1)$ and $\mathbf{r} = 2\mathbf{i} - \mathbf{j} + t(\mathbf{i} + 3\mathbf{j})$

6 The position vectors of A, B and C are $\mathbf{i} + \mathbf{j}$, $2\mathbf{i} + 3\mathbf{j}$ and $4\mathbf{i} - \mathbf{j}$ respectively. Find vector equations for the line BC, and for the perpendicular from A to BC. Find where this perpendicular meets BC.

7 Find the angles between the following pairs of lines.

a) $\mathbf{r} = 2\mathbf{i} + \mathbf{j} + t(\mathbf{i} + \mathbf{j})$ and $\mathbf{r} = -\mathbf{i} - \mathbf{j} + s(2\mathbf{i} - \mathbf{j})$

b) $\mathbf{r} = \mathbf{i} - 3\mathbf{j} + t(-2\mathbf{i} + 3\mathbf{j})$ and $\mathbf{r} = \mathbf{i} - 2\mathbf{j} + s(\mathbf{i} + 2\mathbf{j})$

c) $\mathbf{r} = \mathbf{i} + \mathbf{j} + \mathbf{k} + t(\mathbf{i} - \mathbf{j} + \mathbf{k})$ and $\mathbf{r} = \mathbf{i} + \mathbf{j} + \mathbf{k} + t(-\mathbf{i} + \mathbf{j} - \mathbf{k})$

d) $\dfrac{x+2}{1} = \dfrac{y+1}{2} = \dfrac{z+2}{-1}$ and $\dfrac{x-1}{1} = \dfrac{y-0}{0} = \dfrac{z-2}{2}$

LONGER EXERCISE

Centres of a triangle

The Longer exercise of Chapter 4 concerned three centres of a triangle. Here we return to the same problems, but using vector notation. We shall see that the same results can be found much more easily.

The vertices of the triangle are A, B and C, with position vectors **a**, **b** and **c**.

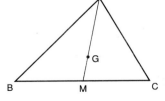

Centroid

Find the midpoint M of BC. Find the point G which divides AM (a median of the triangle) in the ratio $2:1$. Simplify your expression as far as possible. What you can say about the other medians of the triangle?

Fig. 15.28

Orthocentre

Suppose the perpendiculars from A to BC and from B to AC meet at D, with position vector **d**, so we have

$$(\mathbf{d} - \mathbf{a}) \cdot (\mathbf{c} - \mathbf{b}) = 0 \text{ and } (\mathbf{d} - \mathbf{b}) \cdot (\mathbf{c} - \mathbf{a}) = 0$$

Show that $(\mathbf{d} - \mathbf{c}) \cdot (\mathbf{b} - \mathbf{a}) = 0$.

What can you conclude about the perpendicular from C to AB?

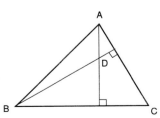

Fig. 15.29

Circumcentre
Suppose that the origin of coordinates is at the circumcentre of ABC. Then A, B and C are equidistant from O, so $|\mathbf{a}| = |\mathbf{b}| = |\mathbf{c}|$. Show that the orthocentre has position vector $\mathbf{a} + \mathbf{b} + \mathbf{c}$. What is the relationship between the three centres?

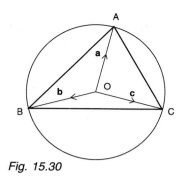

Fig. 15.30

EXAMINATION QUESTIONS

1 The position vectors of the vertices of triangle ABC are $\mathbf{a}$, $\mathbf{b}$ and $\mathbf{c}$ respectively. P is the point on AB such that $AP:PB = 3:2$, Q is the point on AC such that $AQ:QC = 3:1$.

(i) Write down the position vectors of P and Q in terms of $\mathbf{a}$, $\mathbf{b}$ and $\mathbf{c}$.

(ii) R is the point on PQ such that $PR:RQ = 4:5$. Find the position vector of R in terms of $\mathbf{a}$, $\mathbf{b}$ and $\mathbf{c}$, expressing your answer in as simple a form as possible.

SMP 1991

2 The line l_1 is parallel to the vector $\mathbf{i} + 2\mathbf{k}$ and passes through the point A with position vector $3\mathbf{i} + 2\mathbf{j} + 5\mathbf{k}$ relative to an origin O. Find the position vector of the point P on l_1 such that OP is perpendicular to l_1. Hence find the perpendicular distance from O to l_1.

A second line l_2 is parallel to the vector $\mathbf{i} + \mathbf{j} + 3\mathbf{k}$ and passes through the point B with position vector $2\mathbf{i} + 3\mathbf{j} + p\mathbf{k}$, where p is a constant. Given that the lines l_1 and l_2 meet at C, find the value of p and the position vector of C.

Write down expressions for the vectors $\overrightarrow{AC}$ and $\overrightarrow{BC}$ and hence show that

$$\cos A\widehat{C}B = \frac{7}{\sqrt{55}}$$

AEB 1991

3 The points A, C have position vectors $\mathbf{a}$, $\mathbf{c}$ with respect to an origin O. D is a point on AC which divides AC in the ratio $\lambda : 1 - \lambda$, where $0 < \lambda < 1$. Show from first principles that D has position vector

$$(1 - \lambda)\mathbf{a} + \lambda\mathbf{c}$$

B is a further point such that OB is a diagonal of the parallelogram OABC. The point E divides AB in the ratio $1:3$ and the line ED meets OA at F.

(i) Show that E has position vector $\mathbf{a} + \frac{1}{4}\mathbf{c}$.

Denoting the position vector of F by $\mu\mathbf{a}$, show that

(ii) $\overrightarrow{FD} = (1 - \lambda - \mu)\mathbf{a} + \lambda\mathbf{c}$ (iii) $\overrightarrow{FE} = (1 - \mu)\mathbf{a} + \frac{1}{4}\mathbf{c}$.

By noting that $\overrightarrow{FD}$ is a scalar multiple of $\overrightarrow{FE}$, show that

$$\mu = \frac{1 - 5\lambda}{1 - 4\lambda}$$

(iv) Find the value of λ such that F is the mid-point of OA.

(v) Find the range of values of λ such that F lies between O and A.

W 1992

Summary and key points

1 A vector is a quantity with direction as well as magnitude. Vectors can be added and multiplied by scalars. Two vectors are parallel if one is a scalar multiple of the other. The position vector of a point is the vector from the origin to the point.

2 Vectors can be represented in terms of components. Vectors are added by adding the components. A vector can be multiplied by a scalar, by multiplying each of the components by the scalar.

3 The modulus of a vector is given by

$$\left| \begin{pmatrix} x \\ y \end{pmatrix} \right| = \sqrt{x^2 + y^2}$$

A unit vector has modulus 1. The unit vectors **i**, **j** and **k** are parallel to the x, y and z-axes respectively.

4 The scalar product of two vectors **a** and **b** is

$$\mathbf{a} \cdot \mathbf{b} = |\mathbf{a}|\,|\mathbf{b}|\cos\theta$$

where θ is the angle between them.

$$\begin{pmatrix} x \\ y \end{pmatrix} \cdot \begin{pmatrix} a \\ b \end{pmatrix} = xa + yb$$

The component of **u** in the direction **v** is $\dfrac{\mathbf{u} \cdot \mathbf{v}}{|\mathbf{v}|}$.

If two vectors are perpendicular, then their scalar product is zero.

5 If P divides the points with position vectors **a** and **b** in the ratio $\lambda : \mu$, then its position vector is $\dfrac{(\mu\mathbf{a} + \lambda\mathbf{b})}{(\lambda + \mu)}$.

If a line is parallel to **u** and passes through a point with position vector **a**, then its vector equation is

$$\mathbf{r} = \mathbf{a} + t\mathbf{u}$$

where t is a scalar.

If $\mathbf{a} = \begin{pmatrix} a \\ b \\ c \end{pmatrix}$ and $\mathbf{u} = \begin{pmatrix} u \\ v \\ w \end{pmatrix}$, the Cartesian equations of the line are

$$\frac{x - a}{u} = \frac{y - b}{v} = \frac{z - c}{w}$$

Graphs

Throughout your mathematical career, you will have drawn many graphs. The procedure is to draw up a table, with values of x and the corresponding values of y. The points (x, y) are then plotted on graph paper and the points are joined up with a smooth curve. This procedure is known as **plotting** the graph.

Often we want to know the general shape of a graph rather than the exact points through which it passes. Then we just want a curve which picks out the important features of the graph. The procedure of drawing this curve is called **sketching** the graph.

By sketching the graph of a function, we can see how that function behaves.

16.1 Sketching graphs

When sketching a graph, we are interested in showing its general behaviour. The following features are significant.

1　Are there any values of x for which the function is not defined?
2　What happens as x tends to plus or minus infinity?
3　Where does the graph cross the axes?
4　Is there any symmetry of the graph?
5　Are there stationary points of the graph?

Notes
1　A function cannot be evaluated if it involves dividing by zero or taking the square root of a negative number. So $f(x) = \frac{1}{x}$ is not defined at $x = 0$, and $f(x) = \sqrt{x}$ is not defined if x is negative.

2　If $f(x)$ is a polynomial function, then its behaviour as x tends to infinity depends on the greatest power of x. If n is even, x^n

tends to plus infinity when x tends to plus or minus infinity, Fig. 16.1. If n is odd, x^n tends to infinity when x tends to infinity, and to minus infinity when x tends to minus infinity, Fig. 16.2. If x^n is multiplied by a constant a, the sign of a must be taken into account.

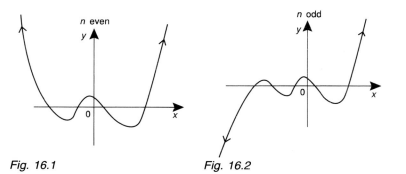

Fig. 16.1   *Fig. 16.2*

Fig. 16.3

3 The graph crosses the y-axis when $x = 0$, at B in Fig. 16.3. It crosses the x-axis when $y = 0$, at A and C. Finding the co-ordinates may involve solving an equation.

4 If $f(x)$ involves only even powers of x, $f(-x) = f(x)$. The graph will be symmetric about the y-axis. If it involves only odd powers of x, $f(-x) = -f(x)$. The graph will have rotational symmetry about the origin of order 2. See Fig. 16.4.

5 To find the exact position of the stationary points, differentiate and put $f'(x) = 0$. Often we can find the approximate position of the stationary points by considering the other four features.

EXAMPLE 16.1
Sketch the graph of $y = x(x-1)(x-2)$.

Solution
Take the features listed above in turn.

1 y is defined for all x.

2 When the brackets are expanded, the greatest power of x will be x^3. This will tend to plus infinity when x tends to plus infinity, and to minus infinity when x tends to minus infinity.

3 The graph crosses the x-axis at $x = 0$, $x = 1$ and $x = 2$. It crosses the y-axis at $(0, 0)$.

4 There is no obvious symmetry of the graph.

5 The graph crosses the x-axis at $x = 0$, $x = 1$ and $x = 2$. Hence there will be stationary points between these values. We do not need to find their exact position.

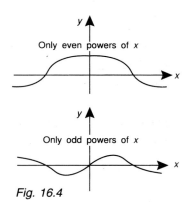

Only even powers of x

Only odd powers of x

Fig. 16.4

Now to sketch the graph. It comes up from minus infinity, goes through the origin, turns down to cross the x-axis again at $(1, 0)$, turns up to cross again at $(2, 0)$, then goes up to infinity. A sketch is shown in Fig. 16.5.

EXAMPLE 16.2

Sketch the graph of $y = 2 + \dfrac{1}{x^2}$.

Solution

The graph is not defined for $x = 0$. When x is very small, $\dfrac{1}{x^2}$ will be very large. It does not cross either of the axes. It is symmetrical about the y-axis. When x tends to infinity $\dfrac{1}{x^2}$ tends to 0, and hence y tends to 2. Differentiating, $\dfrac{dy}{dx} = -\dfrac{2}{x^3}$, which is never zero. Hence there are no stationary points.

The curve will come from $y = 2$ and shoot up to infinity as x approaches 0. The graph will be symmetrical about the y-axis. The graph is sketched in Fig. 16.6.

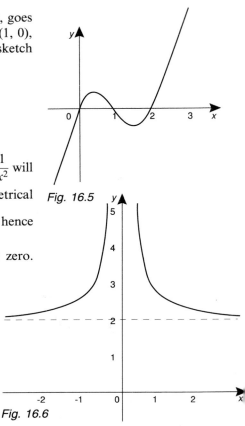

Fig. 16.5

Fig. 16.6

EXERCISE 16A

1 Sketch the graphs of the following functions.

a) $y = (x - 1)(x + 2)$

b) $y = x(x + 3)$

c) $y = x(x - 1)(x + 2)$

d) $y = x(4 - x)$

e) $y = x^2(x - 3)$

f) $y = (2 - x)x$

g) $y = (1 + x)^2(x - 1)$

h) $y = 1 + \dfrac{1}{x}$

i) $y = 3 + \dfrac{2}{x}$

j) $y = \dfrac{1}{x - 1}$

k) $y = 2 - \dfrac{1}{x - 1}$

l) $y = \dfrac{1}{x^2}$

m) $y = 2 - \dfrac{1}{(x - 1)^2}$

n) $y = \sqrt{x + 3}$

o) $y = \sqrt{x - 3}$

p) $y = \sqrt{2 + x}$

q) $y = \sqrt{2 - x}$

*2 Draw graphs with the following features.

a) Through $(-1, 0)$, $(1, 0)$, with a minimum at $(0, -1)$, symmetric about the y-axis, tending to plus infinity as x tends to infinity.

b) Symmetrical about the y-axis, with a maximum at $(0, 2)$, never crossing the x-axis, tending to 0 as x tends to infinity.

c) Not defined at $x = -1$, crossing the y-axis at $(0, 1)$, never crossing the x-axis, tending to zero as x tends to plus or minus infinity.

3 The graph of $y = \dfrac{ax + b}{x + 2}$ is drawn. When x tends to infinity y tends to 2. The graph goes through the point $(1, 4)$. Find a and b.

16.2 Standard graphs

The graphs of linear, quadratic and reciprocal functions follow fixed patterns.

Linear functions

These are functions of the form $y = mx + c$, where m and c are constant. The graph of such a function is a **straight line**, with gradient m, crossing the y-axis at $(0, c)$.

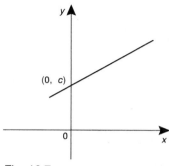

Fig. 16.7

Quadratic functions

These are functions of the form $y = ax^2 + bx + c$, where a, b and c are constant. The graph of such a function is a **parabola**, as shown.

These graphs can be sketched by completing the square, as described in Chapter 2. If the function is of the form $y = a(x - k)^2 + j$, then the following facts hold.

There is a stationary point at (k, j), which is a minimum if $a > 0$, or a maximum if $a < 0$.

The line of symmetry of the graph is $x = k$.

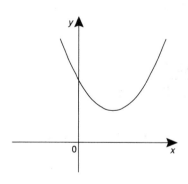

Fig. 16.8

Reciprocal functions

These are functions of the form $y = a + \dfrac{b}{x - c}$. For $x = c$, y is not defined. As x tends to infinity in either direction the graph tends to $y = a$.

The vertical line $x = c$ and the horizontal line $y = a$ are **asymptotes**. The graph approaches these lines but never meets them.

EXAMPLE 16.3

Sketch the graph of $y = x^2 - 4x - 2$.

Solution

Complete the square for this quadratic expression.

$$y = (x - 2)^2 - 6$$

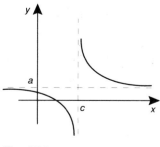

Fig. 16.9

The minimum value of this graph is -6, achieved when $x = 2$. Hence there is a minimum at $(2, -6)$. It crosses the y-axis at $(0, -2)$. The graph is sketched in Fig. 16.10.

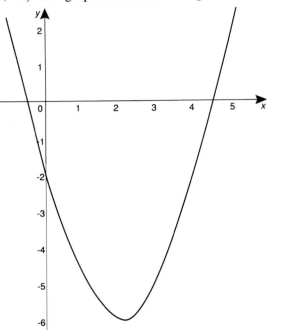

Fig. 16.10

EXAMPLE 16.4

Find the equation of the graph shown, given that it is of the form

$$y = a + \frac{1}{x - b}$$

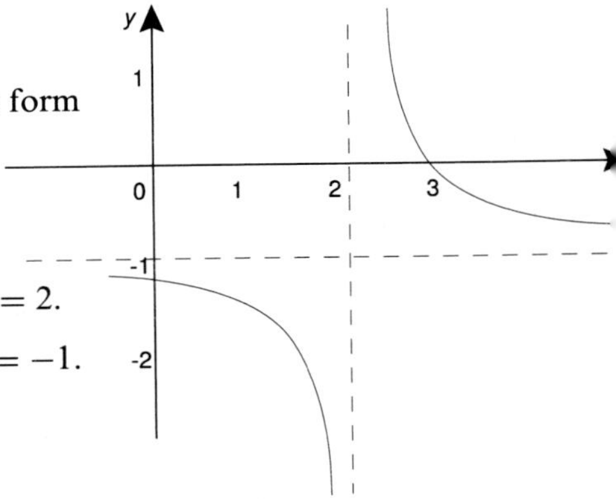

Solution

Notice that the function is not defined for $x = 2$. Hence $b = 2$.

Notice also that as x increases, y approaches -1. Hence $a = -1$.

The equation is $y = -1 + \dfrac{1}{x - 2}$.

Fig. 16.11

EXERCISE 16B

1 Sketch the graphs of the following functions.

  a) $y = 2x - 1$ **b)** $y = 3x + 2$ **c)** $y = 1 - 3x$ **d)** $y = 4 - \dfrac{1}{2}x$

  e) $y = x^2 - 2x - 2$ **f)** $y = x^2 + 4x - 1$ **g)** $y = x^2 + x - 3$

  h) $y = 4 + 2x - x^2$ **i)** $y = 3 - 8x - 2x^2$ **j)** $y = \dfrac{1}{x - 1}$

  k) $y = 4 + \dfrac{2}{x - 3}$ **l)** $y = 1 - \dfrac{1}{x + 1}$ **m)** $y = \dfrac{3}{x + 1} - 2$

2 Find the equations of the graphs below, given that they are of the form $y = x^2 + bx + c$.

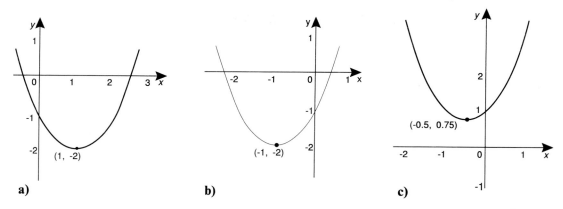

a) **b)** **c)**

3 Find the equations of the graphs below, given that they are of the form $y = a + \dfrac{1}{x - b}$.

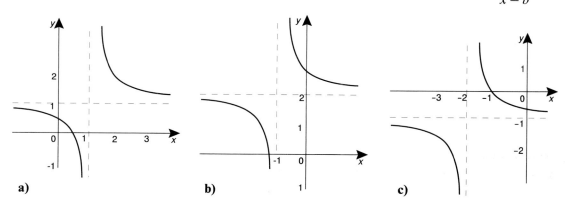

a) **b)** **c)**

*4 Find the equations of the quadratic functions which have the following graphs:

a) with a minimum at $(1, -4)$, through $(0, 6)$

b) with a maximum at $(1, 2)$, through $(0, 0)$.

*5 Find the functions of the form $a + \dfrac{b}{(x - c)}$ for which the graph

a) is not defined at $x = 3$, tends to 4 as x tends to infinity and goes through $(0, 6)$

b) is not defined for $x = -1$, tends to -1 as x tends to infinity and goes through $(0, 1)$.

16.3 Transformations of graphs

Suppose we are given the graph of $y = x^3$ and want to draw the graph of a similar cubic, such as $y = 2x^3$, or $y = (x + 1)^3$. It is possible to draw the new graph by making a suitable transformation of the old graph.

This applies to other functions besides cubics. In the following a is a constant.

1 The graph of $y = f(x) + a$ is drawn by shifting the original curve a units upwards. If a is negative the graphs will be moved downwards.

2 The graph of $y = f(x + a)$ is drawn by shifting the original curve a units to the left (because everything happens a units earlier). If a is negative the graph will be shifted to the right.

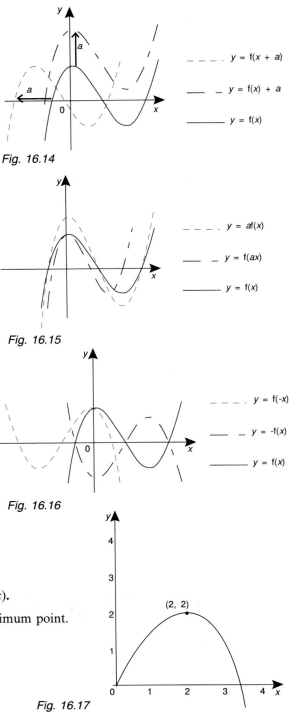

- - - - $y = f(x + a)$

— — $y = f(x) + a$

——— $y = f(x)$

Fig. 16.14

3 The graph of $y = af(x)$ is drawn by stretching the original graph vertically by a factor of a.

4 The graph of $y = f(ax)$ is drawn by compressing the original graph horizontally by a factor of a.

- - - - $y = af(x)$

— — $y = f(ax)$

——— $y = f(x)$

Fig. 16.15

5 The graph of $y = -f(x)$ is drawn by reflecting the original graph in the x-axis.

6 The graph of $y = f(-x)$ is drawn by reflecting the original graph in the y-axis.

- - - - $y = f(-x)$

— — $y = -f(x)$

——— $y = f(x)$

Fig. 16.16

EXAMPLE 16.5

Figure 16.17 shows the graph of $y = f(x)$.

Make sketches of **a)** $y = f(x - 2)$ **b)** $y = f(2x)$.

In both cases indicate the position of the maximum point.

(2, 2)

Fig. 16.17

Solution

a) Shift the graph to the left by -2 units, i.e. to the right by 2 units.

The result is shown.

Note that the maximum point is at (4, 2).

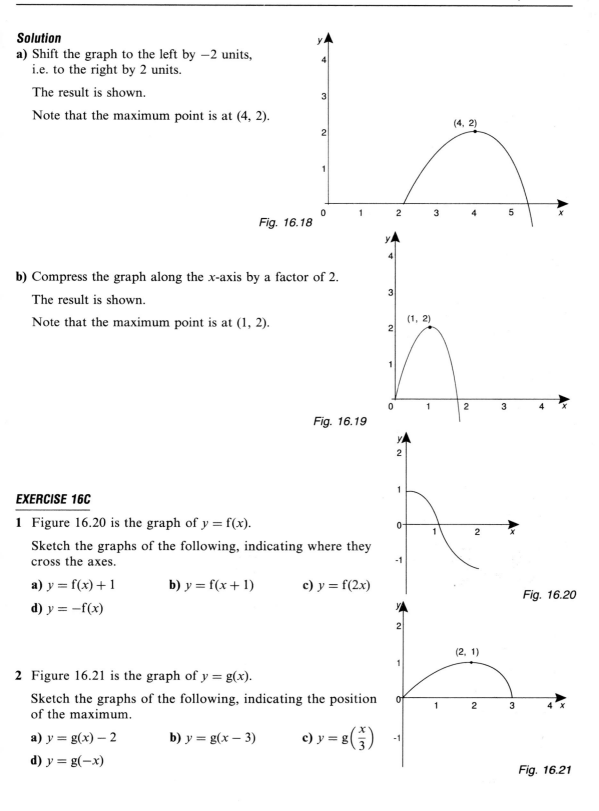

Fig. 16.18

b) Compress the graph along the *x*-axis by a factor of 2.

The result is shown.

Note that the maximum point is at (1, 2).

Fig. 16.19

EXERCISE 16C

1 Figure 16.20 is the graph of $y = f(x)$.

Sketch the graphs of the following, indicating where they cross the axes.

a) $y = f(x) + 1$ **b)** $y = f(x + 1)$ **c)** $y = f(2x)$

d) $y = -f(x)$

Fig. 16.20

2 Figure 16.21 is the graph of $y = g(x)$.

Sketch the graphs of the following, indicating the position of the maximum.

a) $y = g(x) - 2$ **b)** $y = g(x - 3)$ **c)** $y = g\left(\dfrac{x}{3}\right)$

d) $y = g(-x)$

Fig. 16.21

3 Figure 16.22 shows the graphs of **a)** $y = f(x)$ **b)** $y = f(x + b)$ **c)** $y = f(cx)$
Find b and c.

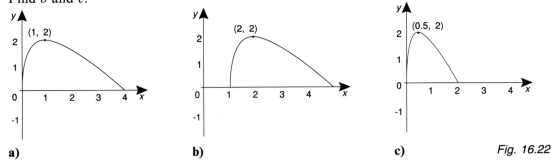

a)   b)   c)   *Fig. 16.22*

4 Describe the transformations which take the graph of $y = x^2$ to the graph of

a) $y = x^2 + 1$ **b)** $y = (x - 2)^2$ **c)** $y = -x^2$

5 Describe the transformations which take the graph of $y = \dfrac{1}{x}$ to the graph of

a) $y = \dfrac{3}{x}$ **b)** $y = \dfrac{1}{x + 1}$ **c)** $y = -\dfrac{1}{x}$

*6 Describe each set of successive transformations which take the graph of $y = x^2$ to the graphs of

a) $y = (x + 1)^2 - 3$ **b)** $y = 2(x + 1)^2 - 1$ **c)** $y = 3x^2 - 12x + 7$

*7 Describe each set of successive transformations which will take the graph of $y = \dfrac{1}{x}$ to the graphs of

a) $y = 2 + \dfrac{3}{x}$ **b)** $y = 1 - \dfrac{1}{2x}$ **c)** $y = \dfrac{x + 3}{x + 1}$

16.4 Reduction of graphs to linear form

The simplest sorts of graph are those which can be drawn with a ruler. These are linear functions of the form $y = mx + c$. The value of m is the gradient of the line, and the value of c is the intercept on the y-axis.

Suppose x and y are related by a non-linear function, with constants which we do not know. Suppose that we are given approximate values of x and y. It can be helpful to reduce a graph to a linear form by a transformation of the variables. Then we can plot a straight line graph and, from the gradient and intercept, find the unknown constants of the function.

Some examples of transformations are below.

$y = ax^2 + c$

Letting $X = x^2$, the graph of y against X will be the linear one of $y = aX + c$.

$y = ax^n$

Take logs of both sides, to obtain

$$\log y = \log(ax^n) = \log a + n \log x$$

Let $X = \log x$ and $Y = \log y$. The graph of Y against X will be the linear one of $Y = \log a + nX$.

$y = ab^x$

Take logs of both sides.

$$\log y = \log(ab^x) = \log a + x \log b$$

Let $Y = \log y$. The graph of Y against x will be the linear one of $Y = \log a + x \log b$.

EXAMPLE 16.6

The variables x and y are related by the equation $y = ax^2 + bx$.

Values of x and y are given in the table below. By plotting $\frac{y}{x}$ against x find the constants a and b.

x	1	2	3	4	5	6	7	8
y	1	7	20	40	64	93	130	170

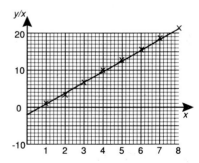

Fig. 16.23

Solution

Write in an extra row for the values of $\frac{y}{x}$.

$\frac{y}{x}$	1	3.5	6.7	10	12.8	15.5	18.6	21.3

Let $Y = \frac{y}{x}$. The graph of Y against x is shown in Fig. 16.23.

Notice that the gradient is 2.9 and the intercept is -2.

The equation of the straight line is $Y = 2.9x - 2$.

This gives values of $a = 2.9$ and $b = -2$.

EXAMPLE 16.7

It is thought that y is proportional to a power of x. A table of values of x and y is given below. By a suitable trasformation change the relationship to a linear one. Plot the transformed values, and hence find the power and the equation giving y in terms of x.

x	2	3	4	5	6	7
y	0.11	0.25	0.46	0.73	1.08	1.49

Solution

Let us suppose that the proportionality relationship is $y = ax^n$, where n is the power and a the constant of proportionality. If we take logs of both sides, we obtain the relationship

$$\log y = \log a + n \log x$$

If we write $X = \log x$ and $Y = \log y$, then there is a linear relationship between X and Y. Draw up a new table showing the values of X and Y.

X	0.30	0.48	0.60	0.70	0.78	0.85
Y	−0.96	−0.60	−0.34	−0.14	0.03	0.17

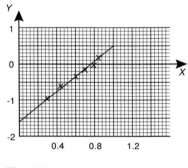

Plot Y against X. Draw a straight line through the points. Notice that the gradient of the graph is 2.1, and the intercept on the Y axis is -1.6. We now have the equation giving Y in terms of X.

$$Y = 2.1X - 1.6$$

Hence $n = 2.1$. Comparing with the above, $\log a = -1.6$, giving $a = 10^{-1.6} = 0.025$.

The power is 2.1, and the equation is $y = 0.025x^{2.1}$.

Fig. 16.24

EXERCISE 16D

1 In each of the cases below, reduce the equation given to linear form. Plot the transformed variables and find the unknown constants.

a) $y = ax^2 + b$

x	1.1	1.4	1.6	1.8	1.9	2.1
y	−0.4	1.9	3.7	5.7	6.8	9.2

b) $y = ax^3 + bx$

x	2	2.1	2.2	2.3	2.4	2.5	2.6
y	5.8	6.3	6.9	7.6	8.2	8.9	9.7

c) $y = ax + \dfrac{b}{x}$

x	1	1.5	2	2.5	3	3.5	4	4.5
y	11	9	10	11	12	13	14	15

d) $y = ax^b$

x	1	2	3	4	5	6	7	8
y	4	15	30	60	90	120	160	210

e) $y = ab^x$

x	1	2	3	4	5	6	7	8
y	8	10	13	17	22	28	37	50

f) $\dfrac{1}{y} = \dfrac{a}{x} + b$

x	1	2	3	4	5	6	7	8
y	0.3	1.8	-2.9	-1.3	-1	-0.8	-0.7	-0.7

2 Suppose a car is travelling at v m s^{-1}. The distance it takes to brake to a halt is s metres, where $s = av^2 + bv$. A table of stopping distances is given below. Transform the variables so that the equation is linear, plot the transformed variables and find a and b.

v	5	10	15	20	25	30
s	8	22	40	60	90	120

LONGER EXERCISE

Kepler's law

'If I have seen further, it is by standing on the shoulders of giants.'

Newton

One of the giants who preceded Newton was Johannes Kepler. His law relating the distance of a planet from the Sun to the length of its year was one of the results which led towards Newton's law of gravity.

Let the distance of the planet be D millions of miles, and the length of its year be L Earth days.

Figures for eight planets are

	Mercury	Venus	Earth	Jupiter	Saturn	Uranus	Neptune	Pluto
D	36	68	94	486	892	1794	2811	3692
L	88	225	365	4329	10753	30660	60150	90670

Kepler's law says that D is proportional to a certain power of L. By the methods of this chapter, find the power and the equation giving D in terms of L.

Mars is 142 000 000 miles from the Sun. How long is the Martian year?

Kepler's law applies to other bodies besides the planets and the Sun. Jupiter has twelve moons. The table below gives the distances in millions of miles of these moons from Jupiter and the length in Earth days of their months. The first four, with names, were discovered by Galileo (another of the giants). The rest, numbered in order of discovery, were found more recently.

	Io	Europa	Ganymede	Callisto	5	6	7	8	9	10	11	12
D	0.264	0.420	0.670	1.18	0.113	7.19	7.34	14.7	14.8	7.34	14.1	13.3
L	0.074	0.148	0.298	0.695	0.021	10.4	10.8	30.8	31.0	10.8	28.8	26.0

Find the law connecting these quantities.

EXAMINATION QUESTIONS

1 In an experiment to determine the relationship between the resistance to motion, R newtons, of a plank towed through water and its speed, V m s^{-1}, the following data were recorded.

V	1.8	3.7	5.5	9.2	11
R	2.01	7.32	15.1	38.9	56.2

Assuming that $R = kV^n$, where n and k are constants,

a) obtain a relation between $\log R$ and $\log V$,

b) by drawing a graph of $\log R$ against $\log V$ estimate, to 2 significant figures, the values of n and k.

L AS 1990

2 Figure 16.25 shows the curve with equation $y = f(x)$ for $0 \le x \le 2a$. It is given that $f(4a - x) = -f(x)$.

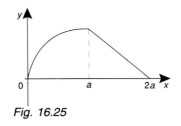

Fig. 16.25

a) Sketch the curve with equation $y = f(x)$ for $0 \le x \le 4a$.

Sketch, on separate axes, the curves with equations

b) $y = f(2x)$, $\quad 0 \le x \le 2a$

c) $y = f(4a - x)$, $\quad 0 \le x \le 4a$

indicating in each case the coordinates of the points where the curve meets the x-axis.

L 1991

3 The variables x and y satisfy the relation

$$y = ab^x$$

where a and b are constants. Reduce this relation to a linear form.

The table shows the combined population of England and Wales at the beginning of each of the last five centuries.

Year	1501	1601	1701	1801	1901
Population (in thousands)	2350	3750	5826	9156	32528

Taking y as the population (in thousands) and x as the number of centuries after 1501, make up a table to show corresponding values of x and $\log y$. Hence plot a suitable graph to investigate whether a relation of the form

$$y = ab^x$$

is consistent with the data in the above table.

Estimate the population in the year 1651.

Comment briefly on the usefulness of the relation $y = ab^x$ for predicting the future population.

J 1991

4 Given that

$$f(x) = (x - \alpha)(x + \beta), \quad \alpha > \beta > 0$$

sketch on separate diagrams the curves with equations

a) $y = f(x)$

b) $y = -f(x + \alpha)$.

On each sketch

c) write the coordinates of any points at which the curve meets the coordinate axes,

d) show with a dotted line the axis of symmetry of the curve and state its equation.

L 1992

Summary and key points

1 The sketch of a graph of a function can be drawn by considering key features of the function.

2 Graphs of linear, quadratic and reciprocal functions have standard shapes.

3 The graph of $y = f(x)$ is transformed as follows.

Shift a units up for $y = f(x) + a$

Shift a to the left for $y = f(x + a)$

Stretch vertically by a for $y = af(x)$

Compress horizontally by a for $y = f(ax)$

Reflect in the x-axis for $y = -f(x)$

Reflect in the y-axis for $y = f(-x)$

4 If y is given in terms of x by a non-linear function, a transformation of variables may convert the function to a linear one. It is often necessary to take logs.

CHAPTER 17

The addition formulae

The trigonometric functions are not simple, linear functions. When we add together two angles we do not simply add their sines to find the sine of the total

$$sin(x + y) \neq sin\ x + sin\ y$$

In this chapter we shall find the correct expansion of $sin(x + y)$, along with similar formulae. From these results many other formulae can be found.

17.1 Formulae for $sin(x + y)$, $cos(x + y)$, $tan(x + y)$

The addition formulae

The formulae which tell us how to find the sine, cosine or tangent of sums of angles are known as **addition formulae**.

$$sin(x + y) = sin\ x \cos y + \cos x \sin y$$

$$cos(x + y) = \cos x \cos y - sin\ x \sin y$$

$$sin(x - y) = sin\ x \cos y - \cos x \sin y$$

$$cos(x - y) = \cos x \cos y + sin\ x \sin y$$

$$tan(x + y) = \frac{tan\ x + tan\ y}{1 - tan\ x \tan y}$$

$$tan(x - y) = \frac{tan\ x - tan\ y}{1 + tan\ x \tan y}$$

Proof

sin(x + y)

Consider the diagram shown in Fig. 17.1. Let OA be 1 unit. We shall consider the length AD. It is built up from AF and FD.

$$AD = AF + FD$$

First find AD.

The angle AOD is $x + y$. Hence

$$AD = \sin(x + y)$$

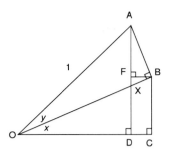

Fig. 17.1

Now find FD.

Consider OAB.

$$OB = \cos y$$

Now consider triangle OBC.

$$BC = OB \sin x = \cos y \sin x$$

FD is equal to BC, hence

$$FD = \cos y \sin x$$

Now find AF.

Angle DXO $= 90° - x$

Angle FAB $= 90° - (90° - x) = x$

Considering triangle OAB,

$$AB = \sin y$$

Considering triangle FAB

$$AF = AB \cos x = \sin y \cos x$$

Now put these results together and use the fact that

$$AD = AF + FD$$

$$\boxed{\sin(x + y) = \sin x \cos y + \cos x \sin y}$$

cos(x + y)

The argument for the expansion of $\cos(x + y)$ is similar.

$$OD = \cos(x + y)$$

$$OC = OB \cos x = \cos y \cos x$$

$$CD = BF = AB \sin x = \sin y \sin x$$

Now use the fact that OD $=$ OC $-$ DC.

$$\boxed{\cos(x + y) = \cos x \cos y - \sin x \sin y}$$

tan(x + y)

For the tan expansion, recall that $\tan P = \dfrac{\sin P}{\cos P}$

$$\tan(x+y) = \frac{\sin(x+y)}{\cos(x+y)} = \frac{\sin x \cos y + \cos x \sin y}{\cos x \cos y - \sin x \sin y}$$

Divide top and bottom by $\cos x \cos y$.

$$\tan(x+y) = \frac{(\sin x/\cos x) + (\sin y/\cos y)}{1 - (\sin x/\cos x)(\sin y/\cos y)} = \frac{\tan x + \tan y}{1 - \tan x \tan y}$$

$$\boxed{\tan(x+y) = \frac{\tan x + \tan y}{1 - \tan x \tan y}}$$

sin(x − y), cos(x − y), tan(x − y)

For the expansions of functions of $x - y$, recall that

$$\sin(-x) = -\sin x$$

$$\cos(-x) = \cos x$$

$$\tan(-x) = -\tan x$$

Substituting these in the above formulae gives

$$\sin(x-y) = \sin(x+-y) = \sin x \cos(-y) + \cos x \sin(-y)$$
$$= \sin x \cos y - \cos x \sin y$$

$$\boxed{\sin(x-y) = \sin x \cos y - \cos x \sin y}$$

$$\cos(x-y) = \cos(x+-y) = \cos x \cos(-y) - \sin x \sin(-y)$$
$$= \cos x \cos y + \sin x \sin y$$

$$\boxed{\cos(x-y) = \cos x \cos y + \sin x \sin y}$$

$$\tan(x-y) = \tan(x+-y) = \frac{\tan x + \tan(-y)}{1 - \tan x \tan(-y)}$$

$$= \frac{\tan x - \tan y}{1 + \tan x \tan y}$$

$$\boxed{\tan(x-y) = \frac{\tan x - \tan y}{1 + \tan x \tan y}}$$

Example 17.1

Find an expression for $\cos(x + 30°)$ in terms of $\sin x$ and $\cos x$.

Solution

Apply the formula for $\cos(x + y)$.

$$\cos(x + 30°) = \cos x \cos 30° - \sin x \sin 30°$$

Use the facts that $\cos 30° = \dfrac{\sqrt{3}}{2}$ and $\sin 30° = \dfrac{1}{2}$.

$$\cos(x + 30°) = \frac{\sqrt{3}}{2}\, \cos x - \frac{1}{2}\, \sin x$$

EXAMPLE 17.2

Prove the identity $\dfrac{\sin(x + y)}{\cos x \cos y} \equiv \tan x + \tan y$.

Solution

Expand $\sin(x + y)$ and divide through by $\cos x \cos y$.

$$\frac{\sin(x + y)}{\cos x \cos y} = \frac{\sin x \cos y + \cos x \sin y}{\cos x \cos y}$$

$$= \frac{\sin x}{\cos x} + \frac{\sin y}{\cos y}$$

This is $\tan x + \tan y$, the right-hand side of the identity.

$$\frac{\sin(x + y)}{\cos x \cos y} \equiv \tan x + \tan y$$

EXERCISE 17A

1 Write down the expansions for the following.

a) $\sin(A + B)$ **b)** $\cos(x - 3y)$ **c)** $\tan(2x + y)$

d) $\sin(x + 45°)$ **e)** $\cos\left(y - \dfrac{\pi}{3}\right)$ **f)** $\tan\left(\dfrac{\pi}{4} + x\right)$

2 Simplify the following.

a) $\cos x \sin 20° + \cos 20° \sin x$ **b)** $\cos 50° \cos 40° + \sin 50° \sin 40°$

3 α and β are acute angles, $\sin \alpha = \frac{3}{5}$ and $\sin \beta = \frac{5}{13}$. Without the use of a calculator find the values of these functions.

a) $\sin(\alpha + \beta)$ **b)** $\cos(\alpha - \beta)$ **c)** $\tan(\alpha + \beta)$

4 Find the greatest values of the following. In each case state the value of x which gives the greatest value.

a) $\sin x \cos 20° + \cos x \sin 20°$ **b)** $\cos x \cos 10° + \sin x \sin 10°$

5 Prove the following identities.

a) $\sin(90° - x) \equiv \cos x$ **b)** $\dfrac{\sin(x + y)}{\sin(x - y)} \equiv \dfrac{\tan x + \tan y}{\tan x - \tan y}$

c) $(\cos A + \cos B)^2 + (\sin A + \sin B)^2 \equiv 2(1 + \cos(A - B))$

d) $\tan(x + 45°) \equiv \dfrac{\cos x + \sin x}{\cos x - \sin x}$

e) $\cos(P + Q) \cos(P - Q) \equiv \cos^2 P - \sin^2 Q$

6 The approximate values of sin 1 and cos 1 are 0.8415 and 0.5403 respectively. Use these values and the approximations $\sin x \approx x$ and $\cos x \approx 1 - \frac{1}{2}x^2$ to find a value for sin 1.01.

7 If $3 \cos(A + B) = 2 \cos(A - B)$, show that $\cos A \cos B = 5 \sin A \sin B$. What is the relationship between $\tan A$ and $\tan B$?

8 A, B and C are the angles of a triangle. Show that

$$\tan A = \dfrac{\tan B + \tan C}{\tan B \tan C - 1}$$

Write down two other similar relationships.

9 Let $\sin x = p$ and $\sin y = q$. Express the following in terms of p and q.

a) cos x **b)** cos y **c)** $\sin(x + y)$ **d)** $\cos(x - y)$

10 A and B are acute angles, $\sin A = \dfrac{1}{\sqrt{10}}$ and $\sin B = \dfrac{1}{\sqrt{5}}$. Without a calculator find surd expressions for the following.

a) cos A **b)** cos B **c)** $\sin(A + B)$ **d)** $\cos(A - B)$ **e)** $\tan(A - B)$

11 If $\tan(x + y) = \frac{4}{7}$ and $\tan x = \frac{2}{5}$ find $\tan y$.

*12 If $\sin(x + y) = 2 \sin(x - y)$ show that $\tan x = 3 \tan y$.

*13 Find an expression for $\cot(x + y)$ in terms of $\cot x$ and $\cot y$.

*14 Find an expression for $\tan(A + B + C)$ in terms of $\tan A$, $\tan B$ and $\tan C$. Arrange your answer so that it is symmetrical in $\tan A$, $\tan B$ and $\tan C$.

17.2 The double angle formulae for sin 2A, cos 2A, tan 2A

Special cases of the addition formulae occur when $x = y$.

$$\sin 2x = 2 \sin x \cos x$$

$$\cos 2x = \cos^2 x - \sin^2 x$$

$$= 2 \cos^2 x - 1$$

$$= 1 - 2 \sin^2 x$$

$$\tan 2x = \dfrac{2 \tan x}{1 - \tan^2 x}$$

Proof

To find sin 2x, use the addition formula for $\sin(x + y)$.

$$\sin 2x = \sin(x + x) = \sin x \cos x + \cos x \sin x$$
$$= 2 \sin x \cos x$$

$$\boxed{\sin 2x = 2 \sin x \cos x}$$

To find cos 2x, use the addition formula for $\cos(x + y)$.

$$\cos 2x = \cos(x + x) = \cos x \cos x - \sin x \sin x$$
$$= \cos^2 x - \sin^2 x$$

$$\boxed{\cos 2x = \cos^2 x - \sin^2 x}$$

Using $\sin^2 x = 1 - \cos^2 x$,

$$\cos 2x = \cos^2 x - (1 - \cos^2 x) = 2 \cos^2 x - 1$$

$$\boxed{\cos 2x = 2 \cos^2 x - 1}$$

Using $\cos^2 x = 1 - \sin^2 x$,

$$\cos 2x = (1 - \sin^2 x) - \sin^2 x = 1 - 2 \sin^2 x$$

$$\boxed{\cos 2x = 1 - 2 \sin^2 x}$$

To find tan 2x, use the addition formula for $\tan(x + y)$.

$$\tan 2x = \tan(x + x) = \frac{\tan x + \tan x}{1 - \tan x \tan x} = \frac{2 \tan x}{1 - \tan^2 x}$$

$$\boxed{\tan 2x = \frac{2 \tan x}{1 - \tan^2 x}}$$

EXAMPLE 17.3

Simplify this expression.

$$\frac{\sin 2x}{1 - \cos 2x}$$

Solution

The expression involves functions of double angles. Use the formulae above. There is a choice of which cos 2x formula to use. If we use the third version we will eliminate the 1.

$$\frac{\sin 2x}{1 - \cos 2x} = \frac{2 \sin x \cos x}{1 - (1 - 2 \sin^2 x)} = \frac{2 \sin x \cos x}{2 \sin^2 x} = \frac{\cos x}{\sin x}$$
$$= \cot x$$

$$\frac{\sin 2x}{1 - \cos 2x} = \cot x$$

EXAMPLE 17.4

Without using a calculator, find a surd expression for $\tan \dfrac{\pi}{8}$.

Solution

Let $\tan \dfrac{\pi}{8} = t$. We know that $2 \times \dfrac{\pi}{8} = \dfrac{\pi}{4}$ and that $\tan \dfrac{\pi}{4} = 1$. Use the formula for $\tan 2x$.

$$1 = \frac{2t}{1 - t^2}$$

Multiply both sides by $(1 - t^2)$ and rearrange as a quadratic.

$$t^2 + 2t - 1 = 0$$

The solutions of this are $t = -1 + \sqrt{2}$ and $-1 - \sqrt{2}$. However, the value of t must be positive.

So the value of $\tan \dfrac{\pi}{8}$ is $\sqrt{2} - 1$.

EXERCISE 17B

1 α is an acute angle for which $\sin \alpha = \frac{3}{5}$. Without a calculator find the following.

a) $\sin 2\alpha$ b) $\cos 2\alpha$ c) $\tan 2\alpha$

2 ABC is a triangle in which $B = x$ and $A = 2x$. Show that $BC = 2\ AC \cos x$.

3 Prove the following identities.

a) $\cos^4 x - \sin^4 x \equiv \cos 2x$ b) $\dfrac{1 - \cos 2x}{1 + \cos 2x} \equiv \tan^2 x$

c) $\cot x - \tan x \equiv 2 \cot 2x$ d) $\sin 4x \equiv 4 \sin x \cos x \cos 2x$

e) $\tan \dfrac{\theta}{2} \equiv \dfrac{\sin \theta}{1 + \cos \theta}$ f) $\dfrac{\tan x + \cot x}{\operatorname{cosec} 2x} \equiv 2$

g) $\cot x - \cot 2x \equiv \operatorname{cosec} 2x$

4 Without the use of a calculator, find surd expressions for the following.

a) $\tan 15°$ b) $\cos 22.5°$ c) $\sin 15°$

5 Show that $\cos 3x = 4 \cos^3 x - 3 \cos x$.

6 Find an expression for $\sin 3x$ in terms of $\sin x$ alone.

*7 Let $t = \tan \frac{1}{2}\theta$. Find expressions for $\tan \theta$, $\sin \theta$ and $\cos \theta$ in terms of t.

8 When θ is small, $\sin \theta$ is approximately θ. Letting $x = \frac{1}{2}\theta$, use a double angle formula for $\cos 2x$ to show that $\cos \theta$ is approximately $1 - \frac{1}{2}\theta^2$.

17.3 Factorising trigonometric expressions

The following four formulae can be obtained by adding or subtracting the addition formulae of Section 17.1.

$$\sin(x + y) + \sin(x - y) = 2 \sin x \cos y \qquad (1)$$

$$\sin(x + y) - \sin(x - y) = 2 \cos x \sin y \qquad (2)$$

$$\cos(x + y) + \cos(x - y) = 2 \cos x \cos y \qquad (3)$$

$$\cos(x + y) - \cos(x - y) = -2 \sin x \sin y \qquad (4)$$

These results can be used to derive the following **factor formulae**.

$$\sin A + \sin B = 2 \sin \frac{A + B}{2} \cos \frac{A - B}{2}$$

$$\sin A - \sin B = 2 \cos \frac{A + B}{2} \sin \frac{A - B}{2}$$

$$\cos A + \cos B = 2 \cos \frac{A + B}{2} \cos \frac{A - B}{2}$$

$$\cos A - \cos B = -2 \sin \frac{A + B}{2} \sin \frac{A - B}{2}$$

(Note the minus sign in this last formula.)

Proof
Let $A = x + y$ and $B = x - y$. Adding and subtracting these equations:

Adding gives $A + B = 2x$.

Subtracting gives $A - B = 2y$.

Hence $\dfrac{A + B}{2} = x$ and $\dfrac{A - B}{2} = y$.

Now substitute into the formula (1) above.

$$\sin A + \sin B = 2 \sin \frac{A + B}{2} \cos \frac{A - B}{2}$$

The other three factor formulae can be obtained by substituting into (2), (3) and (4).

EXAMPLE 17.5
Find the greatest value of $\cos(x + 10°) + \cos(x - 20°)$.

Solution
Rewrite this expression using the factor formula.

$$\cos(x + 10°) + \cos(x - 20°) = 2 \cos \tfrac{1}{2}(2x - 10°) \cos \tfrac{1}{2}(30°)$$

$$= 2 \cos(x - 5°) \cos 15°$$

The greatest value of $\cos(x - 5°)$ is 1.

The greatest value of $\cos(x + 10°) + \cos(x - 20°)$ is $2 \cos 15°$.

EXERCISE 17C

1 Write each of the following as the sum or difference of two trigonometric functions.

   a) $2 \sin 40° \cos 20°$ **b)** $2 \cos 30° \sin 60°$ **c)** $2 \cos 30° \cos 70°$

   d) $2 \sin 70° \sin 10°$ **e)** $2 \sin \dfrac{\pi}{3} \cos \dfrac{\pi}{4}$ **f)** $2 \sin \dfrac{\pi}{3} \sin \dfrac{\pi}{4}$

2 Write the following in terms of the product of two trigonometric functions.

   a) $\sin 60° + \sin 20°$ **b)** $\sin 30° - \sin 16°$ **c)** $\cos 50° + \cos 10°$

   d) $\cos 40° - \cos 80°$ **e)** $\sin \dfrac{\pi}{3} - \sin \dfrac{\pi}{4}$ **f)** $\cos \dfrac{\pi}{2} + \cos \dfrac{\pi}{3}$

3 Without the use of a calculator, show that each of the following is true.

   a) $\sin 70° - \sin 50° = \sin 10°$ **b)** $\cos 50° + \cos 70° = \cos 10°$

   c) $\sin 10° + \sin 80° = \sqrt{2} \cos 35°$

4 Prove the following identities.

   a) $\dfrac{\sin 2x + \sin 2y}{\cos 2x + \cos 2y} \equiv \tan (x + y)$ **b)** $\sin 3x + \sin x = 4 \sin x \cos^2 x$

   c) $\sin x + \sin (x + 120°) + \sin (x + 240°) \equiv 0$

   d) $\dfrac{\sin x + \sin 3x + \sin 5x}{\cos x + \cos 3x + \cos 5x} \equiv \tan 3x$

   e) $\sin x + \sin 2x + \sin 3x \equiv \sin 2x(1 + 2 \cos x)$

5 Find the greatest values of the following expressions, and the values of x at which they are reached.

   a) $\cos (x + 50°) + \cos (x - 20°)$ **b)** $\sin (x + \dfrac{\pi}{3}) + \sin (x + \dfrac{\pi}{4})$

6 Write $2 \cos 3x \cos x$ in the form $\cos A + \cos B$. Hence write $4 \cos 3x \cos 2x \cos x$ in the form $\cos C + \cos D + \cos E + k$

17.4 The expression *a* cos *θ* + *b* sin *θ*

Suppose we have an expression of the form $a \cos \theta + b \sin \theta$, where a and b are constants. Then it can be reduced to a multiple of a single sine function, as follows.

$$a \cos \theta + b \sin \theta = r \sin(\theta + \alpha)$$

where $r = \sqrt{a^2 + b^2}$ and $\alpha = \tan^{-1} \dfrac{a}{b}$.

Proof

We seek to match the expression with the expansion

$$\sin(\theta + \alpha) = \sin \alpha \cos \theta + \sin \theta \cos \alpha$$

It might not be possible to put a and b equal to $\sin \alpha$ and $\cos \alpha$. For example a and b might be greater than 1. Therefore we must divide the expression by a term r, so that $\dfrac{a}{r} = \sin \alpha$ and $\dfrac{b}{r} = \cos \alpha$.

This factor r can be found from the triangle shown in Fig. 17.2. By Pythagoras's theorem, $r = \sqrt{a^2 + b^2}$.

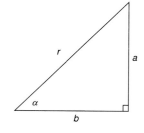

Fig. 17.2

$$a \cos \theta + b \sin \theta = r \left(\frac{a}{r} \cos \theta + \frac{b}{r} \sin \theta \right)$$

$$= r(\sin \alpha \cos \theta + \cos \alpha \sin \theta)$$

$$a \cos \theta + b \sin \theta = r \sin(\theta + \alpha)$$

EXAMPLE 17.6

Find the greatest value of $2 \cos \theta + 3 \sin \theta$, and the value of θ for which this is reached (θ is measured in radians).

Solution

Apply the formula above.

$$2 \cos \theta + 3 \sin \theta = \sqrt{2^2 + 3^2} \sin(\theta + \tan^{-1} \tfrac{2}{3})$$

$$2 \cos \theta + 3 \sin \theta = \sqrt{13} \sin(\theta + \tan^{-1} 0.666\,67)$$

The $\sin \theta$ function reaches its greatest value of 1 at $\theta = \dfrac{\pi}{2}$. So the greatest value of the expression above is obtained when

$$\theta + \tan^{-1} \frac{2}{3} = \frac{\pi}{2}.$$

$$\theta = 0.983$$

The greatest value of $2 \cos \theta + 3 \sin \theta$ is $\sqrt{13}$, achieved at $\theta = 0.983$.

EXERCISE 17D

1 Express each of the following in the form $r \sin(\theta + \alpha)$. Angles are measured in degrees.

a) $\cos \theta + 3 \sin \theta$ b) $4 \cos \theta + 3 \sin \theta$ c) $\sin \theta - \cos \theta$

d) $0.3 \cos \theta + 0.5 \sin \theta$ e) $12 \sin \theta - 10 \cos \theta$ f) $7 \sin \theta + 9 \cos \theta$

2 Find the greatest values of the expressions in Question 1, and the values of θ at which they are reached.

3 Show that $a \cos \theta + b \sin \theta$ can also be expressed as

$r \cos(\theta - \alpha)$

where $r = \sqrt{a^2 + b^2}$ and $\alpha = \tan^{-1} \dfrac{b}{a}$.

4 Find the greatest and least values of $\dfrac{1}{3 \cos x + 4 \sin x + 6}$ and the values of x, in radians, at which they are reached.

5 Express $(\cos x + 3 \sin x)^8$ in terms of the power of a single trigonometric function.

17.5 Practice in solving equations

With the formulae that have been derived in this chapter, a wide range of equations can be solved.

There are often several choices of formula to use. If one formula does not work then try another. It may seem obvious, but one general principle is to look for the formula which makes the expression simpler rather than more complicated.

EXAMPLE 17.7
Find the solutions of the equation

$$\sin(x + 30°) + \sin(x - 40°) = 1.1$$

in the range $0°$ to $180°$.

Solution
We could expand both terms of the left-hand side using the addition formulae. This will make the expression more complicated, with four terms involving x, instead of two. It is simpler to use the factor formula for the left-hand side.

$$2 \sin \tfrac{1}{2}(2x - 10°) \cos \tfrac{1}{2}(70°) = 1.1$$

$$\sin(x - 5°) \cos 35° = 0.55$$

$$\sin(x - 5°) = \frac{0.55}{\cos 35°} = 0.67$$

Now take the $\sin^{-1}$ of the right-hand side.

$$x - 5° = 42.2° \text{ or } x - 5° = 137.8°$$

If we add or subtract $360°$, the result will lie outside the given range.

The solution of the equation is $x = 47.2°$ or $x = 142.8°$.

EXAMPLE 17.8

Find the solutions of the equation $4 \sin x + 3 \cos x = 1$ in the range 0 to 2π.

Solution

Here the left-hand side can be written as a single sine by the method of Section 17.4.

$$4 \sin x + 3 \cos x = r \sin(x + \alpha)$$

where $r = \sqrt{4^2 + 3^2} = 5$, and $\alpha = \tan^{-1} \frac{3}{4} = 0.644$

So $\sin(x + \alpha) = \frac{1}{5} = 0.2$.

$$x + \alpha = \sin^{-1} 0.2 = 0.201$$

$$x + \alpha = 0.201 \text{ or } 2.94 \text{ or } 6.48 \text{ etc.}$$

Note: We include solutions outside the range 0 to 2π, as when we subtract α the result may lie within the range.

$$x = -0.442 \text{ or } x = 2.30 \text{ or } x = 5.84 \text{ etc.}$$

Of these, the first is outside the range.

The solution of the equation is $x = 2.30$ or $x = 5.84$.

EXAMPLE 17.9

Solve the equation $\cos 2\theta + 5 \cos \theta = 2$, in the range 0 to π.

Solution

If we use a double angle formula to express $\cos 2\theta$ in terms of $\cos \theta$, then we will have an equation in $\cos \theta$ alone.

$$2 \cos^2 \theta - 1 + 5 \cos \theta = 2$$

This is a quadratic in $\cos \theta$. Collecting everything on the left, and writing c for $\cos \theta$ gives

$$2c^2 + 5c - 3 = 0$$

$$(2c - 1)(c + 3) = 0$$

So $c = 0.5$ or -3. It is not possible for $\cos \theta$ to be -3. The only solutions come from $\cos^{-1} 0.5$. There is only one in the range given.

$$\cos^{-1} 0.5 = \frac{\pi}{3}$$

The solution is $\theta = \frac{\pi}{3}$.

EXERCISE 17E

In Question 1 the method is suggested. In the remaining questions you have to find the method.

1 Find the solutions of the following equations in the range $0° \le x \le 180°$.

a) (Using addition formulae)

(i) $\sin x \cos 20° + \cos x \sin 20° = 0.5$ (ii) $\cos 50° \cos x + \sin 50° \sin x = 0.8$
(iii) $\sin(x + 45°) = 2 \cos x$ (iv) $\cos(x + 35°) = 2 \sin(x - 25°)$

b) (Using double angle formulae)

(i) $\cos 2x + \sin 2x = 1$ (ii) $\cos x = 1 - \cos \frac{1}{2}x$
(iii) $\tan x \tan 2x = 3$ (iv) $\cos 2x = 2 \sin x \cos x$

c) (Using factor formulae)

(i) $\sin 3x + \sin x = 0$ (ii) $2 \sin x - \sin 2x + 2 \sin 3x = 0$
(iii) $\cos 3x + \cos x = \sin 3x - \sin x$ (iv) $\sin 3x + \sin x = \cos x$

d) (Using the form $a \cos \theta + b \sin \theta$)

(i) $3 \sin x + 2 \cos x = 2.5$ (ii) $4 \cos x + 2 \sin x = 3$
(iii) $3 \cos x - 2 \sin x = 1$ (iv) $6 \cos x + 5 \sin x = -4$

2 Find the solutions of the following equations, in the range $0 \le x \le \pi$.

a) $3 \sin x + 4 \cos x = 2$ b) $\sin x + \sin(x - 0.2) = 0.9$

c) $\sin x + \sqrt{3} \sin 2x + \sin 3x = 0$ d) $3 \sin 2x + \cos x = 0$

3 Solve the following equations in the range $0° \le x \le 360°$.

a) $\sin x \cos 20° + \cos x \sin 20° = 0.3$ b) $\sin(x + 60°) = \sin x$

c) $\tan 2x = 3 \tan x$ d) $\sin 2x + 3 \cos^3 x = 2 \cos x$

4 Solve the following equations in the range $0 \le x \le \pi$.

a) $\sin 2x = \cos x$ b) $\cos 2x + \cos x = 1$

c) $\cos 2x + \sin x = 1$ d) $\cos 2x + \cos^2 x = 1$

5 Find the solutions of the following equations, in the range $0 \le x \le 2\pi$.

a) $\cos(x + 1) - 3 \sin x = 0$ b) $\sin\left(x + \frac{\pi}{3}\right) + \sin\left(x - \frac{\pi}{4}\right) = 0.5$

c) $\cos x = 5 \sin x + 3$ d) $2 \cos 3x = \cos x$

6 Solve the following equations, for $0° \le x \le 180°$.

a) $\sin 4x + \sin 2x = 0$ b) $\sin x + \sin 2x + \sin 3x = 0$

c) $\cos x + 2 \sin 2x = \cos 3x$ d) $\cos 2x = \sin 2x - 1$

*7 For each of the following sets of simultaneous equations, find one pair of solutions.

a) $A + B = 60°$ and $\cos A + \cos B = 1.6$ b) $A - B = \pi$ and $\sin A - \sin B = 0.6$

LONGER EXERCISE

Practical applications

The formulae of this chapter may seem very theoretical, but they have many applications. Below are two.

Beats

Suppose two musical instruments are playing at the same volume. One instrument plays middle C exactly, the other is slightly sharp. The air vibrations caused by the instruments could be modelled by $k\sin(260 \times 2\pi t)$ and $k\sin(261 \times 2\pi t)$, where k represents the **amplitude**.

The total vibration is the sum of these functions. Apply the factor formula to $\sin(260 \times 2\pi t) + \sin(261 \times 2\pi t)$. What sound does this represent?

Electricity supply

The electricity that comes into our homes is connected to a **live** wire and a **neutral** wire. If you touch a live wire the results could be fatal. If you touch a neutral wire you may not feel anything at all. Why?

The electricity supplied by alternating current at 50 cycles per second can be modelled by $k\sin(50 \times 2\pi t)$. The electricity company will send current to three nearby regions at slightly different phases, so that the current in the live wires of the three regions can be modelled by $k\sin(50 \times 2\pi t)$, $k\sin\left(50 \times 2\pi t + \dfrac{2\pi}{3}\right)$, $k\sin\left(50 \times 2\pi t + \dfrac{4\pi}{3}\right)$. The regions share a common neutral connection.

Evaluate the sum of the three expressions above, simplifying as far as possible. How much current returns along the neutral wire?

EXAMINATION QUESTIONS

1 Show that, for a small angle t radians,

$$\sin\left(\frac{\pi}{4} + t\right) \approx \frac{1}{\sqrt{2}}(1 + t - \tfrac{1}{2}t^2)$$

Use your calculator to estimate the error in using this approximation when $t = 0.1$, giving your answer to 5 decimal places.

L 1992

2 Express $3\sin x + 2\cos x$ in the form $r\sin(x + \alpha)$, where r and α are constants, $r > 0$ and $0 < \alpha < 90°$.

Hence, or otherwise, find

a) the least value of $(3\sin x + 2\cos x)^5$

b) the solutions of the equation

$$3\sin x + 2\cos x = 1$$

for $0 < x < 360°$, giving your answers in degrees to 1 decimal place.

L 1989

3 Prove the identity $\dfrac{1}{\sin 2x} + \dfrac{1}{\tan 2x} \equiv \cot x$.

Hence

a) show that $\cot 15° = 2 + \sqrt{3}$

b) solve the equation

$$\frac{1}{\sin 3\theta} + \frac{1}{\tan 3\theta} = 2$$

giving your answers in the interval $0 < \theta < 180°$, to the nearest $0.1°$.

O 1992

4 Consider the expression

$$E = \sin\left(3\theta - \frac{\pi}{4}\right) + \sin\left(\theta + \frac{\pi}{4}\right) \quad \text{where } 0 \le \theta \le \frac{\pi}{2}.$$

(i) Show that E can be written as $2 \sin 2\theta \cos\left(\theta - \frac{\pi}{4}\right)$.

(ii) What value of θ makes $\sin 2\theta$ greatest?

(iii) What value of θ makes $\cos\left(\theta - \frac{\pi}{4}\right)$ greatest?

(iv) Determine the maximum value of E.

In a particular case, $\theta = \dfrac{\pi}{4} + x$ where x is small.

(v) Show that $E = 2 \cos 2x \cos x$ and, by taking a suitable approximation for $\cos x$, show that $E \approx 2 - 5x^2$.

MEI 1992

Summary and key points

1 The addition formulae are

$$\sin(x + y) = \sin x \cos y + \cos x \sin y$$

$$\sin(x - y) = \sin x \cos y - \cos x \sin y$$

$$\cos(x + y) = \cos x \cos y - \sin x \sin y$$

$$\cos(x - y) = \cos x \cos y + \sin x \sin y$$

$$\tan(x + y) = \frac{\tan x + \tan y}{1 - \tan x \tan y}$$

$$\tan(x - y) = \frac{\tan x - \tan y}{1 + \tan x \tan y}$$

2 The double angle formulae are

$$\sin 2x = 2 \sin x \cos x$$

$$\cos 2x = \cos^2 x - \sin^2 x$$

$$= 2 \cos^2 x - 1$$

$$= 1 - 2 \sin^2 x$$

$$\tan 2x = \frac{2 \tan x}{1 - \tan^2 x}$$

3 The factor formulae are

$$\sin A + \sin B = 2 \sin \frac{A + B}{2} \cos \frac{A - B}{2}$$

$$\sin A - \sin B = 2 \cos \frac{A + B}{2} \sin \frac{A - B}{2}$$

$$\cos A + \cos B = 2 \cos \frac{A + B}{2} \cos \frac{A - B}{2}$$

$$\cos A - \cos B = -2 \sin \frac{A + B}{2} \sin \frac{A - B}{2}$$

4 The expression $a \cos x + b \sin x$ can be written as $r \sin(x + \alpha)$, where $r = \sqrt{a^2 + b^2}$ and $\alpha = \tan^{-1} \frac{a}{b}$.

5 The formulae of this chapter can be used to solve a wide range of equations.

When finding the solutions within a given range, you may have initially to consider values outside the range.

chapter 18

Differentiation of other functions

So far the functions you have been able to differentiate are built up from powers of x. In this chapter we extend the range of functions which you will be able to differentiate.

Any function whose graph is smooth can be differentiated. In particular the sine graph, with which you are already familiar, consists of the smooth wave-like curve shown in Fig. 18.1. We shall show that the sine function can be differentiated.

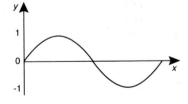

Fig. 18.1

18.1 Finding the derivative of f($ax + b$) from the derivative of f(x)

Suppose we want to differentiate $y = (2x + 1)^6$. We could expand this expression by the binomial theorem, and then differentiate each term, but this would take a long time, and we would be very liable to make a mistake. There is a way to differentiate expressions of this sort very quickly.

Expressions of the form $y = f(x + b)$

Suppose our function is of the form $y = f(x + b)$, where b is a constant. For example, we might have $(x + 3)^6$. This is f($x + 3$), where f(x) = x^6. Then the derivative is given by

$$\frac{dy}{dx} = f'(x + b)$$

Justification

Suppose the graph of $y = f(x)$ is as shown in Fig. 18.2. Then the graph of $y = f(x + b)$ is obtained by shifting this original graph b units to the left. The tangent has also been shifted to the left by b, but its gradient has not been altered.

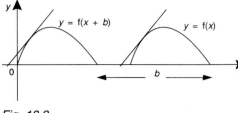

Fig. 18.2

Expressions of the form y = f(ax)

Suppose our function is of the form $y = f(ax)$, where a is a constant. For example we might have $(2x)^6$. This is $f(2x)$ where $f(x) = x^6$. Then the derivative is given by

$$\frac{dy}{dx} = a\,f'(ax)$$

Justification

Suppose the graph of $y = f(x)$ is as shown in Fig. 18.3. Then the graph of $y = f(ax)$ is obtained by compressing this original graph by a factor of a, parallel to the x-axis. Consider what has happened to the tangent. The y-change of the tangent is unchanged, but the x-change has been divided by a. Hence the gradient of the tangent has been multiplied by a.

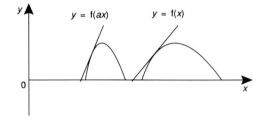

Fig. 18.3

Expressions of the form y = f(ax + b)

Putting these results together, we obtain:

If $y = f(ax + b)$, then $\dfrac{dy}{dx} = af'(ax + b)$.

EXAMPLE 18.1

Find the equation of the tangent to $y = (2x + 1)^4$ at (1, 81).

Solution

Here y is of the form $f(ax + b)$, where $a = 2$, $b = 1$ and $f(x) = x^4$. We know that the derivative of x^4 is $4x^3$. Put this result into the formula above.

$$\frac{dy}{dx} = 2f'(2x + 1) = 2 \times 4(2x + 1)^3 = 8(2x + 1)^3$$

For $x = 1$, $\dfrac{dy}{dx} = 216$.

Hence the tangent at $x = 1$ has equation

$$y = 216x + c$$

where c is a constant.

We know that $y = 81$ when $x = 1$.

$$81 = 216 + c$$

Hence $c = -135$, giving

$$y = 216x - 135$$

The tangent has equation $y = 216x - 135$.

EXAMPLE 18.2
Find the derivative of $\sqrt{10 - 4x}$.

Solution
Here y is of the form $f(ax + b)$, where $a = -4, b = 10$ and $f(x) = \sqrt{x}$. The derivative of $f(x) = \sqrt{x}$ is $f'(x) = \frac{1}{2}x^{-\frac{1}{2}}$. Put this result into the formula above.

$$\frac{dy}{dx} = (-4)\, f'(x) = (-4)\tfrac{1}{2}(10 - 4x)^{-\frac{1}{2}}$$

$$= -2(10 - 4x)^{-\frac{1}{2}}$$

The derivative is $-2(10 - 4x)^{-\frac{1}{2}}$

EXERCISE 18A

1 Find the derivatives of the following functions.

a) $(2x + 3)^5$

b) $(4x - 1)^6$

c) $(1 - 2x)^8$

d) $3(2x + 2)^6$

e) $-4(1 + 2x)^4$

f) $(1 + x)^4 - (2 - x)^5$

g) $(2x + 1)^{\frac{1}{2}}$

h) $(3x + 2)^{-\frac{1}{2}}$

i) $(2x + 5)^{-1}$

j) $(3x - 1)^{-3}$

k) $\sqrt{3x + 2}$

l) $\sqrt[3]{2x - 1}$

m) $\dfrac{1}{2x + 1}$

n) $\dfrac{2}{(3x + 1)^3}$

o) $\dfrac{1}{\sqrt{2x + 1}}$

p) $\dfrac{3}{(3x - 1)^{\frac{1}{3}}}$

2 Find the equation of the tangent to $y = (3x - 1)^6$ at the point $(1, 64)$.

3 Find the equation of the tangent to $y = \sqrt{4x - 3}$ at the point $(3, 3)$.

4 Find the equation of the normal to $y = (2x + 3)^4$ at the point $(-1, 1)$.

5 Find the equation of the normal to $y = \dfrac{4}{x + 1}$ at the point $(1, 2)$.

6 Let $y = (2x + 3)^2$. Find $\dfrac{dy}{dx}$

a) using the methods of this section

b) by expanding the expression and differentiating term by term.

Check that your answers agree.

7 Find the stationary points of the curve $y = x + \dfrac{1}{x + 1}$.

8 Find the stationary points of the curve $y = 2x - 4\sqrt{x+2}$.

*9 Suppose for a function f(x), $f'(x) = \sqrt{1+x^3}$. Find the derivatives of the following.

    a) f(2x)   **b)** f(x + 3)   **c)** f(1 − 2x)

18.2 The derivatives of trigonometric functions

As we noted above, the graph of $y = \sin x$ is smooth. Hence we can find its derivative.

In all trigonometric work involving calculus, the angles are measured in radians. The main results are as follows.

    If $y = \sin x$, then $\dfrac{dy}{dx} = \cos x$.

    If $y = \cos x$, then $\dfrac{dy}{dx} = -\sin x$.

Proof

First we prove the result for $\sin x$.

The formula for the derivative is

$$\frac{dy}{dx} = \lim_{\delta x \to 0} \frac{\delta y}{\delta x} = \lim_{\delta x \to 0} \frac{\sin(x + \delta x) - \sin x}{\delta x}$$

Apply the addition formula to the top line of this equation.

$$\frac{\delta y}{\delta x} = \frac{\sin x \cos \delta x + \cos x \sin \delta x - \sin x}{\delta x}$$

By the result of Chapter 7, when θ is small, $\sin \theta$ is approximately θ and $\cos \theta$ is approximately $1 - \frac{1}{2}\theta^2$. Put these results into the expression, replacing θ by δx.

$\dfrac{\delta y}{\delta x}$ is approximately $\dfrac{\sin x(1 - \frac{1}{2}(\delta x)^2) + \cos x(\delta x) - \sin x}{\delta x}$

$$= \frac{\sin x(-\frac{1}{2}(\delta x)^2) + \cos x\,(\delta x)}{\delta x}$$

Now divide by δx.

    $\dfrac{\delta y}{\delta x}$ is approximately $\sin x(-\frac{1}{2}\delta x) + \cos x$

Finally take the limit.

    $\dfrac{\delta y}{\delta x}$ tends to $\dfrac{dy}{dx}$, and $\sin x \,(-\frac{1}{2}\delta x)$ tends to zero.

    $\dfrac{dy}{dx} = \cos x$

Instead of applying a similar procedure for $y = \cos x$, we shall use the results of the previous section.

$\cos x = \sin\left(\dfrac{\pi}{2} - x\right)$. So $\cos x$ is of the form $f(ax + b)$, where

$a = -1$, $b = \dfrac{\pi}{2}$ and $f(x) = \sin x$.

$$\frac{dy}{dx} = -1\, f'\left(\frac{\pi}{2} - x\right) = -\cos\left(\frac{\pi}{2} - x\right)$$

But $\cos\left(\dfrac{\pi}{2} - x\right) = \sin x$. This gives our result.

$$\frac{dy}{dx} = -\sin x$$

In general,

If $y = \sin(ax + b)$, $\dfrac{dy}{dx} = a \cos(ax + b)$.

If $y = \cos(ax + b)$, $\dfrac{dy}{dx} = -a \sin(ax + b)$.

EXAMPLE 18.3

Find the equation of the tangent to the curve $y = \sin x$ at $\left(\dfrac{\pi}{3}, \dfrac{\sqrt{3}}{2}\right)$.

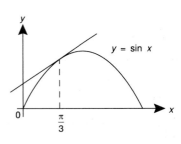

Fig. 18.4

Solution

Differentiating, $\dfrac{dy}{dx} = \cos x$.

At $x = \dfrac{\pi}{3}$, $\cos x = 0.5$.

This gives us the gradient of the tangent.

$$y = 0.5x + c$$

To find the constant c, put $x = \dfrac{\pi}{3}$ and $y = \dfrac{\sqrt{3}}{2}$.

$$\frac{\sqrt{3}}{2} = 0.5\frac{\pi}{3} + c$$

$$c = \frac{\sqrt{3}}{2} - \frac{\pi}{6}$$

$$y = 0.5x + \frac{\sqrt{3}}{2} - \frac{\pi}{6}$$

The equation is $y = 0.5x + \dfrac{\sqrt{3}}{2} - \dfrac{\pi}{6}$.

EXAMPLE 18.4

Find the first stationary point of the curve $y = x + \cos 2x$, and state whether it is a maximum, minimum or point of inflection.

Solution

Differentiating,

$$\frac{dy}{dx} = 1 + 2(-\sin 2x)$$

Put this expression equal to 0.

$$1 - 2 \sin 2x = 0.$$

Hence $\sin 2x = \frac{1}{2}$.

The first solution of this is $2x = \frac{\pi}{6}$, i.e. $x = \frac{\pi}{12}$.

Put this into the expression for y, to obtain

$$\frac{\pi}{12} + \cos \frac{\pi}{6} = \frac{\pi}{12} + \frac{\sqrt{3}}{2}$$

The first stationary point is at $\left(\dfrac{\pi}{12}, \dfrac{\pi}{12} + \dfrac{\sqrt{3}}{2} \right)$.

Differentiating again, $\dfrac{d^2y}{dx^2} = -4 \cos 2x$. This is negative at $x = \dfrac{\pi}{12}$.

The first stationary point is a maximum.

EXERCISE 18B

1 Find the derivatives of the following functions.

a) $2 \sin x$ b) $3 \cos x$ c) $\sin 2x$ d) $\cos 4x$

e) $\sin(2x + 1)$ f) $\sin(3 - 2x)$ g) $\cos(3x - 1)$ h) $\cos\left(2x - \dfrac{\pi}{3}\right)$

2 Find the equations of the tangents to the following curves at the points indicated.

a) $y = 2 \sin x$, at $(0, 0)$ b) $y = 4 \cos x$, at $\left(\dfrac{\pi}{3}, 2\right)$

3 Find the equations of the normals to the following curves at the points indicated.

a) $y = \sin\left(3x + \dfrac{\pi}{6}\right)$, at $\left(\dfrac{\pi}{6}, \dfrac{\sqrt{3}}{2}\right)$ b) $y = \cos x + \sqrt{3} \sin x$, at $\left(\dfrac{\pi}{3}, 2\right)$

4 Find the first positive stationary points of the following curves, and indicate whether they are maxima, minima or points of inflection.

a) $y = 2x - \sin 3x$ b) $y = 4 \cos 2x + 3 \sin 2x$

*5 Prove that the derivative of $\cos x$ is $-\sin x$ by using the addition formula for $\cos(x + \delta x)$.

*6 When finding the derivative of $\sin x$, we used the addition formula. Show that the same result can be obtained by applying the factor formula to $\sin(x + \delta x) - \sin x$.

7 Let $y = \sin x + \cos x$. Show that the second derivative of y is equal to $-y$.

18.3 The derivatives of exponential and logarithmic functions

An exponential function is of the form $y = a^x$, where a is a positive constant. The graphs of these functions are smooth, hence they can be differentiated.

Apply the formula for $\dfrac{\delta y}{\delta x}$.

$$\frac{\delta y}{\delta x} = \frac{a^{x+\delta x} - a^x}{\delta x} = \frac{a^x a^{\delta x} - a^x}{\delta x} = \frac{a^x(a^{\delta x} - 1)}{\delta x} = a^x \frac{a^{\delta x} - 1}{\delta x}$$

Now take the limit as x tends to zero.

$$\frac{dy}{dx} = \lim_{\delta x \to 0} a^x \frac{a^{\delta x} - 1}{\delta x} = a^x \lim_{\delta x \to 0} \frac{a^{\delta x} - 1}{\delta x}$$

The expression $\lim\limits_{\delta x \to 0} \dfrac{a^{\delta x} - 1}{\delta x}$ is not dependent on the value of x.

Hence it is a constant.

$$\text{If } y = a^x, \text{ then } \frac{dy}{dx} = ka^x, \text{ where } k = \lim_{\delta x \to 0} \frac{a^{\delta x} - 1}{\delta x}.$$

Finding k for $a = 2$ and $a = 3$

Now let us find the approximate value of k in the cases $a = 2$ and $a = 3$, by trying small values of δx. First put $a = 2$.

$$\frac{2^{0.1} - 1}{0.1} \approx 0.718 \qquad \frac{2^{0.01} - 1}{0.01} \approx 0.696 \qquad \frac{2^{0.001} - 1}{0.001} \approx 0.693$$

These values seem to be approaching a number near 0.69. So for $a = 2$, k is approximately 0.69.

Now try for $a = 3$.

$$\frac{3^{0.1} - 1}{0.1} \approx 1.16 \qquad \frac{3^{0.01} - 1}{0.01} \approx 1.10 \qquad \frac{3^{0.001} - 1}{0.001} \approx 1.10$$

These values seem to be approaching a number near 1.1. So for $a = 3$, k is approximately 1.1.

Definition of e

For $a = 2$, we found that k is less than 1. For $a = 3$, we found that k is greater than 1. So there is a number, between 2 and 3, for which $k = 1$. This number is called e.

This is one of the most important numbers in mathematics, and it occurs in many other contexts. By definition

If $y = e^x$, then $\dfrac{dy}{dx} = e^x$.

You will find on your calculator a button to evaluate e^x. By putting $x = 1$, we can find the approximate numerical value of e.

$$e^1 = e \approx 2.718$$

Differentiation of e^{ax+b}

Applying the result of Section 18.1, the derivative of e^{ax+b} is ae^{ax+b}.

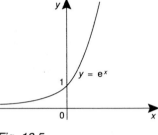

Graph of e^x

When x is large and negative, e^x is very small. When x is large and positive, e^x is very large. The graph of $y = e^x$ is shown in Fig. 18.5. Note that e^x is never negative.

Fig. 18.5

EXAMPLE 18.5
Sketch the graph of the function $y = e^x + 3$. Find the equation of the tangent to the curve at the point where $x = 2$.

Solution
The graph of $y = e^x + 3$ is obtained by shifting the graph of $y = e^x$ up by 3. It is shown in Fig. 18.6.

Differentiating $y = e^x + 3$ gives

$$\frac{dy}{dx} = e^x$$

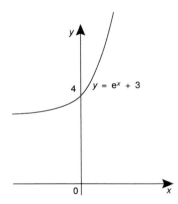

Fig. 18.6

So the gradient at the point where $x = 2$ is e^2. The tangent has equation $y = e^2 x + c$.

The value of y when $x = 2$ is $y = e^2 + 3$.

The point $(2,\ e^2 + 3)$ must lie on the line.

This gives

$$e^2 + 3 = 2e^2 + c$$
$$c = e^2 + 3 - 2e^2 = 3 - e^2$$

Now put these results into the equation.

$$y = e^2 x + 3 - e^2$$

The equation is $y = e^2 x + 3 - e^2$.

EXERCISE 18C

1 Find the derivatives of the following.

    a) $e^x + x^2$ **b)** $3x - 2e^x$ **c)** e^{3x-2} **d)** $2e^{1-x}$

2 Find the equation of the tangent to $y = e^x$ at $(0, 1)$.

3 Sketch the graph of $y = e^{2x-1} - e$. Find the equation of the tangent at $(1, 0)$.

4 Sketch the curve of $y = e^{1-x}$. Find the equation of the normal at $(1, 1)$.

5 Find the stationary point of $y = e^x - e^{3x}$ and identify its nature. Sketch the graph of the function.

6 Find the stationary point of $y = e^{x+3} + e^{-x}$ and identify its nature. Sketch the graph of the function.

7 A function is given by $y = a + be^{-cx}$ where a, b and c are constant. The graph goes through the origin with a gradient of 2, and tends to 2 when x tends to infinity. Find a, b and c.

8 The height of a plant t days after germinating is given by $h = 0.1e^{0.01t}$ centimetres. Find the rate of growth after five days.

9 A radioactive material decays, so that after t hours its mass is $2e^{-0.001t}$ kilograms. Find the rate of decay after 100 hours.

10 Let $y = e^{ax}$. Show that the second derivative of y is a^2y.

Logarithmic function

The logarithmic function is the inverse of the exponential function.

       If $a^x = b$, then $x = \log_a b$.

This is called the log of b to the base a.

Logarithms to the base e are called **natural** logarithms, and are written **ln**.

       $\log_e b = \ln b$

Your calculator will have a button to evaluate the ln function.

Differentiating ln *x*

Let $y = \ln x$.

Then $\dfrac{dy}{dx} = \dfrac{1}{x}$.

Proof
By definition of logarithms

$$x = e^y$$

Differentiate this expression with respect to y.

$$\frac{dx}{dy} = e^y$$

This gives

$$\frac{dy}{dx} = \frac{1}{e^y}$$

Recalling that $e^y = x$, we obtain the result

If $y = \ln x$, $\dfrac{dy}{dx} = \dfrac{1}{x}$.

We can combine this result with that from Section 18.1.

If $y = \ln(ax + b)$, $\dfrac{dy}{dx} = \dfrac{a}{ax + b}$.

Other bases

Logarithms to any other base can be converted to natural logarithms using a formula from Chapter 6.

$$\log_a x = \frac{\log_e x}{\log_e a} = \frac{\ln x}{\ln a}$$

Power functions can also be converted to the exponential function using a formula from Chapter 6.

$$a^x = (e^{\ln a})^x = e^{x \ln a}$$

EXAMPLE 18.6
Find the derivatives of these functions.

a) $y = \log_{10} x$ **b)** $y = 2^x$

Solution
a) Use the formula above to convert $\log_{10}$ to $\log_e$.

$$\log_{10} x = \frac{\log_e x}{\log_e 10} = \frac{\ln x}{\ln 10}$$

Now differentiate.

$$\frac{dy}{dx} = \frac{1}{x \log_e 10}$$

b) Use the formula to convert 2^x.

$$2^x = e^{x \ln 2}$$

Now differentiate:

$$\frac{dy}{dx} = \ln 2 \, e^{x \ln 2} = \ln 2 \times 2^x = 2^x \ln 2$$

Note: A calculator gives $\ln 2$ as 0.693 correct to three decimal places. This confirms the result we obtained when differentiating 2^x at the beginning of this section.

EXAMPLE 18.7

Find the stationary points of $y = x - e^{2x}$ and identify their nature.

Solution

Differentiating gives

$$\frac{dy}{dx} = 1 - 2e^{2x}$$

Put this equal to 0.

$$1 - 2e^{2x} = 0$$

Hence $e^{2x} = \frac{1}{2}$.

This gives $2x = \ln \frac{1}{2}$, so $x = \frac{1}{2} \ln \frac{1}{2} = \ln \frac{1}{\sqrt{2}}$.

Put this in the original formula.

$$y = \ln \frac{1}{\sqrt{2}} - \frac{1}{2}$$

There is a stationary point at $\left(\ln \dfrac{1}{\sqrt{2}}, \ \ln \dfrac{1}{\sqrt{2}} - \dfrac{1}{2} \right)$.

Differentiate again, to obtain

$$\frac{d^2 y}{dx^2} = -4e^{2x}$$

This is negative for all values of x.

The stationary point is a maximum.

EXERCISE 18D

1 Differentiate the following functions.

a) $y = x^2 + \ln x$	**b)** $\ln(x + 1)$	**c)** $\ln(2x + 3)$
d) $\ln(2x - 2)$	**e)** $\ln(2 - 3x)$	**f)** $\log_2 x$
g) $\log_{10}(x + 1)$	**h)** 10^x	**i)** 3^x

2 Find the equations of the tangents to the following curves at the points indicated.

   a) $y = \ln x$ at $(1, 0)$ **b)** $y = \ln(3x + 1)$ at $(0, 0)$

3 Show that the derivatives of $\ln x$ and $\ln 2x$ are equal. Why is this so?

4 Find the stationary points of the following curves and identify their nature.

   a) $y = x + e^{-x}$ **b)** $y = x^2 - 2\ln x$ **c)** $y = 3x - e^{3x}$ **d)** $y = x^2 + x - \ln(x + 1)$

18.4 Integration

We have found how to differentiate several functions. This adds to the range of functions we can integrate.

$$\frac{d}{dx} \sin x = \cos x \qquad \text{Hence} \int \cos x \, dx = \sin x + c$$

$$\frac{d}{dx} \cos x = -\sin x \qquad \text{Hence} \int \sin x \, dx = -\cos x + c$$

$$\frac{d}{dx} e^x = e^x \qquad \text{Hence} \int e^x \, dx = e^x + c$$

$$\frac{d}{dx} \ln x = \frac{1}{x} \qquad \text{Hence} \int \frac{1}{x} \, dx = \ln x + c$$

Note: This last result completes our integration of powers of x.

$$\int x^n \, dx = \frac{x^{n+1}}{n+1} + c \text{ if } n \neq -1 \qquad \int x^{-1} \, dx = \ln x + c$$

In Section 18.1 we showed that the derivative of $f(ax + b)$, where a and b are constant, is $af'(ax + b)$. This can be applied to integrals, provided that a is non-zero.

$$\int \cos (ax + b) \, dx = \frac{1}{a} \sin(ax + b) + c$$

$$\int \sin(ax + b) \, dx = \frac{-1}{a} \cos(ax + b) + c$$

$$\int e^{ax+b} \, dx = \frac{1}{a} e^{ax+b} + c$$

$$\int \frac{1}{ax + b} \, dx = \frac{1}{a} \ln(ax + b) + c$$

$$\int (ax + b)^n \, dx = \frac{1}{a(n + 1)} (ax + b)^{n+1} + c \text{ (provided that } n \neq -1)$$

EXAMPLE 18.8

Find the area under the curve $y = \sin x$ between $x = 0$ and $x = \pi$.

Solution

This region is shown in Fig. 18.7. The area is given by the integral

$$\int_0^\pi \sin x \, dx$$

Evaluate this integral.

$$[-\cos x]_0^\pi = -\cos \pi - (-\cos 0) = -(-1) - (-1)$$

The area is 2.

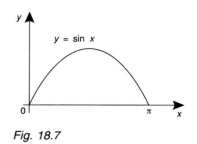

Fig. 18.7

EXERCISE 18E

1 Evaluate the following integrals.

a) $\int (\sin x + \cos x) \, dx$ **b)** $\int (2 \sin x - 3 \cos x) \, dx$ **c)** $\int \sin 2x \, dx$

d) $\int \cos \left(3x + \dfrac{\pi}{3}\right) dx$ **e)** $\int \sin \left(\dfrac{\pi}{6} - 2x\right) dx$ **f)** $\int 2 \cos (\pi - 2x) \, dx$

g) $\int e^{2x} \, dx$ **h)** $\int e^{-x} \, dx$ **i)** $\int e^{2x-3} \, dx$

j) $\int \dfrac{1}{2x + 3} \, dx$ **k)** $\int \dfrac{1}{3 - x} \, dx$ **l)** $\int \dfrac{2}{3 - 3x} \, dx$

m) $\int (2x + 1)^3 \, dx$ **n)** $\int 72(1 - 4x)^5 \, dx$ **o)** $\int (2 - \frac{1}{2}x)^4 \, dx$

p) $\int 6(x + 3)^{-4} \, dx$ **q)** $\int \dfrac{5}{(2x + 3)^6} \, dx$ **r)** $\int (2x + 3)^{\frac{1}{2}} \, dx$

s) $\int \sqrt{3x + 1} \, dx$ **t)** $\int \sqrt[3]{2x + 1} \, dx$ **u)** $\int \dfrac{1}{\sqrt{1 + x}} \, dx$

2 Evaluate the following definite integrals.

a) $\displaystyle\int_0^1 \cos x \, dx$ **b)** $\displaystyle\int_0^{\frac{\pi}{6}} \sin 3x \, dx$ **c)** $\displaystyle\int_0^1 e^x \, dx$

d) $\displaystyle\int_1^2 e^{-2x} \, dx$ **e)** $\displaystyle\int_1^{10} \dfrac{1}{x} \, dx$ **f)** $\displaystyle\int_0^5 \dfrac{1}{2x + 1} \, dx$

g) $\displaystyle\int_{-1}^2 (1 + 2x)^5 \, dx$ **h)** $\displaystyle\int_0^6 \dfrac{1}{(x + 1)^2} \, dx$ **i)** $\displaystyle\int_1^6 \sqrt{4x + 1} \, dx$

3 Find the areas under the following curves between the limits stated.

a) $y = \cos x$, from $x = -\dfrac{\pi}{2}$ to $\dfrac{\pi}{2}$ **b)** $y = \sin 2x$, from $x = 0$ to $x = \dfrac{\pi}{2}$

c) $y = e^{-x}$, from $x = 0$ to $x = 2$ **d)** $y = \dfrac{1}{x}$, from $x = 1$ to $x = 10$

e) $y = \dfrac{1}{2x + 3}$, from $x = 0$ to $x = 5$

4 Show that for all a, $\displaystyle\int_{-a}^a \sin x \, dx = 0$.

*5 In Question 9 of Exercise 18A a function f(x) had the derivative $f'(x) = \sqrt{1 + x^3}$. Find these integrals in terms of f(x).

a) $\displaystyle\int_0^1 \sqrt{1 + x^3} \, dx$
    b) $\displaystyle\int_1^2 \sqrt{1 + 8x^3} \, dx$
    c) $\displaystyle\int_0^1 \sqrt{2 + 3x + 3x^2 + x^3} \, dx$

LONGER EXERCISE

All about e

1 In Section 18.3 we showed by numerical arguments that e lies between 2 and 3. Using similar methods, find e to three decimal places.

This will be very tedious to do on a calculator. The computer investigation on page 384, will set up a spreadsheet to perform this investigation.

2 Investigate what happens to the following expressions when n is large.

a) $\left(1 + \dfrac{1}{n}\right)^n$
    b) $\left(1 - \dfrac{1}{n}\right)^n$
    c) $\left(1 + \dfrac{2}{n}\right)^n$

3 The number e can be found from the following series.

$$e = 1 + 1 + \frac{1}{2!} + \frac{1}{3!} + \frac{1}{4!} + \cdots$$

Use this to find e correct to five decimal places.

*4 Use the definition of part **3** to show that e is irrational. (**Hint:** Suppose that $e = a/b$, and multiply both sides of the expression above by $b!$.)

EXAMINATION QUESTIONS

1 (i) Find the area of the region bounded by the curve with equation $y = (\sin 2x) + 1$ and the lines $x = 0$, $x = \dfrac{\pi}{3}$ and $y = 0$. Leave your answer in terms of π.

   (ii) The gradient at the point (x, y) on a curve is given by $\dfrac{dy}{dx} = e^{3x}$. Given that the curve also passes through the point with coordinates $(0, 1)$ find an equation of this curve.

L AS 1989

2 The equation of the curve C is $y = 2e^x - e^{2x}$.

   (i) Show that C intersects the x-axis at one point whose x-coordinate is ln 2.

   (ii) Find the coordinates of the turning point of C and determine whether it is maximum or a minimum.

   (iii) Sketch the curve C.

   (iv) On your sketch shade the region R for which $x \geq 0$, $y \geq 0$ and $y \leq 2e^x - e^{2x}$.

   (v) Find the area of R.

W AS 1992

3 A water-trough is to be constructed so that its cross-section is a trapezium PQRS in which PQ = RS = 6 cm, QR = 14 cm and ∠SPQ = ∠PSR = θ.

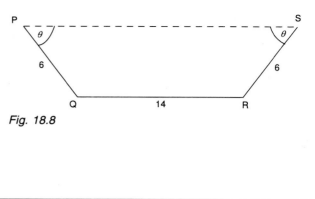

Show that the area A of PQRS is given by

$$A = 84 \sin \theta + 18 \sin 2\theta$$

Fig. 18.8

Prove that A is maximum when $\theta = \cos^{-1} \frac{1}{3}$.

J AS 1991

Summary and key points

1 If a and b are constants, the derivative of $f(ax + b)$ is $af'(ax + b)$.

2 The derivatives of $\sin x$ and $\cos x$ are $\cos x$ and $-\sin x$ respectively.

These results assume that x is measured in radians.

3 The constant e is a number between 2 and 3, such that the derivative of e^x is e^x.

$\log_e x$ is written $\ln x$. The derivative of $\ln x$ is $\frac{1}{x}$.

4 The integrals of $\cos x$ and $\sin x$ are $\sin x$ and $-\cos x$ respectively.

The integrals of e^x and $\frac{1}{x}$ are e^x and $\ln x$ respectively.

If $\int f(x)\, dx = F(x)$, then $\int f(ax + b)\, dx = \frac{1}{a} F(ax + b)$.

Consolidation section D

Chapter 15

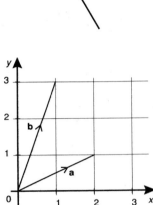

1 The diagram shows vectors **a** and **b**. Make a copy of the diagram, and on it draw the following vectors.

   a) $\mathbf{a} + \mathbf{b}$ **b)** $3\mathbf{b}$ **c)** $\mathbf{b} - \mathbf{a}$ **d)** $2\mathbf{a} - 3\mathbf{b}$

2 **a** and **b** are non-parallel vectors. Solve this equation.

   $(x - 2y)\mathbf{a} + (y + 3x)\mathbf{b} = \mathbf{a} + 10\mathbf{b}$

3 The diagram shows vectors **a** and **b**. Write them in coordinate notation.

4 With **a** and **b** as shown in Question 3, solve the equation $x\mathbf{a} + y\mathbf{b} = 4\mathbf{i} + 7\mathbf{j}$.

5 With **a** and **b** as shown in Question 3, find the moduli of **a**, **b** and **a** + **b**. Find unit vectors parallel to **a** and **b**.

6 Find the angle between the vectors **a** and **b** of Question 3.

7 Find the value of k which will ensure that the lines $\mathbf{r} = \mathbf{i} + t(\mathbf{j} + \mathbf{k})$ and $\mathbf{r} = \mathbf{k} + s(k\mathbf{i} + \mathbf{j} + 2\mathbf{k})$ intersect. With this value of k, find the angle between the two lines.

Chapter 16

8 Sketch the graphs of the following functions.

   a) $y = x(x + 1)(x + 2)$ **b)** $y = \dfrac{3}{x^2}$ **c)** $y = \sqrt{x - 6}$

   d) $y = x^2 - 3x + 4$ **e)** $y = 2 + \dfrac{1}{x}$ **f)** $y = 1 - \dfrac{1}{2 - x}$

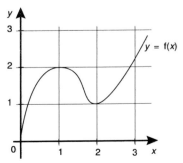

9 The diagram shows the graph of $y = f(x)$. Sketch graphs of the following functions, indicating the positions of the stationary points.

   a) $y = f(x) - 2$ **b)** $y = f(x - 1)$ **c)** $y = f(-x)$

   d) $y = -2f(x)$

10 It is thought there is a relationship between x and y of the form $y = ab^x$. The table below gives values of x and y. Make a suitable transformation to the data, plot a straight line which fits the transformed points, and hence estimate a and b.

x	1	2	3	4	5	6
y	3.9	6.6	11.3	19.2	32.7	55.5

Chapter 17

11 Write down the expansions of these functions.

a) $\sin(x + 60°)$ **b)** $\tan\left(x - \dfrac{\pi}{3}\right)$

12 Factorise these expressions.

a) $\sin 3x + \sin 7x$ **b)** $\cos(x + 10°) - \cos(x - 20°)$

13 Let $f(x) = 2 \sin x + 4 \cos x$. Find the greatest value of $f(x)$. Find the solutions of the equation $f(x) = 3$ in the range $0 \le x \le 2\pi$.

14 Solve each of the following equations in the range indicated.

a) $\sin x + 2 \cos 2x = 1$, for $0° \le x \le 180°$

b) $\cos x - 2 \sin 3x - \cos 5x = 0$, for $0 \le x \le \pi$

Chapter 18

15 Find the derivatives of the following functions.

a) $(2x - 3)^{2.5}$ **b)** $\sqrt{1 - 3x}$ **c)** $\sin(2x - 5)$

d) $3 \cos(2x + 1)$ **e)** e^{2x+1} **f)** $\ln(1 - 3x)$

16 Find the following integrals.

a) $\int (4x - 7)^7 \, dx$ **b)** $\displaystyle\int \frac{1}{\sqrt{2x+1}} \, dx$ **c)** $\int 2 \sin(1 - 2x) \, dx$

d) $\int \cos(0.5x - 2) \, dx$ **e)** $\int e^{3x+1} \, dx$ **f)** $\displaystyle\int \frac{1}{2x - 3} \, dx$

MIXED EXERCISE

1 The first three terms of a geometric progression are $\cos \theta$, $\sin \theta$, $\sin 2\theta$. Find θ, given that it is acute.

2 If θ is not a multiple of π, show that

$\sin \theta + \sin \theta \cos \theta + \sin \theta \cos^2\theta + \cdots = \cot \frac{1}{2}\theta$

3 Let **a** and **b** be defined as in the diagram. Find the values of t for which the modulus of $\mathbf{a} + t\mathbf{b}$ is 5.

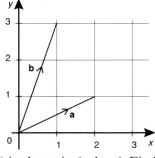

4 Try to find a value of k for which the lines $\mathbf{r} = \mathbf{i} + k\mathbf{j} + t(\mathbf{i} - \mathbf{j} + \mathbf{k})$ and $\mathbf{r} = \mathbf{j} + \mathbf{k} + s(-\mathbf{i} + \mathbf{j} - \mathbf{k})$ intersect. What has gone wrong? Give a geometrical interpretation.

5 Find a unit vector which makes an angle of 60° with the vector $\mathbf{i} + \mathbf{j}$.

6 The point (x, y) divides the line segment between $(1, 0)$ and $(0, 1)$ in the ratio $\theta : 1 - \theta$. Find the x and y coordinates in these cases.

a) $\theta = \frac{1}{3}$ **b)** $\theta = 2$ **c)** $\theta = -1$

What is the geometric interpretation in cases **b)** and **c)**?

7 A line is given by $\mathbf{r} = \mathbf{a} + t\mathbf{u}$, where $\mathbf{u}$ is a unit vector. $\mathbf{r}_0$ is a point not on the line. Show that the point on the line closest to $\mathbf{r}_0$ is given by $t = \mathbf{u} \cdot (\mathbf{r}_0 - \mathbf{a})$.

8 Sketch the graph of $y = \dfrac{1}{x^n}$, where n is a positive integer, in these cases.

a) n is odd   **b)** n is even

9 What transformation will take the graph of $y = \cos x$ to that of $y = \sin x$?

10 Let $f(x) = 3 \sin x + 4 \cos x$. Write $f(x)$ in the form $A \sin(x + \alpha)$. Describe the transformations which will take the graph of $y = f(x)$ to that of $y = \sin x$.

11 Given $f(x) = 3 \sin x + 4 \cos x$, find $f'(x)$

a) directly   **b)** by using the form $A \sin(x + \alpha)$.

Verify that the results are the same.

12 Describe the transformation which transforms the graph of $y = g(x)$ to that of $y = 3g(\frac{1}{3}x)$.

13 The graph of $y = g(x)$ is reflected in the line $x = 2$. Find the equation of the reflected graph.

14 Find surd expressions for the following.

a) $\sin 22.5°$ **b)** $\cos 15°$

15 The diagram shows the graph of $y = \sin x$ for $0 \le x \le \dfrac{\pi}{2}$.

Describe how transformations can be used to extend the graph to the interval $0 \le x \le 2\pi$.

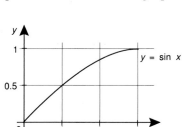

16 Write down the binomial expansion for $\dfrac{1}{1 + x}$ up to the term in x^4. By integrating this expansion, find a series for $\ln(1 + x)$ up to the term in x^5. Use this series to evaluate $\ln 1.2$.

17 The diagram shows three equal squares joined together. Show by trigonometry that $A + B = C$.

This result is a purely geometrical one. See if you can prove it without using trigonometrical functions.

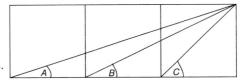

18 Show that the following equation is true for $0 \le \theta \le 45°$.

$$\sqrt{\frac{1 - \sin 2\theta}{1 + \sin 2\theta}} = \frac{1 - \tan \theta}{1 + \tan \theta}$$

LONGER EXERCISE

Cubic equations

1 Show that $\cos 3\theta = 4 \cos^3\theta - 3 \cos \theta$.

2 By substituting $x = \cos \theta$, solve the equation $4x^3 - 3x = 0.5$. (You should get three different solutions, corresponding to different values of 3θ.)

3 Solve the equation $8x^3 - 6x = 1.5$.

4 Consider the equation $x^3 - 3x = 1$. Show that by a suitable choice of k, the substitution $x = ky$ will transform this equation to $4y^3 - 3y = n$, for some n. Hence solve the original equation.

5 Can you use this method to solve these equations?

  a) $4x^3 - 3x = 1.5$ **b)** $x^3 + 3x = 1$

  Why not?

Parabolic telescopes and searchlights revisited

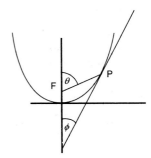

The Longer exercise at the end of Chapter 8 dealt with a mirror in the shape of the curve $y = x^2$. You considered various rays leaving the point F$(0, \frac{1}{4})$, and showed that they were reflected vertically. Now you are in a position to prove this property for all rays.

1 Take a point P(x, x^2) on the curve. Find the gradient of the line joining P to F.

2 Find the gradient of the tangent at P.

3 Let PF make angle θ with the vertical, and the tangent make angle ϕ with the vertical. Use the formula for $\tan 2x$ to show that $\theta = 2\phi$.

4 Show that a ray from F is reflected vertically at P.

Method of least squares

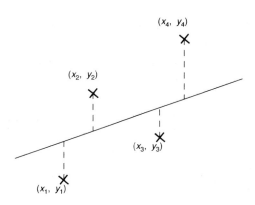

In Section 16.4 of Chapter 16, you fitted straight lines to points, by eye rather than by calculation. There is a more exact method of finding a line of best fit.

Suppose the points are at $(x_1, y_1), (x_2, y_2), ..., (x_n, y_n)$. The points are fixed, so these x and y values are constant. Suppose the line we fit is $y = ax + b$. One way of making this a line of best fit is to minimise the squared differences between the actual y-values and the y-values on the line. This is the **method of least squares**.

1 Show that the sum of the squared differences is $S = \Sigma(ax_i + b - y_i)^2$.

2 Keep b constant, and differentiate S with respect to a. If S is a minimum, show that

$$a\Sigma x_i^2 + b\Sigma x_i = \Sigma x_i y_i$$

3 Keep a constant, and differentiate S with respect to b. If S is a minimum, show that

$$a\Sigma x_i + nb = \Sigma y_i$$

4 The results of parts **2** and **3** can be regarded as simultaneous equations in a and b. Solve these equations to find expressions for a and b in terms of n, Σx_i^2, Σx_i, $\Sigma x_i y_i$ and Σy_i.

5 Plot the values given below on a graph. By eye, draw a line of best fit. Then calculate the line of best fit from the expressions found in part **4**. How close are the two lines?

x	1	2	3	4	5	6
y	12	14	15	16	19	20

EXAMINATION QUESTIONS

1 The figure shows a cross-section of a vertical wall AB of height h m surmounted by a vertical flagpole BC of height H m. D is a point on the same horizontal level as A and AD $= 1$ m. The wall and flagpole subtend equal angles α at D.

Write down expressions for tan α and tan 2α in terms of h and H. By using an appropriate double angle formula, show that

$$H = \frac{h(1 + h^2)}{1 - h^2}$$

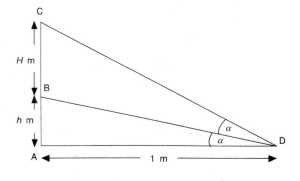

W AS 1991

2 Show that if **a** and **b** are two non-zero vectors, the vector $\mathbf{n} = (\mathbf{b} \cdot \mathbf{b})\mathbf{a} - (\mathbf{a} \cdot \mathbf{b})\mathbf{b}$ is perpendicular to **b**.

Relative to an origin O the points A and B have position vectors $(c\mathbf{i} + c\mathbf{j} + c\mathbf{k})$ and $(c\mathbf{i} - c\mathbf{j} - c\mathbf{k})$ respectively, where c is a constant. Find

a) a vector **m** which is perpendicular to the line OB and is parallel to the plane OAB,

b) a vector equation of the line l_1 which passes through A and is parallel to **m**.

A second line l_2 has equation

$$r = (c\mathbf{i} - c\mathbf{j} - c\mathbf{k}) + t(2\mathbf{i} - \mathbf{j} - \mathbf{k})$$

where t is a parameter. Show that the lines l_1 and l_2 intersect and find the position vector of their point of intersection.

L 1988

3 The variables x, y are related by the equation $y = ax^b$, where a, b are constants.

Approximate measurements of x and y gave the following results.

x	1.22	1.68	2.39	3.52	4.02
y	2.48	5.01	10.86	25.51	34.15

Plot $\ln y$ against $\ln x$ and draw the line of best fit to the plotted points.

Use your line to estimate a and b.

AEB 1991

4 Integrate $\cos\left(\dfrac{x}{3}\right) + e^{2x}$ with respect to x.

Hence evaluate $\displaystyle\int_0^\pi \left[\cos\left(\frac{x}{3}\right) + e^{2x}\right] \, dx$

AEB AS 1992

5 OABC is a trapezium. The position vectors of A, B and C (relative to O as origin) are **a**, **b** and $\mathbf{b} - 3\mathbf{a}$ respectively. Q is the point on AB such that $AQ:QB = 2:1$, R is the point on QC such that $QR:RC = 2:3$ and S is the point on RO such that $RS:SO = 1:5$. Show that the position vector of R is $-\mathbf{a} + \frac{4}{5}\mathbf{b}$, and find the position vector of S.

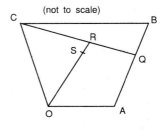

Deduce that SQ is parallel to OA.

SMP 1993

6 a) (i) Show that $(\cos x + \sin x)^2 = 1 + \sin 2x$, for all x.

(ii) Hence, or otherwise, find the derivative of $(\cos x + \sin x)^2$.

b) (i) By expanding $(\cos^2 x + \sin^2 x)^2$, find and simplify an expression for $\cos^4 x + \sin^4 x$ involving $\sin 2x$.

(ii) Hence, or otherwise, show that the derivative of $\cos^4 x + \sin^4 x$ is $-\sin 4x$.

MEI 1992

Progress tests

TEST 1

1. Let $f(x) = 2x^3 + ax^2 - 3x + b$. The remainder when $f(x)$ is divided by $(x+1)$ is -12, and $(x-2)$ is a factor of $f(x)$. Find a and b.

2. A line has equation $x + 2y = 3$. A circle has equation $x^2 + y^2 = 10$. Find the x-coordinates of the points where the line crosses the circle.

3. If $\sin x = 0.3$, find two values of $\cos x$.

4. The numbers $x - 3$, $2x + 1$ and $4x - 4$ are in arithmetic progression. Find the common difference of the progression.

5. Solve these equations.

    a) $9^x = 3^{4-x}$ b) $3^x = 5$ c) $\log_2(2x + 1) - \log_2 x = 3$

6. Find the equation of the tangent to $y = 3x^2 - 8x$ at $(2, -4)$.

7. Find the expansion of $(1 + x)^{\frac{1}{3}}$ up to the term in x^3. Hence find an approximation for $\sqrt[3]{990}$.

8. Find where the curve $y = 3 - 2x - x^2$ crosses the x-axis. Find the area enclosed between the curve and the x-axis.

9. Find the solutions of $\sin x = 0.45$, for values of x between 0 and 4π.

10. Sketch the graph of $y = x + \dfrac{1}{x}$, indicating any stationary points and asymptotes.

11. An Applied Mathematics exam contains six questions on Mechanics and seven on Statistics. Candidates must attempt three questions from each section. How many possible selections are there?

TEST 2

1. A stone is dropped from the top of a cliff. The distance it falls is proportional to the square of the amount of time that passes after it is dropped. After 2 seconds it has dropped 20 m.

    Find an equation giving the distance h m fallen in terms of the time t seconds. Find the time taken to fall 100 m.

2 Find the equation of the line through $(1, 2)$ which is perpendicular to the line with equation $y = 4x - 1$. Find where the two lines meet.

3 Write the following in the form $a + b\sqrt{2}$, where a and b are rational.

  a) $(3 + \sqrt{2})(2 - \sqrt{2})$ **b)** $\dfrac{1 + 3\sqrt{2}}{2 - \sqrt{2}}$

4 A wheel of radius 15 cm is rotating at 20 radians per second. Find the speed of a point on the rim.

The point is initially directly to the right of the centre of the wheel, and the wheel rotates anticlockwise. Find the height of the point above the centre of the wheel 2 seconds after starting.

5 Find the largest angle in a triangle with sides 12 inches, 14 inches and 9 inches. Find the area of the triangle.

6 A cardboard box with square base is to have volume 2000 cm³. The top and bottom are to be made of double thickness cardboard. If x is the side of the base, show that the total area of cardboard to be used is A cm² where

$$A = 4x^2 + \frac{8000}{x}$$

Find the least possible area of cardboard.

7 Find the angle between the vectors $\mathbf{i} + 7\mathbf{j}$ and $2\mathbf{i} - 3\mathbf{j}$.

8 Describe the transformations which will take the graph of $y = x^2$ to those of the following functions.

  a) $y = 3x^2$ **b)** $y = (x - 2)^2$

9 Suppose x and y are angles, for which $\tan x = \frac{3}{5}$ and $\tan y = \frac{1}{8}$. Without the use of a calculator, find the values of $\tan(x + y)$ and $\tan 2x$, leaving your answers as fractions.

10 Evaluate the following.

  a) $\int \sin 3x \, dx$ **b)** $\displaystyle\int_0^7 \sqrt{3x + 4} \, dx$

11 An aunt sends you to the library to borrow two books by the romantic novelist Violet Mullions. On the shelf there are ten books by her. If your aunt has already read four of these books, what is the probability that you will select two books which she has not yet read?

TEST 3

1 The marketing manager of a publisher reckons that if a certain book costs £x, the number sold will be $10\,000 - 1000x$. Show that the total revenue from the book is $10\,000x - 1000x^2$. Write this expression in the form $a - b(x - c)^2$, where a, b and c are constants. Hence find the price which will provide the greatest revenue.

2 $x + 3$, $x + 10$ and $32 - x$ are consecutive terms of a geometric progression. Find the possible values of x.

3 A sector of angle 0.5 radians has area $1.3 \, \text{cm}^2$. Find the radius of the circle from which it is taken.

4 Differentiate the following.

a) $4x^2 + 5x^3$ **b)** $7\sqrt{x} + \dfrac{1}{x^4}$

5 A trapdoor is a square of side 1 m. It is propped open by a stick of length 0.8 m as shown. If the trapdoor makes an angle of $30°$ with the horizontal, find the angle between the stick and the horizontal.

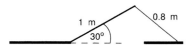

6 Find the stationary points of the curve of $y = x^3 - 2x^2 + x - 3$, stating whether they are maxima, minima or points of inflection. How many roots are there of the equation $x^3 - 2x^2 + x - 3 = 0$?

7 Evaluate $\displaystyle\int_1^3 (x - 1)(x + 3) \, dx$.

8 Solve the equation $6 \sin^2\theta + 7 \cos \theta = 8$, for $0 \le \theta \le 180°$.

9 Find a vector equation of the line joining $(1, 2, 4)$ and $(3, 1, 4)$. Does this line meet the line with vector equation $\mathbf{r} = \mathbf{i} + t(\mathbf{j} + 2\mathbf{k})$?

10 A model of wind resistance gives the resistance F as $av + bv^2$, where v is the speed of the body and a and b are constants. Experiments in a wind tunnel gave the following values of F and v.

v (m s^{-1})	10	20	30	40	50	60
F (newtons)	5.8	13.8	23.5	36.2	51.3	67.2

Using a suitable change of variables, reduce the relationship to linear form. Plot the new variables and draw a straight line through the points. From your straight line estimate the values of a and b.

11 After a car accident, one of the cars drove away. A witness remembers that the registration plate began with F, and had the numbers 5, 6 and 5 and the letters Y, F and L, though not necessarily in those orders. How many possible registration numbers will the police have to check?

TEST 4

1 Two women stand on opposite sides of a tower of height h. They measure the angle of elevation of the top of the tower as α and β. Show that the distance apart of the women is

$h(\cot \alpha + \cot \beta)$

2 Four points are A(1, 1), B(2, 3), C(6, 3) and D(5, 1). Show that ABCD is a parallelogram. M is the midpoint of BC. Find the coordinates of M.

X is the point which divides AM in the ratio 2 : 1. Find the coordinates of X and show that it lies on BD.

3 $1, 2x, 4x^2, 8x^3$ are the first four terms of a geometric progression. If the sum to infinity of the progression exists, find the range of possible values of x. If the sum to infinity is 2, find x.

4 Given that x is a small angle measured in radians, find approximations for the following.

a) $\sin 2x \tan x$　　**b)** $\dfrac{1 - \cos x}{\sin^2 x}$

5 Write down expansions of the following expressions up to the term in x^2.

a) $(4 + 3x)^8$　　**b)** $\sqrt[4]{1 - x}$

6 Find the points on the curve with equation $y = x^3 - 5x^2 + 3x$ where the gradient is 11.

7 Show that for all x, $\dfrac{\tan x + \cot x}{\cot x} = \sec^2 x$.

8 Solve the equation $\sin(x + 0.2) + \sin x = 1.2$, for x in the range $0 \le x \le \pi$.

9 Differentiate the following expressions.

a) $(1 + 3x)^{10}$　　**b)** $\cos 2x - \sin 3x$

10 A curve goes through the point (1, 3), and at the point (x, y) its gradient is $6x^2$. Find the equation of the curve.

11 Every year a man buys a ticket for the state lottery. His chance of winning in any one year is $1/100\,000$. What is the probability that he will win at least one prize in 50 years?

Inequalities

An **equality** or **equation** states that one expression is equal to another. By contrast, an **inequality** states that expressions are not equal, that one expression is **less** than another.

Inequalities involve one of the four signs: $<, \leq, >, \geq$.

$x < 3$	x is less than 3.
$x \leq 4$	x is less than or equal to 4.
$x > 0$	x is greater than 0.
$x \geq -3$	x is greater or equal to -3.

The solution of an equality consists of one or more isolated values. The solution of an inequality consists of a whole range of values.

The solution of an inequality is sometimes shown on a number line, as in Fig. 19.1. The bars indicate the range of values. If the bar ends with a closed or solid dot, the endpoint is included. If it ends with an open or hollow dot, the endpoint is not included.

The top bar represents the inequality $x < 2$. The bar below represents $x \geq -1$.

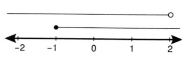

Fig. 19.1

19.1 Linear inequalities in one variable

A linear inequality in x involves only single powers of x. An example of a linear inequality is

$$3x - 2 < 4x - 7$$

The same rules and techniques for solving linear equations apply to linear inequalities, with one additional rule.

When multiplying or dividing by a negative number, the inequality sign **reverses direction**.

If $x > 3$, then $-x < -3$.

If $-2x \leq 6$, then $x \geq -3$.

As an example to justify this, in terms of temperature $20°$ is warmer than $10°$ but $-20°$ is colder than $-10°$.

$20 > 10$, but $-20 < -10$

EXAMPLE 19.1
Solve the inequality $3x + 2 > x - 8$.

Solution
Subtract 2 from both sides, and then subtract x from both sides.

$$3x > x - 10$$

$$2x > -10$$

Divide both sides by 2. This is a positive number, so the inequality sign is unchanged.

$$x > -5$$

The solution is $x > -5$.

EXAMPLE 19.2
Solve the inequality $x(x - 3) < (x + 1)(x - 5)$.

Solution
This seems to be a non-linear inequality, but when we expand the brackets and simplify the non-linear term will cancel.

$$x(x - 3) < (x + 1)(x - 5)$$

$$x^2 - 3x < x^2 - 4x - 5$$

$$-3x < -4x - 5$$

This is now similar to the previous example. Add $4x$ to both sides.

$$x < -5$$

The solution is $x < -5$.

EXERCISE 19A

1 Solve the following inequalities.

a) $3x - 5 < 13$

b) $2x + 3 > 15$

c) $1 - 3x \leq 7$

d) $x + 3 \geq 3x - 9$

e) $4x - 3 < 12 - x$

f) $3 - 2x \leq 5 - 3x$

g) $\frac{1}{2}x + \frac{1}{3}x < 5$ **h)** $\frac{1}{2}x - \frac{1}{3}x \geq 2$ **i)** $\frac{1}{2}x + 1 > \frac{1}{4}x - 1$

j) $2(x + 3) \geq 3(x - 1)$ **k)** $4(1 - 2x) < 3(1 + x)$ **l)** $2(1 - x) \leq 3(2 - x)$

m) $x^2 > (x - 2)(x - 3)$ **n)** $(x - 1)(x - 2) > (x - 3)(x + 4)$

o) $(1 - x)(1 + 4x) \leq (2x - 1)(3 - 2x)$

2 Solve the following inequalities.

    a) $x < x + 1$ **b)** $x < x - 3$

*3 What is the solution of $ax + b < c$, where a, b and c are any numbers?

*4 Which of the following are always true? For those which can be false, give examples of when they are invalid.

    a) If $a < b$ and $c < d$, then $a + c < b + d$ **b)** If $a < b$ and $c < d$, then $ac < bd$

    c) if $a < b$ and $c < d$, then $a - c < b - d$ **d)** If $a < b$ and $c < d$, then $\dfrac{a}{c} < \dfrac{b}{d}$

Two inequalities

Sometimes we are told that a variable satisfies two inequalities. Then the solution set will be the region common to the two solution sets of these inequalities.

EXAMPLE 19.3
Solve the inequalities $3 < 2x - 5 < 7$. Illustrate the solution on a number line.

Solution
Add 5 to the left, middle and right.

    $8 < 2x < 12$

Now divide throughout by 2.

    $4 < x < 6$

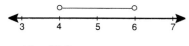

This is illustrated on the number line on the right.

Fig. 19.2

EXAMPLE 19.4
A measurement x is given as 3.2, correct to one decimal place. Express, in terms of inequalities, the range of possible values of x.

Solution
x must be at least 3.15, otherwise it would be rounded to 3.1 or less. Similarly x must be less than 3.25, otherwise it would be rounded to 3.3 or more.

    $3.15 \leq x < 3.25$

So the possible range of x is $3.15 \leq x < 3.25$.

EXERCISE 19B

1 Solve the following inequalities, representing the solutions on a number line.

    a) $3 < x + 1 < 7$ **b)** $5 < 2x - 1 < 13$ **c)** $2 \leq 3x + 5 \leq 14$

2 Express, in terms of inequalities, the range of values of x for each of the following.

    a) $x = 0.34$ correct to two decimal places

    b) $x = 32\,400$ correct to three significant figures

    c) $x = 42\,000$ correct to the nearest thousand

    d) $x = 0.0264$ correct to three significant figures

19.2 Inequalities in two variables

All the inequalities so far have involved only one variable. Their solutions can be represented on a number line. If an inequality involves two variables, then its solution can be represented on a plane.

Suppose the inequality is $x + 2y \leq 3$. First we draw the boundary line $x + 2y = 3$. This line splits the plane into two regions. Above the line, $x + 2y > 3$. Below the line $x + 2y < 3$.

We want $x + 2y$ to be less than 3, so we want the region below the line. It is shown shaded in Fig. 19.3(a).

The inequality $x + 2y \leq 3$ includes the boundary line. If the inequality were $x + 2y < 3$, then the boundary would not be included. It is shown as a broken line in Fig. 19.3(b).

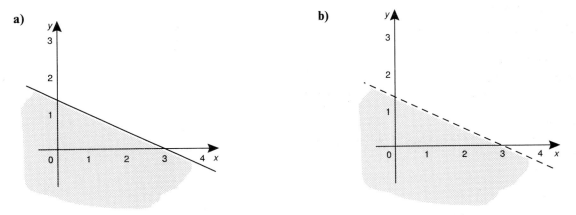

Fig. 19.3

When there are several inequalities to be illustrated simultaneously, it is often clearer to shade out the regions we **don't** want.

EXAMPLE 19.5

A region is bounded by the inequalities $y \geq \frac{1}{2}, y < 3x$ and $y + 4x < 8$. Illustrate these inequalities on a graph, shading out the regions not required. List the points within the region whose coordinates are integers.

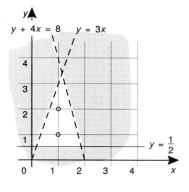

Fig. 19.4

Solution

Draw the lines $y = \frac{1}{2}, y = 3x$ and $y + 4x = 8$. The first line is solid, the second and third lines are broken. We want the region above the first line, to the right of the second and to the left of the third. It is shown unshaded in Fig. 19.4.

The points whose coordinates are integers are circled. They can be listed.

$\qquad$ (1, 1) and (1, 2)

The integer-valued points are (1, 1) and (1, 2)

EXERCISE 19C

1 Illustrate the following inequalities on graphs.

   a) $x + y < 2$ **b)** $2x + 3y \leq 6$ **c)** $x + 2y \geq 4$

   d) $y \geq 2x$ **e)** $y < 3x - 1$ **f)** $2y \leq 3x - 6$

   g) $\frac{1}{2}x + \frac{1}{3}y < 1$ **h)** $2x + 5y > 10$ **i)** $0.1x - 0.5y \leq 1$

2 Identify in terms of inequalities the regions not shaded below.

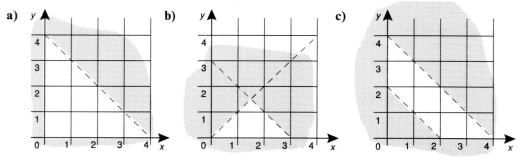

3 Illustrate the following sets of inequalities on graphs, shading out the regions not required.

   a) $x \geq 0, y \geq 0, x + y \leq 3$ **b)** $x > 1, y > 1, 2x + 5y < 10$

   c) $y \geq 0, y \leq x, x + y \leq 4$ **d)** $x \geq 0, y \geq 0, 2x + 3y \leq 12, x + 2y \leq 6$

*4 Illustrate the following inequalities on graphs.

   a) $xy > 0$ **b)** $(x - 1)(y - 2) > 0$ **c)** $(x - y)(x + y) < 0$

19.3 Non-linear inequalities

A non-linear inequality may involve functions of x such as x^2, or $\frac{1}{x}$. Care is needed when solving these inequalities. If an expression involves $\frac{1}{x}$, then we cannot multiply through by x, unless we know whether x is positive or negative. If an expression involves x^2, then we cannot take the square root without considering the negative root.

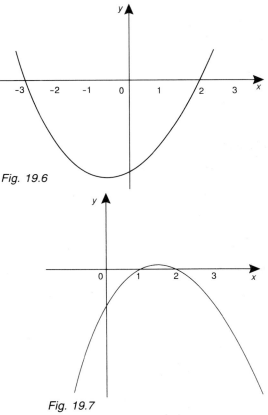

Solution by graph

Suppose we have the inequality $f(x) < 0$. If we know the graph of $y = f(x)$, then we can pick out the negative values of $f(x)$. This is certainly possible when $f(x)$ is a quadratic expression with known roots.

EXAMPLE 19.6

Solve the following inequalities.

a) $(x - 2)(x + 3) < 0$ **b)** $(1 - x)(x - 2) \leq 0$ *Fig. 19.6*

Solution

a) The graph of $y = (x - 2)(x + 3)$ is shown in Fig. 19.6. It crosses the x-axis at $x = -3$ and $x = 2$. The function is negative in between these values.

The solution to the inequality is $-3 < x < 2$.

b) The graph of $y = (1 - x)(x - 2)$ is shown in Fig. 19.7. It crosses the x-axis at $x = 1$ and $x = 2$. The function is negative to the left of 1 and to the right of 2.

The solution to the inequality is $x \leq 1$ or $x \geq 2$.

Fig. 19.7

EXERCISE 19D

1 Solve the following inequalities by sketching their graphs.

    a) $(x + 1)(x - 2) < 0$ **b)** $(x + 2)(x + 5) > 0$ **c)** $x(2x - 3) \leq 0$

    d) $(1 - x)(2 + x) \geq 0$ **e)** $(2 - x)(3 + x) \leq 0$ **f)** $(1 + x)^2 \geq 0$

*2 Solve the inequality $(x - \alpha)(x - \beta) \leq 0$, given that $\alpha < \beta$.

*3 Solve the following inequalities by sketching their graphs.

    a) $(x - 1)(x - 2)(x - 3) < 0$ **b)** $(x + 1)x(x - 2) > 0$ **c)** $x(x^2 + 1) > 0$

Solution by table

One method we can use for solving a non-linear inequality is to collect everything over on one side and factorise. Then we draw up a table and consider whether each of the factors is positive or negative. This table method can be used when we do not know the shape of the graph.

EXAMPLE 19.7
Solve the inequality $x^2 - 4x + 3 > 0$.

Solution
This could be solved by the graph method above, but we shall use it to illustrate the table method. The left-hand side factorises, so we obtain

$$(x - 1)(x - 3) > 0$$

$(x - 1)$ changes sign when x passes 1 and $(x - 3)$ changes sign when x passes 3. Looking at the number line on the right, we see that the points 1 and 3 separate the line into three regions.

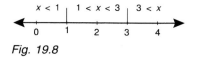

Fig. 19.8

Draw up a table in which the values of x are in regions separated by $x = 1$ and $x = 3$. For each factor, show whether it is positive or negative in each region. For example, the factor $(x - 1)$ is negative up to $x = 1$, and positive thereafter.

	$x < 1$	$1 < x < 3$	$3 < x$
$(x - 1)$	$-$	$+$	$+$
$(x - 3)$	$-$	$-$	$+$
$(x - 1)(x - 3)$	$+$	$-$	$+$

The final row of the table is obtained by seeing whether the factors above are positive or negative. Note that the function is positive in two of the regions.

The solution is $x^2 - 4x + 3 > 0$ for $x < 1$ or $x > 3$.

EXERCISE 19E

1 Solve the following inequalities by setting up a table.

a) $x^2 - 5x + 6 < 0$ **b)** $x^2 + 3x - 10 < 0$ **c)** $x^2 - 8x + 15 \geq 0$

d) $2 - x - x^2 > 0$ **e)** $24 + 2x - x^2 \leq 0$ **f)** $x^2 - 4x + 4 \leq 0$

2 Solve the following inequalities by the table method.

a) $x(x - 1)(x - 2) < 0$ **b)** $(x + 2)(x + 1)(x - 3) > 0$ **c)** $(x + 1)^2(x - 1) > 0$

d) $x(x^2 + 1) \leq 0$ **e)** $x^2(x - 1) \geq 0$ **f)** $x(x - 1)(x - 2)(x - 3) > 0$

Taking square roots

Suppose we have the inequality $x^2 < 4$. When we take the square root of both sides we must be careful with negative values of x. If x is less than -2, its square will be greater than $+4$.

If $x^2 < 4$, then $-2 < x < 2$

In general, for a positive:

If $x^2 < a$, then $-\sqrt{a} < x < \sqrt{a}$.

If $x^2 > a$, then $x < -\sqrt{a}$ or $x > \sqrt{a}$.

Quadratic inequalities are often best solved by completing the square, especially when the quadratic expression does not factorise.

EXAMPLE 19.8

Complete the square of the quadratic $x^2 + 4x - 3$. Hence solve the inequality $x^2 + 4x - 3 > 5$.

Solution

Complete the square using the techniques of Chapter 2.

$$x^2 + 4x - 3 = (x + 2)^2 - 7$$

Now rewrite the inequality.

$$(x + 2)^2 - 7 > 5$$

$$(x + 2)^2 > 12$$

When we take the square root we must consider the negative root of 12 as well as the positive.

$$(x + 2) < -\sqrt{12} \text{ or } (x + 2) > \sqrt{12}$$

The solution is $x < -\sqrt{12} - 2$ or $x > \sqrt{12} - 2$.

EXERCISE 19F

1 Solve the following inequalities.

a) $x^2 < 1$ **b)** $x^2 < 3$ **c)** $2x^2 \leq 3$

d) $x^2 - 1 < 6$ **e)** $x^4 < 3$ **f)** $x^2 > -1$

2 Solve the following inequalities.

a) $(x - 1)^2 < 3$ **b)** $(x + 3)^2 > 5$ **c)** $(x - 2)^2 \leq 5$

d) $(2x - 1)^2 < 2$ **e)** $(3x - 7)^2 > 3$ **f)** $(2x - 5)^2 > 1$

3 Solve the following inequalities by completing the square.

a) $x^2 - 4x + 2 \leq 0$ **b)** $x^2 - 2x - 1 < 0$ **c)** $x^2 + 6x + 2 > 0$

d) $x^2 - 3x - 7 > 0$ **e)** $x^2 - 2x < 5$ **f)** $x^2 \geq 2x + 4$

Fractional inequalities

Suppose an inequality involves algebraic fractions. We cannot multiply both sides by an algebraic expression unless we are confident that it is always positive or always negative.

The usual technique is to collect all the terms on one side and express it as a single fractional expression, then factorise the numerator and denominator of this expression. We are only interested in whether the factors are positive or negative, so it does not matter whether a factor is in the numerator of the fraction or the denominator. We can then use the table method.

EXAMPLE 19.9

Solve the inequality $x > \dfrac{2}{x+1}$.

Solution

Note that we cannot multiply through by $(x + 1)$, as we do not know whether this is positive or negative. Instead collect everything over on the right.

$$0 > \frac{2}{x+1} - x$$

$$0 > \frac{2 - x^2 - x}{x+1}$$

The numerator factorises.

$$0 > \frac{(1-x)(2+x)}{x+1} = f(x)$$

The table we draw up will have three factors: $(1 - x)$ which changes sign at $x = 1$, $(2 + x)$ which changes sign at $x = -2$, and $(x + 1)$ which changes sign at $x = -1$.

	$x < -2$	$-2 < x < -1$	$-1 < x < 1$	$1 < x$
$(x+1)$	$-$	$-$	$+$	$+$
$(2+x)$	$-$	$+$	$+$	$+$
$(1-x)$	$+$	$+$	$+$	$-$
$f(x)$	$+$	$-$	$+$	$-$

Note that $f(x)$ is negative in two of these regions.

$$f(x) < 0 \text{ for } -2 < x < -1 \text{ and for } 1 < x$$

The solution is $x > \dfrac{2}{x+1}$ for $-2 < x < -1$ and for $1 < x$.

EXERCISE 19G

Solve the following inequalities.

1 $\dfrac{x}{x+1} < 2$

2 $\dfrac{x+1}{x-1} > 3$

3 $\dfrac{2x-1}{x} < 4$

4 $\dfrac{x}{x-1} > x$

5 $\dfrac{x+4}{x+1} < x$

6 $\dfrac{2x+3}{x} > x$

7 $\dfrac{1}{x} > \dfrac{1}{x+3}$

8 $\dfrac{2}{x} < \dfrac{1}{x-1}$

9 $\dfrac{1}{2x-1} < \dfrac{1}{x+1}$

10 $\dfrac{x-1}{x+1} > \dfrac{x+3}{x-2}$

11 $\dfrac{x+3}{x+1} < \dfrac{x+1}{x-2}$

19.4 The modulus function and associated inequalities

The **modulus** of a number is its magnitude, regardless of whether it is positive or negative. Modulus is defined in terms of inequalities.

$$|x| = \begin{cases} x \text{ if } x \geq 0 \\ -x \text{ if } x < 0 \end{cases}$$

So the modulus function 'makes a number positive'. If the number is already positive it is unchanged, if the number is negative it is multiplied by -1.

The modulus of a number is sometimes called its **absolute value**. In many computer languages ABS(X) gives the absolute value of X.

The graph of $y = |x|$ is shown in Fig. 19.9. Notice that it consists of two straight line segments: $y = x$ for x positive, and $y = -x$ for x negative.

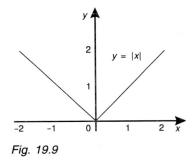

Fig. 19.9

Inequalities involving the modulus function are often dealt with in a similar way to those involving x^2. If $|x| < 1$, x is less than 1 and greater than -1.

If $|x| < 1$, then $-1 < x < 1$.

EXAMPLE 19.10

Solve the inequality $|2x - 3| < 5$.

Solution

Remove the modulus sign, and write the expression inside as greater than -5 and less than $+5$.

$$-5 < 2x - 3 < 5$$

Now add 3 throughout, and then divide by 2.

$$-2 < 2x < 8$$
$$-1 < x < 4$$

The solution is $-1 < x < 4$.

EXAMPLE 19.11
Solve the inequality $|x - 3| \leq |2x + 5|$.

Solution
Both sides are in modulus form. Both sides are positive, so if we square both sides we can remove the modulus signs.

$$(x - 3)^2 \leq (2x + 5)^2$$
$$x^2 - 6x + 9 \leq 4x^2 + 20x + 25$$

Now, all the terms on the right-hand side gives

$$0 \leq 3x^2 + 26x + 16$$

The right-hand side factorises, so that the inequality becomes

$$0 \leq (3x + 2)(x + 8)$$

Solve this by the methods of Section 19.3.

The solution is $x \leq -8$ or $x \geq -\frac{2}{3}$.

EXERCISE 19H

1 Solve the following inequalities.

    a) $|x| < 6$ **b)** $|x| \leq 2$ **c)** $|x| > 5$

    d) $|x| \geq 7$ **e)** $|x + 1| < 5$ **f)** $|x - 3| \leq 4$

    g) $|x + 2| > 0.5$ **h)** $|x - 1| \geq 1$ **i)** $|2x + 1| < 3$

    j) $|3x - 1| \leq 4$ **k)** $|0.5x + 1| < 2$ **l)** $|0.1x + 0.2| < 0.3$

2 Solve the following inequalities.

    a) $|x - 1| \geq |x|$ **b)** $|1 - x| < |2 + x|$ **c)** $|3 + 2x| < |1 - 2x|$

    d) $|2x - 1| < |x - 3|$ **e)** $|3x| \geq |x + 1|$ **f)** $|x + 1| < |2x - 1|$

    g) $\dfrac{1}{|x + 1|} < \dfrac{2}{|x|}$ **h)** $\dfrac{1}{|x + 1|} > \dfrac{1}{|x|}$ **i)** $\dfrac{2}{|x - 1|} \geq \dfrac{3}{|x - 2|}$

3 Express the following statements using modulus signs.

    a) x is within 1 of 5   **b)** x is more than 2 away from 3

    c) x is 510, correct to the nearest 10   **d)** x is 0.34, correct to two decimal places

Solution by graph

Modulus inequalities can sometimes be solved with the help of a graph. This is especially useful when only one part of the inequality is enclosed in modulus signs.

EXAMPLE 19.11

Solve the inequality $|x - 2| > x$.

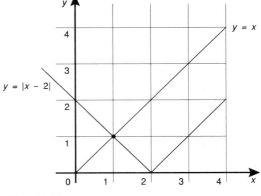

Solution

The graph of $y = |x - 2|$ is obtained by shifting the graph of $y = |x|$ two units to the right. It is shown in Fig. 19.10. The graph of $y = x$ is also shown. Notice that they cross at $(1, 1)$. Up to this point the modulus graph lies above the $y = x$ line.

The solution is $x < 1$.

Fig. 19.10

EXERCISE 19I

Solve the following inequalities using graphs.

1 $|x| > 1 - x$

2 $|x| < x + 1$

3 $|x + 1| > x$

4 $|x - 3| < 2x$

5 $|x| > 2x + 3$

6 $|2x| < x + 1$

19.5 Practice in solving inequalities

There is no single rule for solving inequalities. See what sort of inequality you have been given, then apply the appropriate technique. Below is a set of miscellaneous exercises covering the different sorts of inequality met in this chapter.

EXERCISE 19J

1 Solve the following inequalities.

a) $3x - 4 < 8$

b) $x^2 - 8x + 7 > 0$

c) $|x - 7| > |2x|$

d) $x(x - 3) \leq (x - 2)(x + 7)$

e) $\dfrac{1}{x - 3} > 2$

f) $x(2x - 1)(x - 3) > 0$

g) $|x + 1| < |3 - x|$

h) $\dfrac{1}{|x|} > \dfrac{2}{|x + 2|}$

i) $\dfrac{x + 2}{x - 1} < \dfrac{1}{2}$

j) $2|x - 3| > |x + 1|$

k) $5 + x - x^2 < 0$

l) $(x + 1)^2 \leq 4x$

*2 Solve the equations $x^2 - 17x + 60 = 0$ and $x^2 - 17x - 60 = 0$. Hence solve the inequality $|x^2 - 17x| < 60$.

*3 Show by giving examples that each of the following arguments is false.

a) Suppose $1 - x < 3$. Then $x < -2$.

b) Suppose $x(x - 1) < 0$. Then $x - 1 < 0$, and so $x < 1$.

c) Suppose $\dfrac{1}{x} < 1$. Then $1 < x$.

d) Suppose $|x - 1| < |2x|$. Then $x - 1 < 2x$, and so $-1 < x$.

LONGER EXERCISE

Linear programming

The two-dimensional inequalities of Section 19.2 may be used to represent restrictions, of material or time or labour. The solution region of the inequalities then represents the possibilities available. Often, there is a function that we wish to maximise or minimise, subject to the restrictions, and if the function is a linear one the process is called **linear programming**.

If the restrictions are all linear then the solution region will be a polygon. If the function concerned is linear then the maximum or minimum will be reached at one of the corners of the polygon.

We shall now take a simple linear programming problem.

An amusement arcade proprietor has 15 m^2 of space in which to install new machines. He has a choice between 'Trollslayer', which takes up 2 m^2 and costs £3000, and 'Golden Gloves' which takes up 3 m^2 and costs £2000. He has £15 000 to spend.

1 Suppose he buys x 'Trollslayers' and y 'Golden Gloves'. Show that the following inequalities must be obeyed.

$$x \geq 0, \quad y \geq 0, \quad 2x + 3y \leq 15, \quad 3x + 2y \leq 15$$

2 Illustrate these inequalities on a graph.

3 The weekly profit from each 'Trollslayer' is £200, and from each 'Golden Gloves' it is £290. Write down an expression in x and y for the total weekly profit.

4 Find how many of each machine should be bought to maximise the total weekly profit.

Here is another problem in linear programming.

5 A farmer has a choice of two types of cattle feed, Type A and Type B. A hundredweight of Type A costs £11, and provides 20 units of protein and 30 units of carbohydrate. A hundredweight of Type B costs £15, and provides 20 units of protein and 45 units of carbohydrate. To keep the herd healthy, they must receive at least 200 units of protein and 360 units of carbohydrate per day. What is the cheapest way of feeding the herd?

EXAMINATION QUESTIONS

1 Given that $x \neq 2$, find the complete set of values of x for which

$$\frac{3x + 1}{x - 2} > 2$$

L 1988

2 By sketching appropriate graphs, or otherwise, solve the inequalities.

(i) $x(x-1) \geq 6$ (ii) $|2x-3| < x+1$

W AS 1991

3 Sets of points A, B, C of the xy-plane are defined as follows.

$$A = \{(x, y) : x \geq 2\}$$

$$B = \{(x, y) : x + y \leq 4\}$$

$$C = \{(x, y) : y < -3\}$$

On a sketch, shade the region representing the set of points in A and B but not in C.

L AS 1990

4 In four separate diagrams, illustrate the following four sets of points.

(i) $S = \{(x, y) : y \leq 6, \text{ where } x, y \text{ are real}\}$

(ii) $T = \{(x, y) : y = x^2 - 5x + 6, \text{ where } x, y \text{ are real}\}$

(iii) $U = \{(x, y) : y \geq |x^2 - 5x + 6|, \text{ where } x, y \text{ are real}\}$

(iv) The region common to S and U.

O&C 1992

Summary and key points

1 Linear inequalities in one variable can be solved by the methods of solving linear equations. When multiplying or dividing by a negative number reverse the direction of the inequality sign.

2 Linear inequalities in two variables can be illustrated on a graph. Draw the lines corresponding to the equalities, then shade the appropriate region.

3 Quadratic inequalities are solved either by factorising or by completing the square.

If an inequality involves algebraic fractions then collect all terms on one side.

Do not multiply or divide by a term unless it is always positive.

4 The modulus function ensures that an expression is positive. If $|x| < k$ then $-k < x < k$.

Differentiation of compound functions

So far, you have learned how to differentiate x^n, $\sin x$ and $\cos x$, e^x and $\ln x$. You have differentiated sums of functions, and constant multiples of these functions. You have differentiated functions of linear functions.

Many functions which we have to deal with are compounds of the basic functions. Examples are

$x \sin x$	a product of two basic functions
e^x/x	a quotient of two basic functions
$\sqrt{1 + x^2}$	the composition of two basic functions

These are all **compound** functions. In this chapter we shall see how to differentiate them.

Any compound of the basic functions, however complicated, can be differentiated using the rules of this chapter. It is a routine procedure – indeed it can be done by computer. A computer package such as 'Derive' is able to differentiate any compound function.

20.1 The rule for differentiating products of functions

The product rule

Suppose we have a function obtained by multiplying together two other functions, such as $y = uv$, where u and v are both functions of x. The **product rule** states

$$\frac{dy}{dx} = u\frac{dv}{dx} + v\frac{du}{dx}$$

Proof

Suppose x increases by a small amount δx. Then u will increase by a small amount δu, v by a small amount δv, and y by a small amount δy. Write down the new value of y.

$$y + \delta y = (u + \delta u)(v + \delta v)$$

Expand this.

$$y + \delta y = uv + u\delta v + v\delta u + \delta u \delta v$$

Recall that $y = uv$. Subtract this from both sides.

$$\delta y = u\delta v + v\delta u + \delta u \delta v$$

Divide through by δx.

$$\frac{\delta y}{\delta x} = u\frac{\delta v}{\delta x} + v\frac{\delta u}{\delta x} + \frac{\delta u \delta v}{\delta x}$$

Now let δx tend to zero. $\dfrac{\delta y}{\delta x}$ tends to $\dfrac{dy}{dx}$, and similarly $\dfrac{\delta u}{\delta x}$ and $\dfrac{\delta v}{\delta x}$ tend to $\dfrac{du}{dx}$ and $\dfrac{dv}{dx}$. The last term, $\delta u\dfrac{\delta v}{\delta x}$ tends to $\delta u\dfrac{dv}{dx}$, which tends to zero. The final result is

$$\frac{dy}{dx} = u\frac{dv}{dx} + v\frac{du}{dx}$$

EXAMPLE 20.1

Find the derivative of $x^2 \sin x$.

Solution

Here $u = x^2$ and $v = \sin x$. We know how to differentiate both of these functions.

$$\frac{du}{dx} = 2x \quad \text{and} \quad \frac{dv}{dx} = \cos x$$

Apply the product rule above.

$$\frac{dy}{dx} = u\frac{dv}{dx} + v\frac{du}{dx} = x^2 \times \cos x + \sin x \times 2x$$

The derivative is $x^2 \cos x + 2x \sin x$.

EXERCISE 20A

1 Differentiate the following expressions.

a) $x^2 \cos x$

b) xe^x

c) $e^x \cos x$

d) $(x^2 + 3x + 1)e^x$

e) $(1 + 3x^2 + 4x^3)(2 - 4x^4)$

f) $(\sin x + \cos x)e^x$

g) $\sqrt{x} \sin x$

h) $e^x \ln x$

i) $x \ln x$

j) $(x^2 + 2x - 1)\ln x$ **k)** $\sin x \ln x$ **l)** $\ln x \cos x$

m) $x \sin 2x$ **n)** $\cos(3x - 1)\,e^x$ **o)** $e^{-x}(x + 3)$

p) $x \ln(2x - 1)$ **q)** $x^2 e^{3x-1}$ **r)** $\sin 2x \cos 3x$

s) $x\sqrt{x + 1}$ **t)** $(x^2 + 1)\sqrt{x - 1}$ **u)** $x(2x - 1)^{-\frac{1}{2}}$

2 Write x^5 as $x^2 \times x^3$. Find the derivative of x^5

    a) directly

    b) using the product rule applied to $x^2 \times x^3$.

    Check that the results are the same.

3 Find the stationary point of $y = xe^{-x}$ and identify whether it is a maximum or minimum.

4 Show that there is a stationary point of $y = e^x \cos x$ in the interval $0 < x < \dfrac{\pi}{2}$. Find its coordinates, and identify whether it is a maximum or minimum.

5 Find the equation of the tangent to $y = (x + 1)\cos x$ at $(0, 1)$.

6 Find the equation of the tangent to $y = x^2 \ln x$ at the point (e, e^2).

*7 Find a rule for differentiating a product of three functions, $y = uvw$.

*8 Figure 20.1 shows the cross-section of a gutter. The sides and base are of equal length a. The sides are inclined at an equal angle θ to the horizontal, as shown. Find an expression for the area enclosed by ABCD, and hence find its maximum.

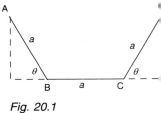

Fig. 20.1

20.2 The rule for differentiating quotients of functions

The quotient rule

Suppose we have a function obtained by dividing one function by another, such as $y = \dfrac{u}{v}$, where u and v are both functions of x. The **quotient rule** states

$$\frac{dy}{dx} = \frac{v\dfrac{du}{dx} - u\dfrac{dv}{dx}}{v^2}$$

Proof

As for the proof of the product rule, let x, y, u and v increase by small amounts δx, δy, δu and δv respectively. Rewriting y gives

$$y + \delta y = \frac{u + \delta u}{v + \delta v}$$

$$\delta y = \frac{u + \delta u}{v + \delta v} - \frac{u}{v} = \frac{v(u + \delta u) - u(v + \delta v)}{(v + \delta v)v}$$

Expand the top and simplify.

$$\delta y = \frac{vu + v\delta u - uv - u\delta v}{(v + \delta v)v} = \frac{v\delta u - u\delta v}{(v + \delta v)v}$$

Divide both sides by δx.

$$\frac{\delta y}{\delta x} = \frac{v\dfrac{\delta u}{\delta x} - u\dfrac{\delta v}{\delta x}}{(v + \delta v)v}$$

Finally, let δx get smaller so $\dfrac{\delta y}{\delta x}$ tends to $\dfrac{dy}{dx}$, and $\dfrac{\delta u}{\delta x}$ and $\dfrac{\delta v}{\delta x}$ tend to $\dfrac{du}{dx}$ and $\dfrac{dv}{dx}$.

The bottom line tends to $v \times v = v^2$.

$$\frac{dy}{dx} = \frac{v\dfrac{du}{dx} - u\dfrac{dv}{dx}}{v^2}$$

EXAMPLE 20.2
Find the derivative of $\dfrac{\sin x}{e^x}$.

Solution
Here $u = \sin x$ and $v = e^x$. Differentiating,

$$\frac{du}{dx} = \cos x \quad \text{and} \quad \frac{dv}{dx} = e^x$$

Put these into the formula of the quotient rule.

$$\frac{e^x \cos x - \sin x \, e^x}{(e^x)^2}$$

We can divide top and bottom by e^x.

The derivative is $\dfrac{\cos x - \sin x}{e^x}$.

EXERCISE 20B

1 Differentiate the following.

a) $\dfrac{\cos x}{x^2}$

b) $\dfrac{e^x}{\sin x}$

c) $\dfrac{x}{e^x}$

d) $\dfrac{x}{x^2+1}$

e) $\dfrac{x+1}{x^2+x+1}$

f) $\dfrac{\ln x}{x}$

g) $\dfrac{\cos x}{\ln x}$

h) $\dfrac{\cos 2x}{\sin 3x}$

i) $\dfrac{e^{2x-1}}{x^2}$

2 Show that the derivative of $\dfrac{\cos x}{\cos x + \sin x}$ is of the form $\dfrac{k}{(\cos x + \sin x)^2}$, where k is constant.

3 Find the stationary points of the following and indicate their nature.

a) $\dfrac{x}{(x+1)^2}$

b) $\dfrac{x^2 - x + 1}{x^2 + x + 1}$

c) $\dfrac{x^3}{1 + 3x^4}$

4 Find the equation of the tangent to $y = \dfrac{x}{x^2+1}$ at the point $(0, 0)$.

Differentiation of tan, cot, sec and cosec

In Chapter 18 we found the derivatives of $\sin x$ and $\cos x$. Now we are in a position to differentiate the other four functions. The results are

$$\frac{d}{dx}\tan x = \sec^2 x \qquad\qquad \frac{d}{dx}\cot x = -\mathrm{cosec}^2 x$$

$$\frac{d}{dx}\sec x = \sec x \tan x \qquad\qquad \frac{d}{dx}\mathrm{cosec}\, x = -\mathrm{cosec}\, x \cot x$$

Proof
We shall prove the results for $\tan x$ and $\sec x$, leaving the other two as exercises.

tan

We know that $\tan x = \dfrac{\sin x}{\cos x}$. We can use the quotient rule, using $u = \sin x$ and $v = \cos x$.

$$\frac{d}{dx}\tan x = \frac{\cos x\dfrac{d}{dx}\sin x - \sin x\dfrac{d}{dx}\cos x}{\cos^2 x}$$

The derivatives of $\sin x$ and $\cos x$ are $\cos x$ and $-\sin x$.

$$\frac{d}{dx}\tan x = \frac{\cos x \cos x - \sin x\,(-\sin x)}{\cos^2 x} = \frac{\cos^2 x + \sin^2 x}{\cos^2 x}$$

Use the facts that $\cos^2 x + \sin^2 x = 1$ and that $\dfrac{1}{\cos x} = \sec x$.

$$\frac{d}{dx}\tan x = \sec^2 x$$

sec

Use the fact that $\sec x = \dfrac{1}{\cos x}$. The quotient rule applies, with $u = 1$ and $v = \cos x$.

$$\frac{d}{dx} \sec x = \frac{\cos x \dfrac{d}{dx} 1 - 1 \dfrac{d}{dx} \cos x}{\cos^2 x}$$

1 is a constant, so its derivative is zero. The derivative of $\cos x$ is $-\sin x$.

$$\frac{d}{dx} \sec x = \frac{+\sin x}{\cos^2 x} = \frac{1}{\cos x} \times \frac{\sin x}{\cos x}$$

$$\frac{d}{dx} \sec x = \sec x \tan x$$

EXAMPLE 20.3
Find the derivative of $y = x \sec x$.

Solution
Apply the product rule, with $u = x$ and $v = \sec x$.

$$\frac{dy}{dx} = \sec x + x \sec x \tan x$$

EXERCISE 20C

1 Show that the derivative of $\cot x$ is $-\text{cosec}^2 x$.

2 Show that the derivative of $\text{cosec } x$ is $-\text{cosec } x \cot x$.

3 Find the derivatives of the following functions.

   a) $\tan 3x$ **b)** $\sec(2x - 1)$ **c)** $\cot\left(4x - \dfrac{\pi}{2}\right)$

   d) $x \tan x$ **e)** $\dfrac{\text{cosec } x}{x}$ **f)** $e^x \cot x$

4 Find the equation of the tangent to $y = \tan x$ at the point $\left(\dfrac{\pi}{4}, 1\right)$.

5 Find the equation of the normal to $y = x \sec x$ at the point $\left(\dfrac{\pi}{3}, \dfrac{2\pi}{3}\right)$.

20.3 The rule for differentiating composite functions

The chain rule

Consider the function $y = \sqrt{\cos x}$. This is a composition of two functions. It can be thought of as an inner and an outer function.

The inner function is applied to x. It is $z = \cos x$. The outer function is applied to $z = \cos x$. It is $y = \sqrt{z}$. To differentiate a function like this we need the **chain rule**.

Suppose we have two functions composed together. Say we have $y = f(z)$ and $z = g(x)$, so $y = f(g(x))$.

$$\frac{dy}{dx} = \frac{dy}{dz} \times \frac{dz}{dx}$$

Proof

Let x, y and z increase by small amounts δx, δy and δz. Then by ordinary algebra

$$\frac{\delta y}{\delta x} = \frac{\delta y}{\delta z} \times \frac{\delta z}{\delta x}$$

As x gets smaller, $\dfrac{\delta y}{\delta x}, \dfrac{\delta y}{\delta z}$ and $\dfrac{\delta z}{\delta x}$ tend to $\dfrac{dy}{dx}, \dfrac{dy}{dz}, \dfrac{dz}{dx}$ respectively.

$$\frac{dy}{dx} = \frac{dy}{dz} \times \frac{dz}{dx}$$

Notes

1 The derivative $\dfrac{dy}{dz}$ is a function of z, which can then be converted back to a function of x.

2 The result of Section 18.1 of Chapter 18 is a special case of the chain rule. If $y = f(ax + b)$, then letting $z = ax + b$

$$\frac{dy}{dx} = \frac{dy}{dz} \times \frac{dz}{dx} = f'(z) \times a = a\, f'(ax + b)$$

EXAMPLE 20.4

Differentiate $y = (\sin x)^2$.

Solution

The inner function here is $z = \sin x$. The outer function is $y = z^2$. Use the chain rule.

$$\frac{dy}{dx} = \frac{dy}{dz} \times \frac{dz}{dx} = 2z \cos x.$$

Write z in terms of x.

$$\frac{dy}{dx} = 2 \sin x \cos x$$

EXERCISE 20D

1 Differentiate the following.

a) $\sin x^2$ b) e^{x^2} c) $\ln x^2$

d) $\sqrt{1-x^2}$

e) $\sqrt{\ln x}$

f) $\sin e^x$

g) $\cos^2 x$

h) $\tan^2 x$

i) $\sqrt{\tan x}$

j) $\ln \cos x$

k) $\ln(\sec x + \tan x)$

l) $(1+x^2)^{10}$

m) $\left(x+\dfrac{1}{x}\right)^8$

n) $\sin(1+\sqrt{x})$

o) $\cos(\cos x)$

2 Find the stationary points of $y = \ln\left(x^2 + \dfrac{1}{x^2}\right)$, describing their nature.

3 Find the stationary points of $y = \sqrt{3 - 2x - x^2}$, identifying their nature.

4 Find the equation of the tangent to $y = \cos(1+x^2)$ at the point $(0, \cos 1)$.

*5 Suppose f(x) is a function for which $f'(x) = \sqrt{1+x^3}$. Find the derivatives of these functions.

a) $f(x^2)$

b) $f(\sqrt[3]{x})$

c) $f(e^x)$

Inverse trigonometric functions

Sometimes we need to find the derivatives of inverse trigono-metric functions.

$$\frac{d}{dx}\sin^{-1}x = \frac{1}{\sqrt{1-x^2}} \qquad \frac{d}{dx}\tan^{-1}x = \frac{1}{1+x^2}$$

Proof

$\sin^{-1}$

Let $y = \sin^{-1} x$.

Then $x = \sin y$.

Differentiate this with respect to y.

$$\frac{dx}{dy} = \cos y$$

Use the identity $\cos^2 y + \sin^2 y = 1$ to write

$$\cos y = \sqrt{1 - \sin^2 y} = \sqrt{1 - x^2}$$

$$\frac{dx}{dy} = \sqrt{1 - x^2}$$

So,

$$\frac{dy}{dx} = \frac{1}{\sqrt{1-x^2}}$$

tan^{-1}

Let $y = \tan^{-1}x$.

Then $x = \tan y$.

Differentiate this with respect to y.

$$\frac{dx}{dy} = \sec^2 y$$

Use the identity $\sec^2 y = 1 + \tan^2 y$.

$$\frac{dx}{dy} = 1 + \tan^2 y = 1 + x^2$$

So,

$$\frac{dy}{dx} = \frac{1}{1 + x^2}$$

EXAMPLE 20.5
Find the derivative of $y = \tan^{-1}x^2$.

Solution
This is a composite function: $y = \tan^{-1}z$, where $z = x^2$. Apply the chain rule.

$$\frac{dy}{dx} = \frac{dy}{dz} \times \frac{dz}{dx} = \frac{1}{1 + z^2} \, 2x$$

The derivative is $\dfrac{2x}{1 + x^4}$.

EXERCISE 20E

1 Differentiate the following expressions.

a) $\sin^{-1}3x$ **b)** $\tan^{-1}2x$ **c)** $\sin^{-1}(2x - 1)$

d) $x \sin^{-1}x$ **e)** $x^2 \tan^{-1}x$ **f)** $\sin^{-1}x^2$

g) $\tan^{-1}\dfrac{1}{x}$ **h)** $\sin^{-1}e^x$ **i)** $(\sin^{-1}x)^2$

2 Find the derivatives of $\cos^{-1}x, \cot^{-1}x, \sec^{-1}x$ and $\csc^{-1}x$.

*3 Find the derivative of $\sin^{-1}(\sin x)$, simplifying your answer as far as possible.

*4 Let $f(x) = \tan^{-1}\left[\dfrac{1 + x}{1 - x}\right]$. Find $f'(x)$, and show that $f(x) = \tan^{-1}x + k$, where k is a constant to be found.

20.4 Practice in differentiation

When you are given a compound function to differentiate, first decide whether it is a product, a quotient or a composition. Then apply the correct rule.

EXAMPLE 20.6

Differentiate $y = \sqrt{1 + x^2}$.

Solution

This is a composite function. The chain rule must be used. The inner function is $z = 1 + x^2$, the outer function is $y = \sqrt{z}$. Apply the rule

$$\frac{dy}{dx} = \tfrac{1}{2}z^{-\frac{1}{2}} 2x$$

$$\frac{dy}{dx} = x(1 + x^2)^{-\frac{1}{2}}$$

EXERCISE 20F

1 Differentiate the following expressions.

a) $x \cos x$ **b)** $\cos x \, e^x$ **c)** $\cos e^x$

d) $\cos x \ln x$ **e)** $\dfrac{\ln x}{\cos x}$ **f)** $\ln(\sin x)$

2 Sometimes we have a choice of which rule to use. Show that $\sec x$ can be expressed either as $\dfrac{1}{\cos x}$ or as $(\cos x)^{-1}$. Differentiate the first form by the quotient rule, and the second form by the chain rule. Show that they give the same result.

Using the rules more than once

Often we have to use the rules more than once.

EXAMPLE 20.7

Differentiate $y = x \sqrt{1 + x^2}$.

Solution

This is a product of x and $\sqrt{1 + x^2}$. The second function is itself a compound function, so we shall have to use both the product rule and the chain rule.

Let $u = x$ and $v = \sqrt{1 + x^2}$. Note that we have already found $\dfrac{dv}{dx}$, in Example 20.6 above.

$$\frac{dy}{dx} = x \times x(1+x^2)^{-\frac{1}{2}} + \sqrt{1+x^2} \times 1$$

$$\frac{dy}{dx} = x^2(1+x^2)^{-\frac{1}{2}} + \sqrt{1+x^2}$$

EXERCISE 20G

1 Differentiate the following expressions.

a) $x^2\sqrt{1+x^2}$ **b)** $\sin\sqrt{1+x^2}$ **c)** $e^{\sqrt{1+x^2}}$

2 Find the derivative of $\ln(1-x^2)$. Hence find the derivative of each of these expressions.

a) $x\ln(1-x^2)$ **b)** $\dfrac{\sin x}{\ln(1-x^2)}$ **c)** $\ln(\ln(1-x^2))$

3 Differentiate the following expressions.

a) $x\cos x^2$ **b)** $\sin\sqrt{\cos x}$ **c)** $\tan^{-1}\sqrt{x^2-1}$

d) $\dfrac{\cos e^x}{x^2}$ **e)** $\sin(\cos x^2)$ **f)** $\ln(\ln(\ln x))$

4 Find the stationary points of $y = \dfrac{(x+1)}{\sqrt{x^2+2}}$ and identify their nature.

*5 An isosceles triangle is drawn inside a circle of radius 1, as shown. The angle at the top is 2θ. Show that the area of the triangle is $4\sin\theta\cos^3\theta$. Find the maximum area of the triangle.

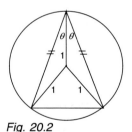

*6 Suppose an object is placed in front of a magnifying lens, so that the image is on the other side of the lens. If u and v are the distances of the object and the image from the lens respectively, the following law holds.

$$\frac{1}{v} + \frac{1}{u} = \frac{1}{f}$$

where f is the constant focal length of the lens.

Find an expression for $u + v$ in terms of u and f, and hence find the least distance apart of the object and its image.

Fig. 20.2

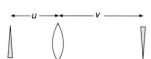

Fig. 20.3

*7 Let a cylinder be cut from a sphere of radius r. Suppose a radius from the centre of the sphere to a point on the rim of the cylinder makes angle θ with the axis of the cylinder. Show that the volume of the cylinder is given by

$$V = 2\pi r^3 \sin^2\theta \cos\theta$$

Find the greatest value of V.

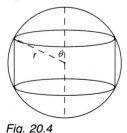

Fig. 20.4

*8 Find the greatest volume of a cone which can be cut from a sphere of radius r.

LONGER EXERCISE

Snell's law

This law tells us how a ray of light is refracted when it passes between two media, such as air and glass or water.

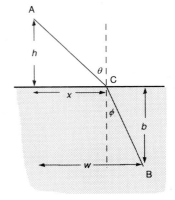

In the diagram the source of light A is h above the glass surface. It reaches a point B b below the surface. The horizontal distance apart of A and B is w.

Let the speed of light in air and glass be v_1 and v_2 respectively. Suppose that the ray strikes a point C on the glass surface which is x horizontally from A.

Show that the time the ray takes to go from A to B via C is

$$T = \frac{\sqrt{x^2 + h^2}}{v_1} + \frac{\sqrt{w^2 + x^2 - 2wx + b^2}}{v_2}$$

Fig. 20.5

Light travels so that it always goes between two points in the shortest possible time. Show that

$$\frac{x}{v_1 \sqrt{x^2 + h^2}} = \frac{w - x}{v_2 \sqrt{w^2 + x^2 - 2wx + b^2}}$$

Do not attempt to solve this equation! Instead rewrite it in terms of the angles θ and ϕ which the ray of light makes with the normal to the glass, before and after refraction. This relationship between the angles is **Snell's law**.

EXAMINATION QUESTIONS

1 The diagram shows a house at A, a school at D and a straight canal BC, where ABCD is a rectangle with $AB = 2\,km$ and $BC = 6\,km$.

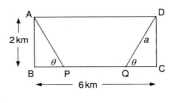

During the winter, when the canal freezes over, a boy travels from A to D by walking to a point P on the canal, skating along the canal to a point Q and then walking from Q to D. The points P and Q being chosen so that the angles APB and DQC are both equal to θ.

Fig. 20.6

Given that the boy walks at a constant speed of $4\,km\,h^{-1}$ and skates at a constant speed of $8\,km\,h^{-1}$, show that the time, T minutes, taken for the boy to go from A to D along this route is given by

$$T = 15\left(3 + \frac{4}{\sin\theta} - \frac{2}{\tan\theta}\right)$$

Show that, as θ varies, the minimum time for the journey is approximately 97 minutes.

L 1988

2 The figure shows a plan view of a thin cable, in two straight sections AP and PB, which is being laid partly underground from A to P, and partly alongside an existing wall from P to B. The plane APB is horizontal and the point C lies on BP produced such that angle ACB = 90°.

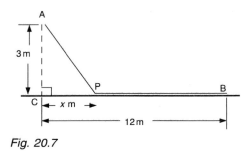

Fig. 20.7

The cost of laying the section AP is £20 per metre; the cost of laying the section PB is £16 per metre. When PC = x metres, the total cost of laying the cable is £y.

a) Show that

$$y = 20\sqrt{(9 + x^2)} + 16(12 - x)$$

b) Determine the positive value of x for which y has a stationary value.

c) Find $\dfrac{d^2y}{dx^2}$.

d) Determine whether the stationary value is a maximum or minimum.

e) Determine the greatest and least costs of laying the cable, justifying your answers.

O 1991

3 Given that $y = e^{-x^2}$ find expressions for

(i) $\dfrac{dy}{dx}$ **(ii)** $\dfrac{d^2y}{dx^2}$

Hence find the x-coordinates of the two points on the graph of y for which $\dfrac{d^2y}{dx^2}$ is equal to zero. Show that these are both points of inflection.

W 1992

---

1 If $y = uv$, then $\dfrac{dy}{dx} = u\dfrac{dv}{dx} + v\dfrac{du}{dx}$

Be sure to get this formula the right way round.

2 If $y = \dfrac{u}{v}$ then $\dfrac{dy}{dx} = \dfrac{v\dfrac{du}{dx} - u\dfrac{dv}{dx}}{v^2}$

Be sure to get u and v the right way round.

3 The derivatives of trigonometric functions are given by:

$$\tan x \rightarrow \sec^2 x$$
$$\cot x \rightarrow -\operatorname{cosec}^2 x$$
$$\sec x \rightarrow \sec x \tan x$$
$$\operatorname{cosec} x \rightarrow -\operatorname{cosec} x \cot x$$

4 If $y = f(z)$ and $z = g(x)$ so $y = f(g(x))$, then

$$\frac{dy}{dx} = \frac{dy}{dz} \times \frac{dz}{dx}$$

In the expression $\frac{dy}{dz}$, y is differentiated as a function of z, not of x.

Write your final answer as a function of x.

5 The derivatives of $\sin^{-1}x$ and $\tan^{-1}x$ are $\dfrac{1}{\sqrt{1-x^2}}$ and $\dfrac{1}{1+x^2}$ respectively.

6 Make sure you use the correct rule when differentiating a compound function. Sometimes it is necessary to use the rules more than once.

Approximations

When we take measurements of distances or weights or temperatures, they are very unlikely to be absolutely accurate. There will usually be some error involved. Even in mathematics, which seems to be an exact science, our results are often only approximate. The number $\sqrt{2}$, for example, cannot be expressed exactly as a terminating decimal or as a fraction. Any numerical value that we give for $\sqrt{2}$ is only an approximation.

When attempting to solve an equation, it is not always possible to guarantee that the solution will be exactly correct. There is often no exact way to solve an equation; instead we provide a method of obtaining an answer as close as we please to the solution.

21.1 Absolute and relative errors

Suppose x is an approximation for a number α. The difference between x and α is the **absolute error** of the approximation.

If the absolute error is h, then we say that

$$\alpha = x \pm h$$

This means that α must lie between $x - h$ and $x + h$.

The **relative error** is the absolute error divided by the true value.

$$\text{Relative error} = \frac{\text{absolute error}}{\alpha}$$

The **percentage error** is the relative error expressed as a pereentage.

$$\text{Percentage error} = \text{relative error} \times 100\%$$

Notes

1 In this chapter we are dealing with approximations, so there is no point in giving the error itself to a high degree of accuracy. It is misleading to speak of a maximum error of 0.012 351 8.

2 Often, we do not know the true value a, so to find the relative error we divide the absolute error by the approximate value x.

3 When we speak of 'the error' without any qualification, we mean the absolute error.

Decimal places and significant figures

When we give a measurement correct to a certain number of decimal places, then we set a bound on the absolute error of the measurement. If a number is expressed as 1.3, correct to one decimal place, then the absolute error is at most 0.05.

When we give a measurement correct to a certain number of significant figures, then we set a bound on the relative error of the measurement.

EXAMPLE 21.1

A weight is given as 3.2 kg, correct to one decimal place. Find the maximum values of the absolute error, the relative error and the percentage error.

Solution

The true weight could lie anywhere between 3.15 and 3.25.

The maximum absolute error is 0.05.

Now we calculate the maximum relative error.

We do not know the true weight, so we divide the absolute error by the approximate weight.

$$\text{Maximum relative error} = \frac{0.05}{3.2} = 0.016$$

The maximum relative error is 0.016.

To find the maximum percentage error, we multiply this relative error by 100.

$$\text{Maximum percentage error} = 0.016 \times 100\%$$

The maximum percentage error is 1.6%.

EXERCISE 21A

1 In each of the following measurements, find the absolute error, the relative error and the percentage error.

a) 4.7 to one decimal place

b) 520 000 to the nearest 10 000

c) 0.235 to three significant figures   **d)** 1 200 000 to two significant figures

e) £123 rounded down to a whole number

f) 415 000 rounded down to a multiple of 1000

g) 5.69 rounded down to two decimal places

2 Find the range of values in the following measurements.

  a) 3.2 ± 0.1 **b)** 0.123 ± 0.0005 **c)** $51\,000 \pm 500$

3 Throughout history, various values have been suggested for π. In each of the following cases find the absolute error and the relative error.

  a) 3 (in the Bible, I Kings, chapter 7 verse 23)

  b) $\frac{22}{7}$ (Archimedes, Greece, 225BC. In fact he proved that π lies between $\frac{22}{7}$ and $\frac{223}{71}$.)

  c) $\frac{355}{113}$ (Tsu-Ch'ung-Chih, China, AD470)

4 In Chapter 7, we found approximations for $\sin x$ and $\cos x$ as follows.

  $$\sin x \approx x \qquad \cos x \approx 1 - \tfrac{1}{2}x^2$$

  Find the absolute and relative errors in using these approximations when $x = 0.1$.

*5 A number is given correct to three significant figures. What is the maximum relative error?

21.2 The arithmetic of errors

When we perform arithmetic on measurements, we compound the errors.

Addition and subtraction

The absolute error in $x + y$ or $x - y$ is the sum of the absolute errors in x and y.

Proof

Suppose our measurements are x and y, with maximum errors of h and k respectively. Suppose the true values are α and β. Then

  $$\alpha - x = \pm h$$
  $$\beta - y = \pm k$$

If x is h too small and y is k too small, then the error in $x + y$ is

  $$(\alpha + \beta) - (x + y) = (\alpha - x) + (\beta - y) = h + k$$

If x is h too small and y is k too *large*, then the error in $x - y$ is

$$(\alpha - \beta) - (x - y) = (\alpha - x) + (y - \beta) = h + k$$

So in both cases, the maximum error in $x \pm y$ is $h + k$.

Multiplying and dividing

Assume that x and y are positive measurements and that the errors are small in comparison with the measurements. Then the maximum relative error in $x \times y$ or $x \div y$ is approximately the sum of the maximum relative errors in x and y.

Proof

As before let x and y be the measurements, α and β the true values, h and k the absolute errors. Then the largest xy could be is

$$(\alpha + h)(\beta + k) = \alpha\beta + h\beta + k\alpha + hk$$

So the absolute error is $h\beta + k\alpha + hk$. Divide by $\alpha\beta$ to find the relative error.

$$\text{Relative error} = \frac{h}{\alpha} + \frac{k}{\beta} + \frac{hk}{\alpha\beta}$$

$\dfrac{h}{\alpha}$ is the maximum relative error in x. $\dfrac{k}{\beta}$ is the maximum relative error in y. $\dfrac{hk}{\alpha\beta}$ is the product of two small quantities, and so can be ignored in comparison with the others.

The largest $x \div y$ could be is

$$\frac{\alpha + h}{\beta - k} = \frac{(\alpha + h)(\beta + k)}{(\beta - k)(\beta + k)} = \frac{\alpha\beta + h\beta + k\alpha + hk}{\beta^2 - k^2}$$

Ignoring k^2 and hk as the products of two small quantities, we obtain

$$\frac{\alpha + h}{\beta - k} = \frac{\alpha}{\beta} + \frac{h}{\beta} + \frac{k\alpha}{\beta^2}$$

The absolute error in $\dfrac{x}{y}$ is therefore $\dfrac{h}{\beta} + \dfrac{k\alpha}{\beta^2}$. Divide by $\dfrac{\alpha}{\beta}$ to obtain the relative error.

$$\text{Relative error} = \frac{h}{\alpha} + \frac{k}{\beta}$$

In both cases, the maximum relative error is the sum of the maximum relative errors in x and y.

EXAMPLE 21.2

Measurements x and y were given correct to two decimal places, as $x = 0.72$ and $y = 0.13$. Find the maximum value of $x - 2y$.

Solution

$x - 2y$ can be written as $x - y - y$. The maximum error in each measurement is 0.005. So the maximum error in $x - 2y$ is 0.015.

Maximum value of $x - 2y = 0.72 - 2 \times 0.13 + 0.015 = 0.475$

The maximum value of $x - 2y$ is 0.475.

EXAMPLE 21.3

A rectangular field is measured as 212 m by 189 m, both distances being given to the nearest metre. What is the maximum error in the area of the field?

Solution

The relative errors in the distances are $\frac{0.5}{212}$ and $\frac{0.5}{189}$. Add these to obtain the relative error in the area.

$$\frac{0.5}{212} + \frac{0.5}{189} = 0.005$$

The maximum relative error in the area is 0.005.

Multiply this by the area of the field.

Maximum error $= 0.005 \times 212 \times 189 = 200.34$

The maximum error in the area is 200 m^2.

EXERCISE 21B

1 Measurements are given as $X = x \pm h$ and $Y = x \pm k$. Express each of the following in the form $a \pm b$.

a) $X + Y$ b) $X - Y$ c) $2X - 3Y$

2 Measurements x, y and z are given correct to one decimal place as $x = 21.2$, $y = 30.4$, $z = 9.7$. Find the maximum errors in the following.

a) $x + y$ b) $x - y$ c) $x + y + z$ d) $2x + 3y - 4z$

e) xy f) $\dfrac{y}{z}$ g) z^2 h) $\dfrac{x + y}{z}$

3 The four stages of a relay race take 21.3, 20.7, 19.5 and 23.1, all measurements being given in seconds correct to one decimal place. What was the least possible total time taken?

4 A distance of 53 miles (to the nearest mile) was driven at 55 m.p.h. (to the nearest 5 m.p.h.). What are the greatest and least values of the time taken?

5 The radius of circle is 6.53 cm, correct to two decimal places. Find the maximum value of the area.

6 The sides of a cuboid are 20.1, 19.3, 25.6, each measurement being given in centimetres correct to one decimal place. Find the maximum values of the volume and the surface area.

7 The height of a cone is 12.8 cm and its base radius is 3.4 cm, both figures being given correct to one decimal place. Find the least value of the volume of the cone.

21.3 Solving equations by iteration

Suppose we are given an equation which we cannot solve directly. Examples of such equations are $x^5 + 3x = 2$, $x = \cos x$ and $e^x = 3 - x$. There are many ways we can solve such equations approximately. One method is by **iteration**.

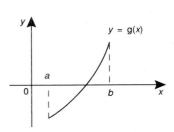

Suppose the equation is $g(x) = 0$. We can find the approximate position of a solution by considering the sign of $g(x)$. Suppose there are x-values a and b, for which $g(a) < 0$ and $g(b) > 0$. Then provided that $g(x)$ is continuous, there is a solution to $g(x) = 0$ between a and b.

Fig. 21.1

Rewrite the equation in the form $x = f(x)$, where f is some function. Pick a starting point x_0, between a and b as found above. Find successive values by

$$x_1 = f(x_0), \quad x_2 = f(x_1), \quad x_3 = f(x_2)$$

and so on. In general

$$x_{n+1} = f(x_n)$$

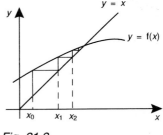

Suppose we want our answer correct to four decimal places. Then we continue the iteration until two successive values agree to four decimal places.

The situation might be as illustrated in Fig. 21.2. The graphs of $y = x$ and $y = f(x)$ intersect at the root. The starting point is x_0. Travel up to the curve of $y = f(x)$. By travelling horizontally to the $y = x$ line, we find $f(x_0) = x_1$. Repeat this process, and we see the terms getting closer to the root.

Fig. 21.2

We cannot be sure what will happen to the sequence x_0, x_1, x_2, ... The terms may approach a fixed value. They may get larger and larger. They may oscillate between two fixed values.

Figure 21.3 shows four possible cases.

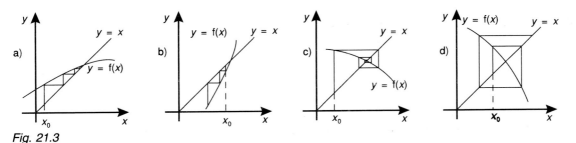

Fig. 21.3

In the 'staircase' diagram of a), the sequence gets closer to the root.

In the 'staircase' diagram of b), the sequence gets further away from the root.

In the 'cobweb' diagram of c), the sequence gets closer to the root.

In the 'cobweb' diagram of d), the sequence gets further away from the root.

If the sequence of numbers approaches a fixed value, then that value will be a solution of the original equation. The sequence **converges** to the solution. If the sequence does not approach a fixed value, then it **diverges**.

EXAMPLE 21.4

Show that there is a solution of $x^3 - 5x + 1 = 0$ between 0 and 1.

Show that the equation can be rewritten in the form $x = \dfrac{x^3 + 1}{5}$,

and hence find the solution correct to two decimal places.

Draw a diagram to illustrate the finding of the solution.

Solution

Put $x = 0$ and $x = 1$.

$x = 0$ gives

$$0^3 - 5 \times 0 + 1 = 1 > 0$$

$x = 1$ gives

$$1^3 - 5 \times 1 + 1 = -3 < 0$$

There is a solution between 0 and 1.

Rearrange the equation by adding $5x$ to both sides, and dividing through by 5.

$$5x = x^3 + 1$$
$$x = \tfrac{1}{5}(x^3 + 1)$$

The equation can be rearranged as $x = \tfrac{1}{5}(x^3 + 1)$.

This rearrangement is now in the form $x = f(x)$. Try the iteration $x_{n+1} = f(x_n)$. Take as starting point halfway between 0 and 1, so that $x_0 = 0.5$. The next values are

$$x_1 = \tfrac{1}{5}(0.5^3 + 1) = 0.225$$
$$x_2 = \tfrac{1}{5}(0.225^3 + 1) = 0.202\,278$$
$$x_3 = \tfrac{1}{5}(0.202\,278^3 + 1) = 0.201\,655$$

The last values agree, correct to two decimal places.

The solution is 0.20.

The sequence decreases towards the root. The situation is shown in the 'staircase' diagram of Fig. 21.4.

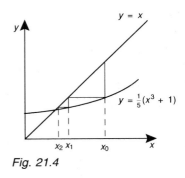

Fig. 21.4

EXERCISE 21C

1 For each of the following equations, find a solution correct to two significant figures. Use the iteration and starting point suggested. Draw a diagram to illustrate the convergence of the sequence.

a) $x^3 - x - 4 = 0$ $\qquad x_{n+1} = \sqrt[3]{x_n + 4}$ $\qquad x_0 = 1.5$

b) $2x = \cos x$ $\qquad x_{n+1} = \frac{1}{2} \cos x_n$ $\qquad x_0 = 0.5$ (*x* is measured in radians.)

c) $\ln x + x = 2$ $\qquad x_{n+1} = 2 - \ln x_n$ $\qquad x_0 = 1.5$

2 For each of the following equations, find a pair of integers between which there is a root. Use the suggested iteration to find the root correct to two significant figures.

a) $x^3 + x - 1 = 0$ $\qquad x_{n+1} = \dfrac{1}{x_n^2 + 1}$ $\qquad$ **b)** $x^3 + 2x^2 - 7 = 0$ $\qquad x_{n+1} = \sqrt[3]{7 - 2x_n^2}$

3 Find a pair of positive integers between which there is a solution of the equation $x^3 - 2x - 5 = 0$. Three iterations are suggested. Find which converges to the solution. In each case, draw a diagram to illustrate the convergence or divergence.

a) $x_{n+1} = \dfrac{5}{x_n^2 - 2}$ $\qquad$ **b)** $x_{n+1} = \frac{1}{2}(5 - x_n^3)$ $\qquad$ **c)** $x_{n+1} = \sqrt[3]{2x_n + 5}$

4 Apply the iteration $x_{n+1} = \dfrac{1}{x_n^2 + 3}$, with a starting value of 0.5, until the values agree, correct to three decimal places. By putting $x_{n+1} = x_n = x$ in the formula, find an equation in *x*. Verify that the limit of the sequence is a solution of this equation.

5 Apply the following iteration schemes until the values agree, correct to three significant figures. Find the equations of which the limit of the sequence is a solution.

a) $x_{n+1} = \dfrac{1}{2}\left(x_n + \dfrac{3}{x_n}\right)$ $\qquad x_0 = 1.5$ $\qquad$ **b)** $x_{n+1} = \dfrac{x_n + 1}{x_n^2 + 10}$ $\qquad x_0 = 0.5$

Newton-Raphson iteration

In all the examples so far, the iteration formula has been provided. Where does it come from? The **Newton-Raphson method** produces an iteration formula which may converge to a solution.

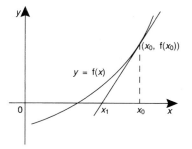

Suppose we want to solve $f(x) = 0$. Figure 21.5 shows a graph of $y = f(x)$ and where it crosses the x-axis, i.e. where the solution is. The formula is

$$x_{n+1} = x_n - \frac{f(x_n)}{f'(x_n)}$$

Fig. 21.5

Justification

Suppose our first approximation for the solution is x_0. On the diagram the tangent to the curve at $(x_0, f(x_0))$ is drawn. This tangent crosses the x-axis at our next approximation, x_1.

The gradient of the tangent is $f'(x_0)$. This gradient can also be given as the ratio of the y-change to the x-change.

$$\text{Gradient} = f'(x_0) = \frac{f(x_0)}{x_0 - x_1}$$

Multiplying both sides by $x_0 - x_1$, and then dividing through by $f'(x_0)$ gives.

$$x_0 - x_1 = \frac{f(x_0)}{f'(x_0)}$$

Rearrange to obtain an expression for x_1.

$$x_1 = x_0 - \frac{f(x_0)}{f'(x_0)}$$

A similar process will give x_2 in terms of x_1.

$$x_2 = x_1 - \frac{f(x_1)}{f'(x_1)}$$

So we have a sequence of approximations, which we hope will tend to the solution of the equation $f(x) = 0$. The sequence is $x_0, x_1, x_2, \cdots, x_n, \cdots$.

The rule for finding successive terms is

$$x_{n+1} = x_n - \frac{f(x_n)}{f'(x_n)}$$

When Newton-Raphson goes wrong

The Newton-Raphson method does not always work. The sequence may converge to the wrong root, or fail to converge at all.

In Fig. 21.6 we see that the first use of Newton-Raphson will take the sequence further away from the root, rather than closer to it.

Provided that we take the initial value close enough to the root, the Newton-Raphson method will usually work.

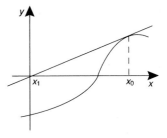

EXAMPLE 21.5
Show that there is a solution of $x^3 + x - 1 = 0$ between $x = 0$ and $x = 1$. Use Newton-Raphson iteration to find this root correct to six decimal places.

Fig. 21.6

Solution
Let $f(x) = x^3 + x - 1$. Then $f(0) = -1$ and $f(1) = 1$. Hence $f(x) = 0$ for some x between 0 and 1.

A suitable first approximation is $x_0 = 0.5$. Write down the Newton-Raphson formula for this equation.

$$x_{n+1} = x_n - \frac{f(x_n)}{f'(x_n)} = x_n - \frac{x_n^3 + x_n - 1}{3x_n^2 + 1}$$

It is a good idea to keep the value of x_n in the memory of your calculator, to save you having to key it in four times in the formula.

$$x_1 = 0.5 - \frac{0.5^3 + 0.5 - 1}{3 \times 0.5^2 + 1} = 0.714\,285\,714$$

We then continue as before.

$$x_2 = 0.683\,179\,724$$

$$x_3 = 0.682\,328\,423$$

$$x_4 = 0.682\,327\,804$$

The last two terms agree, correct to six decimal places.

The solution is 0.682 328.

EXERCISE 21D

1 Use Newton-Raphson iteration to solve the following equations, correct to six decimal places.

a) $x^3 + x - 3 = 0$ **b)** $x = \cos x$ **c)** $7x = e^x$

2 A chord subtends angle θ at the centre of a circle. If the area of the smaller segment cut off by the curve is one third of the area of the circle, show that θ obeys the equation

$$3\theta - 3\sin\theta - 2\pi = 0$$

Starting with $\theta_0 = 1.5$, use Newton-Raphson iteration to solve this equation, correct to four decimal places.

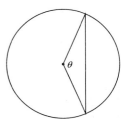

Fig. 21.7

3 Regard $\sqrt{2}$ as the solution of the equation $x^2 - 2 = 0$. Use a Newton-Raphson iteration on this equation to find $\sqrt{2}$, correct to six decimal places.

4 Use Newton-Raphson iteration to find $\sqrt[3]{2}$, correct to six decimal places.

Interval bisection

The Newton-Raphson method is very quick when it does work. Unfortunately there are equations for which it doesn't. A method which is slow but sure is the **interval bisection method**. If a continuous function has a root between two values, this method will always find it.

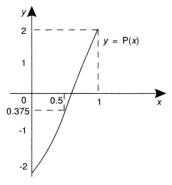

Suppose we want to solve the equation $x^3 + 3x - 2 = 0$. Let $P(x) = x^3 + 3x - 2$. Evaluate P at $x = 0$ and $x = 1$ to find

$$P(0) = -2 < 0$$

$$P(1) = 2 > 0$$

$P(x)$ goes from negative to positive between $x = 0$ and $x = 1$, so there must be some value of x between 0 and 1 for which $P(x) = 0$, i.e. there is a root between 0 and 1. Evaluate P at $x = 0.5$.

Fig. 21.8

$$P(0.5) = -0.375$$

$P(x)$ changes sign between $x = 0.5$ and $x = 1$, so the root lies between 0.5 and 1. Repeat this process, by evaluating $P(0.75)$, and finding whether the root lies between $x = 0.5$ and $x = 0.75$ or between $x = 0.75$ and $x = 1$.

The process can be continued. The method repeatedly bisects the interval within which the root lies, getting closer and closer to the root.

EXAMPLE 21.6

Find a pair of positive integers between which there is a solution of the equation $x^3 + x - 5 = 0$. Use the interval bisection method to find the solution correct to one decimal place.

Solution

Try integer values.

$x = 0$	$x = 1$	$x = 2$
$0^3 + 0 - 5 < 0$	$1^3 + 1 - 5 < 0$	$2^3 + 2 - 5 > 0$

There is a solution between 1 and 2.

Try $x = 1.5$ $1.5^3 + 1.5 - 5 < 0$ So the solution lies between $x = 1.5$ and $x = 2$.

Try $x = 1.75$ $1.75^3 + 1.75 - 5 > 0$ So the solution lies between $x = 1.5$ and $x = 1.75$.

Try $x = 1.625$ $1.625^3 + 1.625 - 5 > 0$ So the solution lies between $x = 1.5$ and $x = 1.625$

Try $x = 1.5625$ $1.5625^3 + 1.5625 - 5 > 0$ So the solution lies between $x = 1.5$ and $x = 1.5625$

Try $x = 1.53125$ $1.53125^3 + 1.53125 - 5 > 0$ So the solution lies between $x = 1.5$ and $x = 1.53125$.

We now know the solution is closer to $x = 1.5$ than to $x = 1.6$, so we have the solution correct to one decimal place.

The solution is 1.5.

EXERCISE 21E

1 Use the interval bisection method to solve the following equations, correct to one decimal place.

  a) $x^5 + x - 1 = 0$ (solution between $x = 0$ and $x = 1$)

  b) $x^3 + 3x^2 + x - 10 = 0$ (solution between $x = 1$ and $x = 2$)

  c) $6 - x - x^3 = 0$ (solution between $x = 1$ and $x = 2$)

2 For the following, find a pair of integers between which the function changes sign. Use the bisection method to find a root between the integers, giving your answer correct to one decimal place.

  a) $x^3 + x + 1$ **b)** $x + \sin x - 2$ **c)** $x - \ln x - 5$

3 Figure 21.9 shows two equal circles, with centres A and B, which meet at C and D. The angle CAD is θ. If the areas of the three regions are equal, show that

$$\theta = \frac{\pi}{2} + \sin \theta$$

Use interval bisection to find θ correct to one decimal place.

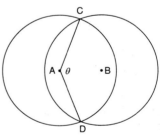

Fig. 21.9

21.4 Solving equations by graphs

We can often solve an equation approximately by seeing where a graph crosses the x-axis, or where two graphs cross each other.

The graph method is especially useful for finding multiple roots of an equation. The Newton-Raphson and bisection methods will find only one solution at a time. The graph method will pick out all the roots.

EXAMPLE 21.7

By drawing the graphs of $y = \cos x$ and $y = x$, solve the equation $x = \cos x$ correct to one decimal place.

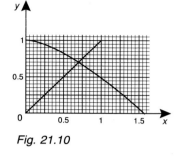

Fig. 21.10

Solution

The graphs are as shown in Fig. 21.10. Notice that they cross at $(0.7, 0.7)$.

The solution is $x = 0.7$.

EXERCISE 21F

1 Use the graph of $y = \cos x$ above to solve the following equations correct to one decimal place.

a) $2x = \cos x$ **b)** $2x - 1 = \cos x$ **c)** $x + 1 = 3 \cos x$

2 Draw the graph of $y = x^3$ for values of x between -2 and 2. Use the graph to solve the following equations.

a) $x^3 = x + 1$ **b)** $x^3 = 3 - x$ **c)** $x^3 = 7x$

3 Draw the graph of $y = e^x$, for x between -1 and 3. Use your graph to solve the following equations correct to one decimal place.

a) $e^x = 2 - x$ **b)** $e^x = 1 + 3x$ ***c)** $e^{2x} = 2 - x$

LONGER EXERCISE

Computer solution of equations

The numerical work in this chapter is very tedious if done by calculator. The use of a computer will make it much easier. Refer to page 383 for an investigation involving computers.

Use of a graphics calculator or graphics program

It takes a long time to draw graphs, and they are usually inaccurate. The use of a graphics calculator or a graphics program will make things easier. Refer to page 384 for an investigation involving these.

EXAMINATION QUESTIONS

1   **a)** Three quantities x, y and z were measured. The results, rounded correct to two significant figures, were

$$x = 3.4, \ y = 9.8, \ z = 9.3$$

Find the upper and lower bounds for the values of

(i) $y - z$ (ii) $\dfrac{x}{y - z}$

b) The volume V of a cylinder is to be calculated from measurements of r, its radius, and h its height. δV denotes the error in the calculation of V produced by errors δr and δh in the measurement of r and h respectively.

Show that, if products of errors can be ignored,

$$\frac{\delta V}{V} = 2\frac{\delta r}{r} + \frac{\delta h}{h}$$

If r and h are measured correct to three significant figures, estimate the largest possible percentage error in the calculated value of V.

O&C 93

2 The tangent at $P(x_n, x_n^2 - 2)$, where $x_n > 0$, to the curve with equation $y = x^2 - 2$ meets the x-axis at the point $Q(x_{n+1}, 0)$. Show that

$$x_{n+1} = \frac{x_n^2 + 2}{2x_n}$$

This relationship between x_{n+1} and x_n is used, starting with $x_1 = 2$, to find successive approximations for the positive root of the equation $x^2 - 2 = 0$. Find x_2 and x_3 as fractions and show that $x_4 = \frac{577}{408}$.

Find the error, correct to one significant figure, in using $\frac{577}{408}$ as an approximation to $\sqrt{2}$.

L 1990

3 The equation $x^3 - 2x - 5 = 0$ has one real root. Show that this root lies between $x = 2$ and $x = 3$.

This root is to be found using the iterative formula $x_{r+1} = f(x_r)$.

Show that the equation can be written as $x = f(x)$ in the three forms

(1) $x = \frac{1}{2}(x^3 - 5)$

(2) $x = 5(x^2 - 2)^{-1}$

(3) $x = (2x + 5)^{\frac{1}{3}}$

Choose the form for which the iteration converges and hence find the root correct to five decimal places.

O&C 1993

Summary and key points

1 The absolute error of an approximation is the difference between it and the true value.

The relative error is the absolute error divided by the value.

2 When approximations are added or subtracted, the absolute errors are added.

When approximations are multiplied or divided, the relative errors are added.

Note that we never subtract, multiply or divide errors.

3 A sequence x_0, x_1, x_2, $\cdots$ which approaches a fixed value converges to that value.

The Newton-Raphson method provides a sequence which may converge to a solution of $f(x) = 0$ by the iteration.

$$x_{n+1} = x_n - \frac{f(x_n)}{f'(x_n)}$$

The interval bisection method involves finding an interval which contains a solution, and then repeatedly cutting the interval in half.

4 Solutions of equations can be found by intersections of graphs.

Handling data

After a survey or experiment, there will be a great mass of data for you to make sense of and interpret. You will want to see quickly what the general trend of the numbers is, and to draw conclusions from them. How can this be done?

We can summarise the data using single numbers, which give the central point of the numbers and their dispersion about this central point.

We can draw a picture, to give a visual impresson of the data at a glance.

22.1 Mean, median, mode

Suppose we want to summarise a collection of numbers using a single number. Such a number is called a **measure of central tendency**.

Mean

The most commonly used measure of a set of numbers is its **mean**. To find the mean of n numbers, we add them up and divide the total by n.

Suppose the numbers are $x_1, x_2, ..., x_n$. The mean $\bar{x}$ is

$$\bar{x} = \frac{x_1 + x_2 ++x_3 + ... + x_n}{n}$$

If the data is given in the form of a frequency table, the mean of the data can still be found. We multiply each value by its frequency, add them all   p, then divide by the sum of all the frequencies. Suppose val x_i occurs with frequency f_i.

$$\bar{x} = \frac{\Sigma x_i f_i}{\Sigma f_i}$$

Suppose the data is given in terms of intervals rather than as exact values. Then we take the midpoint of the intervals for the x_is. This will give an approximation for the mean.

Median

The **median** of a set of numbers is the middle number when the numbers are arranged in order of size. Half the numbers are less than the median and half are greater.

If the data is given in terms of intervals instead of exact values, then we assume that the data is evenly spread within each interval. From this we can find an approximation for the median.

Mode

The **mode** of a set of numbers is the most commonly occurring number.

If the data is given in terms of intervals, then we will not be able to find the mode. The interval containing the greatest density of data is the **modal interval**.

Use of calculator

A scientific calculator can be used to find the mean of data. The method of entering the data varies between models. But do not rely too much on your calculator – you must be familiar with the formula for the mean.

EXAMPLE 22.1

Find the mean and median of the following numbers.

7, 8, 6, 9, 5, 7, 8, 6, 9, 10

Solution

Add these numbers up and divide the total by 10.

The total is $(7 + 8 + 6 + 9 + 5 + 7 + 8 + 6 + 9 + 10) = 75$

The mean is $75 \div 10 = 7.5$

The mean is 7.5.

Arrange the numbers in order.

5 6 6 7 7 8 8 9 9 10

The fifth number is 7 and the sixth is 8. Take the value halfway between these.

The median is 7.5.

EXAMPLE 22.2

The maximum temperature in a town was found every day for 100 days. The results are given in the table below. Estimate the mean and median. What was the modal interval?

Temperature F (°F)	$40 \leq F < 50$	$50 \leq F < 60$	$60 \leq F < 65$	$65 \leq F < 70$	$70 \leq F < 80$
Frequency	16	28	19	21	16

Solution

We do not know how the temperatures are spread within each interval. The best we can do is to assume they are evenly spread. So assume the sixteen temperatures between 40° and 50° have an average of 45°. Make similar assumptions in the other intervals. The mean is approximately

$$\frac{45 \times 16 + 55 \times 28 + 62.5 \times 19 + 67.5 \times 21 + 75 \times 16}{100} = \frac{6065}{100}$$

$$= 60.65$$

The mean is approximately 60.65°.

The median temperature will lie halfway between the 50th and the 51st temperatures. Note that 44 days had temperatures less than 60°, and 63 less than 65°. So the median temperature will lie between 60° and 65°.

There are 19 temperatures in this interval, and we want the first 6.5 of them, to get halfway between the 50th and the 51st. Again assuming that the temperatures are evenly spread, the temperature we want is

$$\frac{6.5}{19} \times 5 + 60$$

$$= 1.71 + 60$$

$$= 61.71$$

The median temperature is approximately 61.7°.

The interval between 50 and 60 has the greatest number of days in it, but this interval is wider than two of the others. We have 28 days in an interval covering 10°. In the interval between 65° and 70°, there are 21 days in an interval covering 5°. This has the greatest **density** of frequencies, and so is the modal interval.

The modal interval is $65 \leq F < 70$.

EXERCISE 22A

1 Find the mean and median of each of the following sets of numbers.

    a) 12, 14, 13, 12, 10, 16, 17   **b)** 105, 109, 99, 113, 120, 118, 106, 95

2 The numbers of children in 90 families are given in the table below. Find the mean and median number of children.

Number of children	0	1	2	3	4	5	6
Frequency	16	21	24	15	8	5	1

3 The number of driving tests taken by 100 motorists before they passed is given in the table below. Estimate the mean number, and find the median number and the mode.

Number of tests	1	2	3	4	5	6	7+
Frequency	40	23	13	11	4	3	6

4 The incomes of 700 employees are given in the table below. Estimate the mean and median incomes. The incomes are given in £1000s.

Income I	$6 \leq I < 8$	$8 \leq I < 10$	$10 \leq I < 12$	$12 \leq I < 14$	$14 \leq I < 16$	$16 \leq I < 18$
Frequency	34	61	103	125	218	159

5 The annual sales of 100 first novels are given in the table below. Estimate the mean and median sales. Comment on your results.

Number of sales S	$0 \leq S < 200$	$200 \leq S < 500$	$500 \leq S < 1000$	$1000 \leq S < 5000$	$5000 \leq S < 10\,000$
Frequency	20	32	36	7	5

6 a) Consider the distribution of income in this country. Which is greater, the mean income or the median?

b) Consider the times taken to run a marathon race. Which is greater, the mean time or the median?

7 The class structure of a country is analysed in the table below. What is the median class? Does it make sense to speak of the mean class?

Class	Upper	Upper middle	Lower middle	Upper working	Lower working
Number in millions	0.3	1.7	5.8	10.6	8.4

8 A questionnaire asked people for their opinion of the Government's housing policy. The responses are given in the table below. What is the median response? Does it make sense to speak of the mean response?

Response	Strongly approve	Approve	Neutral	Disapprove	Strongly disapprove
Frequency	36	58	173	102	283

9 An examination was taken by a class of ten boys and twelve girls. The mean mark for the boys was 63.2, and for the girls 58.5. What was the mean for the class as a whole?

10 The mean age of nine children is 10.3 years. A tenth child joins the group, and the mean age is now 10.5 years. What is the age of the tenth child?

11 An agility test is taken by 10 people. The times they take to complete a task are recorded. Would you use the mean or the median time to give an impression of the performance of the group as a whole? Give reasons.

12 The frequencies of the lifetimes of 100 batteries are given in the table below.

Lifetime in hours L	$2 \leq L < 3$	$3 \leq L < 4$	$4 \leq L < 5$	$5 \leq L < 6$	$6 \leq L < 7$
Frequency	12	35	x	y	4

Given that the mean lifetime is approximately 4.22 hours, find x and y.

*13 A scientific calculator can work out the mean of data entered into it. Why can't it work out the median or the mode?

22.2 Variance and standard deviation

Very often, it is important to know not just the mean of a set of numbers, but also how widely they are spread about the mean. For example, the mean income of all the people in a country will tell us how prosperous the country is. We might also want to know how widely spread the prosperity is – whether or not there is a wide gap between rich and poor. A measure of the spread of data is a **measure of dispersion**.

Suppose we have a set of data $x_1, x_2, x_3, \cdots, x_n$. Let the mean of these numbers be $\bar{x}$.

The difference between x_1 and $\bar{x}$ is $x_1 - \bar{x}$. We want all the differences expressed as positive terms, otherwise the negative differences might cancel the positive differences. To convert $x_1 - \bar{x}$ to a positive term we square it. We do the same to all the data, add the results and divide by n. This gives the **variance**.

$$\text{Variance} = \frac{(x_1 - \bar{x})^2 + (x_2 - \bar{x})^2 + (x_3 - \bar{x})^2 + \cdots + (x_n - \bar{x})^2}{n}$$

The **standard deviation** of a set of figures is the square root of the variance. The standard deviation is measured in the same units as the original data.

The form $\frac{1}{n}\sum x_i^2 - \bar{x}^2$

The formula for variance is $\frac{1}{n}\sum(x_i - \bar{x})^2$. This is equivalent to $\frac{1}{n}\sum x_i^2 - \bar{x}^2$. The second formula is usually easier to use, especially when the mean is not a whole number.

Proof of equivalence

Expand the formula

$$\frac{1}{n}\sum(x_i - \bar{x})^2 = \frac{1}{n}\sum x_i^2 - \frac{1}{n}\sum 2x_i\bar{x} + \frac{1}{n}\sum \bar{x}^2$$

In the second term we can take out a factor of $2\bar{x}$. The third term contains $\bar{x}^2$, repeated n times.

$$\frac{1}{n}\sum(x_i - \bar{x})^2 = \frac{1}{n}\sum x_i^2 - 2\bar{x}\left(\frac{1}{n}\sum x_i\right) + \frac{1}{n}(n\bar{x}^2)$$

Note that $\frac{1}{n}\sum x_i = \bar{x}$

$$\frac{1}{n}\sum(x_i - x)^2 = \frac{1}{n}\sum x_i^2 - 2\bar{x}^2 + \bar{x}^2 = \frac{1}{n}\sum x_i^2 - \bar{x}^2$$

In words this is 'the mean of the squares minus the square of the means'.

Use of calculators

A scientific calculator can evaluate the standard deviation of the data entered into it, but you must know the formulae for variance and standard deviation.

EXAMPLE 22.3

Find the variance of the numbers in Example 22.1.

Solution

We have already found the mean to be 7.5. Subtract this from each of the numbers, square the results and then average these squares. The variance is

$$\frac{(7 - 7.5)^2 + (8 - 7.5)^2 + (6 - 7.5)^2 + (9 - 7.5)^2 + (5 - 7.5)^2 + (7 - 7.5)^2 + (8 - 7.5)^2 + (6 - 7.5)^2 + (9 - 7.5)^2 + (10 - 7.5)^2}{10}$$

$$= \frac{22.5}{10} = 2.25$$

The variance is 2.25.

EXAMPLE 22.4

The weekly wages of the employees of two companies, A and B, are given in the table below. Estimate the means and standard deviations for both companies. What does this tell you about the companies?

Weekly wage W (£)	$100 \leq W < 200$	$200 \leq W < 250$	$250 \leq W < 300$	$300 \leq W < 350$	$350 \leq W < 400$	$400 \leq W < 500$
Frequency A	10	15	58	165	31	12
Frequency B	28	35	42	82	57	50

Solution

Find the means by the method of Section 22.1.

Mean wage for Company A $= £314$

Mean wage for Company B $= £320$

To evaluate the variances, we assume that the numbers in each interval are at the centres of the intervals. So the variance for Company A is

$$\frac{10(150 - 314)^2 + 15(225 - 314)^2 + 58(275 - 314)^2 + 165(325 - 314)^2 + 31(375 - 314)^2 + 12(450 - 314)^2}{291}$$

The variance for Company B is found similarly. In both cases take square roots to find the standard deviation.

The standard deviation for Company A is £53.50, and for Company B it is £87.

Notice that the means are approximately equal, but that the standard deviation for B is greater.

The mean wages are about the same, there is a wider spread of wages in Company B.

EXAMPLE 22.5

Five numbers, in increasing order, are 1, 5, 8, x and y. Their mean is 7, and their variance is 14. Find x and y.

Solution

The sum of the numbers is $5 \times 7 = 35$.

Hence $x + y = 35 - 1 - 5 - 8 = 21$

Using the second version for the variance, we have that

$$\frac{1^2 + 5^2 + 8^2 + x^2 + y^2}{5} - 7^2 = 14$$

Hence $x^2 + y^2 = 225$

Solve the equations $x + y = 21$ and $x^2 + y^2 = 225$.

We find that x and y are 9 and 12. We know that x is less than y.

The solution is $x = 9$ and $y = 12$.

EXERCISE 22B

1 Find the variances of the sets of numbers in Question 1 of Exercise 22A.

2 Two classes A and B took a test. Their results are given in the table below. Find the means and variances of the marks, and comment on the differences.

Mark	0	1	2	3	4	5	6	7	8	9	10
A frequency	0	3	3	6	6	5	1	1	0	0	0
B frequency	0	0	1	7	10	6	1	0	0	0	0

3 Two groups of people ran a 400 m race. Their finishing times, in seconds, were recorded in this table.

Group A	63	68	65	61	68	66	70	71	66	65
Group B	58	66	65	50	69	62	75	83	62	81

For each group find the mean and standard deviation of the times. Comment on the difference.

4 Six numbers in increasing order are 2, 5, x, y, 12, 16. Their mean is 8.5 and their variance is $20\frac{11}{12}$. Find x and y.

5 A class of 12 boys and 13 girls took an examination. The scores for the boys had mean 53 and standard deviation 10. The scores for the girls had mean 55 and standard deviation 12. Find the mean and standard deviation for the scores of the class as a whole.

6 Measurements $x_1, x_2, ..., x_{10}$ are made, and the results are summarised by $\sum x_i = 153$ and $\sum x_i^2 = 4000$. Find the mean and standard deviation of the measurements.

7 Twelve heights are measured. The sum of all the heights is 19 m, and the sum of the squares of the heights is 33 m². What are the mean and the standard deviation of the heights?

*8 When your calculator works out the variance of a set of numbers, which version of the formula does it use? Why?

22.3 Illustrating frequency

There are many ways of illustrating the frequencies of data. Here we discuss **histograms**, **frequency polygons** and **stem and leaf diagrams**.

Histograms

A **histogram** is like a bar chart, but the frequency of each interval is represented by the **area** of the corresponding bar. First we decide upon a standard width for the intervals. In Fig. 22.1 the intervals labelled 2, 3, 6 and 7 are of this standard width. If an interval is wider than the standard width by a factor of k, we must reduce the height of the bar by a factor of k. This reduction was done to the first interval in Fig. 22.1. If an interval is thinner by a factor of k, we increase the height of the bar by a factor of k. This was done to the intervals labelled 4 and 5 in Fig. 22.1.

In all cases the height of the intervals will be proportional to the **density** of data in the interval.

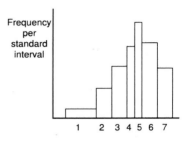

Frequency per standard interval

1 2 3 4 5 6 7

Fig. 22.1

Frequency polygons

If we draw a line through the centres of the tops of the bars of a histogram, we obtain a **frequency polygon**. This is a graph illustrating the frequencies.

Fig. 22.2

Stem and leaf diagrams

A stem and leaf diagram is a type of bar chart, in which the bars consist of the data itself. Suppose the data lies within the range 40 to 79. We could have one bar containing the numbers beginning with 4, another bar with the numbers beginning with 5, and so on. The numbers representing the tens digit, 4, 5, 6 and 7, are the **stems**. The numbers representing the units, which follow the stems, are the **leaves**.

EXAMPLE 22.6

The daily sales in a department store were measured for a period of a year. The results are given in the table below. Draw a histogram and a frequency polygon to illustrate the data.

Sales in £10 000s	0 − 1	1 − 1.5	1.5 − 1.75	1.75 − 2	2 − 2.5	2.5 − 3	3 − 4	4 − 6
Frequency	45	30	22	28	38	49	51	37

Solution

Let us take an interval containing £5000 to be the standard width. When we draw the histogram, we will halve the height of the bars in intervals containing £10 000, and double the height of bars in intervals containing £2500. The height of the last bar will be divided by 4. The histogram is shown in Fig. 22.3.

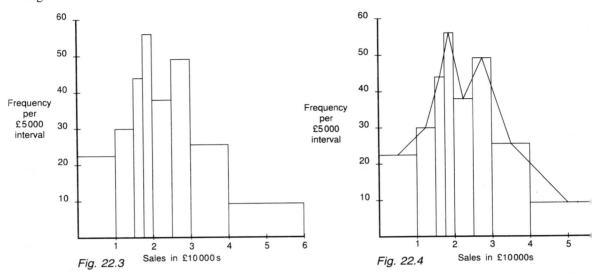

Fig. 22.3   Sales in £10 000 s

Fig. 22.4   Sales in £10 000s

For the frequency polygon, mark points in the centres of the tops of the bars. Join these points up by straight lines, as in Fig. 22.4.

EXAMPLE 22.7

The ages at graduation of 40 Open University students are given below. Construct a stem and leaf diagram to illustrate the data, using as stem values 20, 25, 30, and so on. Describe the data.

24 38 41 29 31   61 51 32 30 28   27 26 35 39 28   31 30 45 39 27

32 36 40 29 49   50 36 26 25 26   33 29 24 25 34   38 33 32 29 31

Solution

(There is more than one method of constructing a stem and leaf diagram. We outline one method here.)

Put the stem values of 20, 25, 30 and so on in a vertical column. There are two people between 20 and 25, both of whom are aged 24. 24 is 4 greater than 20, so put two 4s in the 20 row. Fill in the other rows similarly. The completed diagram is shown in Fig. 22.5.

20	4 4
25	0 0 1 1 1 2 2 3 3 4 4 4 4
30	0 0 1 1 1 2 2 2 3 3 4
35	0 1 1 3 3 4 4
40	0 1
45	0 4
50	0 1
55	
60	1

Fig. 22.5

Notice that most of the people are in the second, third and fourth bars.

Most of the students are in their late twenties or thirties.

EXERCISE 22C

1 A consumer organisation measured the weights, to the nearest gram, of 60 'Whoppaburgers' (which should each weigh 400 grams). The results are shown in the table below. Draw a histogram to illustrate the data, and comment on it.

Weight (grams)	360–379	380–389	390–399	400–404	405–409	410–414	415–419	420–430
Frequency	4	6	11	11	13	6	4	5

2 A consumer group tested 50 batteries by using them until they were exhausted. The lifetimes of the batteries are given in the table below. Draw a histogram and a frequency polygon to illustrate the data.

Lifetime (hours)	2–3	3–3.5	3.5–3.75	3.75–4	4–4.5	4.5–5	5–6
Frequency	3	6	7	8	12	6	8

3 To measure the fuel efficiency of different makes of cars, one gallon of petrol was put into the tank of each of 40 cars, which were then driven over the same route. The distances they travelled before running out of petrol are given in the table below. Draw a histogram to illustrate the data.

Distance (miles)	20–25	25–30	30–32.5	32.5–35	35–40	40–50
Frequency	3	10	8	7	7	5

4 The maximum temperatures over 50 days of summer were measured and the results were as shown in the frequency table below. Draw a histogram and a frequency polygon to illustrate the data.

Temperature (°F)	60–70	70–75	75–80	80–85	85–90	90–95
Frequency	5	8	12	15	8	2

5 The salaries of 30 employees of a firm are given below. The salaries are given in £1000s. Construct a stem and leaf diagram to illustrate the data. Use stems of 15, 20, 25 and so on.

20 23 44 19 16 33 32 21 28 40 17 26 23 21 34

39 31 22 24 20 18 27 24 21 22 16 16 15 28 20

6 The waistlines (in inches) of 20 men were measured as given below. Construct a stem and leaf diagram, using stems of 20, 25, 30 and so on.

28 32 36 34 42   24 29 26 31 37

49 31 30 26 33   41 42 39 34 35

7 A puzzle was given to 40 people, and the times they took, in minutes, to solve it were recorded. Results are given below. Construct a stem and leaf diagram, and comment on the results.

3   7 32   8 29   10 12   7 1 52   32 38 13   3   5   28 25 31 10   6

26 24 42 19 12   9   8 25 2 42   40 27 34 32 41   46 51 24   8 35

22.4 Cumulative frequency

Suppose the grade boundaries for a GCSE exam are 70% for grade A, 60% for grade B, 50% for grade C, etc. We would be interested in finding how many candidates scored less than A, how many scored less than B and so on. These would be the cumulative totals up to 70%, 60% and so on.

Suppose we are given data in a frequency table. For each possible value for the data, a certain number of the items are less than that value. This is the **cumulative frequency** at the value.

A list of the cumulative frequencies is a **cumulative frequency table**. It provides a running total of the ordinary frequencies.

Suppose that b is the endpoint of an interval $a \leq x < b$, and that the cumulative frequency at b is N. This means that N of the data have values less than b. We plot the cumulative frequencies against the right-hand ends of the intervals, i.e. for each b, plot a point at (b, N). Join the points with a smooth curve.

The result is a **cumulative frequency curve**. This curve is sometimes called an **ogive** curve as it resembles the ogive shape often found in Gothic architecture.

Fig. 22.6 A gothic window

Median and quartiles

From the cumulative frequency curve we can read off the median, by finding the value corresponding to half the total frequency.

The **lower quartile** is the value corresponding to a quarter of the total frequency. The **upper quartile** is the value corresponding to three quarters of the total frequency. These quartiles can also be read off from the graph.

In the diagram the total frequency is 50. The median (M or Q_2) is about 3.4. The lower and upper quartiles (LQ and UQ or Q_1 and Q_3) are about 2.2 and 4.2.

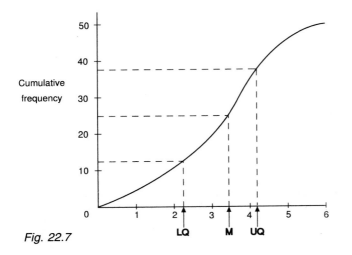

Fig. 22.7

Interquartile range

The difference between the quartiles is the **interquartile range**. Like variance and standard deviation, it is a measure of the dispersion of the data. It tells us the length of the interval which contains the middle half of the data.

In the diagram above the interquartile range is $4.2 - 2.2 = 2$.

Box and whisker diagrams

Once we have found the median and the quartiles of data, we can illustrate them on a **box and whisker** diagram. Draw a number line incorporating a box covering the region between the quartiles. Mark the lower quartile, the median and the upper quartile by Q_1, Q_2 and Q_3.

If there is a very extreme value, by convention, more than 1.5 times the interquartile range from the nearest quartile, then it is represented by a cross. These points are called 'outliers'.

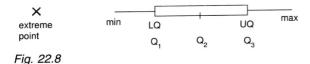

Fig. 22.8

Next we draw lines (the whiskers) from each end of the box, stretching to the maximum and minimum values other than the extreme values.

EXAMPLE 22.8

The times taken by 100 people to run a mile race are given below. Plot a cumulative frequency curve. Find the median and the interquartile range. What proportion of runners took less than 5.3 minutes?

Time (minutes)	4–4.25	4.25–4.5	4.5–4.75	4.75–5	5–5.25	5.25–5.5	5.5–6
Frequency	8	19	28	17	14	8	6

Solution

Write out an extra row for the cumulative frequencies.

Cumulative frequency	8	27	55	72	86	94	100

Plot the graph, putting the points at the right-hand ends of the intervals. The result is as shown.

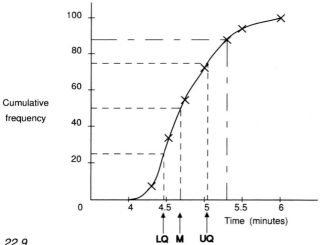

Fig. 22.9

For the median, we find the point which gives a cumulative frequency of 50.

The median time is 4.7 minutes.

The quartiles correspond to cumulative frequencies of 25 and 75. They are 4.45 and 5.05.

The interquartile range is 0.6 minutes.

A time of 5.3 seconds corresponds to a cumulative frequency of 88.

The proportion of 0.88 of the runners took less than 5.3 minutes.

EXAMPLE 22.9

Twenty people were asked to measure how long they slept on a particular night. The results, in hours, are given below. Find the median and the interquartile range. Draw a box and whisker diagram for the data.

8.3 8.8 7.4 9.2 8.9   9.0 8.0 4.2 7.7 7.2

8.1 7.6 9.5 9.1 8.4   10.4 8.1 9.3 7.6 7.1

Solution

Arrange the numbers in increasing order.

4.2 7.1 7.2 7.4 7.6 7.6 7.7 8.0 8.1 8.1 8.3 8.4 8.8 8.9
9.0 9.1 9.2 9.3 9.5 10.4

The median lies between the 10th and 11th entries. This is 8.2.

The lower quartile lies between the 5th and 6th entries. This is 7.6. The upper quartile lies between the 15th and 16th entries. This is 9.05. Find the difference.

$$9.05 - 7.6 = 1.45$$

The median is 8.2 and the interquartile range is 1.45 hours.

Multiply this by 1.5, to obtain 2.175. The least time of 4.2 hours is more than 2.175 hours away from the lower quartile, so we shall consider 4.2 as an extreme value. The greatest time of 10.4 hours is within 2.175 of the upper quartile, so we shall not consider it as extreme.

The box stretches from 7.6 to 9.05. The whiskers stretch from 7.1 to 10.4. The extreme value is marked with a cross. The complete diagram is shown in Fig. 22.10.

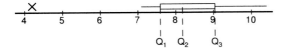

Fig. 22.10

EXERCISE 22D

1 The prices of 120 paperback books are as given in the table below. Draw a cumulative frequency graph for the data. Find the median and the interquartile range. If a book is selected at random, what is the probability it will cost less than £7.20?

Price (£)	4–4.99	5–5.99	6–6.99	7–7.99	8–8.99	9–10
Frequency	9	18	32	38	14	9

2 The majorities obtained by the Feudal Party in 600 parliamentary seats are given in the table below. Negative majorities refer to seats where the Feudal Party lost the election. Draw a cumulative frequency graph for the data. Find the median majority. A seat is classified as **marginal** if the winning party had a majority of fewer than 2000. How many marginal seats are there?

Majority in 5000s	−4 to −3	−3 to −2	−2 to −1	−1 to 0	0 to 1	1 to 2	2 to 3	3 to 4
Frequency	12	37	93	164	198	58	27	11

3 A survey into the length of time that accident patients stayed in hospital produced the data given in the table below. Plot a cumulative frequency curve. Find the interquartile range. What proportion of patients spent longer than five days?

Time (days)	1–3	4–6	7–9	10–15	16–21	22–30
Frequency	35	14	5	8	2	3

4 The ages of a rugby team are given below. Construct a box and whisker diagram for the data.

23.5 22.8 19.7 25.1 20.4

19.2 23.8 26.7 35.1 27.4

21.4 24.3 26.4 25.1 19.4

5 Twenty self-employed people were asked how many hours they worked per week. The results are below. Construct a box and whisker diagram for the data.

45 57 36 41 69   32 43 56 29 41

52 44 61 31 38   49 57 28 21 72

6 A driving school asked 20 of its pupils, who had failed the driving test, how many points they had failed on. The results are below. Construct a box and whisker diagram for the data.

2 3 5 4 2   4 5 3 2 1

5 6 5 10 3   4 3 2 1 2

*7 The process of going from frequency to cumulative frequency is very similar to an operation you have been handling throughout A-level. What is it?

LONGER EXERCISE
American presidential elections

The table below gives details of the American Presidential elections between 1948 and 1984. Analyse the data by some of the methods of this chapter.

1948	Democrat (Truman)	24 104 031
	Republican (Dewey)	21 971 004
1952	Democrat (Stevenson)	27 314 992
	Republican (Eisenhower)	33 937 252
1956	Democrat (Stevenson)	26 035 504
	Republican (Eisenhower)	35 589 471
1960	Democrat (Kennedy)	34 221 341
	Republican (Nixon)	34 108 647
1964	Democrat (Johnson)	43 129 484
	Republican (Goldwater)	27 178 188
1968	Democrat (Humphrey)	31 275 165
	Republican (Nixon)	31 785 480
1972	Democrat (McGovern)	29 168 509
	Republican (Nixon)	47 167 314
1976	Democrat (Carter)	40 827 394
	Republican (Ford)	39 145 977
1980	Democrat (Carter)	27 314 992
	Republican (Reagan)	43 267 486
1984	Democrat (Mondale)	36 930 923
	Republican (Reagan)	56 428 352

EXAMINATION QUESTIONS

1 The figures in the table below are the ages, to the nearest year, of a random sample of 30 people negotiating a mortgage with a bank.

29 26 31 42 38   36 39 49 40 32   33 31 33 52 44

45 35 37 38 38   32 34 27 61 29   32 30 38 42 33

Copy and complete the following stem and leaf diagram. Use the diagram to identify **two** features of the shape of the distribution.

25 4 1

30 1

35

Find the mean age of the 30 people. Given that 18 of them are men and that the mean age of the men is 37.72, find the mean age of the 12 women.

MEI 1991

2 A small business has 12 employees. Their weekly wages, £x, are summarised by

$$\sum x = 2501 \qquad \sum x^2 = 525266.8$$

(i) Calculate the mean and standard deviation of the employees' weekly wages.

A second business has 17 employees. Their weekly wages, £y, have a mean of £273.20 and a standard deviation of £23.16.

(ii) Find $\sum y$ and show that $\sum y^2 = 1\,277\,969$.

(iii) Now consider all 29 employees as a single group. Find the mean and standard deviation of their weekly wages.

MEI 1993

Summary and key points

1 The mean of n items of data is obtained by adding the values and dividing the total by n. The median is obtained by arranging the items in order and taking the middle one. The mode is the most commonly occurring item.

2 The variance of n items of data is

$$\frac{1}{n}\sum(x_i - \bar{x})^2 = \frac{1}{n}\sum x_i^2 - \bar{x}^2$$

The standard deviation is the square root of the variance.

3 In a histogram, the frequency in each interval is shown by the area of its corresponding bar.

If an interval differs from the standard width, then its height is adjusted accordingly.

In a stem and leaf diagram, the bars will contain the data.

4 Cumulative frequency is obtained by adding up frequencies.

When plotting a cumulative frequency graph, the x-values are taken from the right-hand ends of the intervals.

A box and whisker diagram shows the central half of the box, the extreme points by crosses, and the range of values excluding the extreme points by lines.

Computer investigations II

The bisection method

In Chapter 21 we introduced the bisection method for solving equations. If you know that an equation has a solution between two values, then this method will find it, but the method is very slow. If the original interval is one unit wide, then it takes four iterations before you have the solution to correct to one decimal place. It takes seven iterations to get it correct to two decimal places. A computer will do all the iterations automatically.

If you are a programmer, you could write a program in BASIC or any other language to perform the iteration. We shall describe a spreadsheet method of doing it. As well as the ordinary spreadsheet operations of entering formulae and copying ranges of cells, we shall be using the @IF function.

$$@IF\ (condition,\ x,\ y)$$

This will take the value x if the condition is true, and y if it is false.

Suppose we are going to solve the equation $f(x) = x^3 + x - 1 = 0$. We know that $f(0) < 0$ and $f(1) > 0$, so there is a root between these values.

Use the A column for the left-hand values, the C column for the right, and the B columns for the middle values. Put headings in A1, B1 and C1 accordingly. Put 0 into A2, 1 into C2. Put 0.5 into B2.

Use the E column to evaluate the function. In E3 enter the value of the function at B2.

$$+ B2\hat{\ }3 + B2 - 1$$

The value -0.375 should appear.

Now we consider what happens to the A column. Since $f(0.5) < 0$, the left-hand end of the interval will become B2 (which is 0.5). If the value had been positive, it would have stayed at A2 (which is 1). So the formula to enter in A3 is

@IF(E3 < 0,B2,A2)

Similarly, the formula to enter in C3 is @IF(E3 > 0,B2,C2). We want the B column entries to be the midpoint of the A and C entries, so enter (A3 + C3)/2 in B3.

If we had to repeat this process all down the page, there would be no advantage in using a computer, but use the Copy process to copy cells A3 to E3 down the page, and you will see the iteration performed for as many times as you wish.

Amend the spreadsheet to solve other equations. Suggestions are

$$x^5 + x - 1 = 0 \qquad x^3 + 5x - 10 = 0 \qquad x = \cos x$$

Cubic graphs

There is only one possible shape for a quadratic graph, but there are several essentially different shapes that a cubic graph could take. It is very tedious to draw all the graphs by hand. A computer will draw the graphs quickly and accurately.

You could use a graph-drawing program, or the graph drawing facility of a spreadsheet. Try several cubic equations, and find the different shapes that they form. Ones you could try are

$$x^3 - 4x \qquad x^3 - 3x^2 + 3x - 5 \qquad 2 - x - x^3$$
$$1 + x^2 + x^3 \qquad x^3 - 2x^2 + x$$

Finding e

In Chapter 18 we gave reasons for thinking that the number e lies between 2 and 3. Essentially we estimated the limit of the following expression, as $\delta x \to 0$,

$$\frac{a^{\delta x} - 1}{\delta x - 1}$$

and found that for $a = 2$ it was less than 1, and for $a = 3$ it was greater than 1. For e, therefore, somewhere between 2 and 3, the limit was exactly 1.

In A2 enter the first value of δx, for instance, 1. Let the next values of δx be 0.1, 0.01 and so on. So in A3 enter +A2/10. Copy this formula down to A12.

In B1 enter the value of *a*, for instance 2. In B2 enter the expression above. You will be copying this formula down the page, so the expressions for *a* must be fixed. The formula is

(B$1^A2 − 1)/A2

Copy this down the B column to B12. At the bottom of the column there should be 0.693147.

Now just the value of *a*, by changing the entry in B1, until the bottom of the B column is 1. What is the value of e?

The logistic function

This function models the growth of a population living in an area with finite resources. If x_n is the population at year *n*, the population next year is given by

$$x_{n+1} = kx_n(K − x_n)$$

So the population is multiplied by *k*, which depends on the fertility of the species. The factor $(K − x_n)$ accounts for the limited resources. By the time the population has reached *K* it has eaten all the food and the species dies out.

By expressing the population as a proportion of *K*, the equation can be written as

$$x_{n+1} = kx_n(1 − x_n)$$

Start with $k = 1.5$. Enter 1.5 in A1.

Assume that the initial value is $x_0 = 0.5$, half way towards eating all the resources. In A2 enter 0.5.

A3 will hold the value of the next generation. This will be $x_1 = kx_0(1 − x_0)$.

So in A3 enter the formula

+A$1*A2*(1 − A2)

(Make the A1 address absolute, as this holds the value of *k* which will not be changing.)
Copy the formula down the A column, to A40 or further. What does the population tend to?

Try other values of *k*. Does the population always settle down to a steady state? Values of *k* to try are 2, 2.5, 3, 3.5. What are the interpretations in the different cases?

So far, you are recording results for only one value of *k* at a time. You can do it for many different values of *k*. Fill the top row of the spreadsheet with values of k from 1.5 to 4, going up in steps of 0.1. Copy the A column (which has the results for $k = 1.5$) across to the other columns. What is happening? Can you plot a graph to show what is happening for different values of k? The picture on the cover of this book shows part of the graph.

You may have heard of **chaos theory**. The logistic function, thought it seems very simple, provides an example of chaotic behaviour.

Consolidation section E

Chapter 19

1 Solve the following inequalities.

a) $2x + 3 > 5$

b) $3x - 1 \leq x + 7$

c) $1 - 3x > 13$

d) $x^2 - 7x - 30 < 0$

e) $x^2 + 4x \geq 45$

f) $x(x - 3)(2x - 1) > 0$

g) $\dfrac{2 + x}{3 - x} > 0$

h) $\dfrac{2}{x + 1} > 3$

i) $\dfrac{1 + x}{x - 1} < \dfrac{2 + x}{x - 3}$

j) $|x + 1| < 3$

k) $|2x + 3| \geq 3$

l) $|x + 1| < 2|x|$

2 Illustrate the following inequalities on graphs, and list the points with integer values which satisfy all the inequalities.

$$y > 0 \qquad x > 0 \qquad x + 2y > 4 \qquad x + 2y < 8$$

3 By drawing graphs or otherwise, solve the inequality $|2x - 1| > x$.

Chapter 20

4 Differentiate the following functions.

a) $x^3 \sin x$

b) $e^x \sin x$

c) $(x^2 + 3x + 1) \ln x$

d) $\dfrac{\sin x}{x}$

e) $\dfrac{\sqrt{x}}{1 + x}$

f) $\dfrac{x^2 + 2x + 3}{x^2 + x + 5}$

g) $\tan 4x$

b) $\sec -2x$

i) $\operatorname{cosec} 0.5x$

j) $\cos(x^3 + 1)$

k) $\ln(x^2 + x + 1)$

l) $\sqrt{1 + x^4}$

m) $\tan^{-1} 4x$

n) $\sin^{-1}(x + 1)$

o) $\tan^{-1}(2x + 1)$

Chapter 21

5 A measurement is given as 3.26 to three significant figures. Find the absolute error and the relative error.

6 Two quantities are measured as A and B, with maximum errors of a and b respectively. Find the maximum error in $A + B$ and in $2A - 3B$. Find the approximate relative error in AB, $\dfrac{B}{A}$ and A^2.

7 Show that the equation $x^2 - \cos x = 0$ has a root between 0 and $\dfrac{\pi}{2}$. Use interval bisection to find this root correct to one decimal place.

8 Find a pair of positive integers between which there is a root of $e^x = 4x^2 + 3$. Use Newton-Raphson iteration to find this root correct to four decimal places.

9 Show that the equation $4x^2 = e^x$ can be written as either $x = \frac{1}{2}e^{\frac{1}{2}x}$ or $x = 2 \ln 2x$. Write down iterations based on these forms. With starting value $x_0 = 1$, find which iteration converges.

Chapter 22

10 Two golfers went round the same course ten times. Their scores were

A   93   88   95   85   90   91   98   86   96   97

B   90   91   89   90   92   89   88   89   90   89

For each golfer, find the mean and standard deviation of the scores. What conclusions can be reached?

11 In a large company 80 typists had their daily work checked for mistakes. The results are given in the table below.

Number of mistakes	0–9	10–14	15–19	20–29	30–39	40–60
Frequency	5	14	16	21	16	8

a) Estimate the mean and median number of mistakes.

b) What is the modal interval?

c) Draw a histogram to illustrate the information.

d) Draw a cumulative frequency curve to illustrate the information, and from it find the interquartile range.

MIXED EXERCISE

1 Differentiate $y = \sin 2x$

a) directly   **b)** using the identity $\sin 2x = 2 \sin x \cos x$.

Verify that your answers are equal.

2 Let $f(x) = e^x \cos x$. Show that $f'(x) = \sqrt{2}e^x \cos\left(x + \dfrac{\pi}{4}\right)$. Show that the positive x-coordinates of the maxima of $f(x)$ form an arithmetic progression, and the y-coordinates form a geometric progression.

3 A cylindrical log floats with its axis horizontal. The relative density of the wood is 0.9 (so that 0.9 of its volume is submerged under the water). If θ is the angle subtended by the arc above the water, show that

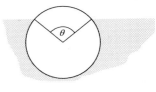

$$\theta = \sin \theta + \frac{\pi}{5}$$

Use Newton-Raphson iteration to solve this equation correct to four decimal places.

Find the ratio of the curved surface area of the log which is above water to the area which is below water.

4 By graphs or otherwise, solve these inequalities.

a) $|2x - 1| < x^2$ **b)** $\frac{1}{x} > |x + 1|$

5 For what values of x is the function $y = \ln(x^2 - 3x + 2)$ not defined?

6 Differentiate the following expressions.

a) $\ln \ln \ln x$ **b)** $\sqrt{1 + \sqrt{1 + \sqrt{x}}}$ **c)** $\ln(x + \sqrt{1 - x^2})$

7 What is wrong with the following argument?

$$x^2 = x + x + x + \cdots + x \quad (x \text{ times})$$
$$2x = 1 + 1 + 1 + \cdots + 1 \quad (\text{differentiating})$$
$$2x = x$$
$$2 = 1 \quad (\text{putting } x = 1)$$

8 Do the following sequences converge?

a) $x_0 = 1$, $x_{n+1} = \sqrt{2x_n}$ **b)** $x_0 = 1$, $x_{n+1} = 0.5e^{x_n}$

9 Draw the graphs of $y = e^x$ and $y = 3x^2$. Find approximate solutions of $3x^2 = e^x$. Solve the inequality $3x^2 > e^x$.

LONGER EXERCISE

1 Finding π – revisited

The Longer exercise of Consolidation section C involved finding π from the binomial expansion of $\sqrt{1 - x^2}$. The following is a much more efficient method.

1 Write down the binomial expansion of $\frac{1}{1 + x^2}$.

2 Integrate this series, to find a series for $\tan^{-1}x$ (called **Gregory's series**).

3 $\tan^{-1}1 = \frac{\pi}{4}$. Put $x = 1$ into the series, to obtain an expression for $\frac{\pi}{4}$.

4 The series converges very slowly for $x = 1$. Instead, by use of double angle formulae, show that

$$4 \tan^{-1} \frac{1}{5} = \frac{\pi}{4} + \tan^{-1} \frac{1}{239}$$

5 Use Gregory's series for $x = \frac{1}{5}$ and $x = \frac{1}{239}$ to find an approximation for π, accurate to six decimal places.

This method is a very practical one. In 1706 John Machin used it to find π to 100 decimal places.

2 Hyperbolic functions

On a scientific calculator there is a button labelled **HYP**. When combined with the sin, cos and tan buttons this enables us to find the **hyperbolic** functions sinh x, cosh x and tanh x.

These are defined by

$$\sinh x = \frac{e^x - e^{-x}}{2} \qquad \cosh x = \frac{e^x + e^{-x}}{2} \qquad \tanh x = \frac{e^x - e^{-x}}{e^x + e^{-x}}$$

1 What are the definitions of sech, cosech and coth?

2 Find the derivatives of cosh x and sinh x.

3 Simplify these expressions.

    a) $\cosh^2 x - \sinh^2 x$ **b)** $\cosh x + \sinh x$ **c)** $\cosh x - \sinh x$

4 Find the derivatives of $\cosh^{-1} x$ and $\sinh^{-1} x$.

5 Draw the graphs of $y = \sinh x$ and $y = \cosh x$.

6 In Chapters 14 and 17 there were many formulae concerning trigonometric functions, such as $\sec^2 x - \tan^2 x = 1$ and $\sin 2x = 2 \sin x \cos x$. What are the corresponding formulae concerning hyperbolic functions? Is there a rule for translating the trigonometric formula to the hyperbolic formula?

3 Elasticity of demand

Suppose a commodity is priced at £p. As p increases, the demand for the commodity will, in general, decrease. The **elasticity of demand** for a commodity tells us the extent to which a change in price alters the demand.

Suppose the demand for the commodity is $c = f(p)$. The elasticity of demand is defined to be

$$E_d = -\frac{dc}{dp} \times \frac{p}{c}$$

1 Suppose $c = 1 - p^2$. Find the elasticity of demand.

2 Suppose c is a linear function. When the price p is 0, i.e. the commodity is free, the demand c is 20 units. When the price p is 40, the demand c is 0, i.e. no one is prepared to buy the commodity at that price. Find the equation giving c in terms of p. Find the elasticity of demand for this commodity.

3 Suppose $c = \dfrac{k}{p}$, where k is constant. Show that the elasticity of demand for the commodity is constant.

4 Find an expression in terms of c and p for the revenue from sales of the commodity. When is the revenue a maximum? How is this related to the elasticity of demand?

5 Suppose at price p, the producers of a commodity will supply s of the commodity. How would you define the **elasticity of supply**?

EXAMINATION QUESTIONS

1 Assuming the identities

$$\sin 3\theta \equiv 3 \sin \theta - 4 \sin^3\theta \quad \text{and} \quad \cos 3\theta \equiv 4 \cos^3\theta - 3 \cos \theta$$

prove that

$$\cos 5\theta \equiv 5 \cos \theta - 20 \cos^3\theta + 16 \cos^5\theta$$

Find the set of all values of θ in the interval $0 < \theta < \pi$ for which $\cos 5\theta > 16 \cos^5\theta$.

AEB 1991 (Part)

2 a) Show that $(x - 2)$ is a factor of $x^3 - 10x^2 + 28x - 24$.

Find the set of values of x for which $x^3 - 10x^2 + 28x - 24 < 0$.

b) The constants m, n, p and q are chosen so that the identity

$$(mt + n)(t^2 - 4t + 5) \equiv 2t^3 + pt^2 + qt + 5$$

is true for all values of t. Find the values of m, n, p and q.

Prove that the equation $2t^3 + pt^2 + qt + 5 = 0$ has only one real root and state its value.

Deduce that $2e^{3u} + pe^{2u} + qe^u + 5 > 0$ for all real u.

AEB 1990

3 State the condition under which the infinite geometric series

$$1 + \left(\frac{2x + 3}{x + 1}\right) + \left(\frac{2x + 3}{x + 1}\right)^2 + \cdots$$

has a sum and obtain the set of values of x for which this condition holds.

Assuming that this condition holds, state the sum of the series in terms of x.

O&C 1992

4 Given that $f(\theta) = \theta - \sqrt{(\sin \theta)}$, $0 < \theta < \dfrac{\pi}{2}$, show that

a) the equation $f(\theta) = 0$ has a root lying between $\dfrac{\pi}{4}$ and $\dfrac{3\pi}{10}$

b) $f'(\theta) = 1 - \dfrac{\cos \theta}{2\sqrt{(\sin \theta)}}$

c) Taking $\dfrac{3\pi}{10}$ as a first approximation to this root of the equation $f(\theta) = 0$, use the Newton-Raphson procedure once to find a second approximation, giving your answer to 2 decimal places.

d) Show that $f'(\theta) = 0$ when $\sin \theta = \sqrt{5} - 2$.

L 1990

Integration as sum

In Chapter 13 integration was defined as the opposite operation to differentiation. The integral of $2x$ is $x^2 + c$, because the derivative of $x^2 + c$ is $2x$.

In fact, integration can be regarded as the limit of a sum. We have already used integration to find areas, and it has many other uses. Integration can be used to find volumes, centres of mass, probabilities and so on.

It is not always possible to integrate a function by treating it as the opposite of differentiation. In these cases, integration is achieved by a numerical procedure of finding a sum which approximates to the integral.

23.1 Integration defined as the limit of a sum

Suppose we have a region under a curve $y = f(x)$ between $x = a$ and $x = b$ as shown in Fig. 23.1. Suppose we wish to find its area. If we divide the region into thin strips, of width δx, the height of the left-hand side of the strip is $f(x)$, the value of the function at this point.

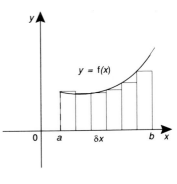

The strip is approximately a rectangle, of height $f(x)$ and width δx. The area of the rectangle is $f(x)\delta x$.

The total area of all the rectangles is

$$\sum f(x)\delta x$$

The total area of all the rectangles is approximately the area of the region. As we let the strips get thinner, the total area of the rectangles approaches the area of the region. The area of the region is therefore the **limit** of the area of the strips.

Fig. 23.1

$$\int_a^b f(x)\,dx = \lim_{\delta x \to 0} \sum f(x)\delta x$$

This is the definition of integration as the **limit of a sum**.

EXAMPLE 23.1

A pyramid has a square base of side a, and height h. Express its
volume as an integral, and hence evaluate its volume.

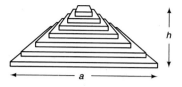

Solution

Divide the pyramid into thin slices as shown in Fig. 23.2. If a slice is x
down from the apex, then it is approximately a cuboid with height δx.

Its side is the base side reduced in the ratio $\dfrac{x}{h}$. This is $\dfrac{ax}{h}$.

Fig. 23.2

$$\text{Volume of slice} \approx \left(\frac{ax}{h}\right)^2 \delta x$$

$$\text{Volume of pyramid} \approx \sum \left(\frac{ax}{h}\right)^2 \delta x$$

Now take the limit as δx tends to 0. The sum tends to an integral,
which is the volume of the pyramid.

$$\text{Volume} = \lim_{\delta x \to 0} \sum \left(\frac{ax}{h}\right)^2 \delta x = \int_0^h \left(\frac{ax}{h}\right)^2 dx$$

To evaluate the integral, expand the brackets.

$$\text{Volume} = \int_0^h \frac{a^2 x^2}{h^2}\,dx = \left[\frac{a^2 x^3}{3h^2}\right]_0^h = \frac{a^2 h^3}{3h^2} = \frac{a^2 h}{3}$$

The volume is $\frac{1}{3}a^2 h$.

EXERCISE 23A

1 A cone has height h and base radius a. Express its volume as an integral and hence evaluate
   it.

2 A rod of length L cm has variable density. At distance x cm from one end, the density is $1 + \dfrac{x}{L}$
   grams per cm. Show that the mass of the rod is $\displaystyle\int_0^L \left(1 + \frac{x}{L}\right) dx$, and evaluate the integral.

3 A sphere of radius a cm has variable density. At distance x cm from the centre, its density is
   $(a - x)$ grams per cm^3. By splitting the sphere into thin shells, express the mass of the sphere
   as an integral. Evaluate this integral.

4 A solid uniform sphere has radius 10 cm. The sphere is cut into two parts by a plane 5 cm
   from the centre. Find the volume of the smaller part.

5 An elastic string is of length L when unstretched. If it is extended by x, the tension in it is kx.

The work required to stretch it a further δx is approximately $kx\,\delta x$. Find the total work needed to stretch the string out to a length of $2L$.

6 At a distance x metres below the surface of the sea, the water pressure is $10000x$ N per square metre. Find the total force on a vertical harbour wall, which is d metres deep and w metres wide.

7 Oil flows through a pipe with circular cross-section of radius a cm. Because of the viscosity of the oil, the rate of flow x cm from the centre is $k(a^2 - x^2)$ cm per second. Find the total flow through the pipe.

23.2 Volumes of revolution

Integration can also be used to find volumes.

Suppose we have a curve $y = f(x)$ as shown in Fig. 23.3. Then let the region between the curve and the x-axis be rotated about the x-axis. The region will sweep out a three-dimensional volume, called a **volume of revolution**.

Many simple shapes are volumes of revolution. A cone is the volume obtained by rotating a straight line which passes through the origin. A sphere is obtained by rotating a semicircle.

A solid object occupying a volume of revolution is a **solid of revolution**.

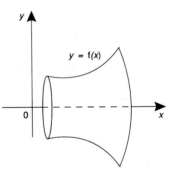

Fig. 23.3

To find the area of a region, we divide it into thin strips. When the region is rotated, these thin strips become thin discs. If the strips have height y and thickness δx, then the discs can be approximated by cylinders with radius y and width δx. Hence the volume of each disc is approximately

$$\pi y^2 \delta x$$

The total volume of the discs is found by adding up the volumes of all the discs.

$$\text{Total volume of discs} = \sum \pi y^2 \delta x$$

Now we let δx tend to zero. The discs get thinner and thinner, and in the limit the total volume becomes the volume of revolution. From Section 23.1, the limit of the sum is the integral.

$$\text{Volume of revolution} = \lim_{\delta x \to 0} \sum \pi y^2 \delta x = \int_a^b \pi y^2 \, dx$$

We can now try to evaluate this integral.

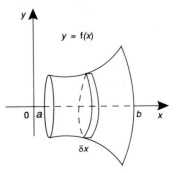

Fig. 23.4

This solid was obtained by rotating a region about the x-axis. If it were rotated about the y-axis, the volume would be

$$\text{Volume of revolution} = \lim_{\delta y \to 0} \sum \pi x^2 \delta y = \int \pi x^2 \, dy$$

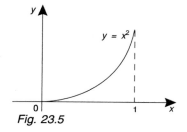

Fig. 23.5

EXAMPLE 23.2

Figure 23.5 shows the curve $y = x^2$ between $x = 0$ and $x = 1$. Find the volume of revolution when the region between this curve and the x-axis is rotated about the x-axis.

Solution

Apply the formula for volume of revolution. The function is $y = x^2$, and the integral is evaluated between $x = 0$ and $x = 1$.

$$\text{Volume} = \int_0^1 \pi (x^2)^2 \, dx = \int_0^1 \pi x^4 \, dx = \left[\frac{\pi}{5} x^5 \right]_0^1$$

The volume is $\dfrac{\pi}{5}$.

EXERCISE 23B

1 Find the volumes obtained when the following regions are rotated about the x-axis.

a) under $y = x^3$, from $x = 0$ to $x = 2$ **b)** under $y = x^2 + 1$, from $x = 1$ to $x = 2$

c) under $y = 4 - x^2$, from $x = 0$ to $x = 2$ **d)** under $y = \sqrt{x}$, from $x = 1$ to $x = 4$

2 Find the volumes obtained when the following regions are rotated about the y-axis.

a) between $y = x^3$ and the y-axis, from $x = 0$ to $x = 1$

b) between $y = \sqrt{x}$ and the y-axis, from $x = 0$ to $x = 4$

c) between $y = x^2 + 1$ and the y-axis, from $x = 1$ to $x = 2$

3 A cone has height h and base radius a. Show that it can be regarded as a solid of revolution, and hence find its volume.

4 A sphere has radius a. Show that it can be regarded as the solid of revolution obtained by rotating the region under $y = \sqrt{a^2 - x^2}$ about the x-axis. Hence find the volume of the sphere.

5 Figure 23.6 shows the curves $y = x^2$ and $y = 0.9x^2 + 0.1$. The region between the curves is rotated about the y-axis, to form the shape of a wine cup.

a) How much wine will the cup hold?

b) What volume of glass will be used to make the cup?

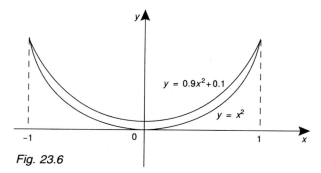

Fig. 23.6

6 Figure 23.7 shows an **ellipse**, with equation

$\dfrac{x^2}{a^2} + \dfrac{y^2}{b^2} = 1$, where $a > b$.

a) If the ellipse is rotated about the x-axis, we obtain a **prolate ellipsoid** (like a rugby ball). Find the volume of this ellipsoid.

b) If the ellipse is rotated about the y-axis, we obtain an **oblate ellipsoid** (like a curling stone). Find the volume of this ellipsoid.

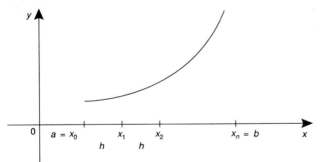

Fig. 23.7

*7 A napkin ring is made by taking a solid sphere of wood and drilling a hole along a diameter. Suppose the height of the resulting napkin ring is 4 cm. What volume of wood does it contain?

23.3 The trapezium rule for approximating integrals

Many functions cannot be integrated in terms of standard functions. None of the following, for example, can be written as the derivative of any standard function.

$\sin x^2 \qquad e^{x^2} \qquad \sqrt{1 - x^3}$

If we wish to evaluate definite integrals of these functions, then we must use some numerical approximation. We approximate the function by straight line segments. The integral is then approximated as the sum of the areas of trapezia.

Fig. 23.8

In Fig. 23.8 the interval between a and b is divided into n equal subintervals. The width of each subinterval is $\dfrac{b - a}{n} = h$. The division points are

$x_0 = a, \quad x_1 = a + h \quad x_2 = a + 2h, \quad \ldots, \quad x_n = a + nh = b$

The **trapezium rule** states that

$$\int_a^b f(x)\, dx \approx \frac{h}{2}[f(x_0) + f(x_n) + 2(f(x_1) + f(x_2) + \cdots + f(x_{n-1}))]$$

Justification

Within each subinterval, we can approximate the function by a straight line segment as shown in Fig. 23.9. The region under this straight line segment is a trapezium. The area of a trapezium is found by multiplying the width by the average of the parallel sides.

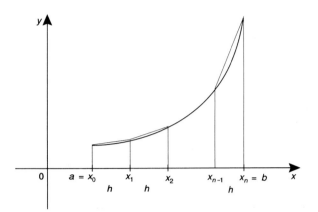

Fig. 23.9

Area of first trapezium $= \frac{1}{2}h[f(x_0) + f(x_1)]$

Area of second trapezium $= \frac{1}{2}h[f(x_1) + f(x_2)]$

$\vdots$

Area of nth trapezium $= \frac{1}{2}h[f(x_{n-1}) + f(x_n)]$

Now we add together these areas, taking out a factor of $\frac{1}{2}h$.

Total area of trapezia

$$= \frac{1}{2}h[f(x_0) + f(x_1) + f(x_1) + f(x_2) + \cdots + f(x_{n-1}) + f(x_n)]$$

Notice that the left and right values each occur once, but that the middle values each occur twice.

Total area $= \frac{1}{2}h[f(x_0) + f(x_n) + 2(f(x_1) + f(x_2) + \cdots + f(x_{n-1}))]$

The total area of the trapezia is approximately equal to the integral.

Use of calculator

In many cases it is possible to use a calculator to work out this formula without having to write anything down. For this you need a button which enables you to add numbers to the memory. One procedure is as follows.

Clear the memory.

Evaluate f(x_1) and add it to the memory. Evaluate f(x_2) and add it to the memory. Continue up to f(x_{n-1}).

Recall the memory. Multiply by 2. Return to the memory.

Evaluate f(x_0) and f(x_n) and add them to the memory.

Recall the memory and multiply by $\frac{1}{2}h$.

EXAMPLE 23.3
Use the trapezium rule with five intervals to find the approximate value of $\int_0^1 \sqrt{1 - x^3}\, dx$.

Solution
The interval between 0 and 1 is divided into five equal parts. The width h of each division is $\frac{1}{5} = 0.2$. The division points are

$$x_0 = 0 \quad x_1 = 0.2 \quad x_2 = 0.4 \quad x_3 = 0.6 \quad x_4 = 0.8 \quad x_5 = 1$$

Apply the formula, evaluating f(x) at each of these division points. This gives

$$\frac{0.2}{2}[\sqrt{1 - 0^3} + \sqrt{1 - 1^3} + 2(\sqrt{1 - 0.2^3} + \sqrt{1 - 0.4^3} + \sqrt{1 - 0.6^3} + \sqrt{1 - 0.8^3})]$$

The value of $\int_0^1 \sqrt{1 - x^3}\, dx$ is approximately 0.809.

EXERCISE 23C

1 Find approximations to the following integrals, using the trapezium rule with the number of intervals indicated.

a) $\int_0^1 \sqrt[3]{1 + x^2}\, dx$ 4 intervals

b) $\int_0^{\frac{\pi}{2}} \sqrt{\sin x}\, dx$ 5 intervals

c) $\int_1^2 e^{-\frac{1}{2}x^2}\, dx$ 8 intervals

d) $\int_0^{\frac{\pi}{2}} \sqrt{\cos x + \sin x}\, dx$ 5 intervals

2 Evaluate $\int_0^{\frac{\pi}{2}} \sin x\, dx$

a) exactly   b) using the trapezium rule with four intervals.

What is the percentage error in using the trapezium rule? Can you explain why your answer to b) is less than your answer to a)?

3 Evaluate $\int_0^1 x^2\, dx$

a) exactly   b) by using the trapezium rule with six intervals.

What is the percentage error? Can you explain why the trapezium rule overestimates the integral?

4 The table below gives the values of a function.

x	0	0.2	0.4	0.6	0.8	1
f(x)	1	1.3	1.5	1.7	1.9	2.2

Use the trapezium rule to estimate $\int_0^1 f(x)\, dx$.

5 The table below gives the values of a function.

x	1	2	3	4	5	6	7
f(x)	21	19	18	18	20	23	29

Estimate $\int_1^7 f(x)\, dx$. Do you think you will overestimate or underestimate the true value?

6 Expand $\sqrt{1+x}$ using the binomial theorem, up to the term in x^3. Evaluate $\int_0^{\frac{1}{2}} \sqrt{1+x}\, dx$

a) exactly   **b)** using the trapezium rule with four intervals

c) by integrating the series found above.

Which of **b)** and **c)** is more accurate?

*7 With f(x) as in Question 4, estimate the following.

a) $\int_0^1 \frac{1}{f(x)}\, dx$ **b)** $\int_0^1 xf(x)\, dx$ **c)** $\int_0^1 (f(x))^2\, dx$

*8 The region under the curve $y = f(x)$, where f(x) is as in Question 4, is rotated about the *x*-axis. Estimate the volume of revolution.

23.4 Simpson's rule for approximating integrals

The trapezium rule is not very accurate. Using the trapezium rule with four intervals to evaluate $\int_0^1 x^2\, dx$, the percentage error is over 3%. To obtain an accuracy of 0.1%, we need at least 23 intervals.

Simpson's rule is a much more accurate way of approximating integrals. For the trapezium rule, we approximate the function by straight line segments. For Simpson's rule, we approximate the function by quadratic segments.

Suppose the interval between *a* and *b* is divided into *n* equal subintervals, where *n* is even. Let $h = \dfrac{b-a}{n}$ be the width of each subdivision, and let the division points be

$$x_0 = a \quad x_1 = a + h \quad x_2 = a + 2h \quad \dots \quad x_n = a + nh = b$$

Then $\displaystyle\int_a^b f(x)\, dx$ is approximately equal to

$$\tfrac{1}{3}h[f(x_0) + f(x_n) + 4(f(x_1) + f(x_3) + \cdots) + 2(f(x_2) + f(x_4) + \cdots)]$$

Notice that the first and last values each occur once, the odd-numbered values each occur four times, and the even-numbered values each occur twice.

Justification

To fit a quadratic $ax^2 + bx + c$ to the curve, we need three points to find the three unknowns a, b and c. Let us take three division points, and for simplicity let them be $-h$, 0 and h.

The area under the function is approximately equal to the area under the quadratic. This area is

$$\int_{-h}^{h} (ax^2 + bx + c)\, dx = \left[\tfrac{1}{3}ax^3 + \tfrac{1}{2}bx^2 + cx\right]_{-h}^{h} = \tfrac{2}{3}ah^3 + 2ch$$

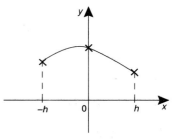

Fig. 23.10

Note that the b terms have cancelled out. Hence we do not need to find an expression for b.

If the quadratic is fitted to the function for these three points, then we have the equations

$$a(-h)^2 + b(-h) + c = f(-h)$$
$$a0^2 + b0 + c = f(0)$$
$$ah^2 + bh + c = f(h)$$

So $c = f(0)$. Adding the first and third equations

$$2ah^2 + 2c = f(-h) + f(h).$$

Hence $a = \dfrac{f(-h) - 2f(0) + f(h)}{2h^2}$

Now substitute for the values of a and c in terms of the function values. The area under the quadratic is

$$\frac{2[f(-h) - 2f(0) + f(h)]}{2h^2} \times \frac{h^3}{3} + 2f(0)h$$

Simplify the fraction, take out a factor of $\tfrac{1}{3}h$ and the expression becomes

$$\tfrac{1}{3}h[f(-h) - 2f(0) + f(h) + 6f(0)] = \tfrac{1}{3}h[f(-h) + 4f(0) + f(h)]$$

So the area is $\frac{1}{3}h$ times (left value $+ 4 \times$ middle value $+$ right value). This will hold, even if the two intervals are not on either side of the origin. So we apply this rule to all the subintervals, taken in pairs. The total area under all the quadratics is

$$\tfrac{1}{3}h[f(x_0) + 4f(x_1) + f(x_2)] + \tfrac{1}{3}h[f(x_2) + 4f(x_3) + f(x_4)] + \cdots + \tfrac{1}{3}h[f(x_{n-2}) + 4f(x_{n-1}) + f(x_n)]$$

Take out a factor of $\frac{1}{3}h$ and collect like terms. The formula becomes

$$\tfrac{1}{3}h[f(x_0) + 4f(x_1) + 2f(x_2) + 4f(x_3) + 2f(x_4) + \cdots + 2f(x_{n-2}) + 4f(x_{n-1}) + f(x_n)]$$

So $f(x_0)$ and $f(x_n)$ occur once. The odd-numbered values each occur four times. The even-numbered values each occur twice, at the end of one pair of intervals and at the beginning of the next. This gives Simpson's rule.

Use of calculator

A procedure to evaluate the formula without writing anything down is as follows.

Clear the memory.

Evaluate all the odd-numbered terms, adding them to the memory.

Recall the memory, and multiply by 2. Return to the memory.

Evaluate all the even-numbered terms, adding them to the memory.

Recall the memory, and multiply by 2. Return to the memory.

Evaluate $f(x_0)$ and $f(x_n)$, adding them to memory.

Recall the memory, and multiply by $\frac{1}{3}h$.

EXAMPLE 23.4
Find an approximation to $\int_0^1 \sin x^2 \, dx$, using Simpson's rule with four intervals.

Solution
Here $h = \frac{1}{4} = 0.25$. The division points are

$$x_0 = 0 \quad x_1 = 0.25 \quad x_2 = 0.5 \quad x_3 = 0.75 \quad x_4 = 1$$

Apply the rule. The integral is approximately

$$\tfrac{0.25}{3}[\sin 0^2 + \sin 1^2 + 4(\sin 0.25^2 + \sin 0.75^2) + 2(\sin 0.5^2)]$$

The integral is approximately $0.309\,94$.

EXERCISE 23D

1 Estimate the following integrals, using Simpson's rule with the number of intervals suggested.

a) $\displaystyle\int_0^{\frac{\pi}{2}} \sqrt{\sin x + \cos x}\ dx$ 4 intervals

b) $\displaystyle\int_0^1 \sqrt[3]{1 + x^2}\ dx$ 2 intervals

c) $\displaystyle\int_{\frac{1}{2}}^1 \sqrt{1 - x^3}\ dx$ 6 intervals

d) $\displaystyle\int_0^{\frac{\pi}{2}} \sqrt{\sin x}\ dx$ 8 intervals

2 Evaluate $\displaystyle\int_0^{\frac{\pi}{2}} \sin x\ dx$

a) exactly   **b)** using Simpson's rule with four intervals.

What is the percentage error? How does your answer compare with that of Question 2, Exercise 23C?

3 Show that Simpson's rule is exact when applied to a quadratic function.

4 The table below gives the values of a function.

x	1	1.5	2	2.5	3
$f(x)$	0.8	1.2	1.7	2.3	3.0

Estimate $\displaystyle\int_1^3 f(x)\ dx$ using Simpson's rule.

5 Let $f(x) = \dfrac{1}{1 + x}$. Use the binomial theorem to expand $f(x)$ up to the x^3 term. Evaluate $\displaystyle\int_0^{\frac{1}{2}} f(x)\ dx$

a) exactly   **b)** by Simpson's rule, using four intervals

c) by integrating the series found above.

Which of **b)** or **c)** is more accurate?

*6 With $f(x)$ as in Question 4, estimate the following.

a) $\displaystyle\int_1^3 \frac{1}{f(x)}\ dx$ **b)** $\displaystyle\int_1^3 xf(x)\ dx$ **c)** $\displaystyle\int_1^3 (f(x))^2\ dx$

*7 Show that Simpson's rule is exact for cubic functions.

LONGER EXERCISE

Consumer's surplus

Figure 23.11 shows a demand curve of a commodity. At a price p_0, the consumer will buy c_0 of the commodity. Notice that the consumer will pay p_0 for the whole quantity he buys, though he would have been prepared to pay a higher price for a smaller part of it. The amount the consumer gains by buying the whole quantity at the lower price is the **consumer's surplus**.

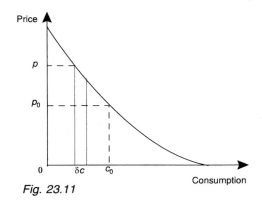

Fig. 23.11

Consider a small amount δc of the commodity. To buy this, the consumer would have been prepared to pay p. Instead, he only pays p_0. So his gain is $(p - p_0)\delta c$.

1 Express the consumer's surplus as a definite integral.

2 Suppose the demand curve is $p = 10e^{-c} - 2$. If the price is 5 units, how much would the consumer buy, and what is his surplus?

3 Suppose the demand curve is linear, that the consumer will obtain nothing at a price of £10, and that he will obtain 40 units if the commodity is free. Sketch the demand curve, and find its equation. Find the consumer surplus if the price is £5.

*4 Where does the surplus come from? It does not come from the producer. There is a similar phenomenon called the **producer's surplus**. See if you can define the producer's surplus in terms of a definite integral.

EXAMINATION QUESTIONS

1 The curve with equation

$$y = \sqrt{x} + \frac{3}{\sqrt{x}}$$

is sketched for $x > 0$.

The region R, shaded in Fig. 23.12, is bounded by the curve, the x-axis and the lines $x = 1$ and $x = 4$.

Determine

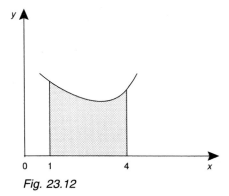

Fig. 23.12

a) the area of R

b) the volume of the solid formed when R is rotated through $360°$ about the x-axis.

O 1992

2 A jug of variable cross-section is 16 cm high inside and its internal radius is measured at equal intervals from the bottom.

Height h (cm)	0	4	8	12	16
Radius r (cm)	5	6.8	7.6	4.6	4.9

Use Simpson's rule to estimate the volume of liquid V, given by

$$V = \pi \int_0^{16} r^2 \, dh$$

which the jug will hold if filled to the brim (in litres to two decimal places).

NI 1992

3 The region R is bounded by the x-axis, the y-axis, the line $y = 12$ and the part of the curve whose equation is $y = x^2 - 4$ which lies between $x = 2$ and $x = 4$.

(i) Copy the sketch graph and shade the region R.

The inside of a vase is formed by rotating the region R through 360° about the y-axis. Each unit of x and y represents 5 cm.

(ii) Write down an expression for the volume of revolution of the region R about the y-axis.

(iii) Find the capacity of the vase in litres.

(iv) Show that when the vase is filled to $\frac{5}{6}$ of its internal height it is three-quarters full.

MEI 1992   *Fig. 23.13*

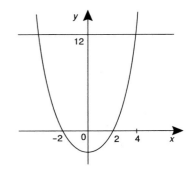

Summary and key points

1 An integral can be regarded as the limit of a sum.

$$\int f(x) \, dx = \lim_{\delta x \to 0} \sum f(x) \, \delta x$$

2 If a region under the curve $y = f(x)$, between $x = a$ and $x = b$, is rotated about the x-axis, the volume of revolution is

$$\int_a^b \pi y^2 \, dx$$

Note that we square y and then integrate, not the other way round.

3 If the interval between a and b is divided into n subintervals of width h, then the area under the curve $y = f(x)$ is approximately

$$\tfrac{1}{2}h[f(a) + f(b) + 2(f(a + h) + f(a + 2h) + \cdots + f(b - h))]$$

(the trapezium rule).

Note that the intervals must be **equal**.

4 Suppose the division for the trapezium rule has an even number of subintervals, i.e. n is even. The area is approximately

$$\tfrac{1}{3}h[f(a) + f(b) + 4(f(a + h) + f(a + 3h) + \cdots)$$
$$+ 2(f(a + 2h) + f(a + 4h) \cdots)]$$

(Simpson's rule).

Distances and angles

In previous chapters we used trigonometry and Pythagoras's theorem to find distances and angles within plane figures. The techniques used can be extended to three-dimensional figures.

Geometry is often more systematic if it is done in terms of coordinates. Here we show that distances and angles can be found using vectors.

24.1 Distances and angles within solids

Suppose we want to find distances and angles within a three-dimensional solid. If we make a flat cut through the solid, a cut surface, called a **section**, will be exposed. This surface will be two-dimensional, and ordinary trigonometry and Pythagoras's theorem can be used for this surface.

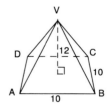

Fig. 24.1

EXAMPLE 24.1

A pyramid has a square base ABCD of side 10 cm. The vertex V is 12 cm above the centre of ABCD. Find

a) AC

b) the inclination of VA to the horizontal

c) the inclination of the plane VAB to the horizontal.

Fig. 24.2

Solution

a) The pyramid is shown in Fig. 24.2. To find AC, make a separate diagram of the base ABCD, Fig. 24.3.

We see that AC is the diagonal of a square of side 10 cm. Pythagoras's theorem now gives

$$AC = \sqrt{10^2 + 10^2} = \sqrt{200} = 14.1 \text{ cm}$$

AC is 14.1 cm.

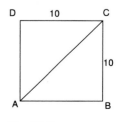

Fig. 24.3

b) To find the inclination of VA, imagine a vertical cut through the pyramid, exposing the surface VAC. This is an isosceles triangle. If M is the midpoint of AC, then the length MA is half the length of AC. Ordinary trigonometry gives

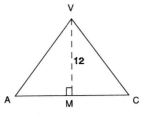

$$V\widehat{A}M = \tan^{-1}\frac{VM}{MA}$$

$$= \tan^{-1}\frac{12}{\frac{1}{2}\sqrt{200}} = 59.5°$$

The inclination of VA is 59.5°.

Fig. 24.4

c) To find the inclination of VAB, imagine a vertical cut through V, which cuts AB and CD in half. This is also an isosceles triangle, in which the base side is 10 cm.

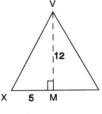

$$V\widehat{X}M = \tan^{-1}\frac{VM}{MX}$$

$$= \tan^{-1}\frac{12}{5} = 67.38°$$

The inclination of VAB is 67.4°.

Fig. 24.5

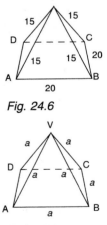

EXERCISE 24A

1 VABCD is a pyramid, with a square base ABCD of side 20 cm. VA = VB = VC = VD = 15 cm. Find

a) AC

Fig. 24.6

b) the height of the pyramid

c) the inclination of VA to the base

d) the inclination of the side VAB to the base.

2 A pyramid VABCD has a square base ABCD of side a and the lengths of VA, VB, VC and VD are also equal to a. Find the inclinations of VA and VAB to the base.

Fig. 24.7

3 Two boys are watching a balloon. One sees the balloon due north, and the other sees it due west. They both measure its angle of elevation as 30°. If the distance between the boys is 200 m, find the height at which the balloon is flying.

4 Two girls are respectively due south and due east of a hovering bird. They measure its angle of elevation as α and β respectively. If the girls are d apart, show that the height of the bird is

$$h = \frac{d}{\sqrt{\cot^2\alpha + \cot^2\beta}}$$

5 ABCDEFGH is a cuboid in which ABCD is the horizontal base. AB = 5 cm, AD = 4 cm and AE = 7 cm. Find the length of AG and its inclination to the horizontal.

6 A disc of radius 10 cm is supported by three strings of length 30 cm equally spaced round its rim. The strings are joined together at a point on the ceiling. Find the depth of the ring below the ceiling. Find the angle each string makes with the vertical, and the angles between the strings.

*7 A plane is at angle α to the horizontal. A line L makes angle β with a line of greatest slope of the plane. Show that the inclination of L to the horizontal is

$$\sin^{-1}(\sin \alpha \cos \beta)$$

*8 A high vertical wall runs north–south. A vertical flagpole is 4 m high and 2 m from the base of the wall. The Sun is in the south-west, at an elevation of 45°. Calculate the length of the shadow of the flagpole on the wall.

*9 A regular hollow tetrahedron made out of thin material has edges of length 2 cm. Find the radius of the largest sphere which can be put inside the tetrahedron.

Find the radius of the smallest sphere which can be placed around the tetrahedron.

Use of sine and cosine rules

Sometimes a three-dimensional problem requires the use of the sine or cosine rule.

EXAMPLE 24.2
Three spheres with centres A, B and C have radii 3 cm, 4 cm and 5 cm respectively. They rest on a horizontal plane in contact with each other, touching the ground at D, E and F respectively. Calculate the angle $D\widehat{E}F$.

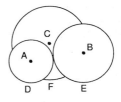

Fig. 24.8

Solution
The situation is shown in Fig. 24.8. Looking at the spheres from one side, the centres A and B are 7 cm apart, and 1 cm apart vertically. We can use Pythagoras's theorem to find their horizontal distance apart.

$$\text{Horizontal distance apart} = \sqrt{7^2 - 1^2} = \sqrt{48}$$

Hence DE = $\sqrt{48}$.

Similarly, EF = $\sqrt{80}$ and FD = $\sqrt{60}$.

Now use the cosine rule in triangle DEF.

$$\cos D\widehat{E}F = \frac{48 + 80 - 60}{2 \times \sqrt{48} \times \sqrt{80}}$$

$$D\widehat{E}F = \cos^{-1}\frac{68}{123.935} = 56.7°$$

$D\widehat{E}F = 56.7°$.

EXERCISE 24B

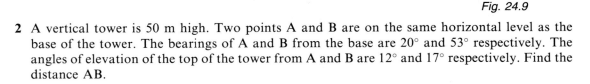

1 In Fig. 24.9, ABC is a horizontal triangle with $AB = 3$ cm, $BC = 4$ cm and $AC = 5$ cm. V is 2 cm vertically above B. Calculate

a) VA and VC   **b)** angle $A\widehat{V}C$.

Fig. 24.9

2 A vertical tower is 50 m high. Two points A and B are on the same horizontal level as the base of the tower. The bearings of A and B from the base are 20° and 53° respectively. The angles of elevation of the top of the tower from A and B are 12° and 17° respectively. Find the distance AB.

3 From the top of a cliff 100 m high two ships P and Q can be seen, at angles of depression 3° and 4° respectively. The distance apart of the two ships is 1000 m. If P is due North of the cliff, find the bearing of Q, given that it is East of P.

*4 A and B are on the same horizontal plane as the foot F of a tower. A is due North of the tower and B is on a bearing of 070°. The angles of elevation of the top T of the tower from A and B are 15° and 20° respectively. If A and B are 50 m apart, find the height of the tower.

*5 Three spheres of radii 3 cm, 3 cm and 4 cm rest on a horizontal plane in contact with each other. Find the inclination of the plane through the centres of the spheres to the horizontal.

24.2 The angle between a line and a plane

Suppose we have a plane Π and a line L going through Π, cutting it at X. What is the angle between the line and the plane? The problem is that there are infinitely many lines in the plane, each making a different angle with L.

In the previous section many questions involved finding the angle between a line and the horizontal. This is the angle between the line and the horizontal plane. Our definition of the angle between a line and a plane follows this.

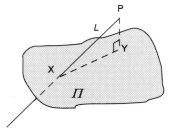

Suppose P is a point on L. Drop a perpendicular from P to the plane Π, meeting the plane at Y. Then the angle between L and Π is $P\widehat{X}Y$.

Fig. 24.10

Notes

1 The angle defined is the **least** angle between L and a line in Π.

2 When finding such angles, it is often best to imagine that the plane Π is horizontal. Then the perpendicular from P is vertical.

EXAMPLE 24.3

In the cuboid ABCDEFGH, AB = 8, AD = 7 and AE = 5 as shown. Find the angle between BH and the plane AEHD. (All measurements in cm.)

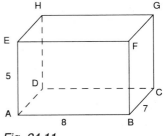

Fig. 24.11

Solution

Here Π is AEHD and L is BH. The point on the line is B. If we drop a perpendicular from B to AEHD, it will cross the plane at A. So the angle we want is $\widehat{\text{BHA}}$.

BHA is a triangle which is right-angled at A. We know that AB = 8. Use Pythagoras's theorem to find the length of HA.

$$HA = \sqrt{AD^2 + DH^2} = \sqrt{7^2 + 5^2} = \sqrt{74} = 8.602$$

Use trigonometry in triangle BHA.

$$\tan \widehat{\text{BHA}} = \frac{AB}{HA} = \frac{8}{\sqrt{74}} = 0.930$$

$$\widehat{\text{BHA}} = \tan^{-1} 0.930 = 42.922°$$

The angle between BH and AEHD is 42.9°.

EXERCISE 24C

1 In the cuboid of Example 24.3 above, find the angles between

a) AG and DCGH   **b)** EC and FGCB.

2 Equal squares are hinged together along an edge, at an angle of 30° to each other. Find the angle between one square and the diagonal of the other.

3 ABCDEFGH is a cube in which ABCD is a square directly below EFGH. Find the angle between AG and ABFE.

4 ABCDEFGH is a cuboid, in which ABCD is a rectangle directly below EFGH. AB = 5 cm, AD = 7 cm and AE = 3 cm. Find the angles AG makes with the faces of the cuboid.

5 The diagram shows a wedge in which ABCD is a square of side 10 cm. Points E and F are 3 cm above A and B respectively. Find the angles between

a) CE and ABCD   **b)** CE and ADE.

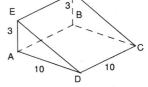

Fig. 24.12

24.3 The angle between two planes

Suppose we have two planes Π and Π'. What is the angle between them? The problem is that there are infinitely many possible angles between lines in Π and lines in Π'. In fact, the angles range from 0° to 180°.

Section 24.1 contained many questions about the inclination of a plane. This is the angle between a tilted plane and the horizontal plane. We define the angle between two planes in accordance with this.

Suppose the planes Π and Π' meet in a line L. Draw lines m and m', in Π and Π' respectively, which are at right-angles to L. The angle between Π and Π' is the angle between m and m'.

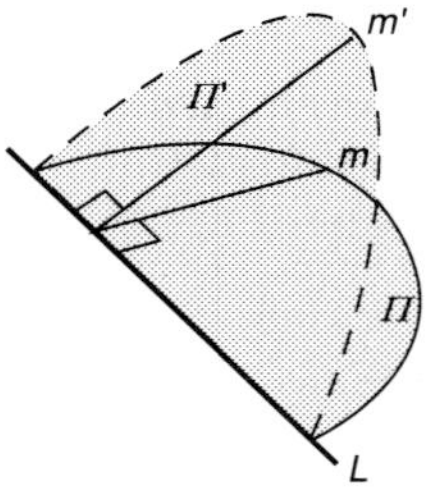

Fig. 24.13

Note
It is sometimes best to imagine that one of the planes is horizontal, and to find the steepest route up the other plane.

EXAMPLE 24.4
Find the angle between two faces of a regular tetrahedron.

Solution
Let the tetrahedron be ABCD. Without loss of generality we may assume that each edge has length 2 units. As the tetrahedron is regular, it does not matter which faces we take. Let us find the angle between ABD and CBD.

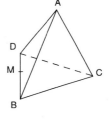

Fig. 24.14

The line of intersection of these planes is BD. Let M be the midpoint of BD. Then AM and CM will be perpendicular to BD. We want to find the angle between these lines, the angle $A\widehat{M}C$.

By Pythagoras's theorem, in triangle ABD, the length of AM is given by

$$AM = \sqrt{2^2 - 1^2} = \sqrt{3}$$

Similarly, $CM = \sqrt{3}$.

AC is one of the edges of the tetrahedron, and hence has length 2 units.

We can now use the cosine rule in triangle AMC.

$$\cos A\widehat{M}C = \frac{3 + 3 - 4}{2 \times \sqrt{3} \times \sqrt{3}} = \frac{1}{3}$$

$$A\widehat{M}C = \cos^{-1}\tfrac{1}{3} = 70.529°$$

The angle between faces is 70.5°.

EXERCISE 24D

1 ABCDEFGH is the cube shown in Fig. 24.15. Find the angle between HFBD and EHDA.

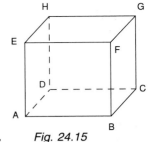

Fig. 24.15

2 In the cube of Question 1, find the angle between HFA and HGFE.

3 A pyramid VABCD has a square base ABCD of side 10 cm. VA, VB, VC and VD are all equal to 9 cm. Find the angle between adjacent triangular faces of the pyramid.

4 Three lines *L*, *M* and *N* are at right angles to each other, and meet at O. Points A, B and C lie on *L*, *M* and *N* respectively, with OA = OB = OC. Find the angle between the plane of ABC and the plane containing *L* and *M*.

5 Two squares ABCD and ABEF are at 30° to each other. Find the angle between AC and AE.

6 XABC is a tetrahedron, in which ABC is an equilateral triangle of side 8 cm. Sides XA, XB, XC are all equal to 11 cm. Find the angle between XAB and XBC.

7 VABCD is a pyramid, in which the base ABCD is a square of side *a*. Each triangular face makes an angle of 60° with the base. Calculate

a) the height of the pyramid   **b)** the angle between adjacent triangular faces.

24.4 Use of vectors

Calculations of distances and angles within solids can often be made using three-dimensional vectors.

The distance between two points can be found as the modulus of the vector joining the points. The angle between two lines can be found as the angle between vectors parallel to the lines.

EXAMPLE 24.5

A pyramid VABCD has a square base ABCD of side 2*a* and is *a* high. Find the length of the sloping edge BV and the angle it makes with BA.

Solution

Let the axes OX, OY and OZ of a coordinate system be along AB, AD and vertically upwards as shown in Fig. 24.16. The vertices of the pyramid are at

A(0, 0, 0)   B(2*a*, 0, 0)   C(2*a*, 2*a*, 0)   D(0, 2*a*, 0)   V(*a*, *a*, *a*)

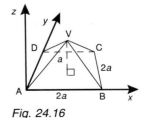

Fig. 24.16

The vector $\overrightarrow{BV}$ is $-a\mathbf{i} + a\mathbf{j} + a\mathbf{k}$. The length BV is the modulus of this vector.

The length of BV is $\sqrt{(-a)^2 + a^2 + a^2} = \sqrt{3}a$.

The angle between $\overrightarrow{BV}$ and $\overrightarrow{AB}$ can be found from the scalar product of $\overrightarrow{BV}$ and $\overrightarrow{BA}$.

$$\overrightarrow{BV} \cdot \overrightarrow{AB} = (-a\mathbf{i} + a\mathbf{j} + a\mathbf{k}) \cdot (-2a\mathbf{i}) = 2a^2$$

The angle is $\cos^{-1} \dfrac{2}{2\sqrt{3}} = 54.7°$.

The angle is $54.7°$.

EXERCISE 24E

1 A cuboid ABCDEFGH is such that AB $= 3$ cm, AD $= 4$ cm, AE $= 7$ cm. Put the x, y and z axes along AB, AD and AE respectively. Find the vectors $\overrightarrow{AE}$, $\overrightarrow{EC}$ and $\overrightarrow{EB}$. Find the length of the space diagonal of the cuboid. Find the angle between $\overrightarrow{EC}$ and $\overrightarrow{EB}$.

2 A wedge ABCDEF is such that ABCD is a rectangle with AB $= 4$ cm and AD $= 7$ cm. AF and DE are perpendicular to ABCD, with AF $=$ DE $= 3$ cm.

Put the x, y and z axes along AB, AD and AF respectively. Find the vectors $\overrightarrow{AE}$, $\overrightarrow{BE}$ and $\overrightarrow{CE}$. Find the angle between CE and AC.

3 A pyramid has a square base with vertices at $(0, 0, 0)$, $(1, 0, 0)$, $(1, 1, 0)$ and $(0, 1, 0)$. The vertex is 2 units above the centre of the base. Find the coordinates of the vertex. Use vectors to find the angle between a sloping edge and the base. Find the angle between a sloping face and the base.

4 Two rectangles ABCD and ABEF are joined along AB and are at right angles to each other. AB $=$ AD $= a$, AF $= 2a$. Let the coordinate axes lie along AB, AD and AF. By use of vectors or otherwise find
a) $B\hat{D}F$ **b)** the angle between DE and CF.

5 A tetrahedron VABC is such that the sides VA, VB and VC are of equal lengths a and are at right angles to each other. By letting the coordinate axes lie along VA, VB and VC, find

a) the angle between BC and VB

b) the angle between the planes ABC and VAB

c) the volume of the tetrahedron

d) the length of the perpendicular from V to ABC.

6 OABC is a square of side a, with O at the origin and OA and OC along the x and y axes. Points D and E are b along the z-direction from A and C respectively. Find the area of triangle ODE in terms of a and b. Find a relationship between $\cos D\hat{O}E$ and $\cos C\hat{E}B$.

LONGER EXERCISE

Platonic solids

The Platonic solids are those for which all the faces are equal regular polygons. There are five possible solids.

Tetrahedron   (4 triangular faces)

Cube   (6 square faces)

Octahedron   (8 triangular faces)

Dodecahedron   (12 pentagonal faces)

Icosahedron   (20 triangular faces)

In Example 24.4 we showed that the angle between two faces of a tetrahedron is $\cos^{-1}\frac{1}{3}$. The angle between adjacent faces of a cube is 90°. What about the other three solids?

1 Octahedron

Let our adjacent faces be AEB and AED. Their line of intersection is AE. Let M be the midpoint of AE. Then MB and MD will be perpendicular to AE. Follow the method of Example 24.4 to find the angle BMD.

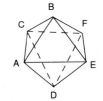

Fig. 24.17

*2 Icosahedron

At each vertex of an icosahedron, five triangular faces meet. Imagine yourself looking straight down at vertex X. You will see the five triangles laid out to make a regular pentagon as shown.

Suppose we find the angle between AXB and AXC. Follow the method of Example 24.4 to find this angle.

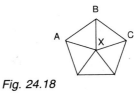

Fig. 24.18

**3 Dodecahedron

At each vertex of a dodecahedron, three pentagonal faces meet. Imagine yourself looking straight down at vertex X. You will see the three pentagons laid out as shown. Find the angle between adjacent faces.

Fig. 24.19

EXAMINATION QUESTIONS

1 An aeroplane P is observed simultaneously from two points A and B on horizontal ground, A being a distance $10\sqrt{2}$ km due North of B. From A the bearing of P is 45° East of South at an elevation of 60°, and from B the bearing of P is 30° East of North.

(i) Carefully draw a sketch to model the bearings and angles of elevation of P from A and B.

(ii) Calculate the elevation of P from B.

(iii) Find the height of P above the ground.

NI 1992

2 The tetrahedron VABC is such that ABC is horizontal and V is 10 cm vertically above A.
Given that angle $V\hat{B}A$ is $50°$, $AC = 5$ cm, $BC = 6$ cm, calculate:

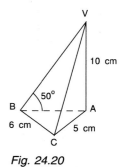

a) the length of AB, to the nearest 0.01 cm

b) the size of angle $A\hat{C}B$, to the nearest degree.

Hence, show that the perpendicular distance from A to the line through B and C is approximately 4.94 cm, and determine the size of the acute angle between the planes VBC and ABC, to the nearest $0.1°$.

O 1992

Fig. 24.20

3 Figure 24.21 shows a plastic waste-paper basket. The base ABCD and the top EFGH are each square in shape, with AB = 26 cm and EF = 30 cm. Each sloping face (e.g. ABFE) is a trapezium, and the lengths AE, BF, CG and DH are each 34 cm. Calculate

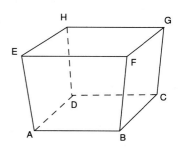

 (i) the perpendicular distance between the planes ABCD and EFGH, giving your answer to the nearest 0.1 cm

(ii) the angle between the planes ABFE and BCGF, giving your answer to the nearest $0.1°$.

C 1992

Fig. 24.21

Summary and key points

1 Distances and angles within solids can be found by considering two-dimensional sections.

2 If a line cuts a plane at X, and a perpendicular is dropped from the line to Y on the plane, the angle between the line and the plane is the angle between the line and XY.

3 If two planes meet in a line L, and lines are drawn in the planes at right-angles to L, the angle between the planes is the angle between these lines.

4 Angles and distances in solids can sometimes be found by using the scalar product and the moduli of vectors.

Algebra II

In this chapter we shall extend some of the algebraic techniques of previous chapters. We shall show how to split up rational functions into simpler functions, and how to sum certain series. We shall investigate the relationships between a quadratic equation and its roots.

25.1 Partial fractions

In Chapter 1 we discussed how to add algebraic fractions to obtain a single rational expression. Now we can go in the other direction: given a complicated rational function, we show how to split it into simpler algebraic fractions.

For example, we showed that the following identity is true for all values of x.

$$\frac{1}{x+1} + \frac{2}{x+2} \equiv \frac{3x+4}{(x+1)(x+2)}$$

In Chapter 1 we went from the left-hand side of the identity to the right. Here we go from the right to the left. The fractions on the left, consisting of constants divided by linear expressions, are examples of **partial fractions**.

A **proper numerical fraction** is one in which the numerator is less than the denominator. Similarly, a **proper algebraic fraction** is one in which the degree of the numerator is less than the degree of the denominator. The technique of partial fractions works for proper rational functions, and the partial fractions themselves must be proper.

Linear factors

In all cases the denominator of the original expressions should be factorised. If these factors are different and linear, then write the expressions as the sum of constants divided by the factors. By putting in special values for x we shall be able to find the constants.

EXAMPLE 25.1

Express $\dfrac{3x - 1}{(x + 1)(x - 3)}$ in partial fractions.

Solution

The denominator is already factorised. Write the expression as the sum of two partial fractions.

$$\frac{3x - 1}{(x + 1)(x - 3)} \equiv \frac{A}{x + 1} + \frac{B}{x - 3}$$

Multiply both sides by $(x + 1)(x - 3)$.

$$3x - 1 \equiv A(x - 3) + B(x + 1)$$

Note that this is an identity, true for all values of x, so we can put in any value of x we please. The most sensible values to put in are those which make either of the linear factors zero.

Put $x = 3$.

$$8 = B(3 + 1) = 4B$$

Hence $B = 2$.

Put $x = -1$.

$$-4 = A(-1 - 3) = -4A$$

Hence $A = 1$.

$$\frac{3x - 1}{(x + 1)(x - 3)} = \frac{1}{x + 1} + \frac{2}{x - 3}$$

EXERCISE 25A

1 Express the following in partial fractions.

a) $\dfrac{2}{(x + 1)(x - 1)}$ **b)** $\dfrac{2x + 1}{x(x + 1)}$ **c)** $\dfrac{2x}{(x + 1)(x - 1)}$

d) $\dfrac{3x}{(x + 1)(x - 2)}$ **e)** $\dfrac{x - 4}{(x - 1)(x - 2)}$ **f)** $\dfrac{8x + 34}{(x + 5)(x + 3)}$

g) $\dfrac{7x - 4}{(2x - 1)(x - 1)}$

h) $\dfrac{5x + 3}{(3x + 2)(2x + 1)}$

i) $\dfrac{3x^2 - 12x + 11}{(x - 1)(x - 2)(x - 3)}$

2 Express the following in partial fractions.

a) $\dfrac{1}{(x + \alpha)(x - \alpha)}$

b) $\dfrac{x}{(x + \alpha)(x - \alpha)}$

Quadratic factors

So far we have looked at cases when the denominator could be factorised into linear factors. Sometimes the denominator contains a quadratic factor, such as $x^2 + x + 2$, which cannot be factorised into linear factors.

Recall that partial fractions must be proper, with the degree of the numerator less than that of the denominator. If the denominator is a quadratic (degree 2), then the numerator could be linear (degree 1). Then the numerator of the partial fraction is of the form $Bx + C$.

For a quadratic factor of the form $x^2 + x + 2$ we include a partial fraction of the form

$$\frac{Bx + C}{x^2 + x + 2}$$

EXAMPLE 25.2

Express $\dfrac{5x + 7}{(x - 1)(x^2 + x + 2)}$ in partial fractions.

Solution
Write the expression in partial fractions.

$$\frac{5x + 7}{(x - 1)(x^2 + x + 2)} \equiv \frac{A}{x - 1} + \frac{Bx + C}{x^2 + x + 2}$$

Multiply both sides by the original denominator.

$$5x + 7 \equiv A(x^2 + x + 2) + (Bx + C)(x - 1)$$

As in the previous example, put $x = 1$.

$$12 = A(1 + 1 + 2) = 4A$$

Hence $A = 3$.

The quadratic expression does not factorise, so there are no more obvious values of x to use. But notice that the expression above is an identity. The left-hand side has no x^2 term, and so the right-hand side cannot have one. The x^2 terms on the right-hand side must cancel each other. This process is called **equating the coefficients** of x^2.

$$0 = A + B$$

Hence $B = -3$.

The constant term must be the same on both sides of the identity.

$$7 = A(2) + C(-1)$$

Hence $C = -1$.

$$\frac{5x + 7}{(x - 1)(x^2 + x + 2)} \equiv \frac{3}{x - 1} - \frac{3x + 1}{x^2 + x + 2}$$

EXERCISE 25B

1 Express the following in partial fractions.

a) $\dfrac{1}{x(x^2 + 1)}$

b) $\dfrac{3x^2 + 3}{(x - 1)(x^2 + 2)}$

c) $\dfrac{2x + 1}{(x + 1)(x^2 + x + 1)}$

d) $\dfrac{2x^2 + x + 1}{x(2x^2 + 1)}$

e) $\dfrac{2 - 5x}{(2x + 1)(x^2 + 2)}$

f) $\dfrac{3 + 2x - x^2}{(x - 1)(2x^2 + x + 1)}$

2 Factorise $x^3 - 1$ and $x^3 + 1$. Hence express these in partial fractions.

a) $\dfrac{3}{x^3 - 1}$

b) $\dfrac{3}{x^3 + 1}$

Repeated factors

So far each factor has occurred just once. Sometimes the denominator may include a factor of the form $(x - a)^2$. Then both $(x - a)$ and $(x - a)^2$ are factors of the denominator, so we include partial fractions of the form

$$\frac{A}{x - a} + \frac{B}{(x - a)^2}$$

EXAMPLE 25.3

Express $\dfrac{7x - 13}{(x + 2)(x - 1)^2}$ in partial fractions.

Solution

Write the expression as

$$\frac{7x - 13}{(x + 2)(x - 1)^2} \equiv \frac{A}{x - 1} + \frac{B}{(x - 1)^2} + \frac{C}{x + 2}$$

Multiply both sides by the denominator of the original fraction.

$$7x - 13 \equiv A(x - 1)(x + 2) + B(x + 2) + C(x - 1)^2$$

Put $x = 1$.

$$-6 = B(3)$$

Hence $B = -2$.

Put $x = -2$.

$$-27 = C(-3)^2 = 9C$$

Hence $C = -3$.

Equate the coefficients of x^2.

$$0 = A + C$$

Hence $A = 3$.

$$\frac{7x - 13}{(x + 2)(x - 1)^2} \equiv \frac{3}{x - 1} - \frac{2}{(x - 1)^2} - \frac{3}{x + 2}$$

EXERCISE 25C

1 Express the following in partial fractions.

a) $\dfrac{2x^2 - 5x + 1}{x(x - 1)^2}$

b) $\dfrac{3x^2 + x + 4}{(x + 1)^2(x - 1)}$

c) $\dfrac{2x^2 - 7x - 1}{(x + 1)(x - 1)^2}$

d) $\dfrac{x^2 + 5x + 2}{x^2(2x + 1)}$

e) $\dfrac{24x + 12}{x^2(x - 2)}$

f) $\dfrac{1}{x(x + 1)^2}$

2 Express the following in partial fractions.

a) $\dfrac{1}{x(x - 1)^3}$

b) $\dfrac{1}{x^2(x + 1)^2}$

Dividing numerator by denominator

So far we have emphasised that we are dealing with **proper** rational functions, in which the degree of the numerator is less than that of the denominator.

If the degree of the numerator is greater or equal to that of the denominator, then we must divide through, using the techniques of Chapter 1, before putting into partial fractions.

EXAMPLE 25.4

Express $\dfrac{x^2 + 3x - 1}{(x - 1)(x + 2)}$ in partial fractions.

Solution

Both numerator and denominator are quadratics, so they have the same degree. Expand the bottom and divide it into the top.

$$
\begin{array}{r}
1 \\
x^2 + x - 2 \overline{\smash{)}x^2 + 3x - 1} \\
\underline{x^2 + x - 2} \\
2x + 1
\end{array}
$$

So the quotient is 1 and the remainder is $2x + 1$. Rewrite the expression.

$$
\frac{x^2 + 3x - 1}{(x - 1)(x + 2)} = 1 + \frac{2x + 1}{(x - 1)(x + 2)}
$$

Using the methods above, we put the algebraic fraction on the right into partial fractions.

$$
\frac{x^2 + 3x - 1}{(x - 1)(x + 2)} = 1 + \frac{1}{x - 1} + \frac{1}{x + 2}
$$

EXERCISE 25D

1 Express the following in partial fractions.

a) $\dfrac{x^2 + x + 1}{x(x + 1)}$

b) $\dfrac{2x^2 + 2x - 17}{(x + 3)(x - 2)}$

c) $\dfrac{3x^2 - 20}{(x + 2)(x - 2)}$

d) $\dfrac{x^3 + 2x^2 + 2x - 2}{x(x + 2)}$

e) $\dfrac{x^3 + 2x^2 - x}{(x + 1)(x - 1)}$

f) $\dfrac{x^4 - 3x + 1}{x(x - 1)}$

2 Express the following in partial fractions.

a) $\dfrac{x^2}{x^2 - \alpha^2}$

b) $\dfrac{x^3}{x^2 - \alpha^2}$

25.2 Expansion of rational functions in series

We can now combine the techniques of Section 25.1 with those of the binomial theorem in Chapter 11.

The expansion of $\dfrac{1}{1+x}$ comes either from the binomial expansion of $(1+x)^{-1}$ or from the sum of an infinite geometric series with common ratio $-x$.

$$\frac{1}{1+x} = 1 - x + x^2 - x^3 + \cdots \quad \text{(provided that } -1 < x < 1)$$

The expansion of $\dfrac{1}{(1+x)^2}$ comes from the binomial expansion of $(1+x)^{-2}$.

$$\frac{1}{(1+x)^2} = 1 - 2x + 3x^2 - 4x^3 + \cdots \quad \text{(provided that } -1 < x < 1)$$

If we have a rational function, then we can split it into partial fractions and find the expansion of each of the fractions.

EXAMPLE 25.5

Let $f(x) = \dfrac{5}{(1+2x)(2-x)}$.

Write $f(x)$ in partial fractions. Hence find the expansion of $f(x)$ up to the term in x^2. For what range of values of x is your expansion valid?

Solution

Use the techniques of Section 25.1 to find the partial fractions.

$$f(x) = \frac{2}{1+2x} + \frac{1}{2-x}$$

Write the second fraction so that the expansion can be used, by dividing top and bottom by 2.

$$f(x) = 2\left(\frac{1}{1+2x}\right) + \frac{\frac{1}{2}}{(1-\frac{1}{2}x)}$$

Now expand these, up to the x^2 term.

$$f(x) \approx 2(1 - 2x + (2x)^2) + \tfrac{1}{2}(1 + \tfrac{1}{2}x + (\tfrac{1}{2}x)^2)$$

$f(x)$ is approximately $2.5 - 3.75x + 8.125x^2$.

The expansion is valid if $-1 < 2x < 1$ and $-1 < \frac{1}{2}x < 1$. Take the stronger of these conditions. The expansion is valid for $-\frac{1}{2} < x < \frac{1}{2}$.

EXAMPLE 25.6

Find the expansion of $\dfrac{x^2 + x + 2}{(x + 1)^2(1 - x)}$ up to the term in x^3.

Solution

Use the techniques of Section 25.1 to find the partial fractions.

$$\frac{x^2 + x + 2}{(x + 1)^2(1 - x)} = \frac{1}{1 - x} + \frac{1}{(1 + x)^2}$$

Expand each of these by the binomial theorem.

$$\frac{1}{1 - x} + \frac{1}{(1 + x)^2} = (1 + x + x^2 + x^3 + \cdots) + (1 - 2x + 3x^2 - 4x^3 + \cdots)$$

Combine these, ignoring the terms after the x^3 term.

$$\frac{x^2 + x + 2}{(x + 1)^2(1 - x)} \approx 2 - x + 4x^2 - 3x^3$$

EXAMPLE 25.7

Find the expansion of $\dfrac{x^2 - x + 3}{(1 - x)(x^2 + 2)}$ up to the term in x^2.

Solution

Use the technique of Section 25.1 to find the partial fractions.

$$\frac{x^2 - x + 3}{(1 - x)(x^2 + 2)} = \frac{1}{1 - x} + \frac{1}{2 + x^2}$$

Divide top and bottom of the second fraction by 2, then use the binomial expansion.

$$\frac{1}{1 - x} + \frac{1}{2 + x^2} = \frac{1}{1 - x} + \frac{\frac{1}{2}}{1 + \frac{1}{2}x^2} = (1 + x + x^2 + \cdots) + \tfrac{1}{2}(1 - \tfrac{1}{2}x^2 + \tfrac{1}{4}x^4 - \cdots)$$

Collect terms and ignore those after the x^2 term.

$$\frac{x^2 - x + 3}{(1 - x)(x^2 + 2)} \approx 1\tfrac{1}{2} + x + \tfrac{3}{4}x^2$$

EXERCISE 25E

1 Express the following in partial fractions. Hence find their expansion up to the x^2 term.

a) $\dfrac{x - 3}{(1 - x)(1 + x)}$ **b)** $\dfrac{3 - x}{(1 - x)(2 - x)}$ **c)** $\dfrac{5x - 13}{(2 - x)(3 - x)}$

d) $\dfrac{3x - x^2}{(x+1)(1-x)^2}$

e) $\dfrac{2x^2 + x + 8}{(2-x)(1+x)^2}$

f) $\dfrac{x^2 - 4x + 12}{(1-x)(2+x)^2}$

g) $\dfrac{x+1}{(1-x)(1+x^2)}$

h) $\dfrac{2x^2 - 6x - 2}{(1+2x)(1+2x^2)}$

i) $\dfrac{2x}{(1+x)(1+x^2)}$

2 The first two terms of the expansion of $\dfrac{Ax + B}{(1+x)(1-x)}$ are $2 + 8x$. Find A and B.

3 The first two terms of the expansion of $\dfrac{Ax + B}{(1+x)(2+x)}$ are $1 - \frac{1}{2}x^2$. Find A and B.

25.3 Expansions of functions

The binomial expansion in Chapter 11 provides a series for $(1+x)^n$. In Section 25.2 we found how to find a series expansion of any rational function. Many other functions can be expanded in a series. In particular, there are series for e^x, $\sin x$ and $\cos x$.

Suppose the function is $f(x)$. Differentiate repeatedly, to obtain $f'(x)$, $f''(x)$, $f'''(x)$ and so on. Find the value of $f(x)$ and its derivatives at $x = 0$, as $f(0)$, $f'(0)$, $f''(0)$, $f'''(0)$ and so on.

The **Maclaurin expansion of** $f(x)$ is

$$f(x) = f(0) + f'(0)x + \frac{f''(0)x^2}{2!} + \frac{f'''(0)x^3}{3!} + \cdots$$

The series is a valid expansion if, whenever we take more and more terms, their sum gets closer to $f(x)$. The series **converges** to $f(x)$.

Justification

Suppose we can expand $f(x)$ in terms of powers of x, and that the following series, where a_0, a_1, a_2, a_3 and so on are constant, is a valid expansion for values of x close to 0.

$$f(x) = a_0 + a_1 x + a_2 x^2 + a_3 x^3 + \cdots$$

The function and its expansion must have the same value at $x = 0$, so put $x = 0$.

$$f(0) = a_0 + a_1 0 + a_2 0^2 + a_3 0^3 + \cdots = a_0$$

The function and its expansion must have the same gradient at $x = 0$. Differentiating gives

$$f'(x) = a_1 + 2a_2 x + 3a_3 x^2 + \cdots$$
$$f'(0) = a_1 + 2a_2 0 + 3a_3 0^2 + \cdots = a_1$$

The function and its expansion must have the same second derivative at $x = 0$. Differentiating again gives

$$f''(x) = 2a_2 + 2 \times 3a_3 x + \cdots$$

$$f''(0) = 2a_2 + 2 \times 3a_3 0 + \cdots = 2a_2$$

And so on. Differentiating a third time and putting $x = 0$ will give $f'''(0) = 2 \times 3a_3$. So we have

$$a_0 = f(0) \quad a_1 = f'(0) \quad a_2 = \frac{f''(0)}{2!} \quad a_3 = \frac{f'''(0)}{3!}$$

and so on.

Notes

1 This method is not guaranteed to work. There are many functions which do not have a Maclaurin expansion.

2 The series may only be valid for small values of x. The binomial expansion of $(1 + x)^{-1}$ is an example of a Maclaurin expansion, and it is only valid for $-1 < x < 1$.

3 If the expansion is valid, then by taking the first few terms of the expansion we obtain an approximation for $f(x)$. The more terms we take, the more accurate the approximation.

Standard series

Four standard series are

1 $e^x = 1 + x + \dfrac{x^2}{2!} + \dfrac{x^3}{3!} + \cdots$

2 $\sin x = x - \dfrac{x^3}{3!} + \dfrac{x^5}{5!} - \cdots$

3 $\cos x = 1 - \dfrac{x^2}{2!} + \dfrac{x^4}{4!} - \cdots$

4 $\ln(1 + x) = x - \dfrac{x^2}{2} + \dfrac{x^3}{3} - \cdots$

Proof

1 Let $f(x) = e^x$.

Then $f(0) = e^0 = 1$.

Differentiating gives

$$f'(x) = e^x$$

and so $f'(0) = 1$.

The nth derivative of $f(x)$ is always e^x. So the value of the nth derivative at $x = 0$ is always 1.

Put $f(0)$, $f'(0)$, $f''(0)$ and $f'''(0)$ equal to 1 in the Maclaurin expansion. The series above is obtained.

2 Let $f(x) = \sin x$.

The derivatives of $f(x)$ are

$$f'(x) = \cos x \qquad f''(x) = -\sin x$$

$$f'''(x) = -\cos x \qquad f''''(x) = \sin x$$

We are back to the original function. The fifth derivative will be equal to the first, the sixth derivative equal to the second and so on.

Put $x = 0$.

$$f(0) = \sin 0 = 0 \qquad f'(0) = \cos 0 = 1$$

$$f''(0) = -\sin 0 = 0 \qquad f'''(0) = -\cos 0 = -1$$

The cycle of 0, 1, 0, -1 will be repeated. So $f''''(0) = 0$, $f'''''(0) = 1$ and so on. Put these values into the Maclaurin expansion and the series above will be obtained.

3 Let $f(x) = \cos x$.

Note that $f(x)$ is the derivative of $\sin x$, so we differentiate the series for $\sin x$.

$$\cos x = \frac{d}{dx} \sin x = 1 - 3 \times \frac{x^2}{3!} + 5 \times \frac{x^4}{5!} - \cdots$$

$$= 1 - \frac{x^2}{2!} + \frac{x^4}{4!} - \cdots$$

4 Let $f(x) = \ln(1 + x)$.

Put $x = 0$.

$$f(0) = \ln 1 = 0$$

Differentiating gives

$$f'(x) = \frac{1}{1 + x}$$

The series for $f'(x)$ can be found using the binomial expansion.

$$f'(x) = 1 - x + x^2 - \cdots$$

Integrate this series.

$$f(x) = c + x - \tfrac{1}{2}x^2 + \tfrac{1}{3}x^3 - \cdots$$

We know that $f(0) = 0$. Hence $c = 0$. The series is obtained.

Notes

1 The expansions of e^x, $\sin x$ and $\cos x$ are valid for all x. The expansion of $\ln(1 + x)$ is valid for $-1 < x < 1$.

2 In Chapter 7 we found approximations for $\sin x$ and $\cos x$, as x and $1 - \frac{1}{2}x^2$ respectively. Note that these are consistent with the series expansions above.

EXAMPLE 25.8

Find the Maclaurin expansion of $\ln(2 + x)$ up to the term in x^2.

Solution

Let $f(x) = \ln(2 + x)$.

Differentiating twice gives

$$f'(x) = \frac{1}{2 + x}$$

$$f''(x) = \frac{-1}{(2 + x)^2}$$

Put $x = 0$ in $f(x)$, $f'(x)$ and $f''(x)$.

$$f(0) = \ln 2 \qquad f'(0) = \tfrac{1}{2} \qquad f''(0) = -\tfrac{1}{4}$$

Put these values into the Maclaurin expansion.

The expansion up to the x^2 term is $\ln 2 + \frac{1}{2}x - \frac{1}{8}x^2$.

EXAMPLE 25.9

Find the expansions of e^{2x} and $\sin 3x$, up to the term in x^3. Hence find the expansion of $e^{2x} \sin 3x$, also up to the term in x^3.

Solution

There is no need to differentiate repeatedly. We can substitute into the expansions already found.

$$e^{2x} \approx 1 + (2x) + \frac{(2x)^2}{2!} + \frac{(2x)^3}{3!} = 1 + 2x + 2x^2 + \tfrac{4}{3}x^3$$

$$\sin 3x \approx (3x) - \frac{(3x)^3}{3!} = 3x - \tfrac{9}{2}x^3$$

Multiply these together.

$$e^{2x} \sin 3x \approx (1 + 2x + 2x^2 + \tfrac{4}{3}x^3)(3x - \tfrac{9}{2}x^3)$$

$$= 3x - \tfrac{9}{2}x^3 + 6x^2 - 9x^4 + 6x^3 - 9x^5 + 4x^4 - 6x^6$$

Discard powers of x beyond x^3.

$e^{2x} \sin 3x$ is approximately $3x + 6x^2 + 1.5x^3$.

EXERCISE 25F

1 By repeated differentiation, find the Maclaurin expansions of the following up to the term in x^3.

a) $\ln(3 + x)$ **b)** $\tan x$ **c)** $\ln(0.5 - x)$

2 Use the standard series above to find the Maclaurin expansions of the following, up to the term in x^3.

a) e^{3x} **b)** $\cos 2x$ **c)** e^{x^2}

d) $\ln(1 + x^2)$ **e)** $\ln(1 + 2x)$ **f)** $\ln(1 - \frac{1}{2}x)$

g) $e^x \cos x$ **h)** $e^{-x} \sin x$ **i)** $e^x \ln(1 + x)$

j) $e^{3x} \sin x$ **k)** $e^{-x} \cos 2x$ **l)** $e^{3x} \cos 2x$

m) $\sin 2x \ln(1 + x)$ **n)** $\cos x^2 \sin x$ **o)** $\sin x^2 \cos x^2$

3 For each of the functions of Question 2, find the range of values of x for which the expansion is valid.

4 Let $f(x) = \sin x \cos x$. Find the Maclaurin expansion of $f(x)$ up to the term in x^3

a) by multiplying the expansions for $\sin x$ and $\cos x$

b) by writing $f(x)$ as $\frac{1}{2} \sin 2x$, and expanding this function.

Verify that your answers to **a)** and **b)** are the same.

5 Use the addition formula to write $\sin\left(\frac{\pi}{6} + x\right)$ in terms of $\sin x$ and $\cos x$. Hence find the expansion of $\sin\left(\frac{\pi}{6} + x\right)$ up to the term in x^2.

6 Find the expansion of $(1 + x)e^x$ up to the term in x^2. Hence find an approximate solution to the equation $(1 + x)e^x = 1.2$.

7 Find constants a and b such that the expansions of $\dfrac{1 + ax}{1 + bx}$ and e^{2x} agree for the first three terms.

8 Show that $\ln(2 + x) = \ln 2 + \ln(1 + \frac{1}{2}x)$. Hence find the expansion of $\ln(2 + x)$ up to the term in x^2. Verify that the expansion agrees with the one found in Example 25.8.

9 Use the approximation $\cos x \approx 1 - \frac{1}{2}x^2 + \frac{1}{24}x^4$ and the binomial expansion of $(1 + x)^{-1}$ to find the expansion of $\sec x$, up to the term in x^4.

10 Find the expansions of the following, up to the term in x^2.

a) $\dfrac{e^{-x}}{1 + x}$ **b)** $\dfrac{\sin 3x}{1 + x^2}$ **c)** $\ln\left(\dfrac{(1 + 2x)^2}{1 - x}\right)$

11 The first three terms of the expansion of e^x are $1 + x + \frac{1}{2}x^2$. Find the error in using these terms as an approximation for e^x in the following cases.

a) $x = 0.1$ **b)** $x = 0.5$ **c)** $x = 1$

12 The first three non-zero terms of the expansion of cos x are used as an approximation for cos x. Find the percentage error in the following cases.

a) $x = 0.5$ **b)** $x = 0.1$

13 Write out the first three terms of the expansion of e^{-x}. Hence find an approximation for this integral.

$$\int_0^{\frac{1}{2}} \sqrt{x}\, e^{-x}\, dx$$

*14 Write out the first three non-zero terms of the expansion of $\ln(1 - x)$. By putting $x = \frac{2}{3}$, find a series for $\ln 3$.

Show how the expansion of $\ln(1 + x)$ can be used to find approximations to $\ln k$, for any positive k.

25.4 Number of roots of an equation

The formula for the solution of the quadratic equation $ax^2 + bx + c = 0$ involves taking the square root of the expression $b^2 - 4ac$. The sign of this expression affects the number of solutions.

If $b^2 - 4ac > 0$ it has two distinct square roots, $+\sqrt{b^2 - 4ac}$ and $-\sqrt{b^2 - 4ac}$.

If $b^2 - 4ac = 0$ it has only one square root, 0 itself.

If $b^2 - 4ac < 0$ it has no square roots, as a negative number has no square root.

So the rule for the number of roots is as follows.

If $b^2 - 4ac > 0$ there are two roots.

If $b^2 - 4ac = 0$ there is one root.

If $b^2 - 4ac < 0$ there are no roots.

The three cases are illustrated in Fig. 25.1.

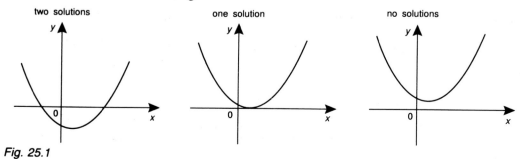

Fig. 25.1

The expression $b^2 - 4ac$ is the **discriminant** of the equation, as it discriminates between these three possible cases.

This discriminant can be used to solve problems in other fields.

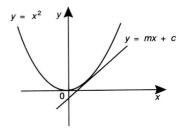

EXAMPLE 25.10
The equation of a curve is $y = x^2$. Find a condition on m and c for the line $y = mx + c$ to be a tangent to the curve.

Solution

Fig. 25.2

When the curve and the line meet, both equations are satisfied.

$$y = x^2 \text{ and } y = mx + c$$

Hence $x^2 = mx + c$.

This is a quadratic equation in x, which can be written as $x^2 - mx - c = 0$. If the line is a tangent to the curve, then this equation will have exactly one solution. The condition for this is that its discriminant is zero.

$$(-m)^2 - 4(-c) = 0$$
$$m^2 + 4c = 0$$

The line is a tangent if $m^2 + 4c = 0$.

EXERCISE 25G

1 Without solving them, find the number of roots for each of the following equations.

a) $x^2 + 10x + 26 = 0$ **b)** $13x^2 - 12x + 2 = 0$ **c)** $0.1x^2 - 0.005x + 0.00007 = 0$

2 Show that the equation $x^2 + 2kx + k^2 - m^2 = 0$ will always have at least one root.

3 Show that the equation $x^2 + kx - m^2 = 0$ will always have at least one root.

4 For what values of k does the equation $x^2 - 4kx + 8 = 0$ have equal roots?

5 Find the value of k which will ensure that the equation $kx^2 - (2k + 1)x + k = 0$ has equal roots.

6 Find the values of k which will ensure that $y = x + k$ is a tangent to the curve $x^2 + 4y^2 = 4$.

7 Find the values of k which will ensure that $y = kx + 2$ is a tangent to $4x^2 + y^2 = 1$.

*8 (The Cauchy-Schwartz inequality)

Suppose **a** and **b** are vectors. Then $\mathbf{a} \cdot \mathbf{b} \leq |\mathbf{a}| \, |\mathbf{b}|$.

Consider $f(x) = (\mathbf{a} + x\mathbf{b}) \cdot (\mathbf{a} + x\mathbf{b})$. Write this as a quadratic in x. Write down the discriminant of this quadratic. Express the fact that $f(x)$ is always positive in terms of the discriminant.

*9 Find the condition on y for there to be solutions of the equation $y = x + \dfrac{1}{x}$. Hence find the range of values taken by the function $y = x + \dfrac{1}{x}$.

LONGER EXERCISE

Parabola of safety

When a bomb explodes, fragments are sent off in all directions. Where are you in danger of being struck by one of the fragments? Suppose the maximum speed of the fragments is u m s^{-1}. There is a curve called the **parabola of safety**, outside which you are not in danger.

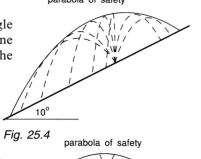

Fig. 25.3

Suppose that a fragment is sent off at a speed of u m s^{-1}, at an angle of θ to the horizontal. It can be shown that the equation of the trajectory of the fragment is

$$y = x \tan \theta - \frac{5x^2}{u^2}\sec^2\theta$$

1 Show that this can be written as $y = xT - \dfrac{5x^2}{u^2}(1 + T^2)$ where $T = \tan \theta$.

2 Suppose you are standing at (x, y). You are safe if the equation above has no solutions. Find the condition that the equation, regarded as a quadratic in T, has no solutions.

3 Find the condition that the equation above has exactly one solution, and show that it reduces to

$$y = \frac{u^2}{20} - \frac{5x^2}{u^2}$$

This is the equation of the parabola of safety. Outside it you cannot be hit by any fragment of the bomb.

4 Let $u = 60$. Suppose the bomb explodes on a slope of angle $10°$. By finding where the parabola of safety meets the line with equation $y = x \tan 10°$, find how far up and down the slope the fragments could reach.

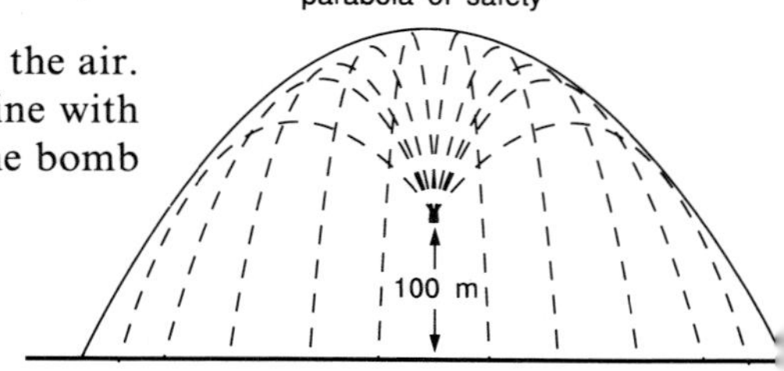

parabola of safety

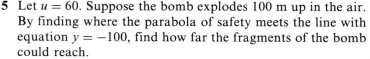

Fig. 25.4

5 Let $u = 60$. Suppose the bomb explodes 100 m up in the air. By finding where the parabola of safety meets the line with equation $y = -100$, find how far the fragments of the bomb could reach.

parabola of safety

Fig. 25.5

EXAMINATION QUESTIONS

1 Express $\dfrac{1 - x - x^2}{(1 - 2x)(1 - x)^2}$ in partial fractions.

Hence, or otherwise, expand this expression in ascending powers of x up to and including the term in x^2.

State the range of values of x for which the full expansion is valid.

Find the coefficient of x^n in the expansion.

J 1991

2 $f(x) \equiv \dfrac{9 - 3x - 12x^2}{(1 - x)(1 + 2x)}.$

a) Given that $f(x) \equiv A + \dfrac{B}{1 - x} + \dfrac{C}{1 + 2x}$

find the values of the constants A, B and C.

b) Given that $|x| < \frac{1}{2}$, expand $f(x)$ in ascending powers of x up to and including the term x^3, simplifying each coefficient.

Hence, or otherwise, find the value of $f'(0)$.

L 1993

Summary and key points

1 To put a rational expression into partial fractions do the following.

(i) If the degree of the denominator is less than or equal to that of the numerator, divide by the denominator and find the remainder.

(ii) Factorise the denominator.

(iii) For a single linear factor $(x - \alpha)$, include a partial fraction $\dfrac{A}{x - \alpha}.$

For a quadratic factor $ax^2 + bx + c$, include a partial fraction $\dfrac{Bx + C}{ax^2 + bx + c}.$

For a repeated factor $(x - \alpha)^2$, include partial fractions $\dfrac{A}{x - \alpha}$ and $\dfrac{B}{(x - \alpha)^2}.$

(iv) Multiply by the denominator. Find the values of the constants by

   a) putting in values of x

   b) equating the coefficients of powers of x.

2 To expand a rational function as a series in x, express the function in partial fractions.

3 The Maclaurin expansion of a function $f(x)$ is

$$f(x) = f(0) + f'(0)x + \frac{f''(0)x^2}{2!} + \frac{f'''(0)x^3}{3!} + \cdots$$

Standard expansions are

$$e^x = 1 + x + \frac{x^2}{2!} + \frac{x^3}{3!} + \cdots \qquad \sin x = x - \frac{x^3}{3!} + \frac{x^5}{5!} - \cdots$$

$$\cos x = 1 - \frac{x^2}{2!} + \frac{x^4}{4!} - \cdots \qquad \ln(1 + x) = x - \frac{x^2}{2} + \frac{x^3}{3!} - \cdots$$

$$(\text{valid for } -1 < x < 1)$$

4 The discriminant of the equation $ax^2 + bx + c = 0$ is $b^2 - 4ac$.

$ax^2 + bx + c = 0$ has two, one or no roots when $b^2 - 4ac$ is positive, zero or negative respectively.

CHAPTER 26

Techniques of integration

The process of integration is opposite to that of differentiation. To integrate a function f(x), we need to find another function F(x) such that $F'(x) = f(x)$.

The standard integrals of basic functions have been found in previous chapters. For example, we know that

$$\int x^n \, dx = \frac{x^{n+1}}{n+1} + c \qquad \text{for } n \neq -1$$

$$\int x^{-1} \, dx = \ln x + c$$

Once we have reduced an integral to one of the standard forms it can be integrated.

Differentiation is a routine procedure. There are fixed rules, such as the product and quotient rules, which we follow to find the derivative of any function. But there is no complete set of rules to tell us how to integrate. There is a variety of techniques for integration, and trial and error often plays a part.

Success is not guaranteed. There are some functions which cannot be integrated in terms of standard functions. One example is $f(x) = e^{x^2}$. There is no function F(x), built up in terms of standard functions, for which $F'(x) = e^{x^2}$.

The function

$$\Phi(x) = \frac{1}{\sqrt{2\pi}} e^{-\frac{1}{2}x^2}$$

is very important in statistics. It cannot be integrated directly, and so there are tables in statistics books, giving the values of the integral. There is a computer investigation on page 505 in which part of these tables can be reproduced.

Below is a list of the integrals of standard functions, found in previous chapters. We shall be using this list throughout the chapter.

$$\int x^n \, dx = \frac{x^{n+1}}{n+1} + c \quad \text{for } n \neq -1$$

$$\int x^{-1} \, dx = \ln x + c \qquad\qquad \int e^x \, dx = e^x + c$$

$$\int \sin x \, dx = -\cos x + c \qquad \int \cos x \, dx = \sin x + c$$

$$\int \tan x \, dx = -\ln \cos x + c \qquad \int \cot x \, dx = \ln \sin x + c$$

In addition, we found how to integrate functions of linear functions. If a and b are constant, and $a \neq 0$

$$\int f'(ax+b) \, dx = \frac{1}{a} f(ax+b) + c$$

26.1 Integration of functions of functions

Suppose the integrand involves a function of a function. There are strategies which we can follow when trying to integrate it.

Substitution

Often an integral can be reduced to one of the standard forms by means of a **substitution**.

When we make a substitution $u = h(x)$, we must be sure to change the dx term as well. Differentiate u to obtain

$$\frac{du}{dx} = h'(x)$$

If δx and δu are small changes in x and u, we can write

$$\delta u \approx h'(x) \, \delta x$$

$$\text{So } \delta x \approx \frac{\delta u}{h'(x)}$$

So in the integral, we can replace dx by $\dfrac{du}{h'(x)}$.

EXAMPLE 26.1
Find the integral $\int x(2x+1)^9 \, dx$.

Solution
If desperate, we could expand $(2x+1)^9$ by the binomial theorem and integrate term by term, but there is a quicker way. If we substitute $u = 2x + 1$, then the term in brackets becomes u^9.

We must also change the x term and the dx.

$$u = 2x + 1$$

so

$$x = \frac{1}{2}u - \frac{1}{2}$$

Now

$$\frac{du}{dx} = 2$$

so

$$dx = \frac{1}{2}du$$

Now we are ready to make the substitution. The integral becomes

$$\int \left(\frac{1}{2}u - \frac{1}{2}\right)u^9 \, \frac{1}{2}du = \int \frac{1}{4}(u^{10} - u^9)du = \frac{1}{44}u^{11} - \frac{1}{40}u^{10} + c$$

The original question was in terms of x, so the final answer should also be in terms of x.

$$\frac{1}{44}u^{11} - \frac{1}{40}u^{10} + c = \frac{1}{44}(2x+1)^{11} - \frac{1}{40}(2x+1)^{10} + c$$

The integral is $\frac{1}{44}(2x+1)^{11} - \frac{1}{40}(2x+1)^{10} + c$.

EXAMPLE 26.2

Use the substitution $u = e^x$ to evaluate $\displaystyle\int_0^1 \frac{e^x}{e^x + 2} \, dx$.

Solution

Rewrite the integral in terms of u.

Change the dx, by using

$$\frac{du}{dx} = e^x = u$$

i.e. replace dx by

$$\frac{1}{u} \, du$$

The limits of the integral are in terms of x. Change them to limits in terms of u.

When $x = 0, u = 1$.

When $x = 1, u = $ e.

The integral becomes

$$\int_1^e \frac{u}{u+2} \frac{1}{u} \, du = \int_1^e \frac{1}{u+2} \, du = \Big[\ln(u+2)\Big]_1^e$$

$$= \ln(e+2) - \ln 3 = 0.4528$$

EXERCISE 26A

1 Find the following integrals.

a) $\int x(x+2)^{15} \, dx$

b) $\int (x+1)(x+2)^{10} \, dx$

c) $\int x\sqrt{x+3} \, dx$

d) $\int \dfrac{x-3}{(x+7)^{15}} \, dx$

e) $\int (x+1)(2x-1)^7 \, dx$

f) $\int (2x-1)\sqrt{x+1} \, dx$

2 Find the following integrals with the substitution indicated.

a) $\int x \cos(x^2 - 1) \, dx$ $(u = x^2 - 1)$

b) $\int \dfrac{x}{x^2+1} \, dx$ $(u = x^2 + 1)$

c) $\int \cos x \sin^2 x \, dx$ $(u = \sin x)$

d) $\int \dfrac{1}{x \ln x} \, dx$ $(u = \ln x)$

e) $\int \dfrac{x^2}{1+x^3} \, dx$ $(u = x^3)$

f) $\int \sin x(1 + \cos x) \, dx$ $(u = 1 + \cos x)$

g) $\int e^x \cos e^x \, dx$ $(u = e^x)$

h) $\int \sec^2 x \tan^2 x \, dx$ $(u = \tan x)$

3 Evaluate the following definite integrals, using the substitutions indicated.

a) $\int_1^2 x\sqrt{x^2 - 1} \, dx$ $(u = x^2 - 1)$

b) $\int_0^{\frac{\pi}{2}} \sin x \cos^3 x \, dx$ $(u = \cos x)$

c) $\int_0^1 e^x\sqrt{1 + 2e^x} \, dx$ $(u = e^x)$

d) $\int_0^1 \dfrac{x^3}{\sqrt{1 - x^4}} \, dx$ $(u = 1 - x^4)$

4 Find the area under the curve $y = \sin x \cos^2 x$ from $x = 0$ to $x = \dfrac{\pi}{2}$.

5 Find the area under the curve $y = x\sqrt{4 - x}$ from $x = 0$ to $x = 4$.

6 Find the volume obtained when the region under $y = \sec x \tan x$, from $x = 0$ to $x = \dfrac{\pi}{4}$, is rotated about the x-axis.

7 Find the volume obtained when the region under $y = \sqrt{x}(1 - \tfrac{1}{2}x)^{\frac{1}{3}}$, from $x = 0$ to $x = 1$, is rotated about the x-axis.

Find the substitution

There is no golden rule for finding the substitution to evaluate an integral, but there are strategies which are often successful. As was stated earlier in the chapter, integration is often a matter of trying different methods until we find one which works.

Linear substitution

In Example 26.1 above, the substitution was $u = 2x + 1$. This sort of substitution is useful when a power of a linear function is involved.

'Spot the derivative'

Consider the integral $\int x \sin x^2 \, dx$. Now $\sin x^2$ is of the form $f(g(x))$, where the inside function $g(x)$ is x^2. The derivative of $g(x)$ is $2x$, which differs from x only by a constant factor. The substitution $u = x^2$ will convert the integral into a simple form.

In general, suppose the function we want to integrate involves something of the form $f(g(x))$, a function of a function. Suppose this is multiplied by $g'(x)$, the derivative of the inside function $g(x)$. Then the substitution $u = g(x)$ is very likely to work.

A particular case of this is when the function is of the form $\dfrac{f'(x)}{f(x)}$. The integral of this is $\ln f(x) + c$, as we shall see in the example below.

Guesswork

Often the quickest way to find an integral is by guesswork. Make a guess for what the integral is, and differentiate this guess. You may be right. You may be out by a constant factor, in which case the guess can be modified to make it correct.

EXAMPLE 26.3

Find $\displaystyle \int \frac{x^3}{1 + x^4} \, dx$.

Solution

Notice that the derivative of $(1 + x^4)$ is $4x^3$, which differs from the numerator only by a constant factor. So make the substitution $u = (1 + x^4)$.

$$\int \frac{x^3}{1 + x^4} \, dx = \int \frac{x^3}{u} \frac{du}{4x^3} = \frac{1}{4} \int \frac{1}{u} \, du = \frac{1}{4} \ln u + c$$

$$\int \frac{x^3}{1+x^4}\, dx = \frac{1}{4}\, \ln(1+x^4) + c$$

EXAMPLE 26.4
Find $\int \sin x \,(1 + \cos x)^9 \, dx$.

Solution
The solution is very likely to involve $(1 + \cos x)^{10}$.

We make a guess of $y = (1 + \cos x)^{10}$ and differentiate.

$$\frac{dy}{dx} = 10 \times (-\sin x)(1 + \cos x)^9$$

This is out by a factor of -10, so adjust the guess and it will be correct.

$$\int \sin x \,(1 + \cos x)^9 \, dx = -\tfrac{1}{10}(1 + \cos x)^{10} + c$$

EXERCISE 26B

1 For the following, write down the substitution you would use. Find the integrals.

a) $\int x \,(x^2 + 8)^{10} \, dx$ **b)** $\int \sin x \, \cos^2 x \, dx$

c) $\int (x + 3)(3x + 2)^{17} \, dx$ **d)** $\int e^{1-4x} \, dx$

e) $\int \cos x \, e^{\sin x} \, dx$ **f)** $\int e^x \sin(e^x + 3) \, dx$

2 With experience, it is possible to do many of the 'Spot the derivative' integrals by guesswork. Try to solve the following integrals by this method. If that fails do them by substitution.

a) $\int x \sin x^2 \, dx$ **b)** $\int x^2 \cos x^3 \, dx$

c) $\int \cos x \, \sqrt{\sin x} \, dx$ **d)** $\int \dfrac{e^x}{e^x + 2} \, dx$

e) $\int \sec^2 x \, e^{\tan x} \, dx$ **f)** $\int \dfrac{x + 2}{x^2 + 4x} \, dx$

26.2 Integration by parts

The product rule tells us how to **differentiate** the product of two functions. There is no infallible rule for **integrating** a product.

Sometimes there is a substitution which will reduce a product to a simpler form. The strategy of 'Spot the derivative' is often successful in this way. If this does not work there is another technique available, called **integration by parts**.

The product rule for differentiation is

$$\frac{d}{dx} uv = u'v + v'u$$

Integrate this equation.

$$uv = \int u'v \, dx + \int v'u \, dx$$

Rewrite as

$$\int u'v \, dx = uv - \int v'u \, dx$$

This is the rule for integration by parts.

Suppose we want to integrate a product of functions, such as $\int wv \, dx$.

Suppose we can integrate one of the functions, and the integral of w is u, i.e. $w = \dfrac{du}{dx}$. Then we can write the integral as

$$\int u'v \, dx$$

Provided we can integrate $v'u$ then we can find the original integral.

EXAMPLE 26.5
Find $\int x \cos x \, dx$.

Solution
We know how to integrate both x and $\cos x$. We have a choice of which part to integrate first. Integrating the $\cos x$ first, we write

$$\frac{du}{dx} = \cos x$$

so that $u = \sin x$ and $v = x$. Apply the formula.

$$\int x \cos x \, dx = x \sin x - \int v' \sin x \, dx$$

Now, $v' = 1$ and the integral of $\sin x$ is $-\cos x$, so we have

$$\int x \cos x \, dx = x \sin x - \int \sin x \, dx = x \sin x + \cos x + c$$

$$\int x \cos x \, dx = x \sin x + \cos x + c$$

Note: We chose to integrate the $\cos x$ and leave the x. If we had made the other decision, we would have obtained

$$\int x \cos x \, dx = \tfrac{1}{2}x^2 \cos x - \int \tfrac{1}{2}x^2(-\sin x) \, dx$$

This formula is correct, but we would now have to integrate $x^2 \sin x$, which is harder than $x \cos x$. If this sort of thing happens go back and try the formula the other way round.

EXAMPLE 26.6
Integrate $\int \ln x \, dx$.

Solution
This does not seem to be a product at all, but write $\ln x$ as $1 \times \ln x$, and then the formula can be used.

There is no choice for which function is to be $\dfrac{du}{dx}$. We know how to integrate 1, but not $\ln x$. If $\dfrac{du}{dx} = 1$, then $u = x$. Put it into the formula.

$$\int 1 \times \ln x \, dx = x \ln x - \int x \frac{d}{dx} \ln x \, dx$$

$\dfrac{d}{dx} \ln x = \dfrac{1}{x}$ which cancels out with the x.

$$\int 1 \times \ln x \, dx = x \ln x - \int 1 \, dx$$
$$= x \ln x - x + c$$
$$\int \ln x \, dx = x \ln x - x + c$$

EXERCISE 26C

1 Find the following.

   a) $\int x \, e^x \, dx$ **b)** $\int x \sin x \, dx$

   c) $\int x^2 \ln x \, dx$ **d)** $\int x \, e^{-x} \, dx$

   e) $\int x^2 \, e^x \, dx$ (For this you will have to integrate by parts twice.)

   f) $\int x^2 \cos x \, dx$ **g)** $\displaystyle\int \frac{\ln x}{x^2} \, dx$

   h) $\int x \sec^2 x \, dx$ **i)** $\int \ln (x + 2) \, dx$

2 Evaluate the following definite integrals.

   a) $\displaystyle\int_0^2 x \, e^{2x} \, dx$ **b)** $\displaystyle\int_0^{\frac{\pi}{2}} x \cos x \, dx$ **c)** $\displaystyle\int_1^{10} x^3 \ln x \, dx$

3 Find the area under the curve $y = x \sin x$ from $x = 0$ to $x = \pi$.

4 Find the area under the curve $y = x \, e^{-x}$ from $x = 1$ to $x = 3$.

5 Find the volume of revolution when the region under the curve $y = \sqrt{x} \, e^{-x}$, from $x = 0$ to $x = 1$, is rotated about the x-axis.

6 Find the volume of revolution when the region under the curve $y = x \, e^x$, from $x = 1$ to $x = 2$, is rotated about the x-axis.

26.3 Strategies for the integration of trigonometric functions

A wide variety of techniques is available for the integration of expressions involving trigonometric functions. Some of them involve the identities from Chapters 14 and 17. The integrals of the basic functions are given in the list at the beginning of this chapter.

Products of functions

Suppose we need to integrate $\sin 4x \cos x$. Then we can use the identity given on page 288.

$$2 \sin A \cos B = \sin (A + B) + \sin (A - B)$$

So $\sin 4x \cos x = \frac{1}{2}(\sin 5x + \sin 3x)$.

The expression on the right can be integrated. A similar method can be used for integrating $\cos 4x \cos x$ or $\sin 4x \sin x$.

Powers of functions

Suppose we need to integrate $\sin^3 x$. Then use the substitution $u = \cos x$, and the Pythagorean formula

$$\sin^2 x = 1 - \cos^2 x$$

This method will work provided we have an odd power of $\sin x$ or $\cos x$. The substitution involves the other function.

To integrate $\sin^2 x$ or $\cos^2 x$, use one of these formulae.

$$\cos 2\theta = 2 \cos^2 \theta - 1 \qquad \cos 2\theta = 1 - 2 \sin^2 \theta$$

Then we can write $\sin^2 x$ or $\cos^2 x$ in terms of $\cos 2x$, which can be integrated. This method will work when we have even powers of $\sin x$ or $\cos x$.

EXAMPLE 26.7
Find these integrals.

a) $\int \cos 3x \cos 2x \, dx$ **b)** $\int \sin^2 x \cos^3 x \, dx$ **c)** $\int \cos^2 3x \, dx$

Solution
a) Apply the formula for the product of two cosines.

$$2 \cos A \cos B = \cos (A + B) + \cos (A - B)$$
$$\int \cos 3x \cos 2x \, dx = \int \frac{1}{2} (\cos 5x + \cos x) \, dx$$

$\int \cos 3x \cos 2x \, dx = \frac{1}{10} \sin 5x + \frac{1}{2} \sin x + c$

b) Note that the power of $\cos x$ is odd. Make the substitution $u = \sin x$, giving $du = \cos x \, dx$. Use the formula $\cos^2 x = 1 - \sin^2 x$ for the remaining powers of $\cos x$.

$\int \sin^2 x \cos^3 x \, dx = \int \sin^2 x \cos^2 x \cos x \, dx = \int u^2(1 - u^2) \, du$

$\int (u^2 - u^4) \, du = \frac{1}{3}u^3 - \frac{1}{5}u^5 + c$

$\int \sin^2 x \cos^3 x \, dx = \frac{1}{3} \sin^3 x - \frac{1}{5} \sin^5 x + c$

c) Use the formula to rewrite $\cos^2 3x$ in terms of $\cos 6x$.

$\cos 6x = 2 \cos^2 3x - 1$

Hence $\cos^2 3x = \frac{1}{2} \cos 6x + \frac{1}{2}$

$\int \cos^2 3x \, dx = \int \left(\frac{1}{2} \cos 6x + \frac{1}{2}\right) dx = \frac{1}{12} \sin 6x + \frac{1}{2}x + c$

EXERCISE 26D

1 Find the following integrals.

a) $\int \sin 3x \cos x \, dx$

b) $\int \sin 2x \cos 5x \, dx$

c) $\int \cos 3x \cos x \, dx$

d) $\int \cos 4x \cos 6x \, dx$

e) $\int \sin 3x \sin x \, dx$

f) $\int \sin 6x \sin 7x \, dx$

g) $\int \sin^3 x \, dx$

h) $\int \cos^3 x \, dx$

i) $\int \cos^5 x \, dx$

j) $\int \sin^3 x \cos^2 x \, dx$

k) $\int \sin^3 x \cos^3 x \, dx$

l) $\int \dfrac{\sin^3 x}{\cos^2 x} \, dx$

m) $\int \sin^2 x \, dx$

n) $\int \sin^2 2x \, dx$

o) $\int \cos^2 4x \, dx$

2 Use the formula $\sin 2A = 2 \sin A \cos A$ to find $\int 8 \sin^2 x \cos^2 x \, dx$.

3 Use the $\cos 2A$ formula twice to find $\int \cos^4 x \, dx$.

*4 Find $\int \sin x \sin 3x \sin 5x \, dx$.

*5 What is $\int \sin nx \cos mx \, dx$ in the following cases?

a) $n \neq m$

b) $n = m$

The integrals of $\dfrac{1}{\sqrt{a^2 - x^2}}$ **and** $\dfrac{1}{a^2 + x^2}$

The integrals of these two functions are $\sin^{-1} \dfrac{x}{a} + c$ and $\dfrac{1}{a} \tan^{-1} \dfrac{x}{a} + c$ respectively.

Proof

We know the following from Chapter 17.

$$\int \frac{1}{\sqrt{1-x^2}}\, dx = \sin^{-1}x + c \qquad \int \frac{1}{1+x^2}\, dx = \tan^{-1}x + c$$

For $\int \frac{1}{\sqrt{a^2 - x^2}}\, dx$, make the substitution $x = au$. Then the integral becomes

$$\int \frac{1}{\sqrt{a^2 - a^2 u^2}}\, a\, du = \int \frac{1}{\sqrt{1 - u^2}}\, du = \sin^{-1}u + c$$

Substitute back for x.

$$\int \frac{1}{\sqrt{a^2 - x^2}}\, dx = \sin^{-1}\frac{x}{a} + c$$

For $\int \frac{1}{a^2 + x^2}\, dx$ again make the substitution $x = au$. The integral becomes

$$\int \frac{1}{a^2 + a^2 u^2}\, a\, du = \frac{1}{a} \int \frac{1}{1 + u^2}\, du = \frac{1}{a}\tan^{-1}u + c$$

$$\int \frac{1}{a^2 + x^2}\, dx = \frac{1}{a}\tan^{-1}\frac{x}{a} + c$$

EXAMPLE 26.8

Find the following integrals.

a) $\int \dfrac{dx}{\sqrt{4 - x^2}}$ **b)** $\int \dfrac{dx}{1 + 4x^2}$ **c)** $\int \dfrac{dx}{(x+1)^2 + 9}$

Solution

a) This fits the form above, putting $a = 2$.

$$\int \frac{dx}{\sqrt{4 - x^2}} = \sin^{-1}\frac{x}{2} + c$$

b) Divide top and bottom of the integrand by 4. Then use the formula, with $a = \frac{1}{2}$.

$$\int \frac{dx}{1 + 4x^2} = \int \frac{1}{4}\frac{dx}{(\frac{1}{4} + x^2)} = \frac{1}{4} \times \frac{1}{\frac{1}{2}} \times \tan^{-1}\frac{x}{\frac{1}{2}} + c$$

$$\int \frac{dx}{1 + 4x^2} = \frac{1}{2}\tan^{-1}2x + c$$

c) Make the substitution $u = x + 1$. We know that $\dfrac{du}{dx} = 1$, so we can change dx to du.

$$\int \frac{dx}{(x+1)^2 + 9} = \int \frac{du}{u^2 + 9} = \frac{1}{3} \tan^{-1} \frac{u}{3} + c$$

$$\int \frac{dx}{(x+1)^2 + 9} = \frac{1}{3} \tan^{-1} \frac{x+1}{3} + c$$

EXERCISE 26E

1 Find the following.

a) $\displaystyle\int \frac{dx}{\sqrt{9 - x^2}}$

b) $\displaystyle\int \frac{dx}{\sqrt{1 - 4x^2}}$

c) $\displaystyle\int \frac{dx}{\sqrt{4 - 9x^2}}$

d) $\displaystyle\int \frac{dx}{4 + x^2}$

e) $\displaystyle\int \frac{dx}{1 + 9x^2}$

f) $\displaystyle\int \frac{dx}{9 + 4x^2}$

2 Use the substitution $u = (x + 2)$ to find $\displaystyle\int \frac{dx}{\sqrt{1 - (x+2)^2}}$.

3 Show that $x^2 + 4x + 5 = (x+2)^2 + 1$. Hence find $\displaystyle\int \frac{dx}{x^2 + 4x + 5}$.

26.4 Integration of rational functions by use of partial fractions

In Chapter 25 we saw how to express a rational function in terms of partial fractions. Any partial fraction can be integrated, as follows.

$$\int \frac{A}{x - \alpha} \, dx = A \ln (x - \alpha) + c$$

$$\int \frac{A}{(x - \alpha)^2} \, dx = -A(x - \alpha)^{-1} + c$$

$$\int \frac{Bx + C}{x^2 + k^2} \, dx = \int \frac{Bx}{x^2 + k^2} \, dx + \int \frac{C}{x^2 + k^2} \, dx$$

$$= \frac{1}{2} B \ln (x^2 + k^2) + \frac{C}{k} \tan^{-1} \frac{x}{k} + c$$

EXAMPLE 26.9

Find $\displaystyle\int \frac{2x - 1}{x(x^2 + 1)}\, dx$.

Solution

The integrand is a rational function. The partial fraction technique of Chapter 25 can be used to show that

$$\frac{2x - 1}{x(x^2 + 1)} = \frac{x + 2}{x^2 + 1} - \frac{1}{x}$$

$$\int \frac{2x - 1}{x(x^2 + 1)}\, dx = \int \left(\frac{x}{x^2 + 1} + \frac{2}{x^2 + 1} - \frac{1}{x} \right) dx$$

The first integral is a 'Spot the derivative', the second is a $\tan^{-1}$ integral, and the third is a ln integral.

$$\int \frac{2x - 1}{x(x^2 + 1)}\, dx = \frac{1}{2}\, \ln (x^2 + 1) + 2 \tan^{-1} x - \ln x + c$$

EXERCISE 26F

Find the following integrals.

1 $\displaystyle\int \frac{dx}{(x + 1)(x + 2)}$

2 $\displaystyle\int \frac{3x\, dx}{(x + 1)(x - 2)}$

3 $\displaystyle\int \frac{dx}{x(x + 1)^2}$

4 $\displaystyle\int \frac{dx}{(x - 1)^2(x + 1)}$

5 $\displaystyle\int \frac{dx}{x^2(x + 1)}$

6 $\displaystyle\int \frac{dx}{(x + 1)(x^2 + 4)}$

26.5 General practice in integration

Deciding which method of integration to use is often a matter of experience. Do not give in too soon – if one method fails, then try another. You will often have to use formulae and results from other areas – to integrate a trigonometric function you may have to use one of the formulae from Chapters 14 and 17, and to integrate a rational function you may have to use the partial fraction technique of Chapter 25. The following exercises contain integrals which require different methods.

EXERCISE 26G

1 Find the following.

a) $\int (x-1)(x+3)^{\frac{1}{3}}\,dx$

b) $\int \cos^3 x \sin x\,dx$

c) $\int \sin^3 2x\,dx$

d) $\int \cos^5 x\,dx$

e) $\int \sin 3x \sin 5x\,dx$

f) $\int \cos 3x \cos x\,dx$

g) $\int \dfrac{1}{(x^2+1)(x+2)}\,dx$

h) $\int \dfrac{1}{x^2(x-1)}\,dx$

i) $\int (\sin x + \cos x)^2\,dx$

2 Find $\int \sin^2 x\,dx$

a) by using the formula for $\cos 2A$ **b)** by using the formula for $\sin A \sin B$.

Check that your answers agree.

3 Find $\int \dfrac{x}{x^2-1}\,dx$

a) by substituting $u = x^2 - 1$ **b)** by partial fractions.

Check that your answers agree.

4 Find $\int \dfrac{2x}{(x+1)^2}\,dx$

a) by parts   **b)** by substituting $u = 1 + x$.

Check that your answers are equivalent.

More than one method

Sometimes we have to apply more than one method to find an integral.

EXAMPLE 26.10
Find $\int \tan^{-1} x\,dx$.

Solution
There is no obvious substitution, so try doing it by parts, by writing $\tan^{-1} x$ as $1 \times \tan^{-1} x$.

$$\int 1 \times \tan^{-1} x\,dx = x \tan^{-1} x - \int x \frac{d}{dx} \tan^{-1} x\,dx$$

The derivative of $\tan^{-1} x$ is $\dfrac{1}{1+x^2}$.

$$\int \tan^{-1} x\,dx = x \tan^{-1} x - \int \frac{x}{1+x^2}\,dx$$

The second integral is now a 'Spot the derivative', as x differs from the derivative of $1 + x^2$ by a constant factor. By guesswork or by substituting $u = 1 + x^2$, the second integral is $\frac{1}{2} \ln(1 + x^2)$.

$$\int \tan^{-1}x \, dx = x \tan^{-1}x - \frac{1}{2} \ln(1 + x^2) + c$$

EXERCISE 26H

1 Find the following.

a) $\int \sin^{-1}x \, dx$

b) $\int 2x \ln(x + 1) \, dx$

c) $\int 4x \cos^2 x \, dx$

d) $\int x \tan^{-1}x \, dx$

e) $\int \dfrac{x}{1 + x^4} \, dx$ (put $u = x^2$)

f) $\int \dfrac{x}{\sqrt{1 - x^4}} \, dx$

g) $\int \sec^4 x \, dx$

h) $\int \dfrac{2x}{(1 + x^2)(4 + x^2)} \, dx$

i) $\int \dfrac{e^x}{2 + e^{2x}} \, dx$

*2 Find the following.

a) $\int \tan^{-1}\sqrt{x} \, dx$

b) $\int \sin^{-1}\sqrt{x} \, dx$

c) $\int x^3 \, e^{x^2} \, dx$

d) $\int \ln(1 - x^2) \, dx$

e) $\int x^3 (x^2 - 2)^{\frac{3}{2}} \, dx$

f) $\int x^{\frac{1}{3}}(1 + x^{\frac{2}{3}})^{-1} \, dx$

LONGER EXERCISE

To end this chapter, we have an example which links up integration with the elementary mensuration of the circle. If a circle has radius 1 unit, then its area is $\pi \, 1^2 = \pi$.

If (x, y) lies on a circle with radius 1 unit, then $x^2 + y^2 = 1$. So we can write the equation of the circle as $y = \sqrt{1 - x^2}$.

The area of the circle is $2 \displaystyle\int_{-1}^{1} \sqrt{1 - x^2} \, dx$.

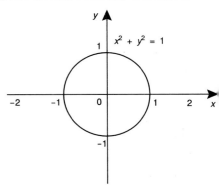

Fig. 26.1

1 Make the substitution $x = \sin u$, and show that this integral becomes

$$2 \int_{-\frac{\pi}{2}}^{\frac{\pi}{2}} \cos^2 u \, du$$

2 Use the identity $\cos^2 u \equiv \frac{1}{2} + \frac{1}{2} \cos 2u$ to evaluate the integral, and so to find the area of the circle. Does this agree with the answer above?

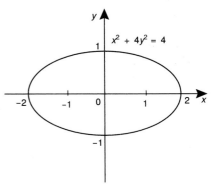

3 An ellipse has equation $x^2 + 4y^2 = 4$. Its graph is shown in Fig. 26.2. Adapt the reasoning for the circle area to find the area of the ellipse. Could your answer have been obtained without integrating?

Fig. 26.2

EXAMINATION QUESTIONS

1 Using the substitution $u^2 = 1 + x^2$ or otherwise, evaluate

$$\int_0^{\sqrt{3}} \frac{x}{\sqrt{1 + x^2}} \, dx$$

The figure shows the region R bounded by the curve with equation $y = \dfrac{x}{\sqrt{1 + x^2}}$ and the lines $x = \sqrt{3}$ and $y = x$.

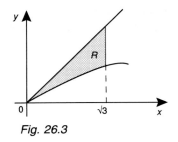

Fig. 26.3

a) Find the area of R.

b) Find the volume of the solid formed when R is rotated through 2π radians about the x-axis.

O 1990

2 **(i)** If $I = \displaystyle\int_{\frac{\pi}{3}}^{\frac{\pi}{2}} \operatorname{cosec} \theta \, d\theta$, use the substitution $x = \cos \theta$ to show that

$$I = \int_0^{\frac{1}{2}} \frac{dx}{1 - x^2}$$

(ii) By using the method of partial fractions, show that $I = \frac{1}{2} \log_e 3$.

NI 1991

3 **a)** Express $\dfrac{1}{x^2(2x - 1)}$ in the form $\dfrac{A}{x} + \dfrac{B}{x^2} + \dfrac{C}{2x - 1}$. Hence, or otherwise, evaluate

$$\int_1^2 \frac{1}{x^2(2x - 1)} \, dx$$

b) Find $\int x^3 \ln(4x) \, dx$.

c) Using the substitution $x = 3 \tan \theta$, evaluate $\displaystyle\int_0^3 \frac{1}{(9 + x^2)^2} \, dx$.

AEB 1992

4 Let $C = \displaystyle\int_0^{\pi} e^{-2x} \cos x \, dx$ and $S = \displaystyle\int_0^{\pi} e^{-2x} \sin x \, dx$. Show by integration by parts, or otherwise, that

$$C = 2S \quad \text{and} \quad S = -2C + 1 + e^{-2\pi}$$

Hence, or otherwise, find the values of C and S.

The finite region bounded by the axes and the curve $y = e^{-x} \cos(\frac{1}{2} x)$, for $0 \le x \le \pi$, is rotated through four right angles about the x-axis. Find the volume of the solid of revolution.

C 1991

Summary and key points

1 Sometimes a substitution enables a function to be integrated. This is especially useful in the following types of integral.

$$\int g'(x)f(g(x)) \, dx = \int f(u) \, du + c \quad \text{(where } u = g(x))$$

$$\int \frac{f'(x)}{f(x)} \, dx = \ln(f(x)) + c$$

When doing a substitution, it is essential to change the dx as well as the function.

In a definite integral, either change the limits or convert the answer back to the original variable.

2 Integration by parts uses the rule

$$\int u'v \, dx = uv - \int uv' \, dx$$

3 When integrating products of sine or cosine, use a formula of the form

$$2 \sin A \cos B = \sin(A + B) + \sin(A - B)$$

When integrating a power of sine or cosine, use $\cos^2\theta + \sin^2\theta = 1$ and substitution for odd powers, and the expansion of $\cos 2\theta$ for even powers.

4 When integrating a rational function, express it in partial fractions.

5 Some integrals require the use of more than one method.

Consolidation section F

Chapter 23

1 A church spire is 40 feet high. At a distance of x feet from the top, the horizontal cross section is a square of side $\dfrac{x^2}{400}$. Find the total volume of the spire.

2 The region under the curve $y = x^2 - x + 1$, for x between 0 and 2, is rotated about the x-axis. Find the volume of revolution.

3 Estimate the integral $\displaystyle\int_0^1 \sqrt{1 + x^3}\ \mathrm{d}x$ using the trapezium rule with five intervals.

4 The values of a function are given in the table below.

x	2	3	4	5	6	7	8
$f(x)$	1.3	1.8	2.2	2.6	2.9	3.2	3.4

Use the trapezium rule to estimate $\displaystyle\int_2^8 f(x)\ \mathrm{d}x$.

5 With $f(x)$ as in Question 4, the area under the curve $y = f(x)$ is rotated about the x-axis. Estimate the volume of revolution.

6 Use Simpson's rule to estimate $\displaystyle\int_2^8 f(x)\ \mathrm{d}x$, where $f(x)$ is as in Question 4.

7 Use Simpson's rule with four intervals to estimate $\displaystyle\int_0^1 \sqrt{1 + x^3}\ \mathrm{d}x$.

Chapter 24

8 A pyramid VABCD has a horizontal base ABCD which is a rectangle of sides 12 m and 15 m. The vertex of the pyramid is 9 m above the centre of the base. Find the length of VA and the angle it makes with the horizontal. Find the angles the faces make with the horizontal.

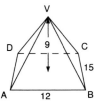

9 ABCDEFGH is a cuboid, with AB $= 4$ cm, AD $= 5$ cm and AE $= 3$ cm. Find the angle between AG and GFBC.

10 For the cuboid of Question 9, find the angle between AHF and CHF.

11 Take the x, y and z axes along AB, AD and AE of the cuboid of Question 9. Find the vectors corresponding to AG, EC and FC. Find the angle between EC and FC.

Chapter 25

12 Express the following in partial fractions.

a) $\dfrac{2x}{(x-1)(x-2)}$
b) $\dfrac{x+3}{x^2(x-1)}$
c) $\dfrac{x^2+x-1}{(x^2+2)(x+1)}$
d) $\dfrac{x^2-x+1}{x^2-3x+2}$

13 Find the series expansions of parts **a)** and **c)** of Question 12, up to the terms in x^3.

14 Find the first three derivatives of $f(x) = \tan\left(x + \dfrac{\pi}{4}\right)$. Hence find the Maclaurin expansion of $f(x)$, up to the term in x^3.

15 Find the expansions of the following, up to the terms in x^2. In each case, find the range of values of x for which the expansion is valid.

a) e^{-3x}
b) $\ln(1 + \tfrac{1}{3}x)$
c) $e^{-2x}\cos 2x$

16 Find the condition on k for the equation $x^2 + 2kx + k + 20 = 0$ to have just one root.

Chapter 26

17 Find the following integrals.

a) $\int (x+1)(x-1)^9 \, dx$
b) $\int x \sec^2 x^2 \, dx$
c) $\int \ln(x-3) \, dx$

d) $\int \sin 2x \cos x \, dx$
e) $\int \sin^3 2x \, dx$
f) $\int \cos^2 5x \, dx$

g) $\displaystyle\int \dfrac{dx}{\sqrt{9 - 4x^2}}$
h) $\displaystyle\int \dfrac{2}{(x-2)(x-3)} \, dx$
i) $\displaystyle\int \dfrac{x+3}{x^2(x+2)} \, dx$

MIXED EXERCISE

1 A pyramid has a square base. Its volume is fixed at 100 cm^3. Find the height which will ensure that the area of the triangular faces is a minimum.

2 The expansion of $\dfrac{1}{(1+ax)(1+bx)}$ agrees with that of $(1 - 5x)^{-\frac{1}{3}}$, up to the term in x^2.

Find a and b.

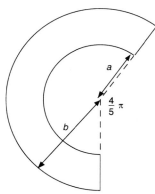

*3 A lampshade is to be made out of the shape shown in the diagram, which is the region between two circles of radii a and b, with a sector of angle $\frac{4}{5}\pi$ removed. The height of the lampshade will be 20 cm, and the surface area will be 525π cm^2. Find a and b.

4 Find an approximation for $\int_0^{\frac{1}{2}} e^{-x^2}\, dx$

    a) using the trapezium rule with four intervals

    b) using the expansion of e^{-x^2} up to the term in x^4.

5 Find the area of the ellipse $x^2 + 4y^2 = k^2$ where k is a constant.

    A vessel is made so that its cross-section z from the bottom is an ellipse with equation $x^2 + 4y^2 = k^2$ where $k = z^2$. Find the volume of the vessel if its height is 2 units.

6 Find the volume of revolution when the region under $y = \sin x$, for x between 0 and π, is rotated about the x-axis.

*7 A sheep is tethered at the edge of a circular building of radius a. The sheep is on a rope of length πa, so it can just reach the far side of the building. Show that the sheep can graze an area of $\frac{5}{6}\pi^3 a^2$.

8 Show that $\cos rx \sin x = \frac{1}{2}[\sin(r+1)x - \sin(r-1)x]$. Hence show that

$$\sum_{r=1}^{n} \cos rx = \frac{\sin(n+1)x + \sin nx - \sin x}{2\sin x}$$

9 A sphere of radius a is divided into two parts by a plane ka away from the centre. Find the volume of the smaller part.

    If the volumes of the parts are in the ratio $1:2$, show that $3k^3 - 9k + 2 = 0$. Use Newton-Raphson iteration to solve this equation.

LONGER EXERCISES

1 Harmonic triangle

The arrangement of fractions below is known as **Leibnitz's harmonic triangle**.

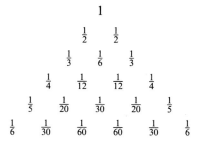

Each slanting row consists of the differences of the slanting row to the left. For example, $\frac{1}{30} = \frac{1}{12} - \frac{1}{20}$.

1 Continue the triangle for two more horizontal rows.

2 The triangle can be used to find the sum of series. For example, taking the third slanting row

$$\frac{1}{2} = \frac{1}{3} + \frac{1}{12} + \frac{1}{30} + \frac{1}{60} + \cdots$$

Verify that the sum of the series is tending to $\frac{1}{2}$. Can you show why it is true?

3 Find some other series within the triangle and verify that they converge to the correct value.

2 Painting strips

You need to paint the strip shown, which is the region under the curve $y = x^{-0.75}$, from $x = 1$ on the left to indefinitely far on the right.

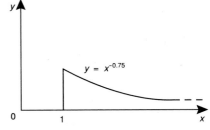

1 If the strip extends to infinity, express the area as an integral. Evaluate this integral, and show that it is infinite. It will require an infinite amount of paint.

2 When you explain the problem to the assistant at the paintshop, he suggests that you rotate the curve about the x-axis, and inject paint into the hollow object formed.

Express the volume of revolution of the region as an integral. Evaluate the integral, and show that it is finite.

3 If the area is infinite, how can you paint it using a finite volume of paint?

3 Integration from first principles

Without the Fundamental Theorem of Calculus, we would have to find integrals as the limit of a sum. This procedure is known as **integrating from first principles**.

We shall find $\int_0^1 x^2 \, dx$ from first principles. The diagram shows the area divided into n vertical strips.

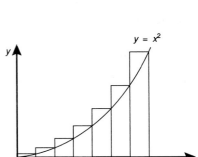

1 Show that the height of the rth strip is $\left(\dfrac{r}{n}\right)^2$.

2 Show that the area of the rth strip is $\dfrac{r^2}{n^3}$.

3 Show that the total area of the strips is $\displaystyle\sum_{r=1}^{n} \frac{r^2}{n^3} = \frac{1}{n^3} \sum_{r=1}^{n} r^2$.

4 Use the summation $\sum_{r=1}^{n} r^2 = \frac{1}{6}n(n+1)(2n+1)$ to find an expression for the total area of the strips.

5 Take more and more strips, by letting n tend to infinity. What happens to the expression found above? Does it give the area under the curve?

6 Evaluate $\int_0^1 x^3 \, dx$ from first principles. You will need the result $\sum_{r=1}^{n} r^3 = \frac{1}{4}n^2(n+1)^2$.

In Question 8 of the Mixed exercise above, you obtained the identity

$$\sum_{r=1}^{n} \cos rx \equiv \frac{\sin(n+1)x + \sin nx - \sin x}{2 \sin x}$$

This can be used to evaluate $\int_0^{\frac{\pi}{2}} \cos x \, dx$ from first principles.

7 Divide the region 0 to $\frac{\pi}{2}$ into n thin strips of width δx, where $\delta x = \frac{\pi}{2n}$. Show that the height of the rth strip is $\cos r \, \delta x = \cos \frac{r\pi}{2n}$.

8 Write down the area of the rth strip.

9 Show that the total area of the strips is $\sum_{r=1}^{n} \cos \frac{r\pi}{2n} \delta x = \frac{\pi}{2n} \sum_{r=1}^{n} \cos \frac{r\pi}{2n}$.

10 Put $x = \frac{\pi}{2n}$ into the identity above, to find an expression for the total area.

11 Let n tend to infinity. The total area of the strips tends to the area under the curve. Does this agree with the value found by ordinary integration?

12 Find $\int_0^1 \sin x \, dx$ by this method.

EXAMINATION QUESTIONS

1 Let $f(x) = \dfrac{1}{(x+1)(2x+1)}$.

  a) Express $f(x)$ in partial fractions.

  b) Show that the graph of $y = f(x)$ has a single stationary point. Find its coordinates and identify it as a maximum or minimum. Sketch the graph of $f(x)$.

  c) Using your results in **a)**, show that

$$[f(x)]^2 = \frac{1}{(x+1)^2} + \frac{4}{(x+1)} - \frac{8}{(2x+1)} + \frac{4}{(2x+1)^2}$$

d) The region enclosed by the graph of $y = f(x)$, the x and y axes and the line $x = 1$ is rotated completely about the x-axis. Show that the volume generated is $\pi(\frac{11}{6} - 4 \ln \frac{3}{2})$.

AEB 1991

2 A piece is cut from one corner of a rectangular block of cheese making a new triangular face ABC as shown in the figure. The lengths PA, PB and PC are 3.5 cm, 2.0 cm and 4.0 cm respectively. Use **one** of the following two methods to find the size of angle ABC.

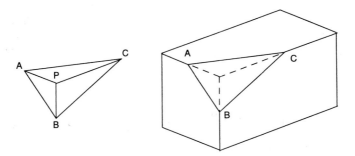

Either (i) **a)** Find the lengths of the sides of triangle ABC.

           b) Use the cosine formula to find angle ABC.

Or   (ii) **a)** Write down the components of vectors $\overrightarrow{BA}$ and $\overrightarrow{BC}$ with respect to suitable coordinate axes.

           b) Use the scalar product of these vectors to find angle ABC.

SMP 1992

3 Find the complete set of values of p for which the roots of the equation $2x^2 + px + 3p - 10 = 0$ are real.

L 1989

4 The flow of water over a weir was monitored over a 15-day period. The flow was measured on Sunday (day 0) and subsequently at noon on every third day. The results were as follows.

Day	0	3	6	9	12	15
Rate of flow (m³/day)	48 000	53 000	54 000	52 000	46 000	38 000

Plot these points on graph paper and draw a smooth curve through them. You may now assume that this smooth curve is a good representation of the actual rate of flow.

By applying the Trapezium Rule, estimate the total volume of water flowing over the weir in this 15-day period.

State, giving your reason, whether your result is likely to be an over-estimate or an under-estimate of the true value.

Why is Simpson's Rule, though generally more accurate, inapplicable in this case?

AEB AS 1991

5 The function f is defined by $f(x) \equiv \sin 2x + 2\cos^2 x$ where x is real and measured in radians.

a) Prove that $1 - \sqrt{2} \le f(x) \le 1 + \sqrt{2}$.

b) The figure shows part of the curve with equation $y = \sin 2x + 2\cos^2 x$.

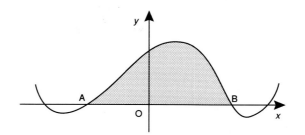

Find the area of the shaded region, giving your answer in terms of π.

AEB 1990 (part)

Functions and graphs

Throughout this course we have been dealing with **functions**. We have considered the trigonometric functions $\sin x$, $\cos x$ and so on, exponential and logarithmic functions such as e^x and $\ln x$, polynomial and rational functions.

In this chapter we discuss the general behaviour of functions. This will be linked up with the shapes of their graphs.

27.1 The domain and range of a function

A function $y = f(x)$ must be unambiguous — the expression $y = \pm\sqrt{x}$ is not a function. However, a function does not have to be defined for all values of x, and need not take all values of y.

Definitions

Some functions are not defined for all possible values. For example, $\dfrac{1}{x}$ is not defined for $x = 0$ and $\sqrt{x}$ is not defined when x is negative.

The set of values of x for which $f(x)$ is defined is the **domain** of $f(x)$.

Some functions do not take all possible real values. For example, x^2 will never take negative values and $\sin x$ will not take values greater than 1 or less than -1.

The set of values taken by $f(x)$ is the **range** of $f(x)$.

Graphs

This graph in Fig. 27.1 is of $y = \dfrac{1}{x}$. The domain is the set of all the
x-values for which the function is defined. In this case it is
$\{x : x \neq 0\}$.

The range is the set of all possible y-values. In this case it is
$\{y : y \neq 0\}$

The graph in Fig. 27.2 is of $y = \sqrt{x - 1}$. As we cannot take the
square root of a negative number, the domain is $\{x : x \geq 1\}$. The
square root of a number is always positive, so the range is
$\{y : y \geq 0\}$

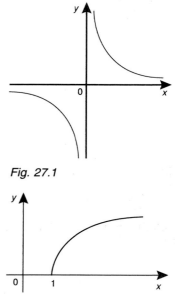

Fig. 27.1

Restricted domains

Sometimes we may want to restrict the domain of a function.
Suppose f(*t*) represents the height of a ball *t* seconds after it has
been thrown up in the air. Then the domain of the function will
be $\{t : t \geq 0\}$, as when *t* was negative the ball had not been
thrown.

Fig. 27.2

Notation

Another notation for functions uses the → symbol

$\qquad$ f : $x \rightarrow x^2 - 1$ means the same as f(x) = $x^2 - 1$.

EXAMPLE 27.1

Find the domain and range of the function f(x) = $\dfrac{1}{x - 3}$. Sketch
the graph of $y = $ f(x).

Solution

We can evaluate this function provided we do not divide by zero,
so *x* can take any value except 3.

The domain is $\{x : x \neq 3\}$.

We can rewrite the function as

$\qquad x - 3 = \dfrac{1}{\text{f}(x)}$

We cannot divide by zero, so f(x) can take any value
except 0.

The range is $\{y : y \neq 0\}$.

The graph is shown in Fig. 27.3. Notice that the graph
does not cross the vertical line $x = 3$ nor the horizontal
line $y = 0$.

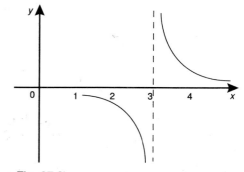

Fig. 27.3

EXAMPLE 27.2

A function is defined by $f : x \rightarrow x^2 - 2x - 2$ for $x \geq 2$. Find the range of the function.

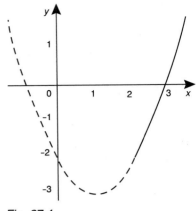

Solution

A sketch of the graph is shown in Fig. 27.4. The domain of the function is restricted to $\{x : x \geq 2\}$, which is the filled-in part of the graph. The range is the set of y-values of this section of the graph.

The range is $\{y : y \geq -2\}$.

Fig. 27.4

EXERCISE 27A

1 Find the domain and range of each of the following functions.

a) $f(x) = x^2$

b) $f(x) = (x - 1)^2$

c) $f(x) = 1 - x^2$

d) $f(x) = 2x^2 + 1$

e) $f(x) = x^2 - 2$

f) $f(x) = 2 - (x - 1)^2$

g) $f(x) = \dfrac{1}{x}$

h) $f(x) = 1 + \dfrac{2}{x}$

i) $f(x) = 3 - \dfrac{1}{x}$

j) $f(x) = \dfrac{1}{x - 2}$

k) $f(x) = 1 + \dfrac{1}{2 + x}$

l) $f(x) = 1 - \dfrac{1}{2x - 1}$

m) $f(x) = \sqrt{x + 3}$

n) $f(x) = 1 + \sqrt{x}$

o) $f(x) = \sqrt{1 - x}$

p) $f(x) = 1 - \sqrt{x}$

q) $f(x) = |x|$

r) $f(x) = 1 + |2x|$

s) $f(x) = e^x$

t) $f(x) = 2 - e^x$

u) $f(x) = \log_{10} x$

v) $f(x) = \ln(x + 3)$

w) $f(x) = \sin x$

x) $f(x) = \tan x$

2 Suppose x and y are related by $xy + 3x - y + 1 = 0$. Rearrange this as a function giving y in terms of x and find the domain and range of the function.

3 The domains are restricted for the following functions. In each case find the range.

a) $f(x) = x^2$ for $x > 1$ **b)** $f(x) = \dfrac{1}{x}$ for $x \geq 2$ **c)** $f(x) = \sqrt{x}$ for $0 \leq x < 16$

d) $f(x) = 1 - x^2$ for $x < 1$ **e)** $f(x) = \sin x$ for $0 \leq x \leq \pi$ **f)** $f(x) = 2^x$ for $x > 0$

*4 A stone is thrown upwards, and t seconds later its height h m is given by $h = 40t - 5t^2$. What would be a sensible domain for this function? What is its range?

One to one functions

Some functions take the same value more than once. For example, the function $f(x) = x^2$ takes the value 4 when $x = 2$ and when $x = -2$. The function $f(x) = \sin x$ takes the value 0.5 at $x = \dfrac{\pi}{6}$ and at $x = \dfrac{5\pi}{6}$.

Other functions, though, take any value at most once. They are called **one to one** functions.

> $f(x)$ is one to one if whenever $a \neq b$, then $f(a) \neq f(b)$.

> Equivalently, $f(x)$ is one to one if, whenever $f(a) = f(b)$, then $a = b$.

This is illustrated in Fig. 27.5. For **a)**, we see that different values in the domain are sent to the same value in the range. For **b)**, different values in the domain always go to different values in the range.

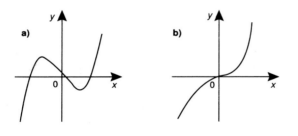

Fig. 27.5

The graph of a function will tell us whether it is one to one. If a horizontal line crosses the graph at more than one point, as in **a)**, then the function is not one to one. If every horizontal line crosses the graph at most once, as in **b)**, then the function is one to one.

EXAMPLE 27.3

Which of the following are one to one functions?

a) $f(x) = x^2 + 3$ **b)** $f(x) = 2x - 3$

Solution

a) Put $x = -1$.

$$f(-1) = (-1)^2 + 3 = 4$$

Similarly $f(1) = 1^2 + 3 = 4$

$f(x) = x^2 + 3$ is not one to one.

b) If $f(a) = f(b)$, then

$$2a - 3 = 2b - 3$$

Hence $2a = 2b$ and so $a = b$.

If the y-values are the same, then the x-values are the same.

$f(x) = 2x - 3$ is a one to one function.

EXERCISE 27B

1 Which of the following are one to one functions? Prove your claims.

  a) $f(x) = 3x + 1$ **b)** $f(x) = x^2 - 1$ **c)** $f(x) = 1 - 2x$

  d) $f(x) = x^3$ **e)** $f(x) = \cos x$ **f)** $f(x) = e^x$

2 Show that $f(x) = x^2 - 4$ is not a one to one function. Suggest a limitation of the domain of $f(x)$ which would make it one to one.

3 Suggest a limitation on the domain of $f(x) = \sin x$ which would make it a one to one function.

4 Which of the functions illustrated below are one to one?

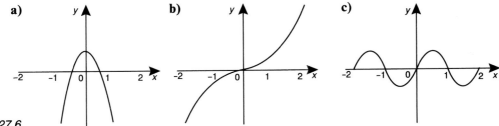

Fig. 27.6

5 For the functions of Question 4 which are not one to one, suggests a limitation of the domain which would make them one to one.

27.2 The composition of functions

If we apply one function to another, the resulting function is the **composition** of the two original functions.

The composition of $f(x)$ and $g(x)$ can be written as $fg(x)$ or as $f \circ g(x)$. We shall use $fg(x)$. It is defined by

$$fg(x) = f(g(x))$$

Note: The order of the functions is important. In general, $fg(x) \neq gf(x)$.

EXAMPLE 27.4

Let $f(x) = 2x - 3$ and $g(x) = 4x + 1$. Find

a) fg(2)   **b)** gf(2)   **c)** fg(x).

Solution

a) First evaluate g(2).

$$g(2) = 4 \times 2 + 1 = 9$$

Apply f to this.

$$fg(2) = f(9) = 2 \times 9 - 3 = 15$$

$$fg(2) = 15.$$

b) First evaluate f(2).

$$f(2) = 2 \times 2 - 3$$

$$= 1$$

Apply g to this.

$$gf(2) = g(1)$$

$$= 4 \times 1 + 1$$

$$= 5$$

$$gf(2) = 5.$$

Note: fg(2) is not equal to gf(2).

c) Apply f to the general expression for g(x).

$$fg(x) = f(4x + 1) = 2 \times (4x + 1) - 3$$

$$= 8x - 1$$

$$fg(x) = 8x - 1.$$

EXERCISE 27C

1 Functions f and g are defined as $f(x) = 2x - 3$ and $g(x) = 4x + 1$. Find the following composite functions.

   a) fg(3)   **b)** gf(3)   **c)** gf(x)

2 Let $f : x \rightarrow 1 + 2x$ and $g : x \rightarrow x^2$. Find the following.

   a) gf(1)   **b)** fg(1)   **c)** gf(x)   **d)** fg(x)

3 With f and g defined as in Question 2 above, solve the equation fg(x) = gf(x).

4 Let $f : x \rightarrow 1 - x$ and $g : x \rightarrow e^x$. Find the following.

   a) fg(0)   **b)** gf(1)   **c)** fg(x)   **d)** gf(x)   **e)** ff(x)

5 Let $f : x \rightarrow \dfrac{1}{x}$ and $g : x \rightarrow x - 1$. Find the following.

a) fg(2)   **b)** gf(2)   **c)** fg(x)   **d)** gf(x)

6 Let $f : x \rightarrow ax + b$ and $g : x \rightarrow cx + d$ where a, b, c and d are constants. Find the following.

a) fg(x)   **b)** gf(x)   **c)** ff(x)   **d)** gg(x)

7 In Question 6, suppose that $a = 1$ and that fg(x) = gf(x) for all x. What can you say about b, c and d?

*8 a) Suppose f and g are both one to one functions. Show that fg is also a one to one function.

   b) Suppose fg is a one to one function. Show that g is a one to one function.

   c) Suppose fg is a one to one function. Show by example that f need not be one to one.

27.3 Identity and inverse functions

The identity function

The **identity** function i(x) leaves x as it is.

    i(x) = x for all x

Inverse functions

Let f(x) be a function. The **inverse** function of f(x), written $f^{-1}(x)$, is such that the composition of f and f^{-1} is the identity.

    $f^{-1}f(x) = ff^{-1}(x) = i(x) = x$ for all x

The graph of $y = f^{-1}(x)$ can be obtained from the graph of $y = f(x)$ by reflecting the graph in the line $y = x$.

Because x and y have been interchanged, the domain of f^{-1} is the range of f and vice versa.

The inverse function of $y = f(x)$ can be found by making x the subject of the formula.

In order for a function to have an inverse, it must be one to one. Suppose f is not one to one, and that f(a) = f(b) = k. Then we would not know whether $f^{-1}(k)$ was a or b.

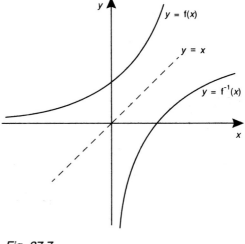

Fig. 27.7

EXAMPLE 27.5

Let f(x) = $2x - 3$. Find $f^{-1}(x)$. Sketch the graphs of f(x) and $f^{-1}(x)$.

Solution

Put $y = f(x) = 2x - 3$.

Make x the subject of the formula by adding 3 to both sides and dividing through by 2.

$$x = \frac{y+3}{2}$$

This gives x as a function of y.

$$x = f^{-1}(y) = \frac{y+3}{2}$$

We want to give the answer as a function of x. Convert to a function of x by replacing y by x.

$$f^{-1}(x) = \frac{x+3}{2}$$

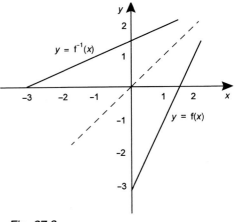

Fig. 27.8

The graphs of $f(x)$ and $f^{-1}(x)$ are shown. Notice that the inverse graph is obtained by reflecting the first graph in the line $y = x$.

EXAMPLE 27.6

A function is defined by $f : x \rightarrow x^2 + 1$ for $0 \leq x$. Show that f is one to one on this domain. Find the range of f and the inverse of f, stating its domain and range.

Solution

The graph of $y = x^2 + 1$ for non-negative x is shown in Fig. 27.9. Notice that for each y value there is only one positive x value.

f is one to one on this domain.

y could be any number greater than or equal to 1.

The range of f is $\{y : y \geq 1\}$.

If $y = x^2 + 1$ then $y - 1 = x^2$ and so $\sqrt{y-1} = x$.

This gives f^{-1}.

The domain of f^{-1} is the range of f and vice versa.

The inverse is $f^{-1} : x \rightarrow \sqrt{x - 1}$.

The domain is $\{x : x \geq 1\}$ and the range is $\{x : x \geq 0\}$.

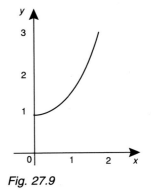

Fig. 27.9

EXERCISE 27D

1 Find the inverse functions of the following:

    a) $f(x) = 3x + 5$ **b)** $f(x) = 2 - 3x$ **c)** $f(x) = \frac{1}{2}x - 2$

    d) $f(x) = 2(x - 1)$ **e)** $f(x) = \frac{1}{3}(x + 2)$ **f)** $f(x) = \frac{1}{2}(2 - x)$

2 Let $f : x \rightarrow x^3 - 1$. Find the inverse function f^{-1}. Sketch the graphs of $f(x)$ and $f^{-1}(x)$.

3 Let $f(x) = e^x$. Find the range of f. Find the inverse function $f^{-1}(x)$ and state its domain. Sketch the graphs of the two functions.

4 Let $f(x) = (x - 1)^2 - 2$ for $x \geq 1$. Find the range of f. Find the inverse function $f^{-1}(x)$, stating its domain and range.

5 Let $f : x \rightarrow \sqrt{x + 1} - 2$ for $x \geq -1$. Find the range of f. Find the inverse function f^{-1}, stating its domain and range.

6 A function is defined by $f(x) = (x - 1)^2 - 1$ for $x \geq k$. Find the least value of k for f^{-1} to exist. With this value of k, find f^{-1}, stating its domain and range.

7 For each of the graphs below, sketch the graphs of the corresponding inverse function.

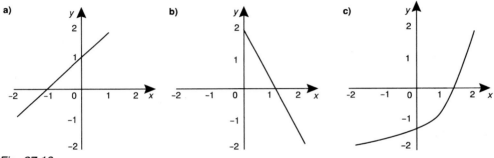

Fig. 27.10

*8 A function f is **self-inverse** if $f(x) = f^{-1}(x)$ for all x in its domain. Show that $f(x) = 1 - x$ and $g(x) = \dfrac{1}{x}$ are both self-inverse. Can you find any other self-inverse functions?

Inverses of trigonometric functions

The trigonometric functions are not one to one. For example $\sin 0 = \sin \pi = 0$. So the inverse trigonometric function take a restricted range of values, to ensure that no ambiguity arises. These are the **principal values** of the inverse trigonometric functions.

sine
The principal values of $\sin^{-1} x$ are in the range $-\dfrac{\pi}{2} \leq x \leq \dfrac{\pi}{2}$.
The graphs of $y = \sin x$ and $y = \sin^{-1} x$ are shown.

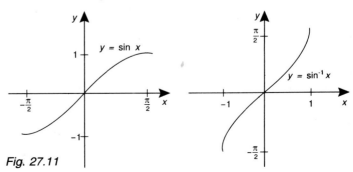

Fig. 27.11

cosine

The principal values of $\cos^{-1}x$ are in the range $0 \le x \le \pi$. The graphs of $y = \cos x$ and $y = \cos^{-1}x$ are shown.

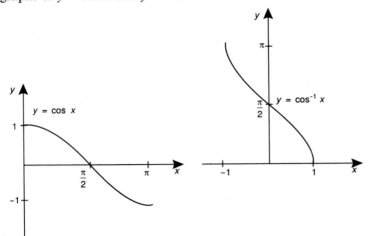

Fig. 27.12

tangent

The principal values of $\tan^{-1}x$ are in the range $-\dfrac{\pi}{2} < x < \dfrac{\pi}{2}$. The graphs of $y = \tan x$ and $y = \tan^{-1}x$ are shown.

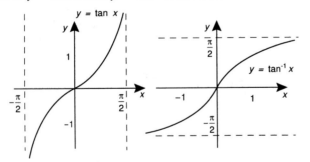

Fig. 27.13

EXERCISE 27E

1 Check that the inverse trigonometric function buttons on your calculator always give the principal values.

2 If $\sin^{-1}(\sin x) = x$ what range of values does x lie in?

3 If $\cos^{-1}(\cos x) = -x$ what range of values does x lie in?

4 If $\tan^{-1}(\tan x) = x - 2\pi$ what range of values does x lie in?

5 Show that $\sin^{-1}x + \cos^{-1}x$ is constant and find its value.

27.4 Odd, even and periodic functions

Some functions possess the following properties.

A function $f(x)$ is **odd** if $f(-x) = -f(x)$ for all x.

The graph of an odd function is symmetrical about the origin.

A function $f(x)$ is **even** if $f(-x) = f(x)$ for all x.

The graph of an even function is symmetrical about the y-axis.

A function $f(x)$ is **periodic** if there is $a > 0$ for which $f(x + a) = f(x)$ for all x.

The least such a is the **period** of $f(x)$.

The graph of a periodic function repeats itself every a along the x-axis.

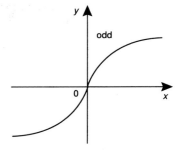

Fig. 27.14

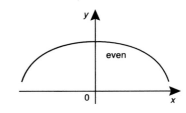

Fig. 27.15

EXAMPLE 27.7
Which of the following are odd, even or periodic? Find the periods of those that are periodic.

a) $f(x) = \sin x$ **b)** $g(x) = x^2 + 3$

c) $h(x) = x^2 + 2x - 1$ **d)** $j(x) = \sin 2x + \cos 2x$

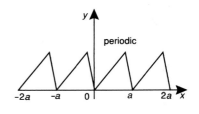

Fig. 27.16

Solution
a) $f(x) = \sin x$

Changing x to $-x$ and using a result from Chapter 3, we have

$$\sin(-x) = -\sin x$$

We know that $\sin(x + 2\pi) = \sin x$.

$\sin x$ is odd, and periodic with period 2π.

b) $g(x) = x^2 + 3$

Changing x to $-x$ we have

$$g(-x) = (-x)^2 + 3 = x^2 + 3 = g(x)$$

$g(x)$ increases as x tends to infinity, hence it cannot be periodic.

$x^2 + 3$ is even.

c) $h(x) = x^2 + 2x - 1$

Changing x to $-x$ we have

$$h(-x) = x^2 - 2x - 1$$

This is neither h(x) nor $-$h(x). For the same reason as in **b**), h(x) is not periodic.

$x^2 + 2x - 1$ is not odd, even or periodic.

d) j(x) $= \sin 2x + \cos 2x$

Changing x to $-x$ we have

$$j(-x) = \sin 2(-x) + \cos 2(-x) = -\sin 2x + \cos 2x$$

This is neither j(x) nor $-$ j(x).

Both $\sin x$ and $\cos x$ functions repeat themselves after 2π.

Hence both $\sin 2x$ and $\cos 2x$ are periodic with period $\dfrac{2\pi}{2} = \pi$.

$$j(x + \pi) = \sin(2x + 2\pi) + \cos(2x + 2\pi)$$
$$= \sin 2x + \cos 2x = j(x)$$

$\sin 2x + \cos 2x$ is periodic with period π.

EXERCISE 27F

1 For the following functions find whether they are odd, even or periodic. Give the periods of those which are periodic.

a) $f(x) = x^3$ **b)** $f(x) = 2x^2 + 13$ **c)** $f(x) = x^3 - 1$

d) $f(x) = |x|$ **e)** $f(x) = e^x$ **f)** $f(x) = \dfrac{1}{x}$

g) $f(x) = \dfrac{1}{1 + x^2}$ **h)** $f(x) = \dfrac{x + x^3}{1 + x^4}$ **i)** $f(x) = \dfrac{2}{2 + \cos x}$

j) $f(x) = x \cos x$ **k)** $x \sin x$ **l)** $\sin 3x$

m) $\cos \frac{1}{3} x$ **n)** $\cos \pi x + \sin \pi x + 2$ **o)** $\tan 2x$

2 The diagram below shows the graphs of three functions. Say whether each of them is odd, even or periodic.

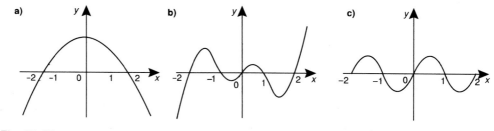

Fig. 27.17

3 For any function f, show that $f(x^2)$ is even.

4 Figure 27.18 shows the graph of $y = f(x)$ for $0 \le x \le 1$. Given that $f(x)$ is even with period 2, sketch the graph for $-2 \le x \le 2$.

5 Figure 27.19 shows the graph of $y = f(x)$ for $0 \le x \le a$ where a is constant. Given that $f(x)$ is odd with period $2a$, sketch the graph for $-2a \le x \le 2a$.

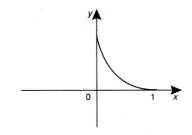

Fig. 27.18

6 Let $f(x)$ and $g(x)$ be even functions, and $j(x)$ and $k(x)$ odd functions. What can you say about the following?

a) $f(x) + g(x)$ **b)** $j(x) + k(x)$ **c)** $f(x)g(x)$

d) $j(x)k(x)$ **e)** $fg(x)$ **f)** $fj(x)$

g) $jk(x)$

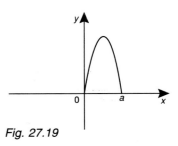

Fig. 27.19

7 The function $f(x)$ is both odd and even. Show that $f(x) = 0$ for all x.

8 Let $f(x)$ be any function. Show that $g(x) = f(x) + f(-x)$ is an even function.

*9 Let $f(x)$ be any function. Show that $f(x)$ can be written as the sum of an odd and an even function.

*10 Let $f(x)$ and $g(x)$ be periodic, with periods a and b respectively. What is the period of $f(x) + g(x)$ in these cases?

a) a is a factor of b **b)** a and b are whole numbers with no common factor

*11 Let k be a positive number. Find a function with period k.

LONGER EXERCISE

Let $f(x) = 1 - x$ and $g(x) = \dfrac{1}{x}$. Find as many functions as you can by taking compositions of functions, starting with $f(x)$ and $g(x)$. What are the inverses of these functions?

EXAMINATION QUESTIONS

1 The diagram shows part of the graph of $y = f(x)$, which is a periodic function with period less than 10.

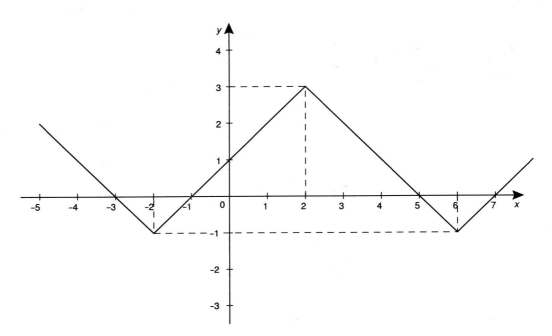

Fig. 27.20

(i) State the period of f.

(ii) If $f(x + a)$ is an even function, state two possible values of a.

(iii) If $f(x) + b$ is an odd function, state the value of b. *SMP 1992*

2 The functions f and g are defined by

$$f: x \rightarrow x - 2, \ x > 2$$

$$g: x \rightarrow e^x, \ x > 0.$$

Given that the function h is defined by

$$h = gf, \ x > 2$$

state

a) the range of h **b)** the domain and range of h^{-1}.

Sketch the curves with equations

c) $y = h(x)$ **d)** $y = h^{-1}(x)$. *L 1989*

3 The function f is defined on the domain $x > 0$ by

$$f(x) = 1 + \frac{1}{x}$$

State the range of f.

Explain why the inverse function f^{-1} exists and obtain an expression for $f^{-1}(x)$ in terms of x.

The composite function g is defined as f∘f.

(i) Show that

$$g(x) = 2 - \frac{1}{1 + x}$$

(ii) State the range of g.

W 1991

Summary and key points

1 The domain of $f(x)$ is the set of x-values for which the function is defined. The range of $f(x)$ is the set of values taken by the function.

A function is one to one if, whenever $a \neq b$ then $f(a) \neq f(b)$.

2 The composition of $f(x)$ and $g(x)$ is defined by

$$fg(x) = f(g(x)).$$

Composition is not the same as multiplication.

$$fg(x) \neq f(x)g(x).$$

The order of the functions is important. In general,

$$fg(x) \neq gf(x).$$

3 The identity function is such that $i(x) = x$.

The inverse $f^{-1}(x)$ of $f(x)$ is such that

$$f^{-1}f(x) = ff^{-1}(x) = i(x).$$

The principal values of the inverse trigonometric functions are as follows.

$$\sin^{-1}: -\frac{\pi}{2} \leq x \leq \frac{\pi}{2} \qquad \cos^{-1}: 0 \leq x \leq \pi$$

$$\tan^{-1}: -\frac{\pi}{2} < x < \frac{\pi}{2}$$

4 An odd function is such that $f(-x) = -f(x)$. An even function is such that $f(-x) = f(x)$.

A function is periodic with period a if a is the least positive number such that $f(x + a) = f(x)$ for all x.

Applications of the chain rule

The chain rule, introduced in Chapter 20, tells us how to differentiate a function of a function. Suppose that y is a function of z, and that z is a function of x. We have

$$\frac{\mathrm{d}y}{\mathrm{d}x} = \frac{\mathrm{d}y}{\mathrm{d}z} \times \frac{\mathrm{d}z}{\mathrm{d}x}$$

The chain rule greatly extends the range of functions we can differentiate. It also provides many extra uses of differentiation.

28.1 Rates of change

Suppose two physical quantities P and Q are changing over time. If the quantities are connected to each other, then their rates of change are also connected. By the chain rule, the connection is given by

$$\frac{\mathrm{d}P}{\mathrm{d}t} = \frac{\mathrm{d}P}{\mathrm{d}Q} \times \frac{\mathrm{d}Q}{\mathrm{d}t}$$

EXAMPLE 28.1

Air is being pumped into a spherical balloon at a rate of 10 cm^3 per second. What is the rate of change of the radius when it is 20 cm?

Solution

The volume V and the radius r of a sphere are connected by

$$V = \tfrac{4}{3}\pi r^3$$

Use the chain rule to connect their rates of change.

$$\frac{\mathrm{d}V}{\mathrm{d}t} = \frac{\mathrm{d}V}{\mathrm{d}r} \times \frac{\mathrm{d}r}{\mathrm{d}t} = 4\pi r^2 \frac{\mathrm{d}r}{\mathrm{d}t}$$

We know that $\dfrac{\mathrm{d}V}{\mathrm{d}t} = 10$ and want to know $\dfrac{\mathrm{d}r}{\mathrm{d}t}$ when $r = 20$. Put these values into the equation.

$$10 = 4\pi(20)^2 \frac{\mathrm{d}r}{\mathrm{d}t}$$

The radius is changing at $\dfrac{\mathrm{d}r}{\mathrm{d}t} = \dfrac{10}{4\pi(20)^2} = 0.00199$ cm per second.

EXERCISE 28A

1 The area of a circle is changing at 2 cm^2 per second. Find the rate of change of the radius when it is 10 cm.

2 The radius of a sphere is changing at 0.1 cm per second. Find the rate of change of the surface area, at the instant when the radius is 10 cm.

3 The area of a circle is changing at 4 cm^2 per second. Find the rate of change of the circumference at the instant when the radius is 5 cm.

4 The volume of a sphere is changing at 2 cm^3 per second. Find the rate of change of the surface area when the radius is 8 cm.

5 A cube has side 10 cm, which is increasing at a rate of 0.1 cm per second. Find the rates of increase of the volume and the surface area.

6 The **space diagonal** of a cuboid is the line joining opposite corners. Find the rate of increase of the space diagonal of the cube of Question 5.

7 A bar has length L cm and cross-section which is a square of side x cm. When it is heated the length is constrained to remain constant. Find the rate of increase of the surface area of the long sides when the volume is increasing at q cm^3 s^{-1}.

8 A sector of angle θ radians is taken from a circle of radius r cm. If the area of the sector is A cm^2, find an expression linking r, θ and A. The area is held constant, and the radius increases at q cm s^{-1}. Find the rate of decrease of θ.

*9 A cup is formed by rotating the curve $y = x^2$, from $x = -2$ to $x = 2$, about the y-axis. Units along the x-axis and the y-axis are in cm. Find the volume of the cup.

Liquid is poured into the cup at a rate of 5 cm^3 s^{-1}. When the depth of liquid is h cm, find the rate of increase of

a) the depth of liquid   **b)** the surface area of liquid.

10 The diagram shows the cross-section of a drinking trough. The angle between the two sides is 90° and the length of the trough is 1.6 m. Show that when the depth of water is h m, the volume of water is $V = 1.6h^2$ m^3. If water is poured in at 0.01 m^3 per second, find the rate of increase of the depth when it is 0.2 m.

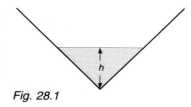
Fig. 28.1

*11 The angle between the sides of a conical champagne glass is 120°. Champagne is poured in at 10 cm^3 per second. Find the rate of increase of the depth of liquid when it is 3 cm.

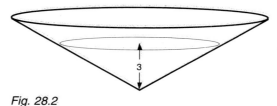
Fig. 28.2

12 A hemispherical bowl has radius 20 cm. When the depth of liquid is x cm, the volume is $\pi x^2 \left(20 - \dfrac{x}{3}\right)$ cm^3. If liquid is dripping out at 1 cm^3 per second, find the rate of decrease of the depth when it is 5 cm.

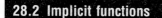
28.2 Implicit functions

Explicit and implicit functions

Most of the functions which you have met so far are similar to those below, in which y is given directly in terms of x.

$$y = 3x^2 + 7x - 2$$

$$y = \sin 2x - 3\cos x$$

$$y = (3x - 2)(4x - 1)$$

Because y is given explicitly in terms of x, these are called **explicit** functions.

Sometimes variables are connected by an equation which does not give one variable in terms of the other. For example, the equation below, which is that of the ellipse in Fig. 28.3, does not give y explicitly in terms of x.

$$x^2 + xy + 3y^2 - 3 = 0$$

This is an **implicit** function.

In general

An explicit function is of the form $y = f(x)$.

An implicit function is of the form $g(x, y) = 0$.

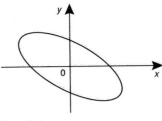

Fig. 28.3

Note: An explicit function can be written as an implicit function. If we have $y = f(x)$, we can write it as $y - f(x) = 0$, which is of the form $g(x, y) = 0$.

When differentiating an implicit function we often need to use the chain rule. Suppose we want to differentiate y^2 with respect to x.

We first differentiate with respect to y, and then multiply by $\dfrac{dy}{dx}$.

$$\frac{d}{dx}y^2 = \frac{d}{dy}y^2 \frac{dy}{dx} = 2y\frac{dy}{dx}$$

We may also need the product rule, for example to differentiate xy.

$$\frac{d}{dx}xy = x\frac{dy}{dx} + y\frac{dx}{dx} = x\frac{dy}{dx} + y$$

EXAMPLE 28.2
Find the gradient of the curve $x^2 + xy + 3y^2 - 3 = 0$, at the point $(1, -1)$.

Solution
Differentiate this equation term by term with respect to x. The right-hand side of the equation is constant, so its derivative is zero.

$$2x + x\frac{dy}{dx} + y + 3(2y)\frac{dy}{dx} = 0$$

Put $x = 1$ and $y = -1$ into this equation.

$$2 + \frac{dy}{dx} + -1 + 3(-2)\frac{dy}{dx} = 0$$

$$-5\frac{dy}{dx} + 1 = 0$$

$$\frac{dy}{dx} = \frac{1}{5}$$

At $(1, -1)$ the gradient is $\dfrac{1}{5}$.

EXERCISE 28B

1 Differentiate the following with respect to x.

a) y^3 b) e^y c) $\sin 3y$

d) $\sqrt{y}$ e) $x^2 y$ f) $y\sqrt{x}$

g) xy^2 h) $x^3 y^2$ i) $\sin x \; e^y$

2 Find $\dfrac{dy}{dx}$ for the following implicit functions at the points indicated.

a) $x^2 + 3y^2 - 4 = 0$ at $(1, 1)$ b) $4x^2 - 3y^2 = 13$ at $(2, 1)$

c) $\sqrt{x} + \sqrt{y} = 3$ at $(4, 1)$ d) $\sin x + \sin y = 1$ at $(0, \frac{\pi}{2})$

e) $2x^2 + 3xy + 2y^2 = 4$ at $(2, -1)$ f) $x^2 + 3xy - 5y^2 = 5$ at $(2, 1)$

g) $x \, e^y = 1$ at $(1, 0)$ h) $2 \sin x \sin y = 1$ at $(\frac{\pi}{4}, \frac{\pi}{4})$

3 For each of the functions of Question 2, find an expression for $\dfrac{dy}{dx}$ at the general point (x, y).

4 If $e^x y = \cos x$, show that $e^x(y + y') = -\sin x$, and that $e^x(y + 2y' + y'') = -\cos x$. Hence show that $y'' + 2y' + 2y = 0$.

5 Let $e^{-x} y = \sin x$. Show that $y'' - 2y' + 2y = 0$.

6 Find the equation of the tangent to the curve $x^2 + 3y^2 = 7$ at the point $(2, 1)$.

7 Find the equation of the tangent to the curve $4x^2 + 3xy - y^2 = 14$ at the point $(-3, 2)$.

8 Find the equation of the normals to the curves of Questions 6 and 7, at the same points.

9 Find $\dfrac{dy}{dx}$ for the curve $x^2 + xy + 2y^2 = 4$. Show that where the curve has gradient $\dfrac{1}{3}$, $x + y = 0$. Hence find the points for which $\dfrac{dy}{dx} = \dfrac{1}{3}$.

10 Find the points on the curve of Question 9 for which the tangent is parallel to the x-axis.

11 Find the maximum value of x for the points on the curve of Question 9.

12 For the curve with equation $2x^2 - 2xy + 5y^2 = 10$, find

a) the maximum value of x b) the maximum value of y.

13 The variables x and y are related by $x^2 y^3 = k$, where k is a constant. Find an expression for $\dfrac{dy}{dx}$ in a form not involving k.

*14 Let $y = x^x$. By taking logs and differentiating, show that $\dfrac{1}{y} \dfrac{dy}{dx} = 1 + \ln x$. Hence write $\dfrac{dy}{dx}$ as a function of x.

*15 Suppose $x = y^x$. Find $\dfrac{dy}{dx}$ in terms of x and y.

28.3 Parametric differentiation

When y and x are given in terms of each other, the equation linking them is a **Cartesian equation**.

Sometimes two quantities x and y are not directly connected, but are both given in terms of a third quantity t. This third quantity is called a **parameter**. For example, suppose a stone is thrown up in the air. Let y be its vertical height, and x its horizontal distance from the point where it was thrown. Both x and y are functions of time t.

The rate of change of y with respect to x can be found from the chain rule.

$$\frac{dy}{dx} = \frac{dy}{dt}\frac{dt}{dx} = \frac{dy}{dt} \bigg/ \frac{dx}{dt}$$

EXAMPLE 28.3

A stone is thrown upwards. After t seconds, the vertical distance y m and horizontal distance x m are given by $y = 40t - 5t^2$ and $x = 30t$.

Find $\dfrac{dy}{dx}$. Hence find the angle at which it is travelling after 2 seconds.

Plot the curve which the stone follows from $t = 0$ and $t = 8$.

Solution

$$y = 40t - 5t^2 \qquad\qquad x = 30t$$

Differentiate both these expressions.

$$\frac{dy}{dt} = 40 - 10t \qquad\qquad \frac{dx}{dt} = 30$$

Divide $\dfrac{dy}{dt}$ by $\dfrac{dx}{dt}$.

$$\frac{dy}{dx} = \frac{40 - 10t}{30} = \frac{4}{3} - \frac{1}{3}t$$

When $t = 2$, $\dfrac{dy}{dx} = \dfrac{4}{3} - \dfrac{2}{3} = \dfrac{2}{3}$.

This gives the gradient of the curve. The gradient of the curve is the tangent of the angle at which the stone is travelling.

So the angle is $\tan^{-1}\dfrac{dy}{dx} = \tan^{-1}\dfrac{2}{3} = 33.69°$.

It is travelling at 33.7° to the horizontal.

Taking values of t from 0 to 8, the following values of x and y are found.

t	0	1	2	3	4	5	6	7	8
x	0	30	60	90	120	150	180	210	240
y	0	35	60	75	80	75	60	35	0

The graph is shown in Fig. 28.4.

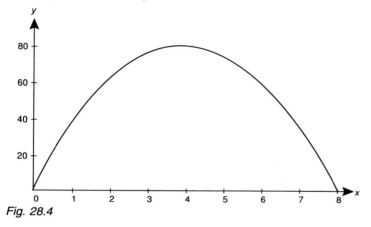

Fig. 28.4

EXAMPLE 28.4

A curve is given parametrically by $x = 2 \cos \theta$, $y = 3 \sin \theta$. Find the equation of the tangent at the point where $\theta = \dfrac{\pi}{4}$.

Eliminate θ to find the Cartesian equation of the curve. Sketch the curve.

Solution

$$x = 2 \cos \theta \qquad\qquad y = 3 \sin \theta$$

Differentiate both these equations.

$$\frac{dx}{d\theta} = -2 \sin \theta \qquad\qquad \frac{dy}{d\theta} = 3 \cos \theta$$

$$\frac{dy}{dx} = -\frac{3 \cos \theta}{2 \sin \theta} = -\frac{3}{2} \cot \theta$$

At $\theta = \dfrac{\pi}{4}$, $x = \dfrac{2}{\sqrt{2}}$, $y = \dfrac{3}{\sqrt{2}}$ and $\dfrac{dy}{dx} = -\dfrac{3}{2}$.

The equation of the tangent is

$$y - \frac{3}{\sqrt{2}} = -\frac{3}{2}\left(x - \frac{2}{\sqrt{2}}\right)$$

The equation of the tangent is $2y + 3x = \dfrac{12}{\sqrt{2}}$.

Divide the equation for x by 2 and that for y by 3 to obtain

$$\frac{x}{2} = \cos\theta \quad \text{and} \quad \frac{y}{3} = \sin\theta$$

Now use the identity

$$\cos^2 x + \sin^2\theta \equiv 1$$

$$\frac{x^2}{4} + \frac{y^2}{9} = 1$$

The Cartesian equation is $\dfrac{x^2}{4} + \dfrac{y^2}{9} = 1$.

To sketch the curve, note that x goes from -2 to 2, and y from -3 to 3. The curve is the oval shape shown.

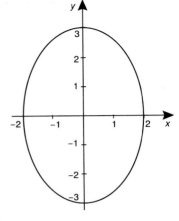

Fig. 28.5

EXERCISE 28C

1 In each of the following, find $\dfrac{dy}{dx}$ in terms of t.

a) $y = t^2,\ x = t^3$ **b)** $y = 3 + t,\ x = 5 - 2t$ **c)** $y = 1 + 2t,\ x = 1 + 3t$

d) $y = \sqrt{t},\ x = t + 3$ **e)** $x = \sqrt{t+1},\ y = \sqrt{t-1}$ **f)** $y = 2 + t,\ x = \dfrac{5}{t}$

g) $x = \dfrac{1}{t+1},\ y = \dfrac{1}{t+2}$ **h)** $x = 3\cos\theta,\ y = 4\sin\theta$ **i)** $y = \sin\theta,\ x = 2\cos\theta$

j) $x = \sec\theta,\ y = \tan\theta$ **k)** $x = 3\tan\theta,\ y = 4\sec\theta$ **l)** $x = e^t,\ y = e^{-2t}$

m) $x = \frac{1}{2}(e^t + e^{-t}),\ y = \frac{1}{2}(e^t - e^{-t})$

2 Find the equations of the following tangents.

a) to Question 1(a) at $t = 1$ **b)** to Question 1(d) at $t = 4$

c) to Question 1(g) at $t = 0$ **d)** to Question 1(h) at $\theta = \dfrac{\pi}{3}$

e) to Question 1(j) at $\theta = \dfrac{\pi}{4}$ **f)** to Question 1(m) at $t = 0$

3 Find the equations of the normals to the curves at the points indicated in Question 2.

4 A stone is thrown upwards and after t seconds it is $30t - 5t^2$ metres high and has travelled $20t$ metres horizontally. Find the angle at which it is travelling after 1 second. Plot the path that the stone follows, from $t = 0$ to $t = 6$.

5 An object is projected in a resisting medium, and t seconds later its height h m and horizontal distance s m travelled are given by

$$h = 10(1 - t) + 10e^{-t} \qquad s = 10(1 - e^{-t})$$

Find the angle at which it is travelling after 1 second.

6 A curve is given parametrically by $x = t^2 - t$, $y = t + 1$. Find the tangent at the point where $t = 1$. Find the Cartesian equation of the curve. Plot the curve for values for t from -2 to 2.

7 A curve is given parametrically by $x = 3t + 1$, $y = 2t^2$. Find the normal to the curve at $t = -1$. Find the Cartesian equation of the curve. Plot the curve for values of t from -2 to 2.

8 Find the normal to the curve given by $x = t^2$, $y = t^3$ at the point where $t = 2$. Plot the curve for values of t from -2 to 2.

9 A curve is given parametrically by $x = t^3 - t$, $y = t^2 + t$. By finding $\dfrac{x}{y}$, obtain the Cartesian equation of the curve. Plot the curve for values of t from -3 to 3.

10 A curve is given by $x = \dfrac{1}{t}$, $y = t^2$. Find the equation of the tangent at the general point P. This tangent cuts the x-axis at X and the y-axis at Y. Show that $\mathbf{PY} = 2\mathbf{PX}$.

11 A curve is given parametrically by $x = \cos\theta$, $y = 2\sin\theta$. Find the Cartesian equation of the curve and sketch it.

12 A curve is given parametrically by $x = 3\cos\theta$, $y = 4\sin\theta$. Find the Cartesian equation of the curve and sketch it.

*13 The diagram shows a wheel of radius a. A point is initially on the bottom of the wheel. The wheel rolls at an angular speed of ω without slipping along a horizontal line. Show that time t later the height y of the point and its horizontal displacement x are given by $y = a(1 - \cos\omega t)$, $x = a(\omega t - \sin\omega t)$.

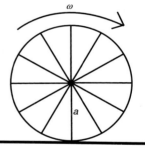

(This curve is called a **cycloid**.) Find the gradient when $t = \dfrac{\pi}{4\omega}$. What is the gradient when $t = 0$?

Fig. 28.6

28.4 Small changes

Suppose x and y are related quantities. The derivative $\dfrac{dy}{dx}$ is defined as the limit of $\dfrac{\delta y}{\delta x}$, where δy and δx are changes in y and x respectively. So if δx and δy are small

$$\frac{\delta y}{\delta x} \approx \frac{dy}{dx}$$

This gives the following relation between the small changes in y and x.

$$\delta y \approx \frac{dy}{dx}\,\delta x$$

EXAMPLE 28.5

A sphere has radius 10 cm. What increase in radius is needed to increase the volume by 5 cm³?

Solution

The volume V and radius r of a sphere are connected by the equation $V = \frac{4}{3}\pi r^3$. Apply the rule above for small changes.

$$\delta V \approx \frac{dV}{dr}\, \delta r = 4\pi r^2\, \delta r$$

We know that $r = 10$ and $\delta V = 5$.

$$5 = 4\pi(10)^2 \delta r \qquad \delta r = \frac{5}{4\pi(10)^2} = 0.003\,98$$

The change in r is approximately $0.003\,98$ cm.

EXAMPLE 28.6

The radius of the sphere of Example 28.5 increases by 1%. What is the approximate percentage change in the volume?

Solution

Take the approximation $\delta V \approx 4\pi r^2 \delta r$ and divide by $V = \frac{4}{3}\pi r^3$.

$$\frac{\delta V}{V} \approx \frac{4\pi r^2 \delta r}{\frac{4}{3}\pi r^3} = \frac{3\delta r}{r}$$

The percentage increase of r is 1%.

So $\dfrac{3\delta r}{r} = 3 \times 0.01 = 0.03$

The percentage increase in V is 3%.

EXAMPLE 28.7

Without the use of a calculator find an approximation for $\sqrt{4.1}$.

Solution

Let $y = \sqrt{x}$. We know that $\dfrac{dy}{dx} = \dfrac{1}{2\sqrt{x}}$.

If we let $x = 4$ and $\delta x = 0.1$, then the formula will apply.

$$\delta y \approx \frac{1}{2\sqrt{4}} \times 0.1 = 0.025$$

The approximate value of $\sqrt{4.1}$ is $2 + 0.025 = 2.025$.

EXERCISE 28D

1 A circle has radius 20 cm. What approximate change in radius is needed to increase the area by 4 cm^2?

2 Let $y = x^3 + x^2 - 2x$. Find the approximate increase in y when x increases from 2 to 2.1.

3 Let $y = \sin x + 2 \cos x$. Find the approximate increase in y when x increases from 1 to 1.05.

4 If a circuit has resistance R ohms, the power P watts it uses is $P = \dfrac{R}{(R+3)^2}$.

Initially $R = 7$. Find the change in resistance necessary to increase the power by 0.002.

5 A surveyor stands 60 m from the base of a tower, and measures the angle of elevation of the top as 0.453 radians. If the measurement of the angle could be out by 0.001, find the approximate error in the calculation of the height of the tower.

6 A child stands on a harbour wall, holding a string attached to a toy boat which is 4 m out to sea. The end of the string is 3 m above sea level. The string is shortened by δ m. What is the approximate change in the boat's distance out to sea?

7 A sphere has radius 15 cm. What approximate increase in radius will increase the surface area by 1 cm^2?

*8 If a cone has base radius r and height h, then its **slant height** is given by $L = \sqrt{r^2 + h^2}$. The base radius is 5 cm and the height 12 cm. If the base radius remains fixed, what approximate change in the slant height will increase the volume by 0.1 cm^3?

9 The radius of a circle increases by 1%. What is the approximate percentage increase in the area?

10 The surface area of a sphere increases by 0.8%. What is the approximate percentage increase in the volume?

11 The volume of a cube increases by 0.6%. What is the approximate percentage increase in the side?

12 If a pendulum has length L m, the time T seconds it takes to swing to and fro is given by

$$T = 2\pi\sqrt{\frac{L}{10}}$$

a) If the time taken increases from 2 seconds to 2.1 seconds, what is the approximate change in the length?

b) If the length changes by 1%, what is the approximate percentage change in the time?

13 The diameter of a sphere changes by $x\%$, where x is small. What is the approximate percentage change in the volume of the sphere?

14 The power P in a circuit is given by $P = I^2 R$, where I is the current and R is the constant resistance. Find the percentage change in P if I changes by $x\%$.

15 Without the use of a calculator, find approximations for the following.

a) $\sqrt{9.3}$ **b)** $\sqrt{99}$ **c)** $\sqrt[3]{8.1}$

*16 Let $y = x^3 + 4x^2$. Check that $y = 768$ when $x = 8$. Use the method of small changes to find the value of x which will give $y = 770$.

17 If the price of a commodity is x pence per kilogram, the national annual production will be $(1000x + x^3)$ tonnes. At the moment $x = 15$. What increase in price would be necessary to increase annual production by 400 tonnes?

LONGER EXERCISE

Solving equations

Repeated use of the method of small changes enables us to solve equations. Suppose we want to solve the equation $x^5 + 2x = 40$.

1 Let $y = x^5 + 2x$. Show that when $x = 2$, $y = 36$.

2 $40 - 36 = 4$. So put $\delta y = 4$ and find the corresponding δx. Find an approximate solution to the equation.

3 Find the value of y corresponding to this approximate solution. Let δy be the difference between this value and 40. Use the method of small changes to find a more accurate solution.

4 Repeat, until successive approximations agree correct to eight decimal places.

5 Use this method to solve the equation $x - \cos x = 0$.

6 This method is very similar to one you may have met already. What is it?

EXAMINATION QUESTIONS

1 A petrol tanker is damaged in a road accident, and petrol leaks onto a flat section of a motorway. The leaking petrol begins to spread in a circle of thickness 2 mm. Petrol is leaking from the tanker at a rate of 0.0084 m^3 s^{-1}. Find the rate at which the radius of the circle of petrol is increasing at the instant when the radius of the circle is 3 m, giving your answer in m s^{-1} to 2 decimal places.

L AS 1990

2 The pressure P and volume V of a gas are related by $PV^n = k$ where k and n are positive constants.

(i) Show that $\dfrac{\mathrm{d}P}{\mathrm{d}V} = \dfrac{-nP}{V}$.

(ii) Find, in terms of n, the approximate percentage decrease in P if V is increased by 1%.

$W\ 1991$

3 The volume V of a sphere is given by the formula $V = \frac{4}{3}\pi r^3$, where r is the radius.

(i) A small increase δr in the radius of a sphere leads to a small increase δV in its volume. Write down an approximate expression for δV in terms of r and δr.

(ii) The radius is initially 24 cm, and then the volume is increased by 500 cm³. Calculate the approximate increase in radius, giving your answer correct to 2 decimal places.

(iii) Show that the percentage increase in V is always approximately three times the percentage increase in r.

(iv) Whilst a spherical balloon is being inflated, its volume is increasing at a rate of 2000 cm³ per second. Calculate the rate at which the radius is increasing when $r = 15$ cm.

$MEI\ 1993$

4 A curve is given parametrically by $x = t^2$, $y = t^3$.

Find $\dfrac{\mathrm{d}y}{\mathrm{d}x}$ in terms of t.

Obtain the equation of the tangent at the point P with parameter t and verify that it cuts the curve again at the point Q with parameter $-\frac{1}{2}t$.

If the line PQ cuts the x-axis at R, show that the value of the ratio $\dfrac{\mathrm{PR}}{\mathrm{RQ}}$ is independent of the position of P on the curve.

$MEI\ 1991$

5 This question is about the ellipse with equation $x^2 + 2xy + 4y^2 = 9$.

(i) Show that the curve cuts the x-axis at the points P$(-3, 0)$ and Q$(3, 0)$, and find the coordinates of the points R and S where it cuts the y-axis.

(ii) Differentiate the equation with respect to x, and hence express $\dfrac{\mathrm{d}y}{\mathrm{d}x}$ in the form

$$\frac{ax + by}{cx + dy}$$

where a, b, c and d are constants.

(iii) There are two points on the curve where the gradient is zero. They both lie on a certain straight line through the origin. Use your answer to part (ii) to find the equation of this line.

(iv) There are two points on the curve where the gradient is infinite. They both lie on another straight line through the origin. Use your answer to part (ii) to find the equation of this second line.

(v) On graph paper, using the same scale on each axis, mark the points P, Q, R and S, and draw the two lines you found in parts (iii) and (iv). [Label these lines clearly with their equations.]

(vi) On your diagram sketch the ellipse $x^2 + 2xy + 4y^2 = 9$.

SMP 1993

Summary and key points

1 Connections between rates of change can be found using $\dfrac{dP}{dt} = \dfrac{dP}{dQ} \times \dfrac{dQ}{dt}$.

If you are given values of P or Q do not substitute them in the equation until after differentiating.

2 Implicit functions can be differentiated.

When differentiating $f(x, y) = 0$, the derivative of the right-hand side is 0. It is not $\dfrac{dy}{dx}$.

3 Functions given in terms of parameters can be differentiated by $\dfrac{dy}{dx} = \dfrac{dy}{dt} \Big/ \dfrac{dx}{dt}$

4 Small changes in related quantities can be found approximately by $\delta y \approx \dfrac{dy}{dx} \, \delta x$.

Differential equations

A **differential equation** is an equation linking a quantity with its derivatives. If y is the quantity, then the equation might involve $\dfrac{dy}{dx}$ and $\dfrac{d^2y}{dx^2}$ as well as y itself. Examples of differential equations are

$$\frac{dy}{dx} + 3y = 2 \qquad x\frac{dy}{dx} + 4y = 0 \qquad 2\frac{d^2y}{dx^2} + 3\frac{dy}{dx} + 7y = 0$$

These sorts of equations occur in many natural situations. Some examples are given below.

The rate of loss of temperature of a body is linked to its temperature.

The acceleration (rate of change of velocity) of a body moving through air is linked to the velocity.

The rate of growth of a population is linked to the size of the population.

So differential equations occur throughout the sciences. Indeed, it has been said that the greater part of scientific knowledge consists of differential equations.

29.1 Modelling a situation by a differential equation

There are often three steps in dealing with a scientific situation.

1 Model the situation in terms of a mathematical problem.

2 Solve the mathematical problem.

3 From the mathematical solution, interpret the situation or predict how it will develop.

The process of forming a differential equation to describe a situation is called **modelling**.

Suppose we are dealing with a quantity for which we cannot directly derive a formula. We might have a law which connects the rate of change of the quantity with other factors, including the quantity itself. This law, translated into mathematical terms, is a differental equation.

Example 29.1
Suppose we have a hot object immersed in a bath of water at a constant temperature of 10°. **Newton's law of cooling** states that

> The rate of loss of temperature of a body is proportional to the difference between its temperature and that of its environment.

Find a differential equation modelling this law.

Solution
Let the temperature of the body be $T°$. The rate of loss of temperature is $-\dfrac{dT}{dt}$, where t represents time in seconds. This rate of loss is proportional to the difference between T and 10, giving

$$-\frac{dT}{dt} \propto T - 10$$

Rewrite this proportionality statement as an equation

$$\frac{dT}{dt} = -k(T - 10)$$

EXERCISE 29A

1 The rate of change of a population is proportional to the population. Letting P be the population, form a differential equation linking $\dfrac{dP}{dt}$ and P.

2 The rate of decay of a radioactive material is proportional to the mass of the material. Write this as an equation linking the mass M with its rate of change.

3 The acceleration (rate of change of velocity) of a moving body is proportional to the square of the velocity. Write this as an equation in the velocity v.

4 A tank contains 1000 litres of sugar solution. This is being fermented, at a rate proportional to the amount left unfermented. Write this as an equation in the amount F litres which has been fermented.

5 A disease is spreading in a population. The rate of infection is proportional to the product of the proportion P of people infected with the proportion of people not yet infected. Write this as an equation in P.

6 A tank is in the shape of a cuboid with base area 1 m². Water is poured into the tank at a rate of p m³ s⁻¹. There is a hole in the bottom of the tank, through which water leaves at a rate numerically equal to q times the volume of water in the tank. Find an equation in the volume V m³ of water in the tank.

7 The gradient of a graph at any point is proportional to the square of the x-coordinate at that point. Express this in terms of a differential equation.

8 The gradient of a graph at any point is proportional to the product of the coordinates of the point. Express this in terms of a differential equation.

9 A car depreciates so that the rate of decline in its value is proportional to its value. Find a differential equation in the value V.

10 A tank initially contains 100 litres of pure water. Two taps lead into the tank; one provides k litres of pure water per second, the other k' litres of pure alcohol. There is a plug at the bottom through which liquid leaves at $(k + k')$ litres per second. Assume the liquid is instantaneously mixed. Find a differential equation involving the volume V litres of alcohol in the tank.

11 The birth rate of a population is proportional to the population. The death rate is proportional to the square of the population. Form a differential equation in the population P.

12 Liquid enters a tank at a constant rate S litres per second. The rate of leakage is proportional to the square of the volume V. If there is a state of equilibrium when $V = V_0$, show that V obeys the equation

$$\frac{\mathrm{d}V}{\mathrm{d}t} = S - \frac{SV^2}{V_0^2}$$

13 The birth rate of a population is proportional to the population P. The death rate is proportional to P^3. If the population is steady when $P = P_0$, show that P obeys an equation of the form

$$\frac{\mathrm{d}P}{\mathrm{d}t} = kP(P_0 - P)(P_0 + P)$$

29.2 General solutions of differential equations

Once we have modelled the situation by a differential equation, we try to solve it. It may not be possible to solve the equation exactly, but there is a whole range of techniques which can be used.

Suppose we have an equation of the form $\dfrac{dy}{dx} = f(x, y)$. Then the equation has **separable variables** if we can write $f(x, y)$ as $g(x)h(y)$.

We might have $\dfrac{dy}{dx} = x^2 \cos y$ which is of this form.

If this is the case, then we can bring all the x-terms to one side of the equation and all the y-terms to the other. Then one side can be integrated with respect to x and the other side with respect to y.

$$\int \frac{dy}{h(y)} = \int g(x) \, dx$$

Notes

1 This chapter is not concerned with the actual integration of these expressions. We assume it can be done, either exactly or by numerical approximation.

2 There are two integrals to be performed, each of which will have a constant of integration. But we can combine these constants into one.

3 If $h(y) = 1$, then our equation is of the form $\dfrac{dy}{dx} = g(x)$. To solve this we need only find the integral $\int g(x) \, dx$.

The solution that we find, involving a constant of integration, is the **general solution** of the equation.

EXAMPLE 29.2

Find the general solution of the equation $\dfrac{dy}{dx} = y \tan x$.

Solution

Here the right-hand side is the product of a function of x with a function of y.

Separate the variables.

$$\int \frac{dy}{y} = \int \tan x \, dx$$

These are standard integrals. The integral of the left-hand side is $\ln y$, and of the right-hand side is $-\ln \cos x$. Put the constant of integration on the right.

$$\ln y = -\ln \cos x + c$$

Exponentiate both sides. Be careful with the right-hand side. Recall that $e^{a+b} = e^a \times e^b$.

$$y = e^{-\ln \cos x + c} = e^{-\ln \cos x} \times e^c$$

$$y = \frac{1}{\cos x} \times e^c$$

It is tidier to rewrite the constant e^c as a constant A.

$$y = \frac{A}{\cos x}$$

EXAMPLE 29.3

In a chemical process substance A is changing into substance B, at a rate which is proportional to the product of the amounts of A and B. Letting y be the proportion of substance B, obtain a differential equation in y. Find the general solution of this equation.

Solution

If y is the proportion of B, then $1 - y$ is the proportion of A and $\frac{dy}{dt}$ is proportional to the product of these two quantities. This gives the equation

$$\frac{dy}{dt} = ky(1 - y)$$

Separate the variables.

$$\int \frac{dy}{y(1 - y)} = \int k \, dt$$

Put the integrand of the left-hand side into partial fractions.

$$\frac{1}{y(1 - y)} = \frac{1}{y} + \frac{1}{1 - y}$$

Now integrate both sides.

$$\ln y - \ln(1 - y) = kt + c$$

$$\ln \frac{y}{1 - y} = kt + c$$

$$\frac{y}{1 - y} = Ae^{kt}$$

Now we make y the subject of the formula. Multiply through by $(1 - y)$ and collect terms.

$$y = Ae^{kt} - yAe^{kt}$$

Hence $y(1 + Ae^{kt}) = Ae^{kt}$.

The general solution is $y = \dfrac{Ae^{kt}}{1 + Ae^{kt}}$.

EXERCISE 29B

1 Find the general solutions of the following equations.

a) $\dfrac{dy}{dx} = y$

b) $\dfrac{dy}{dx} = 4x\sqrt{y}$

c) $(x + 3)\dfrac{dy}{dx} = y$

d) $\dfrac{dy}{dx} = 2x\sqrt{1 - y^2}$

e) $\dfrac{dy}{dx} = e^{x+y}$

f) $(1 + x^2)\dfrac{dy}{dx} = y^2$

g) $\dfrac{dy}{dx} = x(y + 1)$

h) $(x + 2)\dfrac{dy}{dx} = y + 3$

i) $2x\dfrac{dy}{dx} = \dfrac{1}{y} + y$

j) $e^x\dfrac{dy}{dx} = e^{y+2}$

k) $\dfrac{dy}{dx} = \cos^2 y \sin x$

l) $\dfrac{dy}{dx} = e^y \cos x$

m) $\dfrac{dy}{dx} = 2y(y + 1)$

n) $2x\dfrac{dy}{dx} = 1 - y^2$

o) $\dfrac{dy}{dx} = xy(1 - y)$

2 The rate of growth per annum of a population is 0.02 times the population. Find a differential equation expressing this fact, and find its general solution. How long does it take for the population to double?

3 The rate of growth of a plant is proportional to the difference between its height h and the maximum height H. Express this in terms of a differential equation, and find its general solution.

4 Suppose $\dfrac{d^2y}{dx^2} = 5\dfrac{dy}{dx}$. By writing $p = \dfrac{dy}{dx}$, find an expression for $\dfrac{dy}{dx}$ in terms of x. Hence find y in terms of x.

5 Solve these equations.

a) $x\dfrac{d^2y}{dx^2} = 2\dfrac{dy}{dx}$

b) $\dfrac{d^2y}{dx^2} = x\sqrt{\dfrac{dy}{dx}}$

c) $\tan x \dfrac{d^2y}{dx^2} = \dfrac{dy}{dx}$

29.3 Particular solutions of differential equations

Suppose we are modelling a situation by a differential equation, and have found the general solution. The particular situation that we are modelling will have extra conditions relevant to it. For example, we might know the state at time zero. When we put in the extra conditions, we obtain a solution which applies to our particular situation. This is a **particular solution**.

EXAMPLE 29.4

Find the solution of $\dfrac{dy}{dx} = y \tan x$ for which $y = 2$ when $x = 0$.

Solution

The general solution of this equation was found in Example 29.2.

$$y = \frac{A}{\cos x}$$

Put in the given values.

$$2 = \frac{A}{\cos 0} = A$$

The solution is $y = \dfrac{2}{\cos x}$.

EXAMPLE 29.5

A body is immersed in a liquid kept at a constant temperature of $10°$. Initially the body is at $90°$. After 30 minutes it is at $50°$. Find an expression for its temperature after t minutes. Find the time for it to cool to $20°$.

Solution

Newton's law of cooling applies here. The situation is modelled by a differential equation, as in Section 29.1.

$$\frac{dT}{dt} = -k(T - 10)$$

Separate the variables.

$$\int \frac{dT}{T - 10} = \int -k \, dt$$

$$\ln(T - 10) = -kt + c$$

$$T - 10 = e^c \times e^{-kt}$$

Put $e^c = A$.

$$T = Ae^{-kt} + 10$$

This is the general solution. There are two unknown constants, k and A. We have two bits of information: that $T = 90$ when $t = 0$, and that $T = 50$ when $t = 30$.

$$90 = A + 10$$

$$50 = Ae^{-30k} + 10$$

The first equation gives $A = 80$. The second gives

$$40 = 80 \, e^{-30k}$$

Hence $-30k = \ln 0.5$

$$k = -\frac{\ln 0.5}{30} = \frac{\ln 2}{30}$$

The temperature after t minutes is $80\,e^{-t\,(\ln 2)/30} + 10$.

Note: $e^{\ln 2} = 2$, so this solution can also be written as

$$T = 80(e^{\ln 2})^{-t/30} + 10 = 80 \times 2^{-t/30} + 10$$

Now put $T = 20$ into this equation.

$$20 = 80 \times 2^{-t/30} + 10$$

$$2^{-t/30} = \frac{1}{8}$$

$$= 2^{-3}$$

So $-\dfrac{t}{30} = -3$

The temperature will be $20°$ after 90 minutes.

EXERCISE 29C

1 Find the particular solutions of the following differential equations, with the conditions as given. Note that these equations are the ones of Question 1 from Exercise 29B.

a) $\dfrac{dy}{dx} = y$. When $x = 0$, $y = 3$

b) $\dfrac{dy}{dx} = 4x\sqrt{y}$. When $x = 0$, $y = 4$

c) $(x + 3)\dfrac{dy}{dx} = y$. When $x = 0$, $y = 6$

d) $\dfrac{dy}{dx} = 2x\sqrt{1 - y^2}$. When $x = 0$, $y = 1$

e) $\dfrac{dy}{dx} = e^{x+y}$. When $x = 0$, $y = 0$

f) $(1 + x^2)\dfrac{dy}{dx} = y^2$. When $x = 0$, $y = 2$

g) $\dfrac{dy}{dx} = x(y + 1)$. When $x = 0$, $y = 3$

h) $(x + 2)\dfrac{dy}{dx} = y + 3$. When $x = -1$, $y = -1$

i) $2x\dfrac{dy}{dx} = \dfrac{1}{y} + y$. When $x = 4$, $y = 3$

j) $e^x\dfrac{dy}{dx} = e^{y+2}$. When $x = 0$, $y = 5$

k) $\dfrac{dy}{dx} = \cos^2 y \sin x$. When $x = \dfrac{\pi}{2}$, $y = \dfrac{\pi}{4}$

l) $\dfrac{dy}{dx} = e^y \cos x$. When $x = \dfrac{\pi}{6}$, $y = 1.5$

m) $\dfrac{dy}{dx} = 2y(y + 1)$. When $x = 0$, $y = 0.5$

n) $2x\dfrac{dy}{dx} = 1 - y^2$. When $x = 1$, $y = 2$

o) $\dfrac{dy}{dx} = xy(1 - y)$. When $x = 0$, $y = \dfrac{1}{2}$

2 The rate of increase of the price of an item is 0.03 times the price. If the original price is £50, find an expression for the price after time t. When will the price reach £75?

3 A tank is a cylinder with base area 2 m². Water is poured in at 0.002 m³ per second. Water leaks from a hole at the bottom, at a rate equal to 0.0004 times the depth of water. If the depth of water is h m, show that

$$\frac{dh}{dt} = 0.001 - 0.0002h$$

The tank starts empty. Find an expression for the depth of water after t seconds. When will the depth of water be 1 m?

4 A vat contains 500 kg of chemical A, and 500 kg of chemical B. B changes into A at a rate, in kg per second, equal to 0.0001 times the product of the amounts of A and B. If y kg is the amount of A after t seconds, show that

$$10\,000\,\frac{dy}{dt} = y(1000 - y)$$

Find the solution of this equation. When will there be 900 kg of A?

5 A disease is spreading through a population. The rate of increase is proportional to the product of the proportion of people infected and the proportion of people not affected. Initially the proportion infected is 0.1%, and 1 year later it is 10%.

Letting P be the proportion infected, find a differential equation in P. Solve this equation to find when 50% of the population will be infected.

6 In Question 11 of Exercise 29A a population P had a birth rate proportional to P and a death rate proportional to P^2. Suppose that the equilibrium population is 1 000 000. Show that P obeys the differential equation

$$\frac{dP}{dt} = kP(1\,000\,000 - P)$$

Solve this equation, given that the population is initially 500 000 and reaches 750 000 after 1 year.

7 Question 10 of Exercise 29A involved a tank being filled with pure water and alcohol. If $k = 2$ and $k' = 3$, show that the volume of alcohol in the tank obeys the equation

$$\frac{dM}{dt} = 3 - \frac{M}{20}$$

Solve this equation. What will be the volume of alcohol at $t = 10$? After a long time, what will be the proportion of alcohol in the tank?

29.4 Families of solutions

Suppose we have a set of similar functions. Often all the functions will obey the same differential equation. For example, every function of the form $y = x^2 + A$ obeys the equation

$$\frac{dy}{dx} = 2x$$

EXAMPLE 29.6

Find a differential equation, not involving A, which is obeyed by every function of the form

$$y = Ae^{3x}$$

Solution

Differentiating, we find that

$$\frac{dy}{dx} = 3Ae^{3x}$$

Note that this is three times y.

The equation is $\dfrac{dy}{dx} = 3y$.

EXAMPLE 29.7

Find a differential equation, not involving A or B, which is obeyed by every function of the form

$$y = Ae^x + Be^{2x}$$

Solution

Differentiating

$$\frac{dy}{dx} = Ae^x + 2Be^{2x} \quad \text{and} \quad \frac{d^2y}{dx^2} = Ae^x + 4Be^{2x}$$

Suppose our equation is of the form

$$a\frac{d^2y}{dx^2} + b\frac{dy}{dx} + cy = 0$$

Rewrite it as

$$a(Ae^x + 4Be^{2x}) + b(Ae^x + 2Be^{2x}) + c(Ae^x + Be^{2x}) = 0$$

$$Ae^x(a + b + c) + Be^{2x}(4a + 2b + c) = 0$$

So we need $a + b + c = 0$ and $4a + 2b + c = 0$.

Subtracting

$$3a + b = 0$$

Put $a = 1$.

$$b = -3$$

Substituting these values into either of the equations, we obtain $c = 2$.

The equation is $\dfrac{d^2y}{dx^2} - 3\dfrac{dy}{dx} + 2y = 0$.

EXERCISE 29D

1 For each of the following functions, find a differential equation obeyed by the function which does not involve the unknown constant.

a) $y = x^3 + A$ **b)** $y = Ae^{2x}$ **c)** $y = Ax$

2 For each of the following functions, find a differential equation obeyed by the function, which does not involve the unknown constants.

a) $y = Ae^x + Be^{-x}$ **b)** $y = Ae^{2x} + Be^{3x}$ **c)** $y = A + Be^x$

d) $y = A \cos x + B \sin x$ **e)** $y = e^x(A \cos x + B \sin x)$

3 Find differential equations, not involving the unknown constants, obeyed by the following functions. (Implicit differentiation may be required.)

a) $x^2 + y^2 = A$ **b)** $y^2 + yx + A = 0$ **c)** $e^y + e^x = A$

*4 Let $y = x^n$. By taking logarithms or otherwise, find a differential equation, not involving n, obeyed by the function.

Sketching families of solutions

Suppose we have found the general solution of a differential equations. Then for each value of the constant of integration, we will have a different particular solution. The collection of all the particular solutions is a **family of solutions**.

A family of solutions can be illustrated by drawing several of its members.

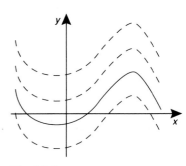

If the equation is of the form $\dfrac{dy}{dx} = f(x)$, then the general solution is $y = F(x) + A$. The graphs of these functions are vertical translations of each other.

Fig. 29.1

EXAMPLE 29.8

Find the general solution of the equation $\dfrac{dy}{dx} = \dfrac{2y}{x}$. Sketch the particular solutions which pass through these points.

a) $(1, 1)$ **b)** $(2, -8)$ **c)** $(2, 0)$

Solution

$$\frac{dy}{dx} = \frac{2y}{x}$$

Separate the variables.

$$\int \frac{dy}{y} = \int \frac{2\, dx}{x}$$

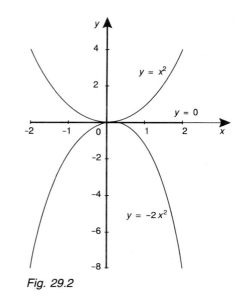

Integrate.

$$\ln y = 2 \ln x + c = \ln x^2 + c$$

Exponentiate both sides, and put $A = e^c$.

$$y = Ax^2$$

The general solution is $y = Ax^2$.

If the curve goes through $(1, 1)$, then $A = 1$.

If the curve goes through $(2, -8)$, then $A = -2$.

If the curve goes through $(2, 0)$, then $A = 0$.

These solutions are shown in Fig. 29.2.

Fig. 29.2

EXERCISE 29E

1 Find the general solution of $\dfrac{dy}{dx} = 5y$. Find the solutions which pass through

a) $(0, 5)$ **b)** $(0, -3)$ **c)** $(1, 4)$.

Sketch the graphs of these solutions.

2 Find the general solution of the equation $\dfrac{dy}{dx} = 4x\sqrt{y}$. Find and sketch the particular solutions which go through these points.

a) $(1, 4)$ **b)** $(4, 1)$ **c)** $(3, 81)$

3 Find the general solution of the equation $\dfrac{dy}{dx} = \sqrt{1 - y^2}$. Find and sketch the particular solutions which go through these points.

a) $(0, 0)$ **b)** $\left(\dfrac{\pi}{2}, 0\right)$ **c)** $(0, 1)$.

4 Find the general solution of the equation $x\dfrac{dy}{dx} = y(1 - x)$. Find an expression for the maximum value of y.

*5 Show that the equation $\left(\dfrac{dy}{dx}\right)^2 + \dfrac{dy}{dx}(e^{-x} - e^x) - 1 = 0$ can be factorised. Hence find the general solution of this equation. Show that when the graphs of two solutions cross, they do so at right angles.

29.5 Numerical solution of differential equations

In many cases, it is not possible to solve a differential equation in terms of standard functions, but there are numerical techniques which we can use to find approximations to the solution.

Suppose water is leaking slowly from a tank, at a rate which depends on the depth of water in the tank. We can find the approximate amount that leaks out in the first minute, by ignoring the change in depth over this period. We can then recalculate the new depth, and find the approximate amount that leaks out in the second minute, and so on – we can find the approximate leakage over any period.

The accuracy of this method can be improved by taking smaller time intervals, say every 15 seconds instead of every minute.

Step-by-step method

The step-by-step method of numerically solving a differential equation follows the procedure above.

Take the differential equation $\dfrac{dy}{dx} = f(x, y)$. Let us suppose we know one particular pair of values, say that $y = k$ when $x = a$. We might want to find the value of y when $x = b$.

Split the interval between a and b into n equal subintervals. Let the initial values of x and y be x_0 and y_0.

$$x_0 = a \qquad y_0 = k$$

Letting δx be the width of each subinterval, i.e.

$$\delta x = \frac{b - a}{n}$$

we have

$$\delta y \approx \frac{dy}{dx} \delta x = f(x_0,\ y_0)\delta x$$

This gives us the next value of x and y.

$$x_1 = x_0 + \delta x \qquad y_1 = y_0 + \delta y$$

This process is repeated n times, until we reach $x_n = b$.

The process is illustrated in Fig. 29.3. It shows curves of functions which are solutions of the differential equation, starting with the one which goes through (a, k). At each stage we take a tangent to one of the curves.

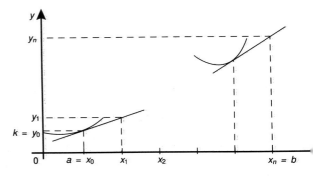

Fig. 29.3

If the solution curve is concave, then we will underestimate the solution, as shown. If it is convex, we will overestimate it.

The whole process can be made more accurate by taking more subintervals.

This operation involves a lot of numerical work, for which a computer is ideal. See page 506 for an investigation using this method.

EXAMPLE 29.9

y is a function of x, for which $\dfrac{dy}{dx} = 2x + y^2$. When $x = 1$, $y = 2$.

Find y when $x = 1.2$, using four subintervals.

Solution

Here $x_0 = 1$ and $y_0 = 2$.

$$\delta x = \frac{0.2}{4} = 0.05$$

Apply the process.

$$\delta y \approx (2 \times 1 + 2^2)\delta x = 6\,\delta x = 6 \times 0.05 = 0.3$$

Hence $y_1 = y_0 + \delta y = 2 + 0.3 = 2.3$.

The next value of x is $x_1 = 1 + \delta x = 1.05$.

Continue the process.

$$\delta y \approx (2 \times 1.05 + 2.3^2) \times 0.05 = 0.3695$$

Hence $y_2 = 2.3 + \delta y = 2.6695$.

$$x_2 = 1.1$$

$$y_3 = 2.6695 + (2 \times 1.1 + 2.6695^2) \times 0.05 = 3.1358$$

$$x_3 = 1.15$$

$$y_4 = 3.1358 + (2 \times 1.15 + 3.1358^2) \times 0.05 = 3.7425$$

$$x_4 = 1.2$$

The approximate value of y when x is 1.2 is 3.7425.

EXERCISE 29F

1 In the following, find the value of y by numerical methods.

a) $\dfrac{dy}{dx} = \sqrt{x + y}$ and $y = 1$ when $x = 0$. Find y when $x = 0.4$, using four subintervals.

b) $\dfrac{dy}{dx} = 3x + y^2$ and $y = 1$ when $x = 1$. Find y when $x = 1.5$, using five subintervals.

c) $\dfrac{dy}{dx} = \sqrt{x^2 + y^2}$ and $y = 1$ when $x = 1$. Find y when $x = 1.1$, using four subintervals.

d) $\dfrac{dy}{dx} = \sqrt{x^2 + 2y^2}$ and $y = 3$ when $x = 1$. Find y when $x = 1.5$, using five subintervals.

2 Suppose $\dfrac{dy}{dx} = y(x + 1)$ and that $y = 1$ when $x = 0$. Find y when $x = 0.5$

  a) exactly, by separating the variables

  b) numerically, using five subintervals.

  What is the percentage error in the numerical method?

3 A tank is a cube of side 1 m. Water is leaking from the tank at a rate equal to $0.01h$ m^3 per minute, where h m is the depth of water in the tank. Initially the depth is 1 m. Form a differential equation in h. Find the depth after 5 minutes

  a) exactly

  b) numerically, using 1 minute intervals.

4 The population P of a colony is growing at a rate equal to $\sqrt{50t + 0.1P}$, where t is time in years after the founding of the colony. The initial population was 10 000. Estimate the population after four years.

LONGER EXERCISE

Simple harmonic motion

If a body moves so that its acceleration (the second derivative of its displacement) is equal to a negative number times its displacement, then it performs **simple harmonic motion**.

This situation occurs frequently in science. Examples of simple harmonic motion occur with weights at the end of springs, objects bobbing up and down in the water and so on.

Let the displacement be y. The differential equation describing simple harmonic motion is

$$\frac{d^2 y}{dt^2} = -\omega^2 y$$

1 Let $\dfrac{dy}{dt} = v$. Write $\dfrac{d^2 y}{dt^2}$ as $\dfrac{dv}{dt} = \dfrac{dy}{dt} \times \dfrac{dv}{dy} = v\dfrac{dv}{dy}$.

  Integrate the equation to find v in terms of y.

2 Integrate a second time to find y in terms of t.

3 Suppose the motion is started with the body at rest, displaced by a, i.e. $\dfrac{dy}{dt} = 0$ and $y = a$. Find the particular solution corresponding to these conditions.

4 Suppose the body is started with $y = a$ and $\dfrac{dy}{dt} = b$. Find the particular solution corresponding to these conditions.

EXAMINATION QUESTIONS

1 In a chemical reaction, hydrogen peroxide is converted into water and oxygen. At time t after the start of the reaction, the quantity of hydrogen peroxide that has **not** been converted is h and the rate at which h is decreasing is proportional to h. Write down a differential equation involving h and t. Given that $h = H$ initially, show that

$$\ln \frac{h}{H} = -kt$$

where k is a positive constant.

In an experiment, the time taken for the hydrogen peroxide to be reduced to half of its original quantity was 3 minutes. Find, to the nearest minute, the time that would be required to reduce the hydrogen peroxide to one tenth of its original quantity.

Express h in terms of H and t, and sketch a graph showing how h varies with t.

JMB 1990

2 Find the solution of the differential equation

$$\frac{dy}{dx} = (1 + x^2)(1 + y^2)$$

which satisfies $y = 1$ when $x = 0$. Give your answer in the form $y = f(x)$.

AEB 1992

3 A curve has the equation $y = (2x + 1)e^{-2x}$.

a) Find $\dfrac{dy}{dx}$ and show that $\dfrac{d^2y}{dx^2} + 4\dfrac{dy}{dx} + 4y = 0$.

b) Calculate the coordinates of the turning point of the curve and determine its nature.

O 1992

4 You are given that $\dfrac{dy}{dx} = \dfrac{y^2}{5 - xy}$ and that $y = 1$ when $x = 0$. The value of y when $x = 0.5$ is required.

The table shows the beginning of a simple step-by-step method of finding the value using steps of 0.25. Complete the calculation, keeping three places of decimals in your working.

x	y	dy/dx	δx	δy
0	1	0.2	0.25	0.05
0.25	1.05			
0.5				

State, with a reason, whether the correct value is greater or less than the value you have calculated.

SMP 1988

Summary and key points

1 If a situation involves rates of change, it can be modelled by a differential equation.

2 If the equation can be written as $\dfrac{dy}{dx} = f(x)g(y)$, then the variables can be separated and both sides integrated.

3 If we know a particular pair of values of x and y, then a particular solution of the equation can be found.

4 The solutions of a differential equation form a family of related curves.

5 Numerical methods can be used to find approximate solutions to differential equations.

Computer investigations III

1 Statistical tables

In Chapter 23 we provided numerical methods, the trapezium rule and Simpson's rule, for evaluating integrals which could not be done exactly. There was a great deal of tiresome calculation involved. A computer can do all the work for us.

Numerical integration is necessary in many fields. In particular, there are many statistical functions which cannot be integrated exactly. The most important statistical function is that of the **normal distribution**, which has the formula

$$\phi(x) = \frac{1}{\sqrt{2\pi}} e^{-\frac{1}{2}x^2}$$

At the end of a statistics textbook you will find tables of the integral of this function.

$$\Phi(x) = \int_{-\infty}^{x} \frac{1}{\sqrt{2\pi}} e^{-\frac{1}{2}x^2} \, dx$$

Here we shall use a computer spreadsheet to evaluate this integral.

We shall use Simpson's rule with ten intervals, to calculate the integral from 0 to 2.

Use A1 to hold the value of x. Enter 2 in A1. Each interval will be a tenth of this, so in A2 enter +A1/10.

The intervals will start at 0 and increase in steps of $\frac{x}{10}$. Enter 0 in A4, and +A4 + A$2 in A5. The $ sign is put in as the formula will be copied down to A14. You should see the values 0, 0.1, 0.2 up to 2 appear in the cells A4 to A14.

Use the B column to enter the value of the function. In B4 enter the formula

$$@EXP(-A4\char`^2/2)/@SQRT(2*@PI)$$

You should see the value 0.398942 appear. Copy this formula down to B14.

Now we apply Simpson's rule. Use C4 for the odd-numbered terms, and C5 for the even-numbered. So enter in these cells

+B5 + B7 + B9 + B11 + B13

+B6 + B8 + B10 + B12

In C6 put the following.

4*C4 + 2*C5 + B4 + B14

This is now multiplied by $\frac{h}{3}$. In C16 enter +A2/3*C6.

This is the integral from 0 to x. You should see the value 0.477248.

To change it to $\Phi(x)$, add 0.5 to this value. You can do this by amending the formula in C16.

Find a table of $\Phi(x)$ in a statistics book. Check that this answer is correct. Try some other values.

How could you amend the spreadsheet to obtain other integrals?

How could you amend the spreadsheet so that it provides the whole $\Phi(x)$ table, instead of just one value?

2 Differential equations

The 'step-by-step' method of solving differential equations was introduced in Chapter 29. It enables us to solve the following sort of equation

$$\frac{dy}{dx} = f(x,\ y)$$

The method involves going from one x-value to another in small steps. If you do it by hand it is very tedious, and unless you take very small steps it is inaccurate. A computer can do all the hard work for you.

If you are a programmer, you could write a program to apply the method. Here we describe how to set up a spreadsheet.

Suppose we are finding the particular solution of the equation $\frac{dy}{dx} = \sqrt{x+3y}$, for which $y = 1$ when $x = 0$. We want the value of y when $x = 2$.

We shall put in the initial x-value, the final x-value, the number of steps and the final y-value. We shall also use the step length. Put labels in the first row: in A1, B1, C1, D1 and E1 of

left value   right value   number   initial y step

For the moment enter 0 in A2, 2 in B2, 4 in C2, 1 in D2. In E2 enter the formula (B2-A2)/C2. 0.5 will appear in E2.

This means that we are starting at the point (0, 1), and finding the value of y at $x = 2$, by means of four intervals of length 0.5.

In the A, B and C columns we shall have the x-values, the y-values and the function values. Enter labels of 'x-value', 'y-value' and 'function' in A4, B4 and C4 respectively.

In A5 we want the initial x-value, so enter +A2. In B5 we want the initial y-value, so enter +D2. In C5 enter the function at the initial values

@SQRT(A5 + 3*B5)

The x-values are to go up in steps of length E2. In A6 enter +A5 + E$2. Copy this down the A column as far as you like.

The y-value increase is equal to $\sqrt{x + 3y}$ multiplied by the step length. So in B6 enter +B5 + C5*E$2. Copy this down the B column.

We already have the formula for f(x, y) in C5. Copy this down the C column.

We can now read off the answer. When the x-value is 2, the y-value is 6.681496.

The accuracy can be greatly improved by taking smaller steps. Alter the number of steps, and find the new value of y.

This spreadsheet can quickly be modified for other starting values, or for other differential equations. Test it on some of the equations of Chapter 29.

3 Random numbers

Computers and scientific calculators have a facility for producing **random numbers**. This is used in simulating probability.

The actions of a computer are determined by the program it is obeying. The numbers produced are only **pseudo** random: they do follow a pattern, but one so complicated that it cannot be easily predicted. There are several ways of producing these pseudo random numbers, and here we shall look at some of them.

1 Modulus method
Start with any two-digit number (called the **seed** of the sequence). Multiply by 13, and discard the multiples of 100, to obtain another two-digit number. Repeat.

If this is done on a spreadsheet, the business of discarding the multiples of 100 can be done by the @MOD function. Enter the seed, say 23, in A1. In A2 enter the formula

@MOD(13*A1,100)

and copy this formula down as far as you please.

Investigate this sequence. Do all two-digit numbers appear in the sequence? Try different seeds, by changing the entry in A1. How long is it before the sequence repeats itself? How could the method be improved to give longer, less predictable sequences?

2 Mid-square method
Start with a seed of a four digit number. Square it, then remove the multiples of 1 000 000 and the last two digits. We will be left with the middle four digits. Repeat.

For example, suppose we start with 5134. Square this, obtaining 26 357 956. Removing the multiples of 1 000 000 and the last two digits leaves 3579.

A spreadsheet formula to perform this is

@INT(@MOD(A1^2,1000000)/100)

Try this sequence with different seeds. Does it ever go wrong?

3 Mid-product method
You may have found that sequences using the Mid-square method often degenerate into a sequence which repeats itself very frequently. The mid-product method is superior. Instead of taking the square of the previous number, take the product of the two previous numbers. We need two four-digit seeds. The spreadsheet formula is

@INT(@MOD(A1*A2,1000000)/100)

Try this method with various seeds. Does it degenerate or repeat itself?

4 Simulation
With any of the random number sequences, you can simulate a simple game which involves the rolling of a die. If the random number in A1 has two digits (as in the modulus method) the following function will give one of the integers between 1 and 6.

@INT(A1*6/100) + 1

Investigate the sequence produced by this method. Is the 'die' fair, or is it biased towards one of the numbers?

Consolidation section G

Chapter 27

1 Find the domain and range of the following functions.

    a) $f(x) = x^2 - 4x - 1$ **b)** $f(x) = \sqrt{2x - 3}$ **c)** $\sec x$ **d)** $\operatorname{cosec} x$

2 The domain of $f(x) = x^2 + 3$ is restricted to $x \geq 0$. Find the range of the function.

3 Which of the following functions are one to one? Show by proof or example.

    a) $f(x) = \sin x$ **b)** $f(x) = 1 - 3x$ **c)** $f(x) = \sin x$ for $0 \leq x \leq \dfrac{\pi}{2}$

4 Let f: $x \to 1 + 2x$ and g: $x \to 1 - 2x$. Find the following.

    a) $fg(2)$ **b)** $gf(2)$ **c)** $fg(x)$ **d)** $gf(x)$ **e)** $ff(x)$ **f)** $f^{-1}(x)$

5 Which of the following are odd and which are even? Give the periods of those which are periodic.

    a) $y = 2x + 3$ **b)** $y = x^4 + 3$ **c)** $y = \sin x + \sin 2x$

Chapter 28

6 The area of an equilateral triangle is $\dfrac{\sqrt{3}}{4} x^2$ cm^2, where x cm is the length of a side. If the area is increasing at 2 cm^2 per second find the rate of increase of the side.

7 Find the gradient of the curve $x^2 - 2xy - 4y^2 = 4$ at the point $(-2, 1)$.

8 Find the equation of the tangent to the curve of Question 7 at the point given.

9 Find the points on the curve of Question 7 at which the tangent is parallel to the y-axis.

10 x and y are given in terms of a parameter t by $x = 1 + t^2$, $y = \sqrt{t}$.

    a) Find $\dfrac{dy}{dx}$ in terms of t.

    b) Find the equation of the tangent to the curve at the point where $t = 4$.

c) Find the Cartesian equation of the curve.

d) Sketch the curve, for values of t from 0 to 4.

11 Consider the equilateral triangle of Question 6. The side is 10 cm. Use the method of small changes to find the increase in side necessary to increase the area by 0.2 cm^2.

Chapter 29

12 A rumour is spreading through a population of size P. The rate of spread is proportional to the product of the number N of people who have heard the rumour with the number of those who have not. Model this situation in terms of a differential equation.

13 Find the general solution of the equation $x^2 \dfrac{dy}{dx} = y + 1$.

14 Find the particular solution of the equation of Question 13, for which $y = 3$ when $x = 1$.

15 Weed is spreading over a lawn at a rate proportional to the area A m^2 already covered by the weed. Initially the area is 1 m^2 and after 10 weeks the area is 5 m^2. Find an equation giving A after t weeks have passed.

16 Find the general solution of the equation $\dfrac{dy}{dx} = y \cot x$. Find and sketch the particular solutions for which

a) $y = 1$ for $x = \dfrac{\pi}{2}$ b) $y = 2$ for $x = \dfrac{3\pi}{2}$ c) $y = 0$ for $x = \dfrac{\pi}{2}$

17 $\dfrac{dy}{dx} = \sqrt{x + 4y}$ and when $x = 0$, $y = 1$. Find y when $x = 0.5$, using a step-by-step method with five intervals.

MIXED EXERCISE

1 A spherical mothball loses weight at a rate which is proportional to its surface area. Show that the rate of decrease of its radius is constant. If the weight is halved after one year, find the time taken for the mothball to disappear completely.

2 Find the period of $f(x) = \sin 2x \sin 3x$.

3 A car depreciates at a rate proportional to its value. It cost £5000 when new, and was reduced to £2000 after three years. When will it be worth £1000?

4 Solve the equation $\dfrac{d^2y}{dx^2} = 3\dfrac{dy}{dx}$. Find a particular solution which is bounded as x tends to infinity.

5 Show that the function $\dfrac{1}{e^x - 1} + \dfrac{1}{2}$ is odd.

*6 If f(x) is even, show that f'(x) is odd.

7 A spherical fungus grows at a rate proportional to its surface area. At 1 o'clock it was 1 mm in radius, and at 2 o'clock it was 1 cm in radius. When was it born?

8 A tank containing liquid is replenished at 1 litre per minute. Liquid evaporates at a rate kV litres per minute, where V litres is the volume of the liquid, and leaks out at a rate kV^2 litres per minute. Write down a differential equation for V in terms of time t.

The equilibrium state is $V = 20$. If the tank is initially empty, find the time taken for it to fill up to 10 litres.

9 Let $f(x) = \ln\left(\dfrac{\frac{1}{2} + x}{\frac{1}{2} - x}\right)$ for $-\frac{1}{2} < x < \frac{1}{2}$. Find the range of f(x). Show that f(x) is odd. Sketch the graph of $y = f(x)$.

10 A tape recorder uses tape which is 0.001 cm thick. Tape is unwound from a spool at 2 cm per second. Find the rate of decrease of the radius of the spool when it is 5 cm.

11 Sand is being poured onto horizontal ground at a rate of 0.2 m³ per second. The sand forms a cone with sides that slope at 5° to the horizontal. Find the rate of increase of the radius of the cone after 200 seconds.

12 On the equator, at the equinox, the Sun will pass directly overhead. Find the rate of increase of the shadow of a man who is 6 foot tall at 5 p.m. on this day.

13 Let $f(x) = x^2 - 2x$ for $0 \le x$. Find the range of x.

14 a) Suppose $f'(x) > 0$ for all x. Show by a diagram that f(x) is one to one.

b) Suppose $f'(x)$ exists for all x and f(x) is one to one. Is it necessarily true that $f'(x) > 0$ for all x or that $f'(x) < 0$ for all x?

LONGER EXERCISE

The acceleration of a particle is given by $\dfrac{dv}{dt} = \dfrac{d^2s}{dt^2}$.

1 Use the chain rule to show that the acceleration can be written as $v\dfrac{dv}{ds}$.

2 (Air resistance) Suppose a body is thrown upwards at initial speed u. There is air resistance, so that its equation of motion is

$$\frac{dv}{dt} = -10 - v$$

a) Find the velocity at time t. Hence find the time when it reaches its greatest height.

b) Use the $v\dfrac{dv}{ds}$ form of the acceleration to find the distance in terms of the velocity. Hence find the greatest height.

3 (Escape from planet Earth) At a distance r m from the centre of the Earth, the acceleration due to gravity is $10\left(\dfrac{R}{r}\right)^2$, where the radius of the Earth R m is $6\,400\,000$ m. (Here we ignore air resistance.)

Suppose an object is fired upwards from the surface of the Earth. Show that r obeys the equation

$$\frac{d^2r}{dt^2} = -10\left(\frac{R}{r}\right)^2$$

Use the $v\dfrac{dv}{ds}$ form of the acceleration to find v in terms of r. Find a condition on the initial speed u for the object to escape completely from the Earth's gravitational field.

EXAMINATION QUESTIONS

1 The functions f and g are defined on the domain $x \geq 0$ by

$$f(x) = x^2 \text{ and } g(x) = 1 - \frac{1}{4}x^2$$

Describe three separate transformations in detail and the order in which they should be applied, whereby the graph of $y = g(x)$ may be obtained from the graph of $y = f(x)$.

Sketch, on the same diagram, the graphs of $y = f(x)$ and $y = g(x)$.

Describe also a single transformation whereby the graph of $y = g^{-1}(x)$ may be obtained from the graph of $y = g(x)$.

Sketch, on the same diagram, the graphs of $y = g(x)$ and $y = g^{-1}(x)$.

State the domain and range of g^{-1}.

Show that the x-coordinate of the point of intersection of the graphs of $y = g(x)$ and $y = g^{-1}(x)$ is given by

$$x^2 + 4x - 4 = 0$$

and find this value correct to two decimal places.

JMB 1990

2 A family of curves is defined by the differential equation

$$\frac{dy}{dx} = (1 - y)^2$$

(i) The **tangent field** is sketched by drawing in short 'needles' which indicate the gradient of the curves at points with integer coefficients. Part of the tangent field is shown on the right.

Copy and complete the sketch of the tangent field for $0 \leq x \leq 4$ and $-2 \leq y \leq 4$.

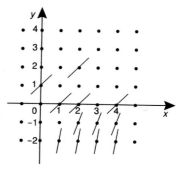

(ii) By solving the differential equation, show that the equation of the curve belonging to this family which passes through $\left(0, \dfrac{2}{3}\right)$ is $y = 1 - \dfrac{1}{x+3}$.

Sketch this curve for $x \geq 0$ on another diagram.

(iii) Another curve of the family passes through $(0, 2)$.

Solve the differential equation with this condition and sketch the curve for $x \geq 0$ on a new diagram.

(iv) A member of the family of curves passes through the point $(0, a)$.

Deduce that, when $x \geq 0$, if $a < 1$ then $a \leq y < 1$; and similarly that, if $a > 1$, then y can take all values except those in the range $1 < y \leq a$.

MEI 1992

3 The graph shows, for some positive values of x, the solution of the differential equation

$$\frac{\mathrm{d}y}{\mathrm{d}x} = (1 + y)\cos x$$

for which $y = -0.5$ when $x = 0$; x is measured in radians.

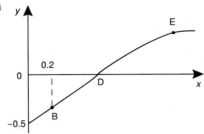

(i) Use a simple step-by-step calculation, with step length $\delta x = 0.1$, to find an approximate value for the y-coordinate of B, where $x = 0.2$. [Work to 3 decimal places.]

(ii) Rewrite the differential equation with the variables separated, and integrate it. Deduce that the graph has equation $y = \frac{1}{2}(e^{\sin x} - 2)$

Hence find the true value for the y-coordinate of B, correct to 3 decimal places.

(iii) Find the coordinates of the point D where the graph first cuts the x-axis.

(iv) By considering how $\sin x$ varies, **write down** the coordinates of the point E where the function first takes its greatest value.

SMP 1989

4 A curve has parametric equations $x = 1 + \sqrt{32}\cos\theta$, $y = 5 + \sqrt{32}\sin\theta$, $0 \leq \theta < 2\pi$.

Show that the tangent to the curve at the point with parameter θ is given by

$$(y - 5)\sin\theta + (x - 1)\cos\theta = \sqrt{32}$$

Find the two values of θ such that this tangent passes through the point A$(1, -3)$. Hence, or otherwise, find the equations of the two tangents to the curve from the point A.

L 1988

Mock examinations

EXAMINATION 1

1 Illustrate the following inequalities on a graph, shading out the regions which do **not** satisfy them.

$$y - 3x \leq 0 \qquad y + 3x \leq 6 \qquad 3y + x \geq 0$$

Find the coordinates of the vertices of the region defined by the inequalities. Find the point in the region which is furthest from the origin, and give its distance from the origin.

2 Functions f and g are defined by

$$f : x \to x^2 \qquad \text{and} \qquad g : x \to x + 1$$

Find the ranges of f and g.

Find, using the same functional notation as above,

a) fg **b)** gf **c)** ff.

Which of f or g has an inverse? Justify your answer with proof or example.

3 A, B, C and D are points with position vectors

$$\mathbf{i} + 2\mathbf{j} \qquad 3\mathbf{i} \qquad 4\mathbf{i} + 5\mathbf{j} \qquad 7\mathbf{i} + 4\mathbf{j}$$

respectively.

a) Show that AB is perpendicular to AC.

b) Show that AC and BD are parallel.

c) Find the angle between BC and BD.

d) AD and BC meet at X. Find the position vector of X.

(No marks will be given for solutions obtained by scale drawing.)

4 The circle shown has centre O and radius R. A point A is taken on the circumference, and an arc is drawn with centre A, cutting the original circle at B and C. The angle $C\hat{A}B$ is 2θ. Find

a) the length of the chord AC

b) the length of the chord CB

c) the area of the sector ACB

d) the area enclosed between the two arcs CB.

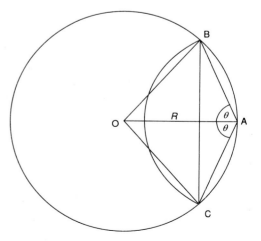

If the area found in **d)** is half the area of the full circle, show that θ obeys the equation

$$\pi + 4\theta \cos 2\theta - 2 \sin 2\theta = 0$$

5 At time $t = 0$ a body leaves the origin with velocity $v = 5$. Its acceleration is given by $\dfrac{dv}{dt} = -e^{-0.1t}$. Find an expression for the velocity after t seconds.

If the displacement from the origin is s, then $v = \dfrac{ds}{dt}$. Show that $s = 100 - 100e^{-0.1t} - 5t$.

Find the greatest distance of the body from the origin.

By Newton-Raphson iteration or otherwise, find correct to two decimal places the time when it returns to the origin.

6 Use the fact that the derivatives of $\sin x$ and $\cos x$ are $\cos x$ and $-\sin x$ respectively to show that the derivative of $\tan x$ is $\sec^2 x$.

Hence or otherwise find the following.

a) $\int x \sec^2 x \, dx$ **b)** $\int \tan^2 x \, dx$ **c)** $\displaystyle\int \frac{1}{\cos 2x + 1} \, dx$

7 A regular octahedron (eight-sided solid) has its vertices at $(\pm 1, 0, 0)$, $(0, \pm 1, 0)$, $(0, 0, \pm 1)$. Sketch the octahedron.

a) Find the distance between adjacent vertices.

b) Find the surface area and the volume of the octahedron.

c) Find the surface area and volume of a regular octahedron for which the distance between adjacent vertices is 1.

d) The vector $\mathbf{i} + \mathbf{j} + \mathbf{k}$ is perpendicular to the face containing $(1, 0, 0)$, $(0, 1, 0)$ and $(0, 0, 1)$. Write down a vector perpendicular to the face containing $(1, 0, 0)$, $(0, 1, 0)$ and $(0, 0, -1)$. Hence or otherwise find the angle between adjacent faces of the octahedron.

8 Find the binomial expansion of $(1 - x)^{-2}$.

A game is played with a six-sided die as follows. The player rolls the die until a six appears. If the six first appears on the first roll, the bank will pay the player £1. If the first six is on the second roll, the bank will pay £2. In general, if the first six is on the nth roll, the bank will pay £n.

Find the probabilities that the player will be paid

a) £1   **b)** £2   **c)** £n.

The **expected gain** of the player will be the sum of all the payments, multiplied by their appropriate probabilities. Write down a series for the expected gain of the player.

If the game is fair, the player will pay an amount equal to the expected gain. Use the binomial expansion at the beginning of this question to find this amount.

9 A territory will support a maximum population of P_0. Let the ratio of the population P to the maximum population be $\dfrac{P}{P_0} = p$.

The rate of change of this ratio p is proportional to the product of p and the difference between p and 1. Write down a differential equation in p.

Show that the growth of population is greatest when P is $\frac{1}{2}P_0$.

The population is initially $\frac{1}{4}P_0$, and reaches $\frac{1}{2}P_0$ after 20 years. When will it reach $\frac{7}{8}P_0$?

EXAMINATION 2

1 a) Show that, for all x and y, $x^2 + y^2 - 2x + 4y + 6$ is greater than 0.

b) Solve the inequality $\dfrac{1}{3+x} > 4$.

2 A balloon consists of a cylinder with hemispheres at both ends. The height of the cylinder is equal to the diameter of the hemisphere.

Letting the height of the cylinder be x cm, show that the volume V cm^3 of the balloon is given by:

$$V = \frac{5\pi}{12}x^3$$

Find an expression for the surface area of the balloon.

Air is pumped in a rate of 10 cm^3 s^{-1}. When the height is 20 cm, find the rate of increase of the height and of the surface area.

3 Let $f(x) = \dfrac{\cos x}{2 - \sin x}$. Find the first two stationary points of the curve $y = f(x)$, describing their nature.

State the range of $y = f(x)$ and sketch the graph of the function.

Evaluate the area under the curve $y = f(x)$ from $x = 0$ to $x = \dfrac{\pi}{2}$.

4 If the price of an item is p the quantity sold per unit of time will be $f(p) = 120 - 7p - p^2$.

How much will be sold if the item is free? At what price will none be sold?

Show that the price at which revenue is maximised is approximately 4.4.

To promote the item, the manufacturer starts selling it at a price of $p = 2$, then gradually increases the price to $p = 4.4$. If the rate of increase of price is $\dfrac{\mathrm{d}p}{\mathrm{d}t} = 0.1$, express as an integral the total revenue from the item during the promotion period. Evaluate this integral.

5 A body starts from the origin, and its velocity is given by the table below.

Time (seconds)	0	1	2	3	4	5	6
Velocity (m/s)	0	6	11	15	18	20	21

Use the trapezium rule to estimate the total distance travelled over these six seconds.

Do you think your answer is an under or an over estimate? Give reasons.

6 'Seppuku' motorcycles were manufactured only during 1989, and were then taken out of production. The original price in the UK on 1 January 1990 was £5000. Because of their scarcity value, since 1990 their price has been rising at 10% per annum. Find the price n years after 1990.

Marianne has set her heart on buying a 'Seppuku' motorcycle. On 1 January 1990 and on 1 January every year since, she has invested £500 at 5% compound interest. Find the total value of all her investments immediately after the nth investment.

If she can afford the motorcycle after n years, show that

$$2 \times 1.05^n - 1.1^n = 2$$

Will she ever be able to afford the motorcycle?

7 A function f(x) is odd and has period $4a$. The graph of $y = \text{f}(x)$ is shown for $0 \le x \le 2a$. Complete the graph for $-4a \le x \le 4a$.

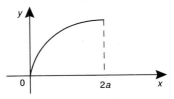

Sketch the graphs of **a)** f($2x$)   **b)** f($x + a$).

Is either **a)** or **b)** odd?

Is either of **a)** or **b)** periodic? If so, give the period.

What is the least k such that f($x + k$) is an even function?

8 Three vertical posts AB, CD and EF have heights 10 m, 14 m and 17 m respectively. They stand on level ground with their feet A, C and E at the vertices of an equilateral triangle of side 20 m. A triangular roof is supported with its vertices at B, D and F. Find the area of this roof.

9 a) Write as a fraction the recurring decimal $0.127\,272\,7\ldots$.

b) Assuming that $\sqrt{3}$ is irrational, prove that $4 - \sqrt{3}$ is irrational.

c) Find an irrational number between $\sqrt{2}$ and $\sqrt{3}$.

10 Twenty numbers in increasing order are

17, 19, 23, 27, 28, x, 33, 34, 38, 39, 40, y, 44, 45, 47, 48, 49, 51, 52, 59

Given that the numbers have mean 38.35 and variance 126.8275, find x and y.

Illustrate these numbers on a box and whisker diagram.

EXAMINATION 3

1 Differentiate the following.

a) $x^{\frac{3}{2}} e^x$ b) $\dfrac{x+7}{x^2 + 3x + 4}$ c) $\sqrt[3]{1 + \sin 2x}$

2 The volume of a cylinder of height h and base radius r is

$$V = \pi r^2 h$$

If r and h are subjected to small changes δr and δh respectively, find the corresponding change δV in the volume, ignoring products of these small changes.

Express the percentage change in the radius in terms of r and δr.

Find the percentage change in the volume in terms of the percentage changes in r and h.

3 Find the stationary points of the function $y = 9x + \dfrac{1}{x}$ and indicate their nature. Sketch the graph of the function.

Find the area of the region under the graph between $x = 1$ and $x = 2$.

Find the volume obtained when this region is rotated about the x-axis.

4 t seconds after a body has started moving, its displacement is s m and its velocity is v m s^{-1}, where $v = \dfrac{ds}{dt}$. Show that the acceleration $\dfrac{dv}{dt}$ can be written as $v\dfrac{dv}{ds}$.

A meteorite is moving through space at a speed of 1000 m s^{-1}. It enters an increasingly thick cloud of gas, so that when it has penetrated s metres, and is travelling at v m s^{-1}, its acceleration $\dfrac{dv}{dt}$ is $-0.000\,01sv$ m s^{-2}.

Find the distance it penetrates the cloud before coming to a halt.

5 Let $f(x) = \dfrac{3x}{(1+x)(1+2x^2)}$. Express $f(x)$ in partial fractions.

Express $f(x)$ as a series in ascending powers of x, up to the term in x^3. For what range of values of x is your expansion valid?

Evaluate $\displaystyle\int_0^2 f(x)\,dx$.

6 Certain packages are to be made with square cross-section. The girth (the perimeter of the cross-section) added to the length cannot be greater than 300 cm. Find the maximum possible volume of the package.

7 The variables x and y are given in terms of a parameter t by

$$x = 3t + 2, \quad y = \frac{1}{t}$$

Find $\dfrac{dy}{dx}$ in terms of t.

Find the tangent to the curve at the point where $t = 1$.

Find the Cartesian equation linking x and y. Sketch the curve of y against x.

Find the area under the curve from $t = 1$ to $t = 3$.

8 The radioactive substance carbon 14 decays so that the rate of decay of its mass is proportional to its mass. Express this fact as a differential equation in terms of the mass M.

The **half-life** of carbon 14, i.e. the time taken for half of it to decay, is 5600 years. If there is originally M_0 kg of the substance, how much will there be after t years?

A boat-shaped object is discovered on Mount Ararat. It is found that the carbon 14 in the wood of the object is 42% of the amount originally present. When was the object built?

9 A bird is 100 m due west of its nest. It can fly at 15 m s^{-1}, but there is a wind of 10 m s^{-1} blowing from the north east, i.e. on a bearing of 225°. Find the direction in which the bird should fly in order to reach its nest, and the time it will take to do so.

10 If a product is priced at £p per kilogram, the amount A (in hundreds of tonnes) that producers will supply of it is given by $A = p + e^{0.1p} - 3$, provided that this is positive. Find the supply when the price is £10.

The price is held fixed at £10, but the Government wishes to increase the supply by 20 tonnes. It can do so by subsidizing the producers at a certain rate per kilogram. Use the method of small changes to find the subsidy necessary.

By Newton-Raphson iteration or otherwise find the price, to the nearest 1p, at which the producers will start to supply the product.

11 The resistances of 100 electrical components are given in the table below.

Resistance in ohms	9–10	10–10.5	10.5–11	11–12	12–14
Frequency	11	17	24	35	13

a) Draw a histogram to illustrate these figures.

b) Estimate the mean resistance. What is the maximum error in your value for the mean?

c) A component is picked at random. Estimate the probability that the resistance of the component is less than 11.5 ohms. What is the greatest possible value for this probability?

Solutions

1 Algebra I

Exercise 1A page 6

1 a) 8   **b)** -7 **c)** $3\frac{1}{2}$

2 a) 19   **b)** 2

3 a) 7   **b)** $z^2 - z + 1$ **c)** $x^6 - x^3 + 1$

4 a) -8 **b)** 32   **c)** $2 - y - y^3$ **d)** $2 + x + x^3$

5 a) $2x^2 + 2$ **b)** $2x$

Exercise 1B page 7

1 a) $3x + 3y$ **b)** $-8x + 12y$ **c)** $x^2 + xy$

  d) $5x - y$ **e)** $-a - 7b$ **f)** $11x - 2y$

  g) $5p - 2q$ **h)** $x^2 + 2xy - y^2$ **i)** $2a^2 - 2ab + b^2$

  j) $2a^2 + 3ab + b^2$ **k)** $6x^2 + 5xy + y^2$ **l)** $p^2 - q^2$

  m) $4a^2 - 9ab - 9b^2$ **n)** $p^2 - p - 6$ **o)** $2q^2 - 3q - 9$

  p) $x^2 + 2xy + y^2$ **q)** $4a^2 + 4ab + b^2$ **r)** $4r^2 - 12rs + 9s^2$

2 a) $3x + 9$ **b)** $3x + 3y + 3$

3 a) $4x^2 + 4x - 1$ **b)** $x^2 + 4x + 2$ **c)** $x^2 + 2xy + y^2 + 2x + 2y - 1$

4 a) $-15 - 14x - 3x^2$ **b)** $-4 + 8x - 3x^2$ **c)** $1 - 2x + 2y - 3x^2 + 6xy - 3y^2$

Exercise 1C page 8

1 a) 2   **b)** 4   **c)** 1   **d)** 5   **e)** 3   **f)** 0

2 $x^2 - 2x - 2;\ 3x^2 + 4x - 4$

3 $-2 - 2x + 2x^2 + x^3;\ -4 - 4x - x^3$

4 $5x^5 + 3x^4 - x^3 - x + 4;\ -5x^5 + 3x^4 + x^3 - x - 8$

5 At most, the maximum of n and m.

Exercise 1D page 9

1 a) $2x^2 + 11x - 21$ **b)** $4x^3 - 15x^2 + 5x + 3$ **c)** $x^3 - 6x^2 - 15x - 8$

  d) $8x^3 + 26x^2 - 26x - 8$ **e)** $x^4 + x^2 + 1$ **f)** $x^5 - x$

  g) $x^2 + 8x + 16$ **h)** $4x^2 + 4x + 1$ **i)** $x^2 - 6x + 9$

  j) $4x^2 - 12x + 9$ **k)** $a^2x^2 - b^2$ **l)** $6a^2x^2 - 19abx + 15b^2$

  m) $a^2x^2 + 2abx + b^2$ **n)** $a^2x^2 - 2abx + b^2$ **o)** $x^3 + 3x^2 + 3x + 1$

2 $a = 5,\ b = -1$ **3** $n + m$ **4 a)** $x^2 - (a + b)x + ab$ **b)** 0

Exercise 1E page 11

1 a) $x^2 + 9x + 40$, 143   **b)** $x^2 - x + 2$, 5   **c)** $x^2 + x$, 1

   d) $x^2 + 2x + 8$, 13   **e)** $4x^2 + \frac{1}{2}x - \frac{7}{4}$, $\frac{3}{4}$ **f)** $4x^3 + 10x^2 + 13x + 27$, 51

   g) $8x^3 + 12x^2 + 18x + \frac{55}{2}$, $\frac{159}{2}$ **h)** $x^4 + x^3 + x^2 + x + 1$, 2

2 $a = 1$ **3** $25\frac{1}{2}$ **4** $a = -2$, $b = -11$

Exercise 1F page 12

1 $3x(1 + 3x)$ **2** $5a(1 + 3b)$ **3** $x^2(1 - x)$

4 $2xy(2 + x - 3y)$ **5** $3p(3 - q^2)$ **6** $z^2(1 + 2z + 3z^2)$

Exercise 1G page 13

1 $(x + 3)(x + 4)$ **2** $(x + 3)(x + 5)$ **3** $(x + 3)^2$

4 $(x - 10)(x - 3)$ **5** $(x - 3)(x - 4)$ **6** $(x - 4)^2$

7 $(x - 5)(x + 8)$ **8** $(x - 10)(x + 5)$ **9** $(x - 8)(x + 1)$

10 $(x - 2)(x + 2)$ **11** $(x - 4)(x + 4)$ **12** $x(x + 8)$

13 $x(x - 12)$

Exercise 1H page 14

1 $(x + 3)(2x + 1)$ **2** $(x + 2)(2x + 1)$ **3** $(x + 3)(6x + 1)$

4 $(2x - 1)(3x - 2)$ **5** $(2x - 1)(5x - 2)$ **6** $(x - 1)(2x - 3)$

7 $(x + 3)(2x - 5)$ **8** $(x - 1)(3x + 2)$ **9** $(x + 2)(3x - 5)$

10 $2(x - 2)(x + 2)$ **11** $3(x - 3)(x + 3)$ **12** $2x(x + 4)$

13 $3x(x + 4)$

Exercise 1I page 16

1 $\dfrac{x + 3}{x - 3}$ **2** $\dfrac{(2x + 1)(x - 3)}{(x - 2)(x + 1)}$ **3** $\dfrac{x - 4}{x - 5}$

4 $\dfrac{2(x - 3)(x - 5)}{(x + 1)(x - 2)}$ **5** $\dfrac{2(x + 1)(x - 3)}{(x + 2)(x - 1)}$ **6** $\dfrac{(2x - 3)(x - 1)}{3(x + 1)(x + 2)}$

7 $\dfrac{(x + 2)(x - 4)}{(x - 2)(x - 1)}$ **8** $\dfrac{(2x + 3)(x + 1)}{(x - 1)(x - 2)}$ **9** $\dfrac{(x - 5)}{(x - 4)}$

Exercise 1J page 17

1 $\dfrac{2x - 5}{(x - 2)(x - 3)}$ **2** $\dfrac{3x - 1}{(2x + 1)(x - 2)}$ **3** $\dfrac{8 - 6x}{(2x - 1)(x + 2)}$

4 $\dfrac{2x}{x - 1}$ **5** $\dfrac{4x - 5}{2(x - 2)}$ **6** $\dfrac{13x - 16}{(x + 2)(x - 5)}$

7 $\dfrac{x - 2}{(x - 1)(x - 3)}$ **8** $\dfrac{4}{(x + 1)(x - 2)}$ **9** $\dfrac{x + 2}{(x + 1)(x - 1)}$

10 $\dfrac{-1}{(x - 1)(x - 2)(x - 3)}$

11 a) $\dfrac{x+3}{x^2-4}$ **b)** $\dfrac{7}{(x-1)(x-2)}$ **c)** $\dfrac{2x+1}{(x^2-4)(x-1)}$

Exercise 1K page 18

1 a) -27 **b)** -13 **c)** 2   **d)** 5   **e)** 4   **f)** $\frac{29}{27}$

2 45   **3** -4 **4** $2\frac{1}{8}$ **5** $a=-5,\ b=-2$ **6** $a=0,\ b=4$

Exercise 1L page 20

1 b) $x-2$ is a factor

2 a) $(x-5)(x+1)(x+3)$ **b)** $(x-1)(x-2)(x+3)$ **c)** $(x+2)$

3 -5 **4** $a=2,\ b=-11$ **5** $a=-1,\ b=-4$

Exercise 1M page 21

1 $(x-2)(x+2)(x+3)$ **2** $(x-2)(x+1)(x+2)$ **3** $(x-4)(x-3)(x-2)$

4 $(x+1)(x-3)(x-4)$ **5** $(x-2)(x-1)^2$ **6** $(x+1)^2(x+3)$

7 $(x-1)(x+1)(2x-1)$ **8** $(x-1)(x+2)(3x+2)$ **9** $(x-2)(2x-1)(3x+1)$

Exercise 1N page 22

1 $y=5x$; 150   **2** 1.25 N   **3** $P=\dfrac{aQ}{b}$

4 $R=\dfrac{20}{S}$; 10; $\dfrac{2}{5}$ **5** 1.25 mins   **6** $P=\dfrac{ab}{Q}$

Exercise 1O page 24

1 $y=4x^2$; 900   **2** 1.25 g   **3** 250 g

4 112.5 N   **5** 1.097 seconds   **6** $Y=\dfrac{20}{X^2}$; $\dfrac{1}{2}$

7 $R=\dfrac{ba^2}{S^2}$ **8** 62.5 candelas   **9** $x\propto\dfrac{1}{y^3}$

Examination questions page 26

1 1

2 $\dfrac{2}{2x-1}$

3 $a=-1,\ b=4$

2 Quadratics

Exercise 2A page 29

1 a) $x=3$ or $x=5$ **b)** $x=3$ or $x=4$ **c)** $x=4$ or $x=-7$ **d)** $x=-5$ or $x=9$

e) $x=\frac{1}{2}$ or $x=2$ **f)** $x=-\frac{1}{3}$ or $x=-\frac{3}{2}$ **g)** $x=-5$ or $x=5$ **h)** $x=0$ or $x=5$

i) $x=-3$ or $x=3$

2 a) $x = 2$ or $x = 3$ **b)** $x = -2$ or $x = 10$ **c)** $x = -1$ or $x = \frac{7}{3}$ **d)** $x = 2$ or $x = 3$

e) $x = \frac{1}{3}$ or $x = 2$ **f)** $x = -1$ or $x = 2$

3 a) $x = 8$ m   **4** 7 m or 8 m   **5** 20 days

Exercise 2B page 30

1 $x = 1$, $x = 2$ or $x = -3$ **2** $x = 1$, $x = -1$ or $x = 6$ **3** $x = -2$, $x = -3$ or $x = -5$

4 $x = -2$, $x = 3$ or $x = 5$ **5** $x = \frac{1}{2}$, $x = -1$ or $x = -2$ **6** $x = \frac{2}{3}$, $x = -2$ or $x = 3$

Exercise 2C page 31

1 a) $(x + 1)^2 - 4$ **b)** $(x - 2)^2 - 2$ **c)** $(x - 4)^2 - 5$

  d) $(x + 2.5)^2 - 7.25$ **e)** $(x - 1.5)^2 - 9.25$ **f)** $(x + 0.5)^2 + 0.75$

2 a) $-4, -1$ **b)** $-2, 2$ **c)** $-5, 4$ **d)** $-7.25, -2.5$ **e)** $-9.25, 1.5$ **f)** $0.75, -0.5$

Exercise 2D page 33

1 a) $3[(x + 1)^2 - \frac{8}{3}]$ **b)** $3[(x - \frac{3}{2})^2 - 1\frac{7}{12}]$ **c)** $2[(x - 1)^2 - 4.5]$

  d) $2[(x + 0.75)^2 - 3\frac{1}{16}]$ **e)** $3[(x - \frac{1}{6})^2 + \frac{35}{36}]$ **f)** $5[(x - 0.2)^2 + 0.56]$

  g) $4 - (x + 1)^2$ **h)** $5.25 - (x + 0.5)^2$ **i)** $-4.75 - (x - 1.5)^2$

  j) $7 - 2(x + 1)^2$ **k)** $5.75 - 3(x - 0.5)^2$ **l)** $7.125 - 2(x + 0.25)^2$

2 a) -8 **b)** -4.75 **c)** -9 **d)** $-6\frac{1}{8}$ **e)** $\frac{35}{12}$ **f)** 2.8

  g) 4 **h)** 5.25 **i)** -4.75 **j)** 7 **k)** 5.75 **l)** 7.125

3 $a\left(x + \dfrac{b}{2a}\right)^2 + c - \dfrac{b^2}{4a}$ **4** 47 m

5 $50 - 2t$, $4t$; 44.7 m   **6** $x(200 - 2x)$; 5000 m^2

Exercise 2E page 34

1 a) $x = 1.45$ or $x = -3.45$ **b)** $x = -0.382$ or $x = -2.62$ **c)** $x = 0.541$ or $x = -5.54$

  d) $x = 0.129$ or $x = 3.87$ **e)** $x = 0.359$ or $x = -8.36$ **f)** $x = -1.23$ or $x = 1.90$

2 $100 - t$, t; 23.5 seconds and 76.5 seconds   **3** 4   **4** $\dfrac{-b \pm \sqrt{b^2 - 4ac}}{2a}$

Exercise 2F page 36

1 a) $x = -0.232$ or $x = -1.43$ **b)** $x = -1.32$ or $x = 5.32$ **c)** $x = 2.54$ or $x = -3.54$

  d) $x = 0.443$ or $x = -1.69$ **e)** $x = -0.459$ or $x = -6.54$ **f)** $x = 0.438$ or $x = 4.56$

2 1.86 seconds and 6.14 seconds   **5** -29 or 30   **7** $\frac{1}{2}p + \sqrt{(\frac{1}{2}p)^2 + q}$

8 30 or 10   **9** $x^2 - 3x + 1$; $x = 1$, $x = 0.382$ or $x = 2.62$

10 a) $x = 1$, $x = -1.14$ or $x = 6.14$ **b)** $x = 2$, $x = -1.79$ or $x = 2.79$

  c) $x = -1$, $x = 1.19$ or $x = -1.69$

Exercise 2G page 38

1 **a)** $x = 0.518$, $x = -0.518$, $x = 1.93$ or $x = -1.93$ **b)** $x = 0.766$ or $x = -1.64$
 c) $x = -1.27$ or $x = 1.77$ **d)** $x = 1$ or $x = 3$ **e)** $x = 0.702$ or $x = 51.3$ **f)** $x = 0$ or $x = 4$

2 **a)** $x = 1.73$ or $x = -1.73$ **b)** $x = -0.414$, $x = -0.732$, $x = 2.41$ or $x = 2.73$
 c) $x = 3$ or $x = -5$ **d)** $x = 0.408$, $x = -0.408$, $x = 0.577$ or $x = -0.577$
 e) $x = 64$ **f)** $x = 0.382$

3 $x = 1$, $x = 1.30$, $x = -2$ or $x = -2.30$ 4 $x = 0.382$ or $x = 2.62$
5 $x = -0.200$ or $x = -4.99$

Exercise 2H page 39

1 **a)** $x = -1.16$, $y = -1.72$ or $x = 5.16$, $y = 0.387$
 b) $x = 0.348$, $y = 1.70$ or $x = -1.15$, $y = -1.30$
 c) $x = 0.509$, $y = 1.24$ or $x = 3.49$, $y = -3.24$
 d) $x = \frac{10}{13}$, $y = 1\frac{6}{13}$ or $x = 2$, $y = -1$
 e) $x = \frac{1}{3}$, $y = 6$ or $x = 4$, $y = \frac{1}{2}$
 f) $x = 1.11$, $y = 4.33$ or $x = 4.43$, $y = -0.639$
 g) $x = 0.252$, $y = 27.7$ or $x = 27.7$, $y = 0.252$
 h) $x = 1.42$, $y = 1.58$ or $x = 4.58$, $y = -1.58$

2 $x = 1.88$, $y = 2.54$ or $x = -1.88$, $y = 2.54$ or $x = 3$, $y = 1$ or $x = -3$, $y = 1$

3 $x = 0.399$, $y = 1.80$ or $x = -1.19$, $y = -1.38$

Examination questions page 40

1 $x = 2$, $y = 3$ or $x = -2$, $y = -3$ or $x = \frac{3}{2}$, $y = 4$ or $x = \frac{-3}{2}$, $y = -4$

2 $10 - (2 + x)^2$; 10

3 (ii) $6x - 4$ (iii) $x = 1$, $x = -0.414$ or $x = 2.41$

3 Trigonometric functions

Exercise 3A page 44

1 **a)** 3.21 **b)** 21.6 **c)** 10.2 **d)** 1.62

2 **a)** 41.4° **b)** 54.7° **c)** 41.8° **d)** 39.4°

3 7.66 cm, 6.43 cm 4 45.6° 5 11.3° 6 7.71 ft 7 72.5°

8 **a)** 10.9 cm **b)** 3.08 cm **c)** 63.6° **d)** 48.6°

11 $\tan P = \dfrac{1}{\tan(90° - P)}$

Exercise 3B page 46

2 **a)** 0 **b)** 1 **c)** 1

Exercise 3C page 47

1 a) 0.675   **b)** 1.47   **c)** 1.05

2 a) 19.5°   **b)** 55.0°   **c)** 56.3°

3 $2, \dfrac{2}{\sqrt{3}}, \sqrt{3}; \sqrt{2}, \sqrt{2}, 1; \dfrac{2}{\sqrt{3}}, 2, \dfrac{1}{\sqrt{3}}$

4 cosec: 0°; sec: 90°; cot: 0°

Exercise 3D page 51

1 a) sin 60°   **b)** −cos 45°   **c)** −tan 82°

    d) −cot 30°   **e)** −sec 4°   **f)** cosec 20°

    g) −cos 70°   **h)** −tan 50°   **i)** −sin 10°

    j) sec 20°   **k)** −cosec 42°   **l)** cot 67°

    m) sin 10°   **n)** −cos 80°   **o)** tan 30°

    p) −sin 32°   **q)** sec 30°   **r)** −cot 70°

    s) sin 60°   **t)** tan 40°   **u)** −cos 60°

2 a) $\dfrac{1}{\sqrt{2}}$ **b)** $-\dfrac{\sqrt{3}}{2}$ **c)** $-\sqrt{3}$

    d) 2   **e)** $-\dfrac{1}{\sqrt{3}}$ **f)** −1

    g) $-\dfrac{1}{\sqrt{2}}$ **h)** $\dfrac{1}{\sqrt{3}}$ **i)** $-\dfrac{\sqrt{3}}{2}$

    j) $-\sqrt{2}$ **k)** −1   **l)** 2

3 4th; 323°   **4 a)** 143°   **b)** 225°   **5** 200°

6 a) 210°, 330°   **b)** 135°, 315°   **c)** 60°, 300°

    d) 19.5°, 160.5°   **e)** 104.5°, 255.5°   **f)** 73.3°, 253.3°

Exercise 3E page 52

2 a) $0° < x < 180°$ **b)** $90° < x < 270°$ **c)** $30° < x < 150°$

3 1.4; 45°   **4** −5; 325°   **5** four

Exercise 3F page 54

2 cot: 0°, 180°, 360°; cosec: 0°, 180°, 360°; sec: 90°, 270°

3 a) $90° < x < 180°$, $270° < x < 360°$ **b)** $45° < x < 90°$, $225° < x < 270°$

4 a) $45° < x < 180°$, $225° < x < 360°$ **b)** $0° < x < 90°$, $270° < x < 360°$

    c) $30° < x < 150°$, $180° < x < 360°$

Exercise 3G page 55

1 a) 180°   **b)** 120°   **c)** 720°

2 a) sin 18x   **b)** $\sin \frac{1}{2}x$ **c)** sin 1.2x

4 0.02 seconds   **5** $\frac{1}{260}$ seconds   **6** $A \sin 187\,200t$

Longer exercise page 56

The chord function

a) 1   **b)** 0.684   **c)** 1.53   **d)** 0.572

e) 0.593   **f)** 29.0°   **g)** 73.7°   **h)** crd $P = 2 \sin \frac{1}{2} P$; $\sin P = \frac{1}{2}$ crd $2P$

Tides

a) 12 hours   **b)** 3 a.m.; 26 ft   **c)** 9 a.m.; 14 ft

d) 1 a.m., 5 a.m.   **e)** draught $\leq$ 14 ft; draught $\geq$ 26 ft

f) between 0 and 7 a.m., after 11 a.m.

Examination questions page 56

1 $0° \leq x < 135°$ and $405° < x \leq 540°$

2 a) 15.6 cm²   **b)** 25.4 cm²

3 (i) 7 m   (ii) 13 m; 29.03

4 (i) 30°, 150°   (ii) 4.11 cm, 13.53 cm

4 Coordinate geometry

Exercise 4A page 59

1 a) $\sqrt{34}$ **b)** $\sqrt{50}$ **c)** $\sqrt{45}$ **d)** $\sqrt{50}$ **e)** $\sqrt{5}$ **f)** $\sqrt{50}$

     g) $\sqrt{170}$ **h)** $\sqrt{274}$ **i)** $\sqrt{(x-1)^2 + (y-1)^2}$ **j)** $\sqrt{(x+2)^2 + (y+3)^2}$

4 $\sqrt{20}, \sqrt{40}, \sqrt{20}$ **5** 5, 10, 15   **6** $\sqrt{90}, \sqrt{90}$; yes   **8** $2 + \sqrt{21}, 2 - \sqrt{21}$

9 a) (i) 8.55 km   (ii) 74.2 km   **b)** 060080

Exercise 4B page 61

1 a) $x^2 + y^2 - 4y - 5 = 0$ **b)** $x^2 + y^2 - 4x + 2y - 11 = 0$ **c)** $x^2 + y^2 + 8x + 2y + 16 = 0$

     d) $x^2 + y^2 - x + y - 3.5 = 0$

2 a) $(2, -1)$, $\sqrt{8}$ **b)** $(-4, -4)$, 6   **c)** $(1.5, -0.5)$, $\sqrt{3.5}$ **d)** $(0.5, 0.5)$, $\sqrt{0.5}$

3 $(0, 0)$, 3; $(3, 4)$, 2; 5 units   **4** $x^2 + y^2 - 4x - 4y + 6 = 0$; $(2, 2)$, $\sqrt{2}$

5 $x^2 + y^2 + 2y - 3 = 0$; $(0, -1)$, 2   **7** $k = 1 \pm 4\sqrt{2}$

Exercise 4C page 63

1 a) 4   **b)** $\frac{1}{2}$ **c)** -2 **d)** 1   **e)** $-\frac{1}{2}$ **f)** 2

3 $y = 7$ **4** no   **5** $x = 3$ **6** $x = -1$

Exercise 4D page 65

1 a) $y = 3x - 1$ **b)** $y = -2x + 5$ **c)** $2y = x + 7$ **d)** $3y = -x + 25$

     e) $3y = x + 2$ **f)** $y = 2x - 4$ **g)** $3y = x + 3$ **h)** $y = -4x + 10$

     i) $y = 4$ **j)** $x = 2$

2 a) 3   **b)** 1.5   **c)** 0.75   **d)** -3 **e)** $\frac{1}{3}$ **f)** -1.5

3 a) $(4,\ 5)$ **b)** $(3,\ 9)$ **c)** $(-0.4,\ 2.2)$

4 $(2,\ 3),\ (1,\ 1)$ and $(7,\ -2)$

6 a) $(-1,\ 3)$ **b)** $y = x + 1$ **c)** $(5,\ 6)$

7 $A(-2,\ 3),\ B(0,\ -3),\ C(2,\ -1)$

Exercise 4E page 67
1 (b) and **(c)** are parallel; **(a)** is perpendicular to **(b)** and **(c)**.

2 a) $y = 3x - 1$ **b)** $3y + x = 11$ **c)** $y = x - 2$ **d)** $2y = x$

3 2 and $-\frac{1}{2}$ **4** $(2,\ 3),\ \sqrt{5}$ **5** $x^2 + y^2 - (a+b)x + ab = 0$ **7** $(0,\ 6),\ (6,\ 2)$

Exercise 4F page 69

1 a) $(2.8,\ 2.4)$ **b)** $(9,\ 10)$ **c)** $\left(\dfrac{1}{1+x},\ \dfrac{1}{1+x}\right)$ **d)** $\left(\frac{5}{9}a,\ \frac{4}{9}b\right)$

2 a) $(2,\ 4)$ **b)** $(1,\ 2)$ **c)** $\left(\frac{1}{2}x,\ \frac{1}{2}y\right)$ **d)** $\left(\frac{1}{2}a,\ \frac{1}{2}b\right)$

3 $(2,\ 7);\ (3,\ 7)$ **4** $\left(\dfrac{y}{1+y},\ \dfrac{1}{(1+x)(1+y)}\right)$ **5** $1:2$

6 $\left(\frac{1}{2}(x_2 + x_3),\ \frac{1}{2}(y_2 + y_3)\right);\ \left(\frac{1}{3}(x_1 + x_2 + x_3),\ \frac{1}{3}(y_1 + y_2 + y_3)\right)$

Exercise 4G page 70
1 a) $63.4°$ **b)** $74.7°$ **c)** $-63.4°$

2 a) $71.6°$ **b)** $76.0°$ **c)** $26.6°$ **d)** $56.3°$ **e)** $-71.6°$ **f)** $-68.2°$

3 a) $8.1°$ **b)** $45°$ **c)** $90°$ **d)** $49.4°$ **e)** $63.4°$ **f)** $8.1°$

4 $57.5°$

Exercise 4H page 72
1 a) $y + x = 5$ **b)** $6y = x - 21$ **c)** $x = 5$ **d)** $y = -1$

2 $(2,\ 2);\ 5$ **3** 230170

Exercise 4I page 73
1 a) 15.5 **b)** 6

Exercise 4J page 74
1 a) $\sqrt{32}$ **b)** $\sqrt{0.4}$ **c)** 1 **d)** $\frac{2}{\sqrt{13}}$ **2** $\sqrt{12.5}$ km

Longer exercise page 74
circumcentre $(1,\ 1)$, orthocentre $(3,\ 2)$, centroid $\left(\frac{5}{3},\ \frac{4}{3}\right)$

Examination questions page 74
1 (i) $y + 7x = -11$ **(ii)** $4y + 3x = 31$ **(iii)** $(-3,\ 10)$

2 a) $9x + 13y + 19 = 0$ **c)** $(7, 11)$

3 a) $(3, 2)$ **b)** $y + 2x = 8$

5 Sequences

Exercise 5A page 79

1 5, 12, 19

2 a) 2, 5, 8 **b)** 3, 6, 12 **c)** 10, 8, 6 **d)** 48, 24, 12

3 $u_1 = 5$, $u_{n+1} = 7 + u_n$ **a)** $u_1 = 2$, $u_{n+1} = 3 + u_n$ **b)** $u_1 = 3$, $u_{n+1} = 2u_n$
c) $u_1 = 10$, $u_{n+1} = u_n - 2$ **d)** $u_1 = 48$, $u_{n+1} = \frac{1}{2}u_n$

4 a) 8, 13, 18 **b)** 2, 5, 10 **c)** 5, 13, 35

5 a) $7 + 9 + 11 + 13 + 15 + 17 = 72$ **b)** $1 + 4 + 9 + 16 + 25 = 55$
c) $2 + 4 + 8 + 16 + 32 = 62$ **d)** $\frac{1}{1} + \frac{1}{2} + \frac{1}{3} + \frac{1}{4} = 2\frac{1}{12}$ **e)** $\frac{1}{2} + \frac{1}{6} + \frac{1}{12} + \frac{1}{20} = \frac{4}{5}$

6 $\sum_{n=1}^{5} (3n - 1)$

7 a) $\sum_{n=1}^{20} n$ **b)** $\sum_{n=1}^{11} 2n$ **c)** $\sum_{n=1}^{10} n^3$ **d)** $\sum_{n=1}^{\infty} (\frac{1}{2})^n$

8 $\sum_{n=1}^{\infty} u_{2n-1}$ **9** $u_1 = u_2 = 1$, $u_{n+2} = u_{n+1} + u_n$

Exercise 5B page 80

1 a) 5, 8, 11 **b)** 3, 10, 17 **c)** 2, 0, −2 **d)** 3, $3\frac{1}{2}$, 4 **e)** 5, 3.5, 2

2 a) $5 + (n - 1)3$ **b)** $3 + (n - 1)7$ **c)** $2 - (n - 1)2$ **d)** $3 + (n - 1)\frac{1}{2}$ **e)** $5 - (n - 1)1.5$

3 −1, 3; $-1 + (n - 1)3$

4 a) 1, 2, $1 + (n - 1)2$ **b)** 0, 4, $(n - 1)4$ **c)** 7, $\frac{1}{2}$, $7 + (n - 1)\frac{1}{2}$ **d)** 5, −1, $5 - (n - 1)$

5 a) 7, 5 **b)** −1, 2 **c)** 0, −3 **d)** 2.1, 0.1 **e)** −2, −2 **f)** 3.5, $-\frac{1}{2}$

6 £3.60 **7** $x = 4$ **8** 11, 2 **9** 2 **10** 332 **11** 22 **13** 21, 28, 36

Exercise 5C page 83

1 a) $\frac{1}{2}n[10 + (n - 1)3]$ **b)** $\frac{1}{2}n[4 + (n - 1)]$ **c)** $\frac{1}{2}n[18 + (n - 1)3]$
d) $\frac{1}{2}n[6 + (n - 1)5]$ **e)** $\frac{1}{2}n[10 + (n - 1)3]$ **f)** $\frac{1}{2}n[12 + (n - 1)\frac{1}{2}]$

2 a) 3925 **b)** 1325 **c)** 4125 **d)** 6275 **e)** 3925 **f)** 912.5

3 $\frac{1}{2}n(n + 1)$ **4** n^2 **5** 0, 2; 2

6 a) 5, 4 **b)** 1, $\frac{1}{2}$ **c)** 2, −2

7 $6 + 2(n - 1)$ **8** of the form $an^2 + bn$

9 a) $S_n - S_{n-1}$ **b)** $S_n + S_{n-2} - 2S_{n-1}$

10 3   **11** 42   **12** 9 109 800

13 2, 5, 2 + $(n-1)3$; $\frac{1}{2}n[4 + (n-1)3]$; 5 storeys   **14** $\frac{1}{2}n(n-1) + 1$ up to $\frac{1}{2}n(n+1)$

15 $\frac{1}{2}n[2p + \frac{1}{2}(n-1)(q-p)]$

Exercise 5D page 85

1   **a)** $2 \times 3^{n-1}$ **b)** $(-2)^{n-1}$ **c)** $8 \times (\frac{1}{2})^{n-1}$

2   **a)** 4374   **b)** -128 **d)** $\frac{1}{16}$

3   12, 3

4   **a)** 2, 2   **b)** $\frac{3}{2}, \frac{1}{2}$ **c)** $-15, -3$

5 $\frac{1}{2}$ or $-\frac{1}{2}$

6   **a)** 5, -5 **b)** $\frac{1}{2}$ **c)** -3 **d)** $\sqrt{2}, -\sqrt{2}$

7   5, 25

Exercise 5E page 87

1   **a)** $\frac{5}{2}(3^n - 1)$ **b)** $8(1 - (\frac{1}{2})^n)$ **c)** $1 - (-2)^n$ **d)** $\frac{15}{2}(1 - (-\frac{1}{3})^n)$

2   **a)** 147 620   **b)** 7.992   **c)** -1023 **d)** 7.4999

3   1, 2

4   **a)** 2, 3   **b)** 5, 2   **c)** 8, 2   **d)** 4, $\frac{1}{2}$ **e)** 13.5, $-\frac{1}{2}$

5 $2^{64} - 1$ **7**   2801   **8**   1, 1, 2, 4, 8; 2^{n-2} (except first term)

Exercise 5F page 88

1   1.1; £1000 $\times 1.1^n$ **2**   10 000 000 $\times 1.02^n$ **3** 0.9^n

4   £8000 $\times 0.85^n$ **5**   £4145   **6**   £174 469

Exercise 5G page 90

1   **a)** yes, 8   **b)** no   **c)** yes, 3

2   0.6   **3**   9   **4** $\frac{2}{3}$

5 $-\frac{1}{2} < x < \frac{1}{2}$, $\frac{1}{1-2x}$

6 $\frac{4}{9}$

7   **a)** $\frac{7}{9}$ **b)** $\frac{2}{11}$ **c)** $\frac{1}{37}$

8   100, 10, 1, 0.1, . . . ; $111\frac{1}{9}$ paces

9   8 seconds; infinitely many   **10** $5 \times 0.8^{n-1}$; 25 cm   **11**   0.25, 7.5

Longer exercise page 91

1   **a)** £205 454   **b)** £x   **c)** £1.08x   **d)** £73.1x   **e)** £2810

2   £70 250   **3** £28 670   **4** £29 450

Examination questions page 91

1 £7.94   2 43.1 km h^{-1}, 92.8 km h^{-1}

3 a) 167 167, 111 445   b) $\dfrac{6}{13}\left(1 - \left(\dfrac{1}{27}\right)^n\right); \dfrac{6}{13}$

Consolidation section A

Extra questions page 94

1 a) $5x^3 + 4x - 6$ b) $-10x^3 + 10x^2 - 27x - 17$ c) $4x^6 + 6x^5 - 9x^4 - 7x^3 - 5x^2 - 52x - 7$

2 $a = 4$, $b = -1$ 3 $x^2 + 7$, 15

4 a) $(x - 7)(x - 1)$ b) $(x - 7)(x + 12)$ c) $(x - 2)(2x - 1)$

5 a) $\dfrac{(2x - 3)(x - 1)}{(x^2 + 1)(x^2 + x + 1)}$ b) $\dfrac{(2x - 3)(x^2 + x + 1)}{(x^2 + 1)(x - 1)}$ c) $\dfrac{3x^3 - 2x^2 - 4}{(x^2 + 1)(x^2 + x + 1)}$

6 20   7 $a = 2$, $b = -3$ 9 $X = 6Y^2$ 10 $x = \dfrac{12}{y}$

11 a) $x = -2$ or $x = 6$ b) $x = 0.186$ or $x = -2.69$ c) $x = -4.76$ or $x = 5.26$

12 a) $(x + 2)^2 - 7$ b) $(x - 1.5)^2 - 3.25$ c) $7 - (x + 2)^2$

13 a) $-7, -2$ b) $-3.25, 1.5$ c) $7, -2$

14 a) $x = \pm\sqrt{5}$ or $x = \pm\sqrt{3}$ b) $x = 4$ or $x = 16$ c) $x = 1, y = 2$ or $x = 2, y = 1$

15 1   a) 0.28   b) 0.96   c) $\dfrac{7}{24}$ d) $\dfrac{24}{7}$ e) $\dfrac{25}{24}$ f) $\dfrac{25}{7}$

16 a) $\dfrac{1}{2}$ b) $\dfrac{1}{\sqrt{2}}$ c) 2   d) 0   e) 0

17 a) $\sin 35°$ b) $-\tan 6°$ c) $-\cos 40°$ d) $-\operatorname{cosec} 80°$
   e) $\cos 30°$ f) $-\sec 30°$ g) $-\cos 40°$ h) $\sin 20°$

18 a) 17.5°, 162.5°   b) 95.7°, 264.3°   c) 14.0°, 194.0°   d) 149.0°, 329.0°

19 a) $\sqrt{37}$ b) $\sqrt{173}$ c) $\sqrt{104}$

20 $x^2 + y^2 - 4x - 2y - 20 = 0$ 21 $(1, -3), \sqrt{22}$

22 a) $y = 3x - 1$ b) $y + 3x = 2$ c) $3y = 2x + 7$
   d) $y + x = 8$ e) $y = 3x - 5$ f) $3y = x + 6$

23 a) $(3, -7)$ b) $(-1, 5)$

24 $y = x$, $(3.5, 3.5)$ 25 $y + 3x = 13$, $(4, 1)$

26 $(6, 5)$

27 a) 63.4°   b) 29.7°

28 a) 8   b) $2 + 3(n - 1)$ c) $\frac{1}{2}n[4 + 3(n - 1)]$

29 a) 4   b) -2 c) $\frac{1}{2}n[8 - 2(n - 1)]$

30 a) 24 $\qquad$ **b)** $81 \times (\frac{2}{3})^{n-1}$ $\qquad$ **c)** $243[1 - (\frac{2}{3})^n]$

31 243 $\qquad$ **32** $\frac{3}{2}$, $16[(\frac{3}{2})^n - 1]$

Mixed exercise page 96

1 $-\frac{1}{2}$ or 1 $\qquad$ **2** $(x+1)(2x+1)(x+2)$ $\qquad$ **3** 54

4 0.2 $\qquad$ **5** 2 or $-\frac{2}{3}$ $\qquad$ **7** $(-1, 4)$, $\sqrt{8}$

9 $\frac{1}{2}$, $\frac{1}{2}$ $\qquad$ **11** $y = x$ $\qquad$ **14** 39.8°

15 70.5° $\qquad$ **16** $\frac{1}{4}$ $\qquad$ **18** $2x^2 - 9x + 10$ $\qquad$ **19** $6\frac{2}{3}$, $(3, 2)$

Longer exercise page 98

1 7.8, 10.1, 16.4 $\qquad$ **2** 6 October, 10 March $\qquad$ **3** 23 December, 23 June

4 23 December $\qquad$ **5** $12 + 4.5 \sin 30t$

Examination questions page 99

1 $(0, 4)$ or $(4, 2)$

2 (ii) $x^2 - 2x + 3$ $\qquad$ (iii) $(x-1)^2 + 2$ $\qquad$ (iv) $x > -\frac{3}{2}$

3 a) 10 sec x, 60 cos x $\qquad$ **c)** $\cos x = \frac{1}{2}$ or $-\frac{1}{3}$, $x = 60°$, $300°$, $109°$, $251°$
$\quad$ **d)** 17.3 m $\qquad$ **e)** 26.5 m

4 a) $d = 5$ $\qquad$ **b)** 26 250

5 $y + 2x = 13$, $(5, 3)$

6 b) 50 to 150 $\qquad$ **c)** $a = 100$, $b = 10\,000$

6 Indices and logarithms

Exercise 6A page 102

1 a) 81 $\qquad$ **b)** 81 $\qquad$ **c)** 125 $\qquad$ **d)** 3 $\qquad$ **e)** 4 $\qquad$ **f)** 1
$\quad$ **g)** $\frac{1}{2}$ $\qquad$ **h)** 0.1 $\qquad$ **i)** $\frac{1}{3}$ $\qquad$ **j)** $\frac{1}{4}$ $\qquad$ **k)** 0.001 $\qquad$ **l)** 2
$\quad$ **m)** 100 $\qquad$ **n)** 2 $\qquad$ **o)** 0.1

2 a) 1 $\qquad$ **b)** 3 $\qquad$ **c)** 1

3 a) $1 + \frac{1}{x}$ $\qquad$ **b)** $\frac{1}{x}$ $\qquad$ **c)** xy

4 a) 3 $\qquad$ **b)** -2 $\qquad$ **c)** $\frac{1}{3}$ $\qquad$ **d)** 0

5 1

Exercise 6B page 104

1 a) 4 $\qquad$ **b)** 8 $\qquad$ **c)** 100 $\qquad$ **d)** $\frac{1}{3}$ $\qquad$ **e)** $\frac{1}{7}$ $\qquad$ **f)** 32

2 a) 2^9 $\qquad$ **b)** 3^{11} $\qquad$ **c)** 5^3 $\qquad$ **d)** 2^{19} $\qquad$ **e)** 3^5 $\qquad$ **f)** 5^{11}

3 a) 2^{6x} $\qquad$ **b)** 2^{10x} $\qquad$ **c)** 10^{2y+x} $\qquad$ **d)** x^9 $\qquad$ **e)** y^{-2} $\qquad$ **f)** z^{a+b-c}

4 a) $x = y^2$ $\qquad$ **b)** $x = y^{\frac{1}{n}}$ $\qquad$ **c)** $x = \dfrac{z^2}{y}$ $\qquad$ **d)** $x = z^3 y$

5 $x\sqrt{\frac{5}{6}}$

6 a) $\frac{1}{x^3}$ **b)** $2^{\frac{1}{2}y}$

Exercise 6C page 105
1 a) 3   **b)** 0   **c)** -3 **d)** 3   **e)** 0   **f)** 3   **g)** $-\frac{2}{3}$ **h)** -5

2 a) $x = 1$ or $x = 2$ **b)** $x = 1$ **c)** $x = 1$ or $x = -1$

3 a) $x = y = \frac{3}{2}$ **b)** $x = 2, \ y = 1$

Exercise 6D page 107
5 a) $\frac{4}{9}$ **b)** $\frac{3}{11}$ **c)** $\frac{41}{333}$ **d)** $\frac{7}{30}$ **e)** $1\frac{14}{55}$ **f)** $23\frac{101}{990}$

6 a) $\frac{a}{9}$ **b)** $\frac{ab}{99}$

Exercise 6E page 109
1 a) $1 + 2\sqrt{2}$ **b)** $4 - \sqrt{2}$ **c)** $\sqrt{3} - 1$ **d)** $5\sqrt{2} - 1$ **e)** -1
   f) $5 - \sqrt{6}$ **g)** $2\sqrt{2} + 3$ **h)** $7 - 4\sqrt{3}$ **i)** $14 - 4\sqrt{6}$

2 a) $3\sqrt{2}$ **b)** $2\sqrt{2}$ **c)** $\sqrt{2}$ **d)** $\frac{1}{\sqrt{2}}$

Exercise 6F page 110
1 a) $6 + 3\sqrt{5} - 2\sqrt{10} - 4\sqrt{2}$ **b)** $\sqrt{10} + \sqrt{15} + 3\sqrt{2} + 3\sqrt{3}$ **c)** $\frac{\sqrt{21}}{18} + \frac{\sqrt{7}}{9} + 5\frac{\sqrt{3}}{18} + \frac{5}{9}$

3 $\sqrt{6}$ **4** $\frac{9\sqrt{3}}{4}$

5 a) $\frac{\sqrt{3}}{2\sqrt{2}}$ **b)** $\sqrt{3} + \frac{7}{4}$ **c)** 2

6 $x^{\frac{1}{2}} + 1$

Exercise 6G page 111
1 a) 2   **b)** 2   **c)** 3   **d)** 2   **e)** 5   **f)** 10   **g)** $\frac{1}{2}$
   h) $\frac{1}{3}$ **i)** $\frac{1}{3}$ **j)** -2 **k)** 0   **l)** $-\frac{1}{2}$ **m)** -3 **n)** $\frac{2}{3}$ **o)** $\frac{3}{4}$

2 a) 2   **b)** 0   **c)** $\frac{1}{2}$ **d)** $\frac{1}{2}$

3 a) $x^z = y$ **b)** $2^2 = 4$ **c)** $9^{\frac{1}{2}} = 3$ **d)** $10^2 = 100$

4 a) 2   **b)** 49   **c)** 9

5 $\frac{1}{2}y$

Exercise 6H page 113
1 a) $\log 3x^2$ **b)** $\log 8x^2$ **c)** $\log 1 = 0$

2 a) 1   **b)** -2 **c)** 1

3 $2 \log 2 + 2 \log 3$

4 a) $\log x + \log y + \log z$ **b)** $2 \log x + \log y$ **c)** $\log x - 2 \log z$
   d) $\frac{1}{2}(\log x + \log y + \log z)$

5 a) $x + y$ **b)** $x + 2y$ **c)** $y - 2x$ **d)** $\frac{1}{2}x$ **e)** $\frac{1}{2}(x + y)$

6 a) 7   **b)** $\frac{5}{4}$ **c)** 3

7 a) $x = 10\,000$, $y = 100$ **b)** $x = 18$, $y = 2$ **c)** $x = 9$, $y = 243$ **d)** $x = 243$, $y = 10$

8 $87 \log_{10} 15$; 103 **9** $65\,050$

Exercise 6I page 115

1 a) 1.58   **b)** 1.21   **c)** 0.792

2 a) 1.16   **b)** 0.631   **c)** −1.66

3 a) 3.82   **b)** 0.518   **c)** 1.51

4 37 years   **5** 13.5 years   **6** 34.3 years

7 a) 25.3%   **b)** 1.53%

Examination questions page 116

1 a) 2^{p+2q+3} **b)** $3p + 2q$

2 a) $\log 3$ **b)** 9.64

3 $x - 1 + \sqrt{3}$

7 Radians

Exercise 7A page 119

1 a) $\dfrac{\pi}{2}$ **b)** $\dfrac{3\pi}{2}$ **c)** π **d)** $\dfrac{\pi}{4}$ **e)** $\dfrac{\pi}{3}$ **f)** $\dfrac{2\pi}{3}$

2 a) $180°$ **b)** $90°$ **c)** $150°$ **d)** $120°$ **e)** $45°$

3 $30°$; 0.5

4 a) 1   **b)** $\dfrac{\sqrt{3}}{2}$ **c)** $\sqrt{3}$

5 0.4; $22.9°$ **6** 0.698; 6.98 cm   **7** $0.000\,291$; $0.000\,004\,85$

Exercise 7B page 120

1 a) 0.841   **b)** 0.825   **c)** −2.19   **d)** 0.775   **e)** 1.27   **f)** 1.11

2 a) 9.97   **b)** 1.12   **c)** 14.1   **d)** 1.05   **e)** 0.896   **f)** 0.589

3 a) 1   **b)** π **c)** $\dfrac{\pi}{6}$

4 a) 0.5   **b)** 0   **c)** 1

Exercise 7C page 121

1 a) 1.6   **b)** 1   **c)** 9π

2 a) 6.25   **b)** $\dfrac{16\pi}{3}$ **c)** $\dfrac{135\pi}{2}$

3 a) 0.446   **b)** 2.57   **c)** 110

4 4.17 cm	**5** 0.163	**6** 5.77 in
7 24.2 cm	**8** 2.28 radians	**10** 8.18 cm^2
11 149 cm^2	**12** 50.0 cm	**13** 0.645 cm^2
14 24.6 cm	**15** 6.28 m	**16** 3438 nautical miles

Exercise 7D page 123

1 2.5 rad s^{-1}	**2** 145 rad s^{-1}	**3** 8.44 rev s^{-1}

4 $\dfrac{30\omega}{\pi}$ **5** $\dfrac{r\pi}{30}$

6 a) 0.000 145   **b)** 0.001 75   **c)** 0.000 000 199   **d)** 0.000 072 7

7 11.7 cm s^{-1} **8** 2.7   **9** 16.7

Exercise 7E page 125

2 a) 0.04   **b)** 0.000 004

3 a) $2x$ **b)** $\frac{1}{2}x$ **c)** $1 - 8x^2$ **d)** $\frac{1}{2}$ **e)** $\frac{1}{2}x$ **f)** $\frac{4}{9}$

4 a) 2   **b)** $\frac{2}{3}$ **c)** $\frac{2}{9}$

5 0.106

6 a) 0.1   **b)** 0.3   **c)** 0.23   **d)** 0.0905

7 a) 0.25   **b)** 0.66

Longer exercise page 126

1 a) 0.099 833 33   **b)** 0.995 004 166   **c)** 0.479 166 67   **d)** 0.877 604 167

2 a) 0.000 000 08   **b)** 0.000 000 001   **c)** 0.0003   **d)** 0.000 02

3 a) 0.100 334 588   **b)** 0.545 994 065

4 0.24   **5** $\frac{1}{3}$

Examination questions page 126

1 0 or $\frac{1}{4}$

2 a) 0.5857   **b) (i)** 4.514 cm   **(ii)** 2.6356   **(iii)** 1.125 cm^2

3 a) 1.18 m   **b)** 0.02

8 Differentiation

Exercise 8A page 129

1 b) 5; 4.5; 4.1; 4.01   **c)** 4   **2** 6

Exercise 8B page 130

3 $2x$

Exercise 8C page 131

1 a) $4x$ **b)** $2x + 2$ **c)** $4x + 1$ **d)** $2x - 2$ **e)** $4x - 3$ **f)** $7 - 2x$

2 a) $2ax$ **b)** $2x + b$ **c)** $2ax + b$

Exercise 8D page 132

1 a) $2x + 5$ **b)** $-2x$ **c)** $6x - 2$

2 $3x^2$ **3** $4x^3$ **4** nx^{n-1}

Exercise 8E page 133

1 $-\dfrac{2}{x^3}$ **2** $\dfrac{1}{2\sqrt{x}}$

Exercise 8F page 134

1 a) $5x^4$ **b)** $11x^{10}$ **c)** $-2x^{-3}$ **d)** $\dfrac{-3}{x^4}$ **e)** $\dfrac{-5}{x^6}$ **f)** $\frac{1}{4}x^{-\frac{3}{4}}$

g) $\frac{1}{3}x^{-\frac{2}{3}}$ **h)** $\frac{1}{4}x^{-\frac{3}{4}}$ **i)** $-\frac{1}{2}x^{-\frac{3}{2}}$

2 a) $5x^4$ **b)** $9x^8$ **c)** $12x^{11}$ **d)** $-3x^{-4}$ **e)** $-3x^{-4}$ **f)** $-12x^{-13}$

g) $1.5x^{0.5}$ **h)** $2\frac{1}{3}x^{1\frac{1}{3}}$ **i)** $1.5x^{0.5}$

3 $(3, 9)$ **4** $(2, 8), (-2, -8)$ **5** $(\frac{1}{2}, 2); (-\frac{1}{2}, -2)$

Exercise 8G page 136

1 a) $10x$ **b)** $48x^3$ **c)** $2.5x^{-\frac{1}{2}}$ **d)** $-21x^{-4}$ **e)** $-\dfrac{8}{x^2}$ **f)** $-\dfrac{1.5}{x^{1.5}}$

2 a) $3x^2 + 5x^4$ **b)** $4x^3 - \dfrac{2}{x^3}$ **c)** $\frac{1}{2}x^{-\frac{1}{2}} - \frac{1}{2}x^{-\frac{3}{2}}$ **d)** $3x^2 + 2x + 1$ **e)** $2x + 1$

3 a) $18x^5 + 21x^2$ **b)** $24x^2 - 20x^3$ **c)** $15x^2 - \dfrac{1}{x^{\frac{1}{2}}}$

d) $-\dfrac{3}{x^2} - \dfrac{4}{x^3}$ **e)** $2x^{-\frac{2}{3}} - x^{-1.5}$ **f)** $8x - 3$

4 a) 5 **b)** 11 **c)** 4 **d)** $\frac{1}{4}$ **e)** -2 **f)** $-15\frac{2}{3}$

5 -3 **6** $5, -5$ **8** $(1, 0), (2, 0); -1, 1$ **9** $(4, 12)$ **10** no

Exercise 8H page 138

1 a) $2x - 7$ **b)** $4x - 5$ **c)** $2x - 7$ **d)** $2x + 6$ **e)** $96x^7$

f) $18x + 6$ **g)** $-\dfrac{1}{x^2} + 3$ **h)** $\dfrac{1}{x^{\frac{1}{2}}} + 2x^{-\frac{3}{2}}$ **i)** $-1.5x^{-2.5} - 2x^{-3}$

Exercise 8I page 139

1 a) $y = 2x - 1$ **b)** $y = 5x - 4$ **c)** $y = -15x - 14$ **d)** $y = 9x - 22$

e) $y = -7x + 11$ **f)** $4y = x + 4$

2 $y = 11x - 16, (-4, -60)$ **3** $(0, -2)$ **4** 2

5 $a = -3, b = 8$ **6** $y = (2ax_0 + b)x + y_0 - 2ax_0^2 - bx_0$

7 $y = -6x + 6, y = 6x - 42; (4, -18)$ **8** $-15, 17$

Exercise 8J page 140

1 a) $12y + x = 98$ **b)** $y = x + 1$ **c)** $y = x + 5$

d) $7y = x + 70$ **e)** $5y = 2x + 16$ **f)** $y = 54x - 485\frac{2}{3}$

2 $3y + x = 8$, $(-1\frac{1}{3}, 3\frac{1}{9})$ **3** $y = 3x + \frac{19}{36}$ **4** $a = \frac{1}{6}$, $c = -\frac{7}{6}$ **5** $-4\frac{11}{16}$

Exercise 8K page 141

1 a) $63.4°$ **b)** $-81.9°$ **c)** $-86.4°$

2 $(0, -3)$, $(2, 3)$; $26.6°$, $7.1°$ **3** $(0, 0)$, $(-2, -2)$, $(2, 2)$; $63.4°$, $38.7°$, $38.7°$

Longer exercise page 141

1 2   **2** $26.6°$ **3** 0.75   **4** $53.1°$

Examination questions page 142

1 $y = -4x + 13$; 16

2 b) $y = 2x + 1$ **c)** $2y + x = 7$ **d)** $\frac{7}{8}$

9 Sine rule and cosine rule

Exercise 9A page 146

1 a) 10.6 cm   **b)** 5.34 cm   **c)** 9.58 cm

2 10.8 cm, 10.1 cm

3 a) 1.96 cm   **b)** 29.9 cm   **c)** 13.2 cm   **d)** 4.16 cm

4 467 m   **5** 192 m   **6** 11.9 m

Exercise 9B page 147

1 a) $39.5°$ **b)** $78.1°$ **c)** $126.4°$

2 $49.6°$, $62.4°$

3 a) $33.9°$ **b)** $48.5°$ **c)** $61.3°$ **d)** $72.9°$

4 $37.8°$ **5** $37.6°$, 37.6 cm   **6** 13.3 cm

Exercise 9C page 149

1 a) $45.8°$ or $134.2°$ **b)** $81.0°$ or $19.0°$ **c)** $36.1°$ or $3.9°$

2 C: $42°$ or $138°$; B: $103°$ or $7°$ **3** $47.9°$

Exercise 9D page 150

1 a) 4.78 cm   **b)** 0.353 cm   **c)** 19.4 cm

2 7.26 cm

3 a) 4.57 cm   **b)** 5.51 cm   **c)** 27.5 cm   **d)** 7.14 cm

4 6.24 inches

Exercise 9E page 151
1 **a)** 57.1°   **b)** 90°   **c)** 92.0°

2 60.4°

3 **a)** 104.5°   **b)** 52.2°   **c)** 82.1°   **d)** 45.7°

6 60°

Exercise 9F page 153
1 **a)** 72.3°   **b)** 8.37 cm   **c)** 12.6 or 1.82 cm

2 60.9°   3 8.93 cm   4 53.2°   5 16.2 or 1.54 cm

7 $b > \dfrac{\sqrt{3}a}{2}$; $b < a$

Longer exercise page 155
(All angles measured with line across river)

1 41.8°, 447 seconds   2 73.1°, 1148 seconds   3 6.78° and 83.2°

Examination questions page 155
1 5.7 cm

2 15.6 miles, 7.7°

3 **a)** $\frac{1}{8}$ **b)** $\sin \theta = \dfrac{3\sqrt{7}}{8}$

10 Probability

Exercise 10A page 158
1 **a)** $\frac{1}{12}$ **b)** $\frac{1}{2}$ **c)** $\frac{1}{6}$ **d)** $\frac{11}{12}$

2 5, $\frac{1}{4}$ 3 **a)** $\frac{1}{8}$ **b)** $\frac{1}{2}$ 4 HH, HT, TH, TT; $\frac{1}{2}$

5 HHH, HHT, HTH, HTT, THH, THT, TTH, TTT; $\frac{3}{8}$

6 WBl, WBr, WG, BlBr, BlG, BrG   **a)** $\frac{1}{6}$ **b)** $\frac{1}{2}$

7 ABC, ACB, BAC, BCA, CAB, CBA; $\frac{1}{6}$

8 **a)** $\frac{1}{36}$ **b)** $\frac{1}{4}$ **c)** $\frac{1}{6}$

9 **a)** $\frac{1}{6}$ **b)** $\frac{1}{3}$ **c)** 0

10 **a)** $\frac{1}{12}$ **b)** $\frac{7}{12}$ **c)** $\frac{1}{3}$

11 $\frac{1}{15}$ 12 $\frac{4}{15}$

Exercise 10B page 160
1 no, no, 0.8   2 **a)** 0.3 **b)** 0.6   3 no, 0.4

4 **a)** xy, $x + y - xy$ **b)** 0, $x + y$ 5 **a)** A&B, C&D, B&D

6 **a)** A&C, B&D   **b)** A&B, A&D, C&B, C&D

Exercise 10C page 162
1 a) 0.000 000 102   **b)** 0.107   **c)** 0.893

2 a) 0.328   **b)** 0.9997

3 0.905   **4** 0.349   **5** 0.868   **6** 0.627

Exercise 10D page 164
1 $\frac{1}{7}$ **2** $\frac{8}{15}$ **3** $\frac{9}{50}$ **4** $\frac{1}{4}$

5 $\frac{3}{10}$ **6** $\frac{11}{450}$ **7** 0.65   **8** $\frac{1}{2}$ **9** 0.512

Exercise 10E page 165
1 a) $\frac{1}{6}$ **b)** $\frac{1}{18}$ **c)** 0

2 0.998   **3** 0.62   **5** 0.643

6 0.6   **7** 0.227   **8** 0.64   **9** 1 in 21

Longer exercise page 166
1 $\frac{1}{4}$ **2** p^2 **3** $\dfrac{2p}{1+p}$ **4** $\dfrac{p}{1+p}$ **5 a)** 0.01   **b)** 0.0099

Examination questions page 166
1 (i) $\frac{76}{153}$ (ii) $\frac{73}{153}$ (iii) $\frac{2}{9}$ (iv) $\frac{34}{73}$

2 (ii) 0.811

3 0.973

Consolidation section B

Extra questions page 174
1 a) 343   **b)** 10   **c)** $\frac{1}{25}$ **d)** 9   **e)** 8

2 a) 3^{2-x} **b)** x^2 **c)** 10^{4x-8}

3 a) $x = -2$ **b)** $x = \frac{1}{4}$ **c)** $x = 1$

4 $\frac{4}{11}$

5 a) 6   **b)** $4 + 2\sqrt{3}$ **c)** $5 + 4\sqrt{2}$

6 a) 3   **b)** 2   **c)** $\frac{1}{4}$ **d)** $-\frac{1}{2}$

7 a) 10   **b)** $\frac{1}{7}$ **c)** 1.29

8 3.60 cm   **9** 0.372 cm^2

10 a) $2x$ **b)** $2x^2$ **11** 0.0202

12 a) $6x^5$ **b)** $\frac{1}{2}x^{-\frac{1}{2}}$ **c)** $-4x^{-5}$ **d)** $6x - 12x^2$

     e) $\frac{1}{3}x^{-\frac{2}{3}} + 6x^{-4}$ **f)** $2x - 7$ **g)** $1 - \dfrac{3}{x^2}$

13 20   **14** $y = 7x - 5$ **15** $y = 3x - 8$

16 97.7°, 37.3°

17 $Q = 45.3°$, $R = 97.7°$; $Q = 134.7°$, $R = 8.3°$

18 79.0°

19 49.7 cm

20 **a)** 4.60 cm

b) 39.1 cm

c) 8.19 cm

21 **a)** 61.9°

b) 111°

c) 86.7°

22 **a)** $\frac{1}{64}$ **b)** $\frac{1}{32}$ **c)** $\frac{1}{8}$, most likely score 9, probability $\frac{1}{8}$

23 0.802

24 a) 0.0467 **b)** 0.107

25 **a)** 0.5 **b)** 0.25

Mixed exercise page 176

1 4, 1.33

2 0.142

3 $\dfrac{\pi}{2}$

4 7

5 0.71 to line across river

6 28.3 ft or 21.2 ft

7 $A = 10x - x^2$, 25 cm²

8 $5a$ cm s^{-1}

9 5.4

10 **a)** $\frac{1}{2}y$ **b)** $x - \frac{1}{2}y$ **c)** $-(\frac{1}{2}y + x)$

11 $\frac{1}{2}(p + q)$, $\frac{1}{2}(p - q)$

12 $x = 2$

14 $\frac{1}{2}r^2\theta$, $\dfrac{\pi - 1}{2}$

16 $2^{\frac{2}{3}} + 2^{\frac{1}{3}} + 1$

18 129.5 cm

19 $a = 2$, $b = 3$, $c = -1$

20 0.3 or 0.7

21 **a)** $\frac{1}{6}$ **b)** $\frac{25}{216}$ **c)** $\frac{625}{7776}$, $\frac{6}{11}$

Examination questions page 179

1 (ii) $k = 3$

2 **a)** $y = x - 1$ **b)** 1:3 **c)** $(1, -3)$, $\sqrt{50}$

3 $x^2 = \frac{5}{4}r^2 - r^2\cos\theta$

4 $3x^2 - 8x + 5$ (i) 4 (ii) 8 (iii) $y = 8x - 20$ (iv) $8y + x = 35$ $x = 0$ or $\frac{8}{3}$

5 **b)** $(6, 9)$ **c)** $(1, -6)$

6 **a)** $(\frac{4}{3}, 0)$ **b)** $(0.8, -0.4)$ **c)** 0.2 **d)** (i) $\frac{14}{9}$ (ii) $\frac{4}{9}$

8 (i) 0.18 (ii) 0.88

11 The binomial theorem

Exercise 11A page 183

1 1 7 21 35 35 21 7 1 1 8 28 56 70 56 28 8 1

2 **a)** $x^5 + 5x^4y + 10x^3y^2 + 10x^2y^3 + 5xy^4 + y^5$
  b) $x^6 + 6x^5y + 15x^4y^2 + 20x^3y^3 + 15x^2y^4 + 6xy^5 + y^6$
  c) $x^7 + 7x^6y + 21x^5y^2 + 35x^4y^3 + 35x^3y^4 + 21x^2y^5 + 7xy^6 + y^7$

Exercise 11B page 185

1 a) $x^4 + 8x^3y + 24x^2y^2 + 32xy^3 + 16y^4$ b) $a^5 - 10a^4b + 40a^3b^2 - 80a^2b^3 + 80ab^4 - 32b^5$

c) $16p^4 + 96p^3q + 216p^2q^2 + 216pq^3 + 81q^4$

d) $729m^6 - 7290m^5n + 30\ 375m^4n^2 - 67\ 500m^3n^3 + 84\ 375m^2n^4 - 56\ 250mn^5 + 15\ 625n^6$

e) $x^6 + 6x^4 + 15x^2 + 20 + \dfrac{15}{x^2} + \dfrac{6}{x^4} + \dfrac{1}{x^6}$

f) $y^7 - 7y^5 + 21y^3 - 35y + \dfrac{35}{y} - \dfrac{21}{y^3} + \dfrac{7}{y^5} - \dfrac{1}{y^7}$

g) $x^5 + 9x^4y + 32x^3y^2 + 56x^2y^3 + 48xy^4 + 16y^5$

h) $a^6 - 11a^5b + 50a^4b^2 - 120a^3b^3 + 160a^2b^4 - 112ab^5 + 32b^6$

i) $16p^5 + 112p^4q + 312p^3q^2 + 432p^2q^3 + 297pq^4 + 81q^5$

2   56   3 $328\sqrt{2}$ 4   1.771 561

5 a) $40x^3y^2$ b) $-270x^2$ c) $600x^2y^2$ d) 70

6   $3\frac{1}{3}$ 7   3.75

Exercise 11C page 186

1 a) 24   b) 720   c) 5040

6 a) 51!   b) $(n+1)!$ c) 9!   d) $(n-1)!$ e) $\dfrac{n+2}{(n+1)!}$

7   3 628 800   8 a)   24   b) 60   c) 120   d) 60

Exercise 11D page 187

1 a) 20   b) 120   c) 42

3   303 600   4   6840   5   7893 600   6   11 232 000

Exercise 11E page 189

1   a) 10   b) 210   c) 210

7   167 960   8   495   9   120

10   66   11   14 700   12   2400

Exercise 11F page 190

1   $\frac{1}{720}$ 2 a)   0.0582   b) 0.207   3   0.102

4   0.008 33   5   0.0655   6   0.0353

Exercise 11G page 192

1 a) $a^6 + 6a^5b + 15a^4b^2 + 20a^3b^3 + 15a^2b^4 + 6ab^5 + b^6$

b) $p^7 - 7p^6q + 21p^5q^2 - 35p^4q^3 + 35p^3q^4 - 21p^2q^5 + 7pq^6 - q^7$

c) $m^5 + 10m^4n + 40m^3n^2 + 80m^2n^3 + 80mn^4 + 32n^5$

d) $32a^5 + 80a^4b + 80a^3b^2 + 40a^2b^3 + 10ab^4 + b^5$

e) $64z^6 + 576z^5w + 2160z^4w^2 + 4320z^3w^3 + 4860z^2w^4 + 2916zw^5 + 729w^6$

f) $16x^4 - 160x^3y + 600x^2y^2 - 1000xy^3 + 625y^4$

2 a) $x^{18} + 18x^{17}y + 153x^{16}y^2$ **b)** $a^{15} + 15a^{14}b + 105a^{13}b^2$

c) $p^{20} - 20p^{19}q + 190p^{18}q^2$ **d)** $x^{17} + 34x^{16}y + 544x^{15}y^2$

e) $4096c^{12} + 73\,728c^{11}d + 608\,256c^{10}d^2$ **f)** $p^{14} - 42p^{13}q + 819p^{12}q^2$

3 a) $792x^5y^7$ **b)** $-56x^3y^5$ **c)** $5280z^4w^7$

d) $41\,472p^7q^2$ **e)** $700\,000x^4y^4$ **f)** $-448w^5z^3$

g) $10x^4y^3$ **h)** $10z^6w^6$ **i)** $112p^{-6}$

4 3432   **5** 1120   **6** 84

7 $1 + 10x + 45x^2$; $1 + 10y + 55y^2$ **8** $1 - 8x + 36x^2$

9 $1 + 21x + 200x^2 + 1140x^3$ **10** $256 + 512x - 256x^2$

Exercise 11H page 194

1 a) $1 + \frac{1}{3}x - \frac{1}{9}x^2 + \frac{5}{81}x^3$ **b)** $1 + x - \frac{1}{2}x^2 + \frac{1}{2}x^3$ **c)** $1 - \frac{3}{2}x - \frac{9}{8}x^2 - \frac{27}{16}x^3$

d) $1 + 6x + 6x^2 - 4x^3$ **e)** $1 - \frac{1}{6}x - \frac{1}{36}x^2 - \frac{5}{648}x^3$ **f)** $1 + 2x + 3x^2 + 4x^3$

g) $1 - \frac{1}{2}x + \frac{3}{8}x^2 - \frac{5}{16}x^3$ **h)** $1 - 4x + 12x^2 - 32x^3$ **i)** $1 + 12x + 96x^2 + 640x^3$

j) $1 - 2x + 3x^2 - 4x^3$ **k)** $1 - 9x + 54x^2 - 270x^3$ **l)** $1 - \frac{1}{3}x + \frac{2}{9}x^2 - \frac{14}{81}x^3$

2 a) $-1 < x < 1$ **b)** $-\frac{1}{2} < x < \frac{1}{2}$ **c)** $-\frac{1}{3} < x < \frac{1}{3}$

d) $-\frac{1}{4} < x < \frac{1}{4}$ **e)** $-2 < x < 2$ **f)** $-1 < x < 1$

g) $-1 < x < 1$ **h)** $-\frac{1}{2} < x < \frac{1}{2}$ **i)** $-\frac{1}{4} < x < \frac{1}{4}$

j) $-1 < x < 1$ **k)** $-\frac{1}{3} < x < \frac{1}{3}$ **l)** $-1 < x < 1$

3 a) $1 + \frac{1}{2}x^2 - \frac{1}{8}x^4$ **b)** $1 + 6x^2 + 24x^4$ **c)** $1 - \frac{1}{2}x^3 + \frac{3}{8}x^6$

4 $1 + x + \frac{1}{2}x^2 + \frac{1}{2}x^3 + \frac{3}{8}x^4 + \frac{3}{8}x^5 + \frac{5}{16}x^6$

5 $1 + \frac{3}{2}x + \frac{15}{8}x^2 + \frac{51}{16}x^3 + \frac{699}{128}x^4 + \frac{2517}{256}x^5$

6 $k = \pm 4$, $a = \mp 6$, $c = \mp 140$ **7** $a = 4$, $n = \frac{1}{2}$ **8** 7

9 $-7.5x^2$; $-\frac{1}{9} < x < \frac{1}{9}$ **10** $a = 3b$ **11** $\frac{1}{2}$; $\frac{1}{8}x^2$

Exercises 11I page 196

1 $x^{12} + 12x^{11}y + 66x^{10}y^2$ **a)** 1.1266   **b)** 2.86   **c)** 0.7864

2 $x^8 + 8x^7y + 28x^6y^2$ **a)** 376.32   **b)** 209.28

3 $1 - 9x + 35x^2$ **a)** 0.9135   **b)** 1.194

4 x^3 term; 1.22   **5 a)** 3.14   **b)** 0.28

6 a) 1.009 95   **b)** 0.989 949   **c)** 1.006 62

d) 4.123 11   **e)** 9.949 87   **f)** 2.0801

7 1.912 95   **8** $1 + \frac{1}{2}x + \frac{1}{2}x^2 + \frac{1}{2}x^3$; $\frac{6633}{2000}$

9 $a = \frac{5}{2}$, $b = \frac{3}{2}$ **10** $a = -1\frac{1}{2}$, $b = -\frac{1}{2}$; $1\frac{577}{1393}$

Examination questions page 197
1 $p = 10$, $q = 45$

2 $1 + nax + \frac{1}{2}a^2n(n-1)x^2$ **b)** $\frac{5}{8}$

3 a) $\frac{21}{16}$ **b)** 165

4 $1 - x + \frac{3}{2}x^2 - \frac{5}{2}x^3$; $\frac{2048}{915}$

12 Differentiation and graphs

Exercise 12A page 201
1 a) $6x$ **b)** 4   **c)** -2 **d)** -2 **e)** $\dfrac{2}{x^3}$ **f)** $-\frac{1}{4}x^{-\frac{3}{2}}$

   g) $24x^{-5}$ **h)** $\dfrac{2}{x^3}$ **i)** $\dfrac{3}{4}x^{-\frac{1}{2}} - \dfrac{1}{4}x^{-\frac{3}{2}}$

2 $x > 0$ **5** $(2, -13)$

Exercise 12B page 203
1 a) $(-2, -7)$, minimum   **b)** $(-0.577, 0.385)$, maximum; $(0.577, -0.385)$, minimum
  c) $(-1, -2)$, maximum; $(1, 2)$, minimum   **d)** $(1, -1)$, point of inflection
  e) $(-\frac{1}{3}, -\frac{1}{27})$, minimum; $(0, 0)$, point of inflection   **f)** $(-1, -4)$, minimum

2 a) $(2, -9)$ **b)** $(1, 0)$

5 a) $6x$, $(0, 0)$, point of inflection   **b)** $12x^2$, $(0, 0)$, minimum   **c)** $-12x^2$, $(0, 1)$, maximum

Exercise 12C page 205
1 a) $(-1, 2)$, maximum; $(0, 0)$ point of inflection; $(1, -2)$, minimum
  b) $(1, -1)$, maximum; $(2, -3)$; point of inflection; $(3, -5)$ minimum
  c) $(-0.5, 5.25)$, minimum; $(0.25, 8.625)$, point of inflection; $(1, 12)$, maximum
  d) $(-2, -16)$, minimum; $(0, 0)$, maximum; $(2, -16)$, minimum; $(-1.15, -8.89)$, point of inflection; $(1.15, -8.89)$, point of inflection
  e) $(-1, 1)$, point of inflection   **f)** $(0, 1)$, point of inflection; $(1, 0)$, minimum

2 $(-1, -2)$, maximum; $(1, 2)$, minimum

3 a) $(0.5, -0.25)$, minimum   **b)** $(0, 0)$, maximum; $(\frac{2}{3}, -\frac{4}{27})$, minimum
  c) $(0, 0)$, point of inflection; $(\frac{3}{4}, \frac{-27}{256})$, minimum

4 $(-2, 16)$, maximum; $(2, -16)$; minimum; $-16 < k < 16$

Exercise 12D page 207
1 £62.50   **2** $\dfrac{1000}{v}$ hours, $\dfrac{1000\,(1 + 0.1v^2)}{v}$ tons; 3.16 m.p.h.

3 $y = mx - 2m + 3$; $(0, 3 - 2m)$ and $(2 - \frac{3}{m}, 0)$; 12   **4** $593\,\text{cm}^3$

5 $x = 5.42$ cm, $h = 10.8$ cm   **6** $514\,\text{cm}^3$ **7** $545\,\text{cm}^2$

8 $\dfrac{4\sqrt{3}\pi a^3}{9}$ **9** $\dfrac{32\pi a^3}{81}$ **10** $272\,\text{cm}^2$

Longer exercise page 208
1 a) $\frac{1}{2}$ **b)** 5.59   **c)** infinity   **2** $(0, \frac{1}{2})$ **3** $p > \frac{1}{2}$

Examination questions page 209
1 $(-1, 5)$ maximum, $(3, -27)$ minimum; $y = -9x$; $(3, -27)$

2 20

3 (i) 0, 0.5   (ii) $x(1 - 2x)^2 \text{ m}^3$ (iii) $\frac{2}{27} \text{ m}^3$ (iv) $\frac{1}{16} \text{ m}^3$

13 Integration

Exercise 13A page 212
1 a) x^3 **b)** $-\dfrac{2}{x}$ **c)** $2x^5$ **d)** $4x^{\frac{1}{2}}$ **e)** $x^2 + 2x$

f) $\dfrac{1}{3}x^3 + \dfrac{1}{2}x^2 + x$ **g)** $-\dfrac{3}{x} + \dfrac{2}{3}x^3$ **h)** $-\dfrac{5}{x} - \dfrac{1}{x^2}$ **i)** $8x^{\frac{1}{2}} - 9x^{\frac{2}{3}}$

Exercise 13B page 213
1 a) $x^4 + 4x^2 + c$ **b)** $\frac{2}{3}x^3 - \frac{3}{5}x^5 + 2x + c$ **c)** $4x - \frac{3}{4}x^4 + c$

d) $\frac{2}{3}x^{\frac{3}{2}} + 2x^2 + c$ **e)** $-\dfrac{1}{x} - \frac{3}{2}x^{-2} + 8x + c$ **f)** $-\dfrac{3}{x} - 4x^{\frac{1}{2}} + c$

2 a) $\frac{1}{3}x^3 + \frac{5}{2}x^2 + 6x + c$ **b)** $\frac{2}{3}x^3 - \frac{5}{2}x^2 + 3x + c$ **c)** $2x - \frac{1}{2}x^2 - \frac{1}{3}x^3 + c$
d) $\frac{1}{4}x^4 + \frac{1}{2}x^2 + c$ **e)** $\frac{1}{5}x^5 - x + c$ **f)** $\frac{1}{3}x^6 + \frac{2}{5}x^5 + \frac{1}{2}x^4 + c$
g) $-\dfrac{1}{x} - \dfrac{1}{2x^2} + c$ **h)** $-\dfrac{3}{x} + x + c$ **i)** $\frac{1}{2}x^2 - x + c$
j) $\frac{2}{5}x^{\frac{5}{2}} - \frac{2}{3}x^{\frac{3}{2}} + c$ **k)** $\frac{4}{5}x^{\frac{5}{2}} - 2x^{\frac{3}{2}} + c$ **l)** $\frac{3}{7}x^{\frac{7}{3}} - 3x^{\frac{1}{3}} + c$
m) $2x^{\frac{1}{2}} + \frac{2}{3}x^{\frac{3}{2}} + c$ **n)** $\frac{2}{3}x^{\frac{3}{2}} + \frac{2}{5}x^{\frac{5}{2}} + c$ **o)** $\frac{2}{3}x^{\frac{3}{2}} + x + c$
p) $\frac{1}{3}x^3 + 3x^2 + 9x + c$ **q)** $\frac{4}{5}x^5 - \frac{4}{3}x^3 + x + c$ **r)** $\frac{1}{3}x^3 + \frac{4}{5}x^{\frac{5}{2}} + \frac{1}{2}x^2 + c$

3 a) $\frac{1}{2}x^2 + c$ **b)** $\frac{2}{3}x^{\frac{3}{2}} + x + c$

4 a) $x^2 - 2x^3 + 3$ **b)** $2x^4 + x^2 + x - 2$ **c)** $4x^{\frac{3}{2}} - 105$

d) $-\dfrac{2}{x} + 9$ **e)** $\frac{1}{3}x^3 + x^2 - 3x + 4$ **f)** $-\dfrac{1}{x} + \frac{1}{2}x^2 + 3\frac{1}{2}$

Exercise 13C page 215
1 a) $2x^2 + c$ **b)** $2x^3 - x^4 + c$ **c)** $2x^{2.5} + c$

2 a) $\int 3x^2 \, dx = x^3 + c$ **b)** $\int -\dfrac{4}{x^2} \, dx = \dfrac{4}{x} + c$

Exercise 13D page 216
1 a) 2   **b)** 12   **c)** 159   **d)** $\frac{1}{2}$
   e) 4   **f)** $\frac{20}{3}$ **g)** 9   **h)** 80
   i) 8   **j)** 1.8   **k)** 90   **l)** 2.9997

2 a) $2b^2 - 2a^2$ **b)** $3(b^2 - a^2 + b - a)$ **c)** 0

Exercise 13E page 218

1 a) 18   **b)** 120   **c)** 6   **d)** 0.9   **e)** 38   **f)** $\frac{8}{3}$

2 $\frac{1}{2}$

3 a) $\frac{14}{3}$ **b)** $\frac{8}{3}$ **c)** $\frac{57}{4}$

Exercise 13F page 219

1 a) 1   **b)** $4\frac{1}{4}$ **c)** $1\frac{1}{3}$ **d)** 8

Exercise 13G page 220

1 $(0, 0)$, $(1, 1)$; $\frac{1}{6}$ **2** $\frac{1}{2}$ **3** $(-1, -3)$, $(1, 1)$; $\frac{4}{3}$

4 $\frac{16}{3}$ **5** $(-2, -3)$, $(2, -3)$; $\frac{64}{3}$ **6** $\dfrac{1}{m+1} - \dfrac{1}{n+1}$

7 $\frac{1}{24}$ **8** $(0, 0)$, $(1, 1)$; $1:7$

9 $y = 2x + 5$, $y = -2x + 13$; $(2, 9)$; $\frac{2}{3}$ **10** $y + 2x = 3$; $(\frac{3}{2}, 0)$; $\frac{11}{12}$ **11** $\frac{1}{3}$

Longer exercise page 221

1 $(-\sqrt{k}, k)$, $(\sqrt{k}, k)$ **2** $\frac{4}{3}k^{\frac{3}{2}}$ **3** $k^{\frac{3}{2}}$

Examination questions page 222

1 b) $(4, 16)$ **c)** $\dfrac{384\sqrt{3}}{5}$

2 a) $A(2, 0)$, $B(-2, 32)$ **b)** $C(-4, 0)$ **c)** 108

3 66

14 Trigonometric identities and equations

Exercise 14A page 224

1 **(a)**, **(d)** and **(e)** are identities   **b)** $x = \frac{3}{2}$ **c)** $x = 90°$ **f)** $x = 0°$

3 $\dfrac{\cos P}{\sin P}$ **4** $\dfrac{\operatorname{cosec} P}{\cot P}$

5 a) $\dfrac{x}{y}$ **b)** $\dfrac{y}{x}$ **c)** $\dfrac{1}{x}$ **d)** $\dfrac{1}{y}$

Exercise 14B page 227

1 $\frac{3}{5}$, $\frac{4}{3}$ **2** $\frac{15}{17}$, $\frac{8}{15}$ **3** $\dfrac{1}{\sqrt{5}}$, $\dfrac{2}{\sqrt{5}}$

4 a) $\sqrt{1 - \cos^2 x}$ **b)** $\sqrt{\sec^2 x - 1}$ **c)** $\sqrt{\tan^2 x + 1}$ **d)** $\sqrt{\cot^2 x + 1}$

5 a) $1 - \sin^2 x + 3 \sin x$ **b)** $\cos^2 x - \cos x + 4$

   c) $\sec x - 3 \sec^2 x + 3$ **d)** $\operatorname{cosec}^2 x + 5 \operatorname{cosec} x - 1$

9 $\dfrac{xX - yY}{yX + xY}$

Exercise 14C page 229

1 a) 23.6°, 156.4°, 383.6°, 516.4°
  c) 71.6°, 251.6°, 431.6°, 611.6°
  e) −30°, −150°, 330°, 210°
  g) 51.3°, −51.3°, 308.7°, −308.7°
  i) −14.5°, 345.5°, 194.5°, 554.5°
  k) 30°, 210°, 390°, 570°

  b) 72.5°, 287.5°, 432.5°, 647.5°
  d) 116.6°, 296.6°, 476.6°, 656.6°
  f) 143°, −143°, 503°, −503°
  h) 73.3°, 253.3°, 433.3°, 613.3°
  j) 45°, 135°, 405°, 495°
  l) 150°, −150°, 210°, −210°

2 a) 0.412, 6.69, 2.73, 9.01
  c) 1.25, 4.39, 7.53, 10.7
  e) $-\frac{\pi}{6}, \frac{7\pi}{6}, \frac{11\pi}{6}, \frac{19\pi}{6}$
  g) 0.896, −0.896, 5.39, −5.39
  i) −0.253, 3.39, 6.03, 9.68
  k) $\frac{\pi}{6}, \frac{7\pi}{6}, \frac{13\pi}{6}, \frac{19\pi}{6}$

  b) 1.27, −1.27, 7.55, −7.55
  d) −1.11, 2.03, 5.18, 8.32
  f) 2.50, −2.50, 3.79, −3.79
  h) 1.28, 4.42, 7.56, 10.7
  j) $\frac{\pi}{4}, \frac{3\pi}{4}, \frac{9\pi}{4}, \frac{11\pi}{4}$
  l) $\frac{5\pi}{6}, -\frac{5\pi}{6}, \frac{7\pi}{6}, -\frac{7\pi}{6}$

3 90°

4 2.50, 8.78

5 1 or −1

Exercise 14D page 232

1 a) 22.2°, 202.2°, 67.8°, 247.8°
  c) 13.3°, 103.3°, 193.3°, 283.3°
  e) 20°, 80°, 140°, 200°, 260°, 320°

  b) 50°, 290°
  d) 95.5°, 304.5°
  f) 81.8°, 178.2°

2 a) $\frac{\pi}{4}, \frac{3\pi}{4}, \frac{5\pi}{4}, \frac{7\pi}{4}$
  c) 0.955, 4.097, 2.186, 5.328
  e) 0.588, 3.730
  g) $\frac{\pi}{2}, \frac{3\pi}{2}$, 0.848, 2.29
  i) $\frac{\pi}{4}, \frac{3\pi}{4}, \frac{5\pi}{4}, \frac{7\pi}{4}$

  b) $\frac{\pi}{6}, \frac{5\pi}{6}, \frac{7\pi}{6}, \frac{11\pi}{6}$
  d) 1.25, 4.39
  f) 0, π, 2π, 1.23, 5.05
  h) 0, π, 2π, 0.340, 2.80

3 a) $\frac{\pi}{6}, \frac{5\pi}{6}$
  d) $\frac{\pi}{6}, \frac{5\pi}{6}$, 0.848, 2.29
  g) $\frac{\pi}{3}$, 1.82

  b) 0.841, π
  e) $\frac{\pi}{6}, \frac{5\pi}{6}$
  h) $\frac{3\pi}{4}$, 0.588

  c) $\frac{\pi}{3}$, π
  f) $\frac{\pi}{3}$, 1.23
  i) 0.340, 2.80

4 a) 45°, 135°, 225°, 315°
  c) 68°, 248°
  e) 60°, 300°

  b) 0°, 180°, 360°
  d) 64°, 350°
  f) 30°, 150°, 270°

Longer exercise page 232

3 $R\sqrt{2 - 2\cos\phi\cos\theta}$

4 $2R\sin^{-1}\sqrt{\frac{1}{2}(1 - \cos\phi\cos\theta)}$

Examination questions page 233

1 30°, 150°, 194°, 346°

2 (i) 2α **(ii)** $\pi + \alpha$; $\frac{\pi}{16}, \frac{\pi}{8}, \frac{5\pi}{16}$

3 (i) −4, 6   **(ii)** 116.6°, 296.6°, 56.3°, 236.3°

Consolidation section C

Extra questions page 235

1 a) $p^5 + 5p^4q + 10p^3q^2 + 10p^2q^3 + 5pq^4 + q^5$ **b)** $81x^4 - 432x^3y + 864x^2y^2 - 768xy^3 + 256y^4$

c) $8x^6 + 36x^3 + \dfrac{27}{x^3} + 54$

2 a) 12	**b)** 35	**c)** 1680
3 a) 19 600	**b)** 117 600	**4** 0.816
5 a) $1 144 066\, x^{10}y^{13}$	**b)** $10\ 264\ 320\, x^8y^4$	**c)** $-960\,740\,352$
6 a) $1 + 2.5x + 1.875x^2$	**b)** $1 + 9x + 54x^2$	**c)** $1 + \frac{1}{4}x - \frac{1}{32}x^2$
7 a) $-1 < x < 1$	**b)** $-\frac{1}{3} < x < \frac{1}{3}$	**c)** $-2 < x < 2$
8 a) 1.01 488	**b)** 0.986 488	**c)** 10.0033
9 a) $12x^2$	**b)** -2	**c)** $\dfrac{36}{x^5}$

10 a) $(3, -10)$, minimum   **b)** $(-1,\ 1)$, maximum; $(-\frac{1}{3},\ \frac{23}{27})$, minimum
c) $(-1, -4)$, maximum; $(1,\ 4)$, minimum

12 2896 cm^2

13 a) $x^4 - 2x^2 + c$ **b)** $4x^{\frac{5}{2}} + c$ **c)** $-\dfrac{4}{x} + c$

d) $6x + \frac{1}{2}x^2 - \frac{1}{3}x^3 + c$ **e)** $\frac{1}{3}x^3 + 4x^2 + 16x + c$ **f)** $\frac{2}{3}x^{\frac{3}{2}} + 4x^{\frac{1}{2}} + c$

14 a) 18	**b)** $\frac{52}{3}$	**c)** $\frac{40}{3}$
15 a) 7	**b)** 16	**c)** $\frac{4}{3}$
17 a) $\frac{\pi}{6}, \frac{5\pi}{6}$	**b)** $\frac{\pi}{3}, \frac{2\pi}{3}$	**c)** $\frac{\pi}{12}, \frac{5\pi}{12}, \frac{3\pi}{4}$

18 a) $54.7°,\ 125.3°$ **b)** $80°,\ 320°$ **c)** 1.11   **d)** 1.23, 0

Mixed exercise page 237

2 $c > a \sin\theta; \ c > a$ **3** $4:9$ **5** $1 + 10x^2 + 45x^4$; 0.310 872

6 0.210, 2.93   **7** $2 + \frac{1}{2}k - \frac{1}{8}k^2 + \frac{1}{16}k^3$; 2.048 81

8 $2 + \frac{3}{2}x + \frac{15}{8}x^2$; $-\frac{1}{2} < x < \frac{1}{2}$ **10** $a = 2,\ b = -8,\ c = 3$

11 $\sqrt{x^4 - x^2 + 1}$; $\dfrac{\sqrt{3}}{2}$ **12** 0.381, π **13** $\dfrac{c^2}{a^2 + b^2}$

14 minimum if n is even; point of inflection if n is odd   **16** $1 + \frac{1}{2}x - \frac{1}{8}x^2$; 0.557 291

17 $\dfrac{-1 - \sqrt{21}}{2}, \dfrac{-1 + \sqrt{21}}{2}; \dfrac{7\sqrt{21}}{2}$ **19** 360; 72

Longer exercise page 238

Finding π
3 $1 - \frac{1}{2}x^2 - \frac{1}{8}x^4 - \frac{1}{16}x^6$ **4** 3.20

Poker hands

2 598 960; **1** 0.001 98 **2** 0.000 24 **3** 0.001 44

4 0.003 55 **5** 0.0211 **6** 0.0475 **7** 0.423

Examination questions page 239

1 $p = 5$, $q = -\frac{12}{5}$, $r = -\frac{72}{125}$

2 (i) $6x^2 - 12x^3$; 0, $\frac{1}{2}$; (0, 4) and $(\frac{1}{2}, \frac{65}{16})$ (ii) $12x - 36x^2$; 0 and -3
(iii) point of inflection (iv) maximum

4 (i) $x^3 - x^2 - x + 1$ (ii) (1, 0); $(-\frac{1}{3}, \frac{32}{27})$

5 (i) $12x^3 + 12x^2$ (ii) (0, 0), $(-1, -1)$ (iii) (0, 0), point of inflection; $(-1, -1)$, minimum

6 45.6°, 314.4°, 91.9°, 268.1°

15 Vectors

Exercise 15A page 244

1 a) $\overrightarrow{AC}$ **b)** $\overrightarrow{DA}$ **b)** $3\overrightarrow{AC}$ **b)** $\overrightarrow{AD}$

3 **(a)**, **(c)** and **(e)** are parallel, **(b)** and **(d)** are parallel

4 a) $x = -3$, $y = 2$ **b)** $x = 2$, $y = -1$ **c)** $x = \frac{3}{8}$, $y = -\frac{1}{4}$

5 13.9 N, bearing 249° **6** 13.9 N, bearing 20.9° **7** 17.8 N, bearing 82.5°

Exercise 15B page 246

1 $\frac{1}{2}\mathbf{a}$, $\frac{1}{2}\mathbf{b}$, $\mathbf{b} - \frac{1}{2}\mathbf{a}$, $\mathbf{a} - \frac{1}{2}\mathbf{b}$, $\frac{1}{2}\mathbf{a} + k(\mathbf{b} - \frac{1}{2}\mathbf{a})$, $\frac{1}{2}\mathbf{b} + m(\mathbf{a} - \frac{1}{2}\mathbf{b})$, $\frac{1}{3}(\mathbf{a} + \mathbf{b})$

2 $\frac{1}{7}\mathbf{a} + \frac{4}{7}\mathbf{b}$ **3** $\mathbf{a} + \mathbf{c}$, $\mathbf{a} + \frac{1}{3}\mathbf{c}$, $\mathbf{c} - \mathbf{a}$; $\frac{3}{4}\mathbf{a} + \frac{1}{4}\mathbf{c}$

4 $\mathbf{a} + \mathbf{c}$, $\frac{1}{2}(\mathbf{a} + \mathbf{c})$, $\frac{1}{2}(\mathbf{a} + 3\mathbf{c})$, $\mathbf{a} + 2\mathbf{c}$

5 a) $\mathbf{a} - \mathbf{e}$ **b)** $\mathbf{a} + \mathbf{e}$ **c)** $2\mathbf{e}$ **d)** $2\mathbf{a} + 2\mathbf{e}$

6 A, B, D **7** $\frac{1}{2}\mathbf{a}$, $\frac{1}{3}\mathbf{a} + \frac{2}{3}\mathbf{b}$, $2\mathbf{b}$

Exercise 15C page 248

1 a) (1, 1) **b)** (2, -1) **c)** (-1, -1)

2 a) $\begin{pmatrix} 7 \\ 2 \end{pmatrix}$ **b)** $\begin{pmatrix} -3 \\ 4 \end{pmatrix}$ **c)** $\begin{pmatrix} 4 \\ 6 \end{pmatrix}$ **d)** $\begin{pmatrix} 19 \\ 3 \end{pmatrix}$ **e)** $\begin{pmatrix} -4 \\ 11 \end{pmatrix}$

3 a) $\begin{pmatrix} 3 \\ -2 \\ 1 \end{pmatrix}$ **b)** $\begin{pmatrix} -1 \\ -2 \\ 3 \end{pmatrix}$ **c)** $\begin{pmatrix} 3 \\ -6 \\ 6 \end{pmatrix}$ **d)** $\begin{pmatrix} -4 \\ -4 \\ 7 \end{pmatrix}$ **e)** $\begin{pmatrix} 7 \\ 2 \\ -6 \end{pmatrix}$

4 a) $x = 2$, $y = 1$ **b)** $\lambda = -2$, $\mu = 1$ **c)** $p = 3$, $q = -1$

5 a) $\begin{pmatrix} 1 \\ 2 \end{pmatrix}$ **b)** $\begin{pmatrix} 2 \\ -2 \end{pmatrix}$ **c)** $\begin{pmatrix} -3 \\ 0 \end{pmatrix}$

6 (a) and (e), (b) and (c)

7 $\begin{pmatrix} 2 \\ 4 \end{pmatrix}, \begin{pmatrix} 3 \\ -1 \end{pmatrix}, \begin{pmatrix} -2 \\ -4 \end{pmatrix}, \begin{pmatrix} -3 \\ 1 \end{pmatrix}$

8 (2, 6)

9 (5, 0), (0, 7), (5, 7)

Exercise 15D page 250
1 a) 5   **b)** 3.61   **c)** 5.10   **d)** 2.45   **e)** 3.74   **f)** 9.90

2 $\begin{pmatrix} 3 \\ 4 \end{pmatrix}, \begin{pmatrix} 5 \\ 0 \end{pmatrix}, \begin{pmatrix} -3 \\ -4 \end{pmatrix}, \begin{pmatrix} -5 \\ 0 \end{pmatrix}$ **3 a)** 3.61   **b)** 3.16 $\lambda = -1$ or $\lambda = -\frac{1}{5}$

Exercise 15E page 251
1 a) $2\mathbf{i} + \mathbf{j}$ **b)** $\mathbf{i} - 2\mathbf{j}$ **c)** $3\mathbf{i}$ **2 a)** $\mathbf{i} + 2\mathbf{j} + 3\mathbf{k}$ **b)** $4\mathbf{i} - \mathbf{j} - \mathbf{k}$ **c)** $2\mathbf{i} + \mathbf{k}$

3 a) $\begin{pmatrix} 1 \\ 3 \end{pmatrix}$ **b)** $\begin{pmatrix} 2 \\ -3 \end{pmatrix}$ **c)** $\begin{pmatrix} 0 \\ 4 \end{pmatrix}$ **4 a)** $\begin{pmatrix} 1 \\ -2 \\ 3 \end{pmatrix}$ **b)** $\begin{pmatrix} 2 \\ 0 \\ -3 \end{pmatrix}$

5 a) $\begin{pmatrix} 3/5 \\ 4/5 \end{pmatrix}$ **b)** $\begin{pmatrix} 2/\sqrt{29} \\ 5/\sqrt{29} \end{pmatrix}$ **c)** $\begin{pmatrix} 1/\sqrt{6} \\ 2/\sqrt{6} \\ -1/\sqrt{6} \end{pmatrix}$

     d) $\dfrac{2}{\sqrt{5}}\mathbf{i} - \dfrac{1}{\sqrt{5}}\mathbf{j}$ **e)** $\dfrac{1}{3}\mathbf{i} + \dfrac{2}{3}\mathbf{j} - \dfrac{2}{3}\mathbf{k}$

6 (a), (c), and (d)

Exercise 15F page 254
1 a) 14   **b)** 17   **c)** -3 **d)** 0   **e)** 1   **f)** 7   **g)** 4

2 a) 8.13°   **b)** 19.7°   **c)** 98.1°   **d)** 17.0°   **e)** 168.5°

    f) 15.8°   **g)** 18.4°   **h)** 112.6°   **i)** 74.2°

3 $|\mathbf{a}|^2 - 4|\mathbf{b}|^2$ **4** 0   **5** 83.4°

Exercise 15G page 255
1 a) $\frac{61}{13}$ **b)** $\frac{11}{5}$ **c)** 2   **d)** $\frac{3}{\sqrt{5}}$

3 a) 3   **b)** 7   **c)** 7.07

4 $(-0.459, 4.702)$

Exercise 15H page 256
1 (a), (d) and (e), (b), (c) and (f)

2 a) -18 **b)** 6   **c)** 1   **d)** -4 **e)** -16 **f)** -1

3 $x = y = 2$

4 a) $\pm \begin{pmatrix} 4/5 \\ -3/5 \end{pmatrix}$ **b)** $\pm \begin{pmatrix} 24 \\ -10 \end{pmatrix}$

c) $\pm \begin{pmatrix} 1/\sqrt{3} \\ -1/\sqrt{3} \\ -1/\sqrt{3} \end{pmatrix}$ **d)** $\begin{pmatrix} 0 \\ 1/\sqrt{2} \\ -1/\sqrt{2} \end{pmatrix}$

7 $\mathbf{a} + \mathbf{c}$, $\mathbf{c} - \mathbf{a}$ **8** They are parallel.

Exercise 15I page 258
1 a) $\frac{5}{8}\mathbf{a} + \frac{3}{8}\mathbf{b}$ **b)** $\frac{2}{3}\mathbf{i} + \frac{4}{3}\mathbf{j}$ **c)** $3\mathbf{j}$

2 a) $\frac{1}{2}\mathbf{i} + \mathbf{j}$ **b)** $2\mathbf{i} + \mathbf{j}$ **c)** $2\mathbf{i} + 2\mathbf{k}$ **d)** $2\mathbf{i} + 2\mathbf{j} - \mathbf{k}$

3 $\frac{1}{2}(\mathbf{a} + \mathbf{b})$, $\frac{1}{2}(\mathbf{b} + \mathbf{c})$, $\frac{1}{2}(\mathbf{c} + \mathbf{d})$, $\frac{1}{2}(\mathbf{d} + \mathbf{a})$

4 $\dfrac{n\mathbf{a} + m\mathbf{b}}{n + m}$, $\dfrac{n\mathbf{a} + m\mathbf{c}}{n + m}$; $\dfrac{n\mathbf{a} + m\mathbf{b} + m\mathbf{c}}{n + 2m}$, $m : n + m$

Exercise 15J page 262
1 a) $\mathbf{r} = 3\mathbf{i} + \mathbf{j} + t(\mathbf{i} + 2\mathbf{j})$ **b)** $\mathbf{r} = -\mathbf{i} + \mathbf{j} + t(2\mathbf{i} - \mathbf{j})$

c) $\mathbf{r} = \mathbf{i} + 2\mathbf{j} + \mathbf{k} + t(\mathbf{i} + 2\mathbf{j} - \mathbf{k})$ $\dfrac{x - 1}{1} = \dfrac{y - 2}{2} = \dfrac{z - 1}{-1}$

d) $\mathbf{r} = \mathbf{i} + \mathbf{j} + t(\mathbf{i} - \mathbf{j})$ **e)** $\mathbf{r} = \mathbf{i} + 2\mathbf{j} + t(\mathbf{i} + 3\mathbf{j})$

f) $\mathbf{r} = \mathbf{i} + 2\mathbf{j} + t(2\mathbf{i} - \mathbf{j})$ **g)** $\mathbf{r} = -\mathbf{i} - 2\mathbf{j} + t(2\mathbf{i} + 3\mathbf{j})$

h) $\mathbf{r} = \mathbf{i} + 2\mathbf{j} + t(2\mathbf{i} - \mathbf{j} + 2\mathbf{k})$ $\dfrac{x - 1}{2} = \dfrac{y - 2}{-1} = \dfrac{z - 0}{2}$

i) $\mathbf{r} = \mathbf{i} + 3\mathbf{j} + t(\mathbf{i} - \mathbf{j})$ **j)** $\mathbf{r} = \mathbf{i} + t(\mathbf{i} - \mathbf{j})$

k) $\mathbf{r} = \mathbf{i} + 2\mathbf{j} + \mathbf{k} + t(\mathbf{i} + 2\mathbf{j} - 3\mathbf{k})$ $\dfrac{x - 1}{1} = \dfrac{y - 2}{2} = \dfrac{z - 1}{-3}$

2 a) $\begin{pmatrix} 2 \\ 0 \end{pmatrix}$ **b)** $\begin{pmatrix} 2 \\ 3 \end{pmatrix}$ **3 a)** $\begin{pmatrix} 3 \\ 2 \\ 3 \end{pmatrix}$ **b)** $\begin{pmatrix} 3 \\ 1 \\ 0 \end{pmatrix}$

4 $x = 2$ **5 a)** $\sqrt{2}$ **b)** $8/\sqrt{10}$

6 $\mathbf{r} = 2\mathbf{i} + 3\mathbf{j} + t(\mathbf{i} - 2\mathbf{j})$; $\mathbf{r} = \mathbf{i} + \mathbf{j} + s(2\mathbf{i} + \mathbf{j})$; $(2.6, \ 1.8)$

7 a) $71.6°$ **b)** $60.3°$ **c)** $0°$ **d)** $79.5°$

Examination questions page 264
1 (i) $\frac{2}{5}\mathbf{a} + \frac{3}{5}\mathbf{b}$, $\frac{1}{4}\mathbf{a} + \frac{3}{4}\mathbf{c}$ **(ii)** $\frac{1}{3}(\mathbf{a} + \mathbf{b} + \mathbf{c})$

2 $0.4\mathbf{i} + 2\mathbf{j} - 0.2\mathbf{k}$, 2.05; $p = 4$, $\mathbf{i} + 2\mathbf{j} + \mathbf{k}$; $-2\mathbf{i} - 4\mathbf{k}$, $-\mathbf{i} - \mathbf{j} - 3\mathbf{k}$

3 (iv) $\frac{1}{6}$ **(v)** $0 < \lambda < \frac{1}{5}$

16 Graphs

Exercise 16A page 268
3 $a = 2$, $b = 10$

Exercise 16B page 270

2 a) $y = x^2 - 2x - 1$ **b)** $y = x^2 + 2x - 1$ **c)** $y = x^2 + x + 1$

3 a) $y = 1 + \dfrac{1}{x-1}$ **b)** $y = 2 + \dfrac{1}{x+1}$ **c)** $y = -1 + \dfrac{1}{x+2}$

4 a) $10(x-1)^2 - 4$ **b)** $2 - 2(x-1)^2$

5 a) $4 - \dfrac{6}{x-3}$ **b)** $-1 + \dfrac{2}{x+1}$

Exercise 16C page 273

1 a) $(0, 2)$, $(2, 0)$ **b)** $(0, 0)$ **c)** $(0, 1)$, $(\frac{1}{2}, 0)$ **d)** $(0, -1)$, $(1, 0)$

2 a) $(2, -1)$ **b)** $(5, 1)$ **c)** $(6, 1)$ **d)** $(-2, 1)$

3 $b = -1$, $c = 2$

4 a) translation 1 up   **b)** translation 2 to right   **c)** reflection in x-axis

5 a) stretch, factor 3, parallel to y-axis   **b)** translation 1 to left   **c)** reflection in y-axis

6 a) translation 1 to left, 3 down

b) translation 1 to left: stretch, factor 2, parallel to y-axis: translation 1 down

c) translation 2 to right: stretch, factor 3, parallel to y-axis: translation 5 down

7 a) stretch, factor 3, parallel to y-axis: translation 2 up

b) compression, factor 2, parallel to x-axis: reflection in y-axis: translation 1 up

c) translation 1 to left: stretch, factor 2, parallel to y-axis: translation 1 up

Exercise 16D page 276

1 a) $y = 3x^2 - 4$ **b)** $y = 0.3x^3 + 1.7x$ **c)** $y = 3x + \dfrac{8}{x}$

d) $y = 4x^{1.9}$ **e)** 6×1.3^x **f)** $\dfrac{1}{y} = \dfrac{5.5}{x} - 2.2$

2 $0.1v^2 + 1.2v$

Longer exercise page 277

Power 2/3 $D = 1.83L^{2/3}$ 680 days $D = 1.5L^{2/3}$

Examination questions page 278

1 a) $\ln R = \ln k + n \ln V$ **b)** $n = 1.8$, $k = 0.67$

3 $y = 2000 \times 1.85^x$ 5030 000

4 c) $(\alpha, 0)$, $(-\beta, 0)$, $(0, -\alpha\beta)$; $(0, 0)$, $(-\beta - \alpha, 0)$ **d)** $x = \frac{1}{2}(\alpha - \beta)$; $x - \frac{1}{2}(-\beta - \alpha)$

17 The addition formulae

Exercise 17A page 284

1 a) $\sin A \cos B + \cos A \sin B$ **b)** $\cos x \cos 3y + \sin x \sin 3y$

c) $\dfrac{\tan 2x + \tan y}{1 - \tan 2x \tan y}$ **d)** $\dfrac{1}{\sqrt{2}} \sin x + \dfrac{1}{\sqrt{2}} \cos x$

e) $\frac{1}{2}\cos y + \frac{\sqrt{3}}{2}\sin y$

f) $\frac{1+\tan x}{1-\tan x}$

2 a) $\sin(x+20°)$

b) $\cos 10°$

3 a) $\frac{56}{65}$ **b)** $\frac{63}{65}$ **c)** $1\frac{23}{33}$

4 a) $1.70°$ **b)** $1.10°$

6 0.8469 **7** $\tan A \tan B = \frac{1}{5}$

9 a) $\sqrt{1-p^2}$ **b)** $\sqrt{1-q^2}$ **c)** $p\sqrt{1-q^2}+q\sqrt{1-p^2}$ **d)** $\sqrt{1-p^2}\sqrt{1-q^2}+pq$

10 a) $\frac{3}{\sqrt{10}}$ **b)** $\frac{2}{\sqrt{5}}$ **c)** $\frac{1}{\sqrt{2}}$ **d)** $\frac{7\sqrt{2}}{10}$ **e)** $-\frac{1}{7}$

11 $\frac{6}{43}$ **13** $\frac{\cot x \cot y - 1}{\cot x + \cot y}$

14 $\frac{(\tan A + \tan B + \tan C - \tan A \tan B \tan C)}{(1 - \tan B \tan C - \tan A \tan B - \tan A \tan C)}$

Exercise 17B page 287

1 a) $\frac{24}{25}$ **b)** $\frac{7}{25}$ **c)** $\frac{24}{7}$

4 a) $2-\sqrt{3}$ **b)** $\sqrt{\frac{1}{2}\left(1+\frac{1}{\sqrt{2}}\right)}$ **c)** $\frac{1}{2}\sqrt{2-\sqrt{3}}$

6 $3\sin x - 4\sin^3 x$

7 $\frac{2t}{1-t^2}, \frac{2t}{1+t^2}, \frac{1-t^2}{1+t^2}$

Exercise 17C page 289

1 a) $\sin 60° + \sin 20°$ **b)** $\sin 90° + \sin 30°$ **c)** $\cos 100° + \cos 40°$

    d) $\cos 60° - \cos 80°$ **e)** $\sin\frac{7\pi}{12} + \sin\frac{\pi}{12}$ **f)** $\cos\frac{\pi}{12} - \cos\frac{7\pi}{12}$

2 a) $2\sin 40° \cos 20°$ **b)** $2\sin 7° \cos 23°$ **c)** $2\cos 30° \cos 20°$

    d) $2\sin 60° \sin 20°$ **e)** $2\sin\frac{\pi}{24}\cos\frac{7\pi}{24}$ **f)** $2\cos\frac{5\pi}{12}\cos\frac{\pi}{12}$

5 a) $1.64; -15°$ **b)** $1.98; \frac{5\pi}{24}$

6 $\cos 4x + \cos 2x;\ \cos 6x + \cos 2x + \cos 4x + 1$

Exercise 17D page 290

1 a) $\sqrt{10}(\sin(\theta + \tan^{-1}\frac{1}{3}))$ **b)** $5(\sin(\theta + \tan^{-1}\frac{4}{3}))$ **c)** $\sqrt{2}(\sin(\theta - \tan^{-1}1))$

    d) $\sqrt{0.34}(\sin(\theta + \tan^{-1}\frac{3}{5}))$ **e)** $\sqrt{244}(\sin(\theta - \tan^{-1}\frac{5}{6}))$ **f)** $\sqrt{130}(\sin(\theta + \tan^{-1}\frac{9}{7}))$

2 a) $\sqrt{10};\ 71.6°$ **b)** $5;\ 36.9°$ **c)** $\sqrt{2};\ 135°$

    d) $\sqrt{0.34};\ 59.0°$ **e)** $\sqrt{244};\ 129.8°$ **f)** $\sqrt{130};\ 37.9°$

4 greatest $1,\ x = 4.07$; least $\frac{1}{11},\ x = 0.927$ **5** $10\,000\sin^8(x + \tan^{-1}\frac{1}{3})$

Exercise 17E page 292

1 a) (i) $10°$, $130°$ (ii) $86.9°$, $13.1°$ (iii) $61.3°$ (iv) $34.9°$

  b) (i) $0°$, $45°$, $180°$ (ii) $77.3°$ (iii) $37.8°$, $142.2°$ (iv) $22.5°$, $112.5°$

  c) (i) $0°$, $90°$, $180°$ (ii) $0°$, $90°$, $75.5°$, $180°$ (iii) $45°$, $135°$ (iv) $90°$, $15°$, $75°$

  d) (i) $10.2°$ (ii) $74.4°$ (iii) $40.2°$ (iv) $160.6°$

2 a) 1.80 **b)** 0.569, 2.77

  c) 0, $\dfrac{\pi}{2}$, $\dfrac{5\pi}{6}$, π **d)** $\dfrac{\pi}{2}$

3 a) $142.5°$, $357.5°$ **b)** $60°$, $240°$ **c)** $0°$, $180°$, $30°$, $210°$, $150°$, $330°$, $360°$

  d) $90°$, $270°$, $199.5°$, $340.5°$

4 a) $\dfrac{\pi}{2}$, $\dfrac{\pi}{6}$, $\dfrac{5\pi}{6}$ **b)** 0.675 **c)** 0, π, $\dfrac{\pi}{6}$, $\dfrac{5\pi}{6}$,   **d)** 0.615, 2.53

5 a) 0.140, 3.28 **b)** 0.292, 2.59

  c) 3.97, 5.85 **d)** 0.361, $\dfrac{\pi}{2}$, 2.78, 3.50, $\dfrac{3\pi}{2}$, 5.92

6 a) $90°$, $0°$, $60°$, $120°$, $180°$ **b)** $0°$, $90°$, $180°$, $120°$

  c) $0°$, $90°$, $180°$ **d)** $90°$, $45°$

7 a) $A = 52.5$, $B = 7.5$ **b)** $A = 2.84$, $B = -0.30$

Examination questions page 294

1 $0.000\,11$

2 $\sqrt{13}\sin(x + 33.7°)$ **a)** -609.3 **b)** $130.2°$, $342.4°$

3 b) $17.7°$, $137.7°$

4 (ii) $\dfrac{\pi}{4}$ (iii) $\dfrac{\pi}{4}$ (iv) 2

18 Differentiation of other functions

Exercise 18A page 299

1 a) $10(2x + 3)^4$ **b)** $24(4x - 1)^5$ **c)** $-16(1 - 2x)^7$ **d)** $36(2x + 2)^5$

  e) $-32(1 + 2x)^3$ **f)** $4(1 + x)^3 + 5(2 - x)^4$ **g)** $(2x + 1)^{-1/2}$ **h)** $-1.5(3x + 2)^{-1.5}$

  i) $-2(2x + 5)^{-2}$ **j)** $-9(3x - 1)^{-4}$ **k)** $1.5(3x + 2)^{-0.5}$ **l)** $\frac{2}{3}(2x - 1)^{-2/3}$

  m) $-2(2x + 1)^{-2}$ **n)** $-18(3x + 1)^{-4}$ **o)** $-(2x + 1)^{-3/2}$ **p)** $-3(3x - 1)^{-4/3}$

2 $y = 576x - 512$ **3** $y = \frac{2}{3}x + 1$ **4** $8y + x = 7$

5 $y = x + 1$ **6 a)** $4(2x + 3)$ **b)** $8x + 12$

7 $(0, 1)$, minimum; $(-2, -3)$, maximum   **8** $(-1, -6)$, minimum

9 a) $2\sqrt{1 + 8x^3}$ **b)** $\sqrt{1 + (x + 3)^3}$ **c)** $-2\sqrt{1 + (1 - 2x)^3}$

Exercise 18B page 302

1 a) $2\cos x$ **b)** $-3\sin x$ **c)** $2\cos 2x$ **d)** $-4\sin 4x$

  e) $2\cos(2x + 1)$ **f)** $-2\cos(3 - 2x)$ **g)** $-3\sin(3x - 1)$ **h)** $-2\sin\left(2x - \dfrac{\pi}{3}\right)$

2 a) $y = 2x$ **b)** $y = -2\sqrt{3}x + 2 + \dfrac{2\sqrt{3}\pi}{3}$

3 a) $y = \dfrac{2}{3}x + \dfrac{\sqrt{3}}{2} - \dfrac{\pi}{9}$ **b)** $x = \dfrac{\pi}{3}$

4 a) $(0.280, -0.185)$, minimum **b)** $(0.322, 5)$, maximum

Exercise 18C page 305

1 a) $e^x + 2x$ **b)** $3 - 2e^x$ **c)** $3e^{3x-2}$ **d)** $-2e^{1-x}$

2 $y = x + 1$ **3** $y = 2ex - 2e$ **4** $y = x$

5 $(-0.549, 0.385)$, maximum **6** $(-1.5, 8.96)$, minimum

7 $a = 2$, $b = -2$, $c = 1$ **8** $0.001\,05$ cm per day **9** $0.001\,81$ kg per hour

Exercise 18D page 307

1 a) $2x + \dfrac{1}{x}$ **b)** $\dfrac{1}{x+1}$ **c)** $\dfrac{2}{2x+3}$

d) $\dfrac{1}{x-1}$ **e)** $\dfrac{-3}{2-3x}$ **f)** $\dfrac{1}{x\ln 2}$

g) $\dfrac{1}{(x+1)\ln 10}$ **h)** $10^x \ln 10$ **i)** $3^x \ln 3$

2 a) $y = x - 1$ **b)** $y = 3x$

4 a) $(0, 1)$, minimum **b)** $(1, 1)$, minimum **c)** $(0, -1)$, maximum **d)** $(0, 0)$, minimum

Exercise 18E page 309

1 a) $-\cos x + \sin x + c$ **b)** $-2\cos x - 3\sin x + c$ **c)** $-\frac{1}{2}\cos 2x + c$

d) $\frac{1}{3}\sin(3x + \frac{\pi}{3}) + c$ **e)** $\frac{1}{2}\cos(\frac{\pi}{6} - 2x) + c$ **f)** $-\sin(\pi - 2x) + c$

g) $\frac{1}{2}e^{2x} + c$ **h)** $-e^{-x} + c$ **i)** $\frac{1}{2}e^{2x-3} + c$

j) $\frac{1}{2}\ln(2x + 3) + c$ **k)** $-\ln(3 - x) + c$ **l)** $-\frac{2}{3}\ln(3 - 3x) + c$

m) $\frac{1}{8}(2x + 1)^4 + c$ **n)** $-3(1 - 4x)^6 + c$ **o)** $-0.4(2 - 0.5x)^5 + c$

p) $-2(x + 3)^{-3} + c$ **q)** $-\frac{1}{2}(2x + 3)^{-5} + c$ **r)** $\frac{1}{3}(2x + 3)^{3/2} + c$

s) $\frac{2}{9}(3x + 1)^{3/2} + c$ **t)** $\frac{3}{8}(2x + 1)^{4/3} + c$ **u)** $2\sqrt{1 + x} + c$

2 a) 0.841 **b)** $\frac{1}{3}$ **c)** $e - 1$

d) 0.0585 **e)** 2.30 **f)** 1.20

g) 1302 **h)** $\frac{6}{7}$ **i)** 19.0

3 a) 2 **b)** 1 **c)** 0.865 **d)** 2.30 **e)** 0.733

5 a) $f(1) - f(0)$ **b)** $\frac{1}{2}(f(4) - f(2))$ **c)** $f(2) - f(1)$

Examination questions page 310

1 (i) $\frac{\pi}{3} + \frac{3}{4}$ **(ii)** $y = \frac{1}{3}e^{3x} + \frac{2}{3}$

2 (ii) $(0, 1)$, maximum **(v)** $\frac{1}{2}$ square unit

Consolidation section D

Extra questions page 312

2 $x = 3$, $y = 1$

3 $\mathbf{a} = \begin{pmatrix} 2 \\ 1 \end{pmatrix}$, $\mathbf{b} = \begin{pmatrix} 1 \\ 3 \end{pmatrix}$

4 $x = 1$, $y = 2$

5 $\sqrt{5}$, $\sqrt{10}$, 5; $\begin{pmatrix} \frac{2}{\sqrt{5}} \\ \frac{1}{\sqrt{5}} \end{pmatrix}$, $\begin{pmatrix} \frac{1}{\sqrt{10}} \\ \frac{3}{\sqrt{10}} \end{pmatrix}$

6 $45°$

7 $k = -1$; $30°$

10 $a = 2.3$, $b = 1.7$

11 a) $\frac{1}{2} \sin x + \frac{\sqrt{3}}{2} \cos x$

b) $\dfrac{\tan x - \sqrt{3}}{1 + \sqrt{3} \tan x}$

12 a) $2 \sin 5x \cos 2x$

b) $-2 \sin (x - 5°) \sin 15°$

13 $\sqrt{20}$; 1.30, 5.91

14 a) $39.8°$, $140.2°$

b) 0, $\dfrac{\pi}{3}$, $\dfrac{2\pi}{3}$, π, $\dfrac{\pi}{4}$

15 a) $5(2x - 3)^{1.5}$

b) $-\frac{3}{2}(1 - 3x)^{-\frac{1}{2}}$

c) $2 \cos (2x - 5)$

d) $-6 \sin (2x + 1)$

e) $2 e^{2x+1}$

f) $\dfrac{-3}{1 - 3x}$

16 a) $\frac{1}{32} (4x - 7)^8 + c$

b) $\sqrt{(2x + 1)} + c$

c) $\cos (1 - 2x) + c$

d) $2 \sin (0.5x - 2) + c$

e) $\frac{1}{3} e^{3x+1} + c$

f) $\frac{1}{2} \ln (2x - 3) + c$

Mixed exercise page 313

1 $51.3°$

3 $t = 1$ or $t = -2$

5 $-0.259\mathbf{i} + 0.966\mathbf{j}$ or $0.966\mathbf{i} - 0.259\mathbf{j}$

6 a) $(\frac{2}{3}, \frac{1}{3})$

b) $(-1, 2)$

c) $(2, -1)$

9 translation $\frac{\pi}{2}$ to right

10 $5 \sin (x + 0.927)$

11 a) $3 \cos x - 4 \sin x$

b) $5 \cos (x + 0.927)$

12 enlargement, scale factor 3, centre $(0, 0)$

13 $y = g(4 - x)$

14 a) $\sqrt{\dfrac{1}{2} - \dfrac{\sqrt{2}}{4}}$

b) $\sqrt{\dfrac{1}{2}\left(1 + \dfrac{\sqrt{3}}{2}\right)}$

16 $1 - x + x^2 - x^3 + x^4$; $x - \frac{1}{2}x^2 + \frac{1}{3}x^3 - \frac{1}{4}x^4 + \frac{1}{5}x^5$; $0.182\,33$

Longer exercise page 315
Cubic equations
2 -0.174, 0.940, -0.766

3 -0.279, -0.692, 0.971

4 1.88, -0.347, -1.53

Parabolic telescopes
1 $\dfrac{x^2 - \frac{1}{4}}{x}$

2 $2x$

Method of least squares
4 $a = \dfrac{n\Sigma x_i y_i - \Sigma x_i \Sigma y_i}{n\Sigma x_i^2 - (\Sigma x_i)^2}$, $b = \dfrac{\Sigma x_i y_i \Sigma x_i - \Sigma x_i^2 \Sigma y_i}{(\Sigma x_i)^2 - n\Sigma x_i^2}$; $y = 1.6x + 10.4$

Examination questions page 316
1 h, $h + H$

2 a) $\mathbf{m} = 2\mathbf{i} + \mathbf{j} + \mathbf{k}$ **b)** $c\mathbf{i} + c\mathbf{j} + c\mathbf{k} + t(2\mathbf{i} + \mathbf{j} + \mathbf{k})$; $-c\mathbf{i}$

3 $a = 1.6$, $b = 2.2$

4 $3 \sin \frac{1}{3}x + \frac{1}{2}e^{2x} + c$; 270

5 $-\frac{5}{6}\mathbf{a} + \frac{2}{3}\mathbf{b}$

6 a) (ii) $2 \cos 2x$ **b)** $1 - \frac{1}{2} \sin^2 2x$

Progress tests

Test 1 page 318

1 $a = 1$, $b = -14$ **2** $\frac{3}{5} - \frac{2}{5}\sqrt{41}$, $\frac{3}{5} + \frac{2}{5}\sqrt{41}$ **3** ± 0.954

4 13   **5 a)** $\frac{4}{3}$ **b)** 1.465   **c)** $\frac{1}{6}$

6 $y = 4x - 12$ **7** $1 + \frac{1}{3}x - \frac{1}{9}x^2 + \frac{5}{81}x^3$; 9.966 55

8 $(1, 0)$, $(-3, 0)$; $10\frac{2}{3}$ **9** 0.467, 2.675, 6.750, 8.958   **11** 700

Test 2 page 318

1 $h = 5t^2$; 4.47 seconds   **2** $4y + x = 9$; $(\frac{13}{17}, 2\frac{1}{17})$ **3 a)** $4 - \sqrt{2}$ **b)** $4 + \frac{7}{2}\sqrt{2}$

4 300 cm s^{-1}; 11.18 cm   **5** 1.436; 53.5 sq. in.   **6** 1200 cm^2

7 138°

8 a) stretch, parallel to y-axis, factor 3   **b)** translation to right by 2

9 $\frac{29}{37}$, $1\frac{7}{8}$ **10 a)** $-\frac{1}{3}\cos 3x + c$ **b)** 26   **11** $\frac{1}{3}$

Test 3 page 319

1 $25\,000 - 1000(x - 5)^2$; £5   **2** $x = 4$ or $x = \frac{1}{2}$ **3** 2.28 cm

4 a) $8x + 15x^2$ **b)** $3.5x^{-\frac{1}{2}} - 4x^{-5}$ **5** 38.7°

6 $(\frac{1}{3}, -2\frac{23}{27})$, maximum; $(1, -3)$, minimum 1   **7** $10\frac{2}{3}$ **8** 60°, 48.2°

9 $\mathbf{r} = \mathbf{i} + 2\mathbf{j} + 4\mathbf{k} + s(2\mathbf{i} - \mathbf{j})$; yes   **10** $\dfrac{F}{v} = a + bv$; $a = 0.47$, $b = 0.01$ **11** 18

Test 4 page 320

2 $(4, 3)$; $(3, \frac{7}{3})$ **3** $-\frac{1}{2} < x < \frac{1}{2}$; $\frac{1}{4}$ **4 a)** $2x^2$ **b)** $\frac{1}{2}$

5 a) $65\,536 + 393\,216\,x + 1032\,192\,x^2$ **b)** $1 - \frac{1}{4}x - \frac{3}{32}x^2$

6 $(4, -4)$ and $(-\frac{2}{3}, -4\frac{14}{27})$ **8** 0.547, 2.39

9 a) $30(1 + 3x)^9$ **b)** $-2 \sin 2x - 3 \cos 3x$ **c)** $12e^{12x}$

10 $y = 2x^3 + 1$ **11** 0.000 4999

19 Inequalities

Exercise 19A page 323

1 a) $x < 6$ **b)** $x > 6$ **c)** $x \geq -2$ **d)** $x \leq 6$ **e)** $x < 3$
  f) $x \leq 2$ **g)** $x < 6$ **h)** $x \geq 12$ **i)** $x > -8$ **j)** $x \leq 9$
  k) $x > \frac{1}{11}$ **l)** $x \leq 4$ **m)** $x > 1.2$ **n)** $x < 3.5$ **o)** $x \geq 0.8$

2 a) always true   **b)** never true

3 $x < \dfrac{c-b}{a}$ if $a > 0$, $x > \dfrac{c-b}{a}$ if $a < 0$

Exercise 19B page 325

1 a) $2 < x < 6$ **b)** $3 < x < 7$ **c)** $-1 \leq x \leq 3$

2 a) $0.335 \leq x < 0.345$ **b)** $32\,350 \leq x < 32\,450$
  c) $41\,500 \leq x < 42\,500$ **d)** $0.026\,35 \leq x < 0.026\,45$

Exercise 19C page 326

2 a) $x \geq 0$, $y \geq 0$, $x+y < 4$ **b)** $y \geq 0$, $x+y < 3$, $y < x$
  c) $x \geq 0$, $y \geq 0$, $x+y > 2$, $x+y < 4$

Exercise 19D page 327

1 a) $-1 < x < 2$ **b)** $x < -5$ or $x > -2$ **c)** $0 \leq x \leq 1.5$
  d) $-2 \leq x \leq 1$ **e)** $x \leq -3$ or $x \geq 2$ **f)** all x

2 $\alpha \leq x \leq \beta$

3 a) $x < 1$ or $2 < x < 3$ **b)** $-1 < x < 0$ or $x > 2$ **c)** $x > 0$

Exercise 19E page 328

1 a) $2 < x < 3$ **b)** $-5 < x < 2$ **c)** $x \leq 3$ or $x \geq 5$
  d) $-2 < x < 1$ **e)** $x \leq -4$ or $x \geq 6$ **f)** $x = 2$

2 a) $x < 0$ or $1 < x < 2$ **b)** $-2 < x < -1$ or $x > 3$ **c)** $x > 1$
  d) $x \leq 0$ **e)** $x \geq 1$ or $x = 0$ **f)** $x < 0$ or $1 < x < 2$ or $x > 3$

Exercise 19F page 329

1 a) $-1 < x < 1$ **b)** $-\sqrt{3} < x < \sqrt{3}$ **c)** $-\sqrt{1.5} \leq x \leq \sqrt{1.5}$
  d) $-\sqrt{7} < x < \sqrt{7}$ **e)** $-\sqrt[4]{3} < x < \sqrt[4]{3}$ **f)** always true

2 a) $-\sqrt{3}+1 < x < \sqrt{3}+1$ **b)** $x < -\sqrt{5}-3$ or $x > \sqrt{5}-3$
  c) $-\sqrt{5}+2 \leq x \leq \sqrt{5}+2$ **d)** $\frac{1}{2}(1 - \sqrt{2}) < x < \frac{1}{2}(1 + \sqrt{2})$
  e) $x < \frac{1}{3}(7 - \sqrt{3})$ or $x > \frac{1}{3}(7 + \sqrt{3})$ **f)** $x < 2$ or $x > 3$

3 a) $2 - \sqrt{2} \leq x \leq 2 + \sqrt{2}$ **b)** $1 - \sqrt{2} < x < 1 + \sqrt{2}$ **c)** $x < -3 - \sqrt{7}$ or $x > -3 + \sqrt{7}$
  d) $x < 1.5 - \sqrt{9.25}$ or $x > 1.5 + \sqrt{9.25}$ **e)** $1 - \sqrt{6} < x < 1 + \sqrt{6}$
  f) $x < 1 - \sqrt{5}$ or $x > 1 + \sqrt{5}$

Exercise 19G page 331

1 $x < -2$ or $x > -1$ 2 $1 < x < 2$ 3 $x < -\frac{1}{2}$ or $x > 0$

4 $x < 0$ or $1 < x < 2$ 5 $-2 < x < -1$ or $x > 2$ 6 $x < -1$ or $0 < x < 3$

7 $x < -3$ or $x > 0$ 8 $x < 0$ or $1 < x < 2$ 9 $-1 < x < \frac{1}{2}$ or $x > 2$

10 $x < -1$ or $-\frac{1}{7} < x < 2$ 11 $-7 < x < -1$ or $x > 2$

Exercise 19H page 332

1 a) $-6 < x < 6$ b) $-2 \le x \le 2$ c) $x < -5$ or $x > 5$

   d) $x \le -7$ or $x \ge 7$ e) $-6 < x < 4$ f) $-1 \le x \le 7$

   g) $x < -2.5$ or $x > -1.5$ h) $x \le 0$ or $x \ge 2$ i) $-2 < x < 1$

   j) $-1 \le x \le \frac{5}{3}$ k) $-6 < x < 2$ l) $-5 < x < 1$

2 a) $x \le \frac{1}{2}$ b) $x > -\frac{1}{2}$ c) $x < -\frac{1}{2}$

   d) $-2 < x < \frac{4}{3}$ e) $x \le -\frac{1}{4}$ or $x > \frac{1}{2}$ f) $x < 0$ or $x > 2$

   g) $x < -2$ or $x > -\frac{2}{3}$ h) $x < -\frac{1}{2}$ i) $-1 \le x \le 1.4$

3 a) $|x - 5| \le 1$ b) $|x - 3| > 2$ c) $|x - 510| < 5$ d) $|x - 0.34| < 0.005$

Exercise 19I page 333

1 $x > \frac{1}{2}$ 2 $x > -\frac{1}{2}$ 3 always true

4 $x > 1$ 5 $x < -1$ 6 $-\frac{1}{3} < x < 1$

Exercise 19J page 333

1 a) $x < 4$ b) $x < 1$ or $x > 7$ c) $-7 < x < \frac{7}{3}$

   d) $x \ge \frac{7}{4}$ e) $3 < x < 3.5$ f) $0 < x < \frac{1}{2}$ or $x > 3$

   g) $x < 1$ h) $-\frac{2}{3} < x < 2$ i) $-5 < x < 1$

   j) $x < \frac{5}{3}$ or $x > 7$ k) $x < 0.5 - \sqrt{5.25}$ or $x > 0.5 + \sqrt{5.25}$ l) $x = 1$

2 $x = 5$ or $x = 12$; $x = -3$ or $x = 20$; $-3 < x < 5$ or $12 < x < 20$

Longer exercise page 334

3 £$(200x + 290y)$ 4 three of each   5 six type A, four type B

Examination questions page 334

1 $x < -5$ or $x > 2$

2 (i) $x < -2$ or $x > 3$ (ii) $\frac{2}{3} < x < 4$

20 Differentiation of compound functions

Exercise 20A page 337

1 a) $2x \cos x - x^2 \sin x$ b) $e^x + xe^x$ c) $e^x \cos x - e^x \sin x$

   d) $(2x + 3)e^x + (x^2 + 3x + 1)e^x$ e) $(6x + 12x^2)(2 - 4x^4) - 16x^3(1 + 3x^2 + 4x^3)$

f) $2e^x \cos x$

g) $\frac{1}{2}x^{-\frac{1}{2}} \sin x + \sqrt{x} \cos x$

h) $e^x \ln x + \dfrac{e^x}{x}$

i) $\ln x + 1$

j) $(2x + 2) \ln x + x + 2 - \dfrac{1}{x}$

k) $\cos x \ln x + \dfrac{\sin x}{x}$

l) $\dfrac{1}{x} \cos x - \sin x \ln x$

m) $\sin 2x + 2x \cos 2x$

n) $-3e^x \sin(3x - 1) + e^x \cos(3x - 1)$

o) $-e^{-x}(x + 3) + e^{-x}$

p) $\ln(2x - 1) + \dfrac{x}{2x - 1}$

q) $2xe^{3x-1} + 3x^2 e^{3x-1}$

r) $2 \cos 2x \cos 3x - 3 \sin 2x \sin 3x$

s) $\sqrt{x + 1} + \dfrac{x}{2\sqrt{x + 1}}$

t) $2x\sqrt{x - 1} + \dfrac{x^2 + 1}{2\sqrt{x - 1}}$

u) $(2x - 1)^{-\frac{1}{2}} - x(2x - 1)^{-1\frac{1}{2}}$

3 $\left(1, \dfrac{1}{e}\right)$, maximum

4 $\left(\dfrac{\pi}{4}, \dfrac{e^{\frac{\pi}{4}}}{\sqrt{2}}\right)$ maximum

5 $y = x + 1$

6 $y = 3ex - 2e^2$

7 $u'vw + uv'w + uvw'$

8 $a^2(1 + \cos \theta) \sin \theta;$ $\dfrac{a^2 3\sqrt{3}}{4}$

Exercise 20B page 339

1 a) $\dfrac{-x \sin x - 2 \cos x}{x^3}$

b) $\dfrac{e^x(\sin x - \cos x)}{\sin^2 x}$

c) $\dfrac{1 - x}{e^x}$

d) $\dfrac{(x^2 + 1) - 2x^2}{(x^2 + 1)^2}$

e) $\dfrac{(x^2 + x + 1) - (x + 1)(2x + 1)}{(x^2 + x + 1)^2}$

f) $\dfrac{1 - \ln x}{x^2}$

g) $\dfrac{-\ln x \sin x - \frac{1}{x} \cos x}{(\ln x)^2}$

h) $\dfrac{-2 \sin 3x \sin 2x - 3 \cos 3x \cos 2x}{\sin^2 3x}$

i) $\dfrac{2xe^{2x-1} - 2e^{2x-1}}{x^3}$

2 $k = -1$

3 a) $(1, \frac{1}{4})$, maximum   **b)** $(-1, 3)$, maximum; $(1, \frac{1}{3})$, minimum

c) $(-1, -\frac{1}{4})$, minimum; $(0, 0)$, point of inflection; $(1, \frac{1}{4})$, maximum

4 $y = x$

Exercise 20C page 341

3 a) $3 \sec^2 3x$

b) $2 \sec(2x - 1) \tan(2x - 1)$

c) $-4 \operatorname{cosec}^2\left(4x - \dfrac{\pi}{2}\right)$

d) $\tan x + x \sec^2 x$

e) $-\dfrac{1}{x^2} \operatorname{cosec} x - \dfrac{1}{x} \operatorname{cosec} x \cot x$

f) $e^x(\cot x - \operatorname{cosec}^2 x)$

4 $y = 2x + 1 - \dfrac{\pi}{2}$

5 $y + 0.178x = 2.28$

Exercise 20D page 342

1 a) $2x \cos x^2$

b) $2xe^{x^2}$

c) $\dfrac{2}{x}$

d) $\dfrac{-x}{\sqrt{1-x^2}}$

e) $\dfrac{1}{2x\sqrt{\ln x}}$

f) $e^x \cos e^x$

g) $-2\cos x \sin x$

h) $2\tan x \sec^2 x$

i) $\frac{1}{2}\sec^2 x(\tan x)^{-\frac{1}{2}}$

j) $-\tan x$

k) $\sec x$

l) $20x(1+x^2)^9$

m) $8\left(1-\dfrac{1}{x^2}\right)\left(x+\dfrac{1}{x}\right)^7$

n) $\dfrac{1}{2\sqrt{x}}\cos(1+\sqrt{x})$

o) $\sin x \sin(\cos x)$

2 $(-1,\ \ln 2)$, minimum; $(1,\ \ln 2)$, minimum   **3** $(-1,\ 2)$, maximum   **4** $y = \cos 1$

5 a) $2x\sqrt{1+x^6}$

b) $\frac{1}{3}x^{-\frac{2}{3}}\sqrt{1+x}$

c) $e^x\sqrt{1+e^{3x}}$

Exercise 20E page 344

1 a) $\dfrac{3}{\sqrt{1-9x^2}}$

b) $\dfrac{2}{1+4x^2}$

c) $\dfrac{2}{\sqrt{4x-4x^2}}$

d) $\dfrac{x}{\sqrt{1-x^2}} + \sin^{-1}x$

e) $\dfrac{x^2}{1+x^2} + 2x\tan^{-1}x$

f) $\dfrac{2x}{\sqrt{1-x^4}}$

g) $-\dfrac{1}{x^2+1}$

h) $\dfrac{e^x}{\sqrt{1-e^{2x}}}$

i) $\dfrac{2\sin^{-1}x}{\sqrt{1-x^2}}$

2 a) $\dfrac{-1}{\sqrt{1-x^2}}$

b) $\dfrac{-1}{1+x^2}$

c) $\dfrac{1}{x\sqrt{x^2-1}}$

d) $\dfrac{-1}{x\sqrt{x^2-1}}$

3 1

4 $\dfrac{1}{1+x^2}$, $k = \dfrac{\pi}{4}$

Exercise 20F page 345

1 a) $\cos x - x\sin x$

b) $e^x(\cos x - \sin x)$

c) $-e^x \sin e^x$

d) $-\sin x \ln x + \frac{1}{x}\cos x$

e) $\dfrac{\frac{1}{x}\cos x + \sin x \ln x}{\cos^2 x}$

f) $\cot x$

Exercise 20G page 346

1 a) $2x\sqrt{1+x^2} + x^3(1+x^2)^{-\frac{1}{2}}$

b) $\dfrac{x}{\sqrt{1+x^2}}\cos\sqrt{1+x^2}$

c) $\dfrac{x}{\sqrt{1+x^2}}e^{\sqrt{1+x^2}}$

2 $\dfrac{-2x}{1-x^2}$ **a)** $\ln(1-x^2) - \dfrac{2x^2}{1-x^2}$

b) $\dfrac{\cos x \ln(1-x^2) + \dfrac{2x\sin x}{1-x^2}}{(\ln(1-x^2))^2}$

c) $\dfrac{-2x}{(1-x^2)\ln(1-x^2)}$

3 a) $\cos x^2 - 2x^2 \sin x^2$

b) $-\frac{1}{2}(\sin x)(\cos x)^{-\frac{1}{2}}\cos\sqrt{\cos x}$

c) $\dfrac{1}{x\sqrt{x^2-1}}$

d) $\dfrac{-e^x \sin e^x}{x^2} - \dfrac{2\cos e^x}{x^3}$

e) $-2x\sin x^2 \cos(\cos x^2)$

f) $\dfrac{1}{x\ln x \ln(\ln x)}$

4 $(2,\ \dfrac{\sqrt{6}}{2})$, maximum

5 $\dfrac{3\sqrt{3}}{4}$

6 $\dfrac{u^2}{(u-f)}$, $4f$

7 $\dfrac{4\sqrt{3\pi r^3}}{9}$

8 $\dfrac{32\pi r^3}{81}$

Examination questions page 347

2 b) 4   **c)** $\dfrac{180}{(x^2 + 9)^{\frac{3}{2}}}$ **d)** minimum   **e)** least £228, greatest £252

3 (i) $-2xe^{-x^2}$ **(ii)** $e^{-x^2}(4x^2 - 2)$, $x = \pm\dfrac{1}{\sqrt{2}}$

21 Approximations

Exercise 21A page 351

1 a) 0.05, 0.01, 1%   **b)** 5000, 0.01, 1%   **c)** 0.0005, 0.002, 0.2%   **d)** 50 000, 0.04, 4%

   e) £1, 0.01, 1%   **f)** 1000, 0.002, 0.2%   **g)** 0.01, 0.002, 0.2%

2 a) 3.1 − 3.3   **b)** 0.1225 − 0.1235   **c)** 50 500 − 51 500

3 a) 0.14, 0.045   **b)** 0.0013, 0.0004   **c)** 0.000 000 3, 0.000 000 08

4 sin: 0.0002, 0.002; cos: 0.000 004, 0.000 004   **5** 0.005

Exercise 21B page 354

1 a) $(x + y) \pm (h + k)$ **b)** $(x - y) \pm (h + k)$ **c)** $(2x - 3y) \pm (2h + 3k)$

2 a) 0.1   **b)** 0.1   **c)** 0.15   **d)** 0.45

   e) 2.6   **f)** 0.02   **g)** 1   **h)** 0.04

3 84.4 seconds   **4** 1 hour, 0.9 hour   **5** 134.2 cm^2

6 10 000 cm^3, 2800 cm^2 **7** 150 cm^3

Exercise 21C page 357

1 a) 1.8   **b)** 0.45   **c)** 1.6

2 a) 0 and 1: 0.68   **b)** 1 and 2; 1.4

3 2 and 3; (c); 2.09   **4** 0.322; $x^3 + 3x - 1 = 0$

5 a) 1.73; $x^2 - 3 = 0$ **b)** 0.111; $x^3 + 9x - 1 = 0$

Exercise 21D page 359

1 a) 1.213 412   **b)** 0.739 085   **c)** 0.169 193

2 2.6053   **3** 1.414 214   **4** 1.259 921

Exercise 21E page 361

1 a) 0.8   **b)** 1.4   **c)** 1.6

2 a) −1 and 0; −0.7   **b)** 1 and 2; 1.1   **c)** 6 and 7; 6.9

3 2.3

Exercise 21F page 362

1 a) 0.5   **b)** 0.8   **c)** 0.9, −3.6, −1.9

2 a) 1.3 **b)** 1.2 **c)** −2.6, 0, 2.6

3 a) 0.4 **b)** 0, 1.9 **c)** 0.3

Examination questions page 362
1 a) (i) 0.6, 0.4 (ii) 8.625, 5.58 **b)** 1.5%

2 $\frac{3}{2}$, $\frac{17}{12}$; 0.000 002

3 2.094 55

22 Handling data

Exercise 22A page 367
1 a) 13.4, 13 **b)** 108.125, 107.5

2 1.97, 2 **3** 2.49, 2, 1 **4** £13 600, £14 250

5 990, 480 **7** Upper working **8** Disapprove

9 60.6 **10** 12.3 **12** $x = 26$, $y = 23$

Exercise 22B page 372
1 a) 5.10 **b)** 66.6

2 A: 3.56, 2.3264 B: 3.96, 0.84 **3** A: 66.3, 2.9 B: 67.1, 9.72

4 $x = 7$, $y = 9$ **5** 54.04, 11.1 **6** 15.3, 12.9

7 1.58 m, 0.493 m

Exercise 22D page 379
1 £7, £1.70, 0.57 **2** −200, 180 **3** 6 days, 0.34

Examination questions page 381
1 36.9 years, 35.6 years

2 (i) £208.42, £18.30 (ii) £4644.4 (iii) £246.39, £38.35

Consolidation section E

Extra questions page 387
1 a) $x > 1$ **b)** $x \le 4$ **c)** $x < -4$

d) $-3 < x < 10$ **e)** $x \le -9$ or $x \ge 5$ **f)** $0 < x < \frac{1}{2}$ or $x > 3$

g) $-2 < x < 3$ **h)** $-1 < x < -\frac{1}{3}$ **i)** $-\frac{1}{3} < x < 1$ or $x > 3$

j) $-4 < x < 2$ **k)** $x \le -3$ or $x \ge 0$ **l)** $x < -\frac{1}{3}$ or $x > 1$

2 (1, 2), (1, 3), (2, 2), (3, 1), (3, 2), (4, 1), (5, 1) **3** $x < \frac{1}{3}$ or $x > 1$

4 a) $3x^2 \sin x + x^3 \cos x$ **b)** $e^x(\sin x + \cos x)$ **c)** $(2x + 3)\ln x + x + 3 + \frac{1}{x}$

d) $\dfrac{x \cos x - \sin x}{x^2}$ **e)** $\dfrac{(1 + x)(\frac{1}{2}x^{-\frac{1}{2}}) - \sqrt{x}}{(1 + x)^2}$

f) $\dfrac{(x^2 + x + 5)(2x + 2) - (x^2 + 2x + 3)(2x + 1)}{(x^2 + x + 5)^2}$

g) $4 \sec^2 4x$ **h)** $-2 \sec(-2x) \tan(-2x)$ **i)** $-0.5 \csc 0.5x \cot 0.5x$

j) $-3x^2 \sin(x^3 + 1)$ **k)** $\dfrac{2x + 1}{x^2 + x + 1}$ **l)** $2x^3(1 + x^4)^{-\frac{1}{2}}$

m) $\dfrac{4}{1 + 16x^2}$ **n)** $\dfrac{1}{\sqrt{1 - (x + 1)^2}}$ **o)** $\dfrac{2}{1 + (2x + 1)^2}$

5   0.005, 0.0015   **6** $a + b$, $2a + 3b$; $\dfrac{a}{A} + \dfrac{b}{B}$, $\dfrac{a}{A} + \dfrac{b}{B}$, $\dfrac{2a}{A}$

7   0.8   **8**   4 and 5, 4.3778   **9**   first

10   A: 91.9, 4.39; B: 89.7, 1.1

11   **a)** 24, 22   **b)** 15 − 19   **d)** 18

Mixed exercise page 388
1   **a)** $2 \cos 2x$ **b)** $2 \cos^2 x - 2 \sin^2 x$ **3**   1.6268; 0.349

4   **a)** $x < -2.4$ or $0.4 < x < 1$ or $x > 1$ **b)** $0 < x < 0.6$ **5** $1 \le x \le 2$

6   **a)** $\dfrac{1}{x \ln x \ln(\ln x)}$ **b)** $\frac{1}{8} x^{-\frac{1}{2}}(1 + \sqrt{x})^{-\frac{1}{2}}(1 + \sqrt{1 + \sqrt{x}})^{-\frac{1}{2}}$

c) $\dfrac{1 - x(1 - x^2)^{-\frac{1}{2}}}{x + \sqrt{1 - x^2}}$

8   **a)** yes, to 2   **b)** no   **9** -0.5, 0.9, 3.7, $x < -0.5$, $0.9 < x < 3.7$

Longer exercise page 389

Finding π
1 $1 - x^2 + x^4 - \cdots$ **2** $x - \frac{1}{3}x^3 + \frac{1}{5}x^5 - \cdots$
Hyperbolic functions

1 $\dfrac{2}{e^x + e^{-x}}$, $\dfrac{2}{e^x - e^{-x}}$, $\dfrac{e^x + e^{-x}}{e^x - e^{-x}}$ **2** $\sinh x$, $\cosh x$

3   **a)** 1   **b)** e^x **c)** e^{-x} **4** $\dfrac{1}{\sqrt{x^2 - 1}}$, $\dfrac{1}{\sqrt{x^2 + 1}}$

Elasticity of demand
1 $\dfrac{2p^2}{1 - p^2}$ **2** $c = 20 - 0.5p$; $\dfrac{p}{40 - p}$ **4** pc; when $c + p\dfrac{dc}{dp} = 0$; when $E_d = 1$

Examination questions page 391
1 $\dfrac{\pi}{3} < \theta < \dfrac{\pi}{2}$, $\theta > \dfrac{2\pi}{3}$

2   **a)** $x < 2$, $2 < x < 6$ **b)** $m = 2$, $n = 1$, $p = -7$, $q = 6$; $t = -\frac{1}{2}$

3 $-1 < \dfrac{2x+3}{x+1} < 1$; $-2 < x < -\frac{4}{3}$; $-\dfrac{x+1}{x+2}$

4 **c)** 0.88

23 Integration as sum

Exercise 23A page 394

1 $\frac{1}{3}\pi a^2 h$ **2** $\dfrac{3L}{2}$ **3** $\frac{1}{3}\pi a^4$

4 $\dfrac{625}{3}\pi$ **5** $\frac{1}{2}kL^2$ **6** $5000\, d^2 w\,\text{N}$

7 $\frac{1}{2}k\pi a^4\ \text{cm}^3$ per second

Exercise 23B page 396

1 **a)** $\dfrac{128}{7}\pi$ **b)** $\dfrac{178}{15}\pi$ **c)** $\dfrac{256}{15}\pi$ **d)** $\dfrac{15}{2}\pi$

2 **a)** $\dfrac{3}{5}\pi$ **b)** $\dfrac{32}{5}\pi$ **c)** $\dfrac{15}{2}\pi$

3 $\frac{1}{3}\pi a^2 h$ **4** $\frac{4}{3}\pi a^3$

5 **a)** $\dfrac{9}{20}\pi$ **b** $\dfrac{1}{20}\pi$ **6 a)** $\frac{4}{3}\pi ab^2$ **b)** $\frac{4}{3}\pi a^2 b$ **7** $\dfrac{32}{3}\pi\ \text{cm}^3$

Exercise 23C page 399

1 **a)** 1.097 **b)** 1.162 **c)** 0.341 **d)** 1.762

2 **a)** 1 **b)** 0.987; 1.3%

3 **a)** $\frac{1}{3}$ **b)** 0.338; 1.4%

4 1.6 **5** 123

6 $1 + \frac{1}{2}x - \frac{1}{8}x^2 + \frac{1}{16}x^3$ **a)** 0.558 08 **b)** 0.557 96 **c)** 0.558 27; b) is better

7 **a)** 0.656 **b)** 0.9 **c)** 2.672 **8** 8.394

Exercise 23D page 403

1 **a)** 1.770 **b)** 1.0948 **c)** 0.3459 **d)** 1.191

2 **a)** 1 **b)** 1.000 135; 0.01% **4** 3.533

5 $1 - x + x^2 - x^3$ **a)** 0.405 465 **b)** 0.405 471 **c)** 0.401 042

6 **a)** 1.305 **b)** 7.8 **c)** 7.057

Longer exercise page 404

2 0.357, 0.503 **3** $4p + c = 40$; 50

Examination questions page 404

1 **a)** $\frac{32}{3}$ **b)** 38π

2 1.82 litres

3 (iii) 47 litres

24 Distances and angles

Exercise 24A page 408

1 a) 28.3 cm   **b)** 5 cm   **c)** 19.5°   **d)** 26.6°

2 45°; 54.7°

3 81.6 m   **5** 9.49 cm; 47.5°

6 28.3 cm;   19.5°;   33.6°   **8** $4 - 2\sqrt{2}$ m   **9** $\dfrac{1}{\sqrt{6}}$ cm; $\sqrt{\dfrac{3}{2}}$ cm

Exercise 24B page 410

1 a) $\sqrt{13}$, $\sqrt{20}$ **b)** 75.6°   **2** 132.5 m   **3** 30.8°

4 13.1 m   **5** 9.1°

Exercise 24C page 411

1 a) 36.6°   **b)** 42.9°   **2** 20.7°   **3** 35.3°

4 33.3°, 50.2°, 19.2°   **5 a)** 12.0°   **b)** 43.8°

Exercise 24D page 413

1 45°   **2** 54.7°   **3** 116.5°   **4** 54.7°   **5** 21.1°   **6** 64.9°

7 a) $\dfrac{a\sqrt{3}}{2}$ **b)** 104.5°

Exercise 24E page 414

1 $7\mathbf{k}$, $3\mathbf{i} + 4\mathbf{j} - 7\mathbf{k}$, $3\mathbf{i} - 7\mathbf{k}$; $\sqrt{74}$; 27.7°

2 $7\mathbf{j} + 3\mathbf{k}$, $-4\mathbf{i} + 7\mathbf{j} + 3\mathbf{k}$, $-4\mathbf{i} + 3\mathbf{k}$; 66.6°

3 $(\frac{1}{2}, \frac{1}{2}, 2)$; 70.5°; 76.0°   **4 a)** 71.6°   **b)** 48.2°

5 a) 45°   **b)** 54.7°   **c)** $\frac{1}{6}a^3$ **d)** $\dfrac{1}{\sqrt{3}}a$

6 $\frac{1}{2}\sqrt{a^4 + 2a^2b^2}$; $\cos \widehat{DOE} = \cos^2\widehat{CEB}$

Longer exercise page 415

1 109.5°   **2** 138.2°   **3** 116.6°

Examination questions page 415

1 (ii) 50.8°   (iii) 12.7 km

2 a) 8.39 cm   **b)** 99°; 63.7°

3 (i) 33.9 cm   (ii) 90.2°

25 Algebra II

Exercise 25A page 418

1 a) $\dfrac{1}{x-1} - \dfrac{1}{x+1}$ **b)** $\dfrac{1}{x+1} + \dfrac{1}{x}$ **c)** $\dfrac{1}{x+1} + \dfrac{1}{x-1}$

d) $\dfrac{2}{x-2}+\dfrac{1}{x+1}$

e) $\dfrac{3}{x-1}-\dfrac{2}{x-2}$

f) $\dfrac{5}{x+3}+\dfrac{3}{x+5}$

g) $\dfrac{1}{2x-1}+\dfrac{3}{x-1}$

h) $\dfrac{1}{3x+2}+\dfrac{1}{2x+1}$

i) $\dfrac{1}{x-1}+\dfrac{1}{x-2}+\dfrac{1}{x-3}$

2 a) $\dfrac{1}{2\alpha(x-\alpha)}-\dfrac{1}{2\alpha(x+\alpha)}$

b) $\dfrac{1}{2(x-\alpha)}+\dfrac{1}{2(x+\alpha)}$

Exercise 25B page 420

1 a) $\dfrac{1}{x}-\dfrac{x}{x^2+1}$

b) $\dfrac{2}{x-1}+\dfrac{x+1}{x^2+2}$

c) $\dfrac{-1}{x+1}+\dfrac{x+2}{x^2+x+1}$

d) $\dfrac{1}{x}+\dfrac{1}{2x^2+1}$

e) $\dfrac{2}{2x+1}-\dfrac{x+2}{x^2+2}$

f) $\dfrac{1}{x-1}-\dfrac{3x+2}{2x^2+x+1}$

2 a) $\dfrac{1}{x-1}-\dfrac{2+x}{x^2+x+1}$

b $\dfrac{1}{x+1}+\dfrac{2-x}{x^2-x+1}$

Exercise 25C page 421

1 a) $\dfrac{1}{x}+\dfrac{1}{x-1}-\dfrac{2}{(x-1)^2}$

b) $\dfrac{1}{x+1}+\dfrac{2}{x-1}-\dfrac{3}{(x+1)^2}$

c) $\dfrac{2}{x+1}-\dfrac{3}{(x-1)^2}$

d) $\dfrac{2}{x^2}+\dfrac{1}{x}-\dfrac{1}{2x+1}$

e) $\dfrac{15}{x-2}-\dfrac{6}{x^2}-\dfrac{15}{x}$

f) $\dfrac{1}{x}-\dfrac{1}{x+1}-\dfrac{1}{(x+1)^2}$

2 a) $\dfrac{1}{(x-1)^3}-\dfrac{1}{(x-1)^2}+\dfrac{1}{x-1}-\dfrac{1}{x}$

b) $\dfrac{1}{(x+1)^2}+\dfrac{2}{x+1}+\dfrac{1}{x^2}-\dfrac{2}{x}$

Exercise 25D page 422

1 a) $1+\dfrac{1}{x}-\dfrac{1}{x+1}$

b) $2-\dfrac{1}{x-2}+\dfrac{1}{x+3}$

c) $3-\dfrac{2}{x-2}+\dfrac{2}{x+2}$

d) $x-\dfrac{1}{x}+\dfrac{3}{x+2}$

e) $x+2+\dfrac{1}{x-1}-\dfrac{1}{x+1}$

f) $x^2+x+1-\dfrac{1}{x}-\dfrac{1}{x-1}$

2 a) $1+\dfrac{\alpha}{2(x-\alpha)}-\dfrac{\alpha}{2(x+\alpha)}$

b) $x+\dfrac{\alpha^2}{2(x+\alpha)}+\dfrac{\alpha^2}{2(x-\alpha)}$

Exercise 25E page 424

1 a) $-\dfrac{1}{1-x}-\dfrac{2}{1+x}$; $-3+x-3x^2$

b) $\dfrac{2}{1-x}-\dfrac{1}{2-x}$; $\dfrac{3}{2}+\dfrac{7}{4}x+\dfrac{15}{8}x^2$

c) $-\dfrac{2}{3-x}-\dfrac{3}{2-x}$; $-\dfrac{13}{6}-\dfrac{35}{36}x-\dfrac{97}{216}x^2$

d) $\dfrac{1}{(1-x)^2}-\dfrac{1}{1+x}$; $3x+2x^2$

e) $\dfrac{2}{2-x}+\dfrac{3}{(1+x)^2}$; $4-\dfrac{11}{2}x-\dfrac{37}{4}x^2$

f) $\dfrac{1}{1-x}+\dfrac{8}{(2+x)^2}$; $3-x+\dfrac{5}{2}x^2$

g) $\dfrac{1}{1-x}+\dfrac{x}{1+x^2}$; $1+2x+x^2$ **h)** $\dfrac{1}{1+2x}-\dfrac{3}{1+2x^2}$; $-2-2x+10x^2$

i) $\dfrac{1+x}{1+x^2}-\dfrac{1}{1+x}$; $2x-2x^2$

2 $A=8,\ B=2$ **3** $A=3,\ B=2$

Exercise 25F page 429

1 a) $\ln 3+\frac{1}{3}x-\frac{1}{18}x^2+\frac{1}{81}x^3$ **b)** $x+\frac{1}{3}x^3$ **c)** $\ln 0.5-2x-2x^2-\frac{8}{3}x^3$

2 a) $1+3x+\frac{9}{2}x^2+\frac{9}{2}x^3$ **b)** $1-2x^2$ **c)** $1+x^2$

d) x^2 **e)** $2x-2x^2+\frac{8}{3}x^3$ **f)** $-\frac{1}{2}x-\frac{1}{8}x^2-\frac{1}{24}x^3$

g) $1+x-\frac{1}{3}x^3$ **h)** $x-x^2+\frac{1}{3}x^3$ **i)** $x+\frac{1}{2}x^2+\frac{1}{3}x^3$

j) $x+3x^2+\frac{13}{3}x^3$ **k)** $1-x-\frac{3}{2}x^2+\frac{11}{6}x^3$ **l)** $1+3x+\frac{5}{2}x^2-\frac{3}{2}x^3$

m) $2x^2-x^3$ **n)** $x-\frac{1}{6}x^3$ **o)** x^2

3 **a), b), c), g), h), j), k), l), n)** and **o)** are valid for all x

    d), i) and **m)** are valid for $-1<x<1$ **e)** $-\frac{1}{2}<x<\frac{1}{2}$ **f)** $-2<x<2$

4 $x-\frac{2}{3}x^3$ **5** $\dfrac{1}{2}\cos x+\dfrac{\sqrt{3}}{2}\sin x$; $\dfrac{1}{2}+\dfrac{\sqrt{3}}{2}x-\dfrac{1}{4}x^2$

6 $1+2x+\frac{3}{2}x^2$; 0.093 **7** $a=1,\ b=-1$ **9** $1+\frac{1}{2}x^2+\frac{5}{24}x^4$

10 a) $1-2x+\frac{5}{2}x^2$ **b)** $3x$ **c)** $5x-\frac{7}{2}x^2$

11 a) $0.000\,17$ **b)** 0.024 **c)** 0.22

12 a) 0.0025 **b)** $0.000\,000\,1$ **13** $1-x+\frac{1}{2}x^2$; 0.178

14 $-x-\frac{1}{2}x^2-\frac{1}{3}x^3$; $\frac{2}{3}+\frac{1}{2}\left(\frac{2}{3}\right)^2+\frac{1}{3}\left(\frac{2}{3}\right)^3+\cdots$

Exercise 25G page 431

1 a) 0 **b)** 2 **c)** 0

4 $\pm\sqrt{2}$ **5** $-\frac{1}{4}$ **6** $\pm\sqrt{5}$

7 $k=\pm\sqrt{12}$ **9** $y\le-2$ or $y\ge2$

Longer exercise page 432

4 302 m, 429 m   **5** 449 m

Examination questions page 433

1 $\dfrac{1}{1-2x}+\dfrac{1}{(1-x)^2}-\dfrac{1}{1-x}$; $1+3x+6x^2$; $-\frac{1}{2}<x<\frac{1}{2}$; 2^n+n

2 a) $A=6,\ B=-2,\ C=5$ **b)** $9-12x+18x^2-42x^3$; $f'(0)=-12$

26 Techniques of integration

Exercise 26A page 438

1 a) $\frac{1}{17}(x+2)^{17} - \frac{1}{8}(x+2)^{16} + c$ b) $\frac{1}{12}(x+2)^{12} - \frac{1}{11}(x+2)^{11} + c$

c) $\frac{2}{5}(x+3)^{5/2} - 2(x+3)^{3/2} + c$ d) $\frac{5}{7}(x+7)^{-14} - \frac{1}{13}(x+7)^{-13} + c$

e) $\frac{1}{36}(2x-1)^9 + \frac{3}{32}(2x-1)^8 + c$ f) $\frac{4}{5}(x+1)^{5/2} - 2(x+1)^{3/2} + c$

2 a) $\frac{1}{2}\sin(x^2-1) + c$ b) $\frac{1}{2}\ln(x^2+1) + c$ c) $\frac{1}{3}\sin^3 x + c$

d) $\ln(\ln x) + c$ e) $\frac{1}{3}\ln(1+x^3) + c$ f) $-\frac{1}{2}(1 + \cos x)^2 + c$

g) $\sin e^x + c$ h) $\frac{1}{3}\tan^3 x + c$

3 a) $\sqrt{3}$ b) $\frac{1}{4}$ c) 3.71   d) $\frac{1}{2}$

4 $\frac{1}{3}$ 5 $\frac{128}{15}$ 6 $\frac{\pi}{3}$ 7 1.19

Exercise 26B page 440

1 a) $x^2 + 8$; $\frac{1}{22}(x^2+8)^{11} + c$ b) $\cos x$; $-\frac{1}{3}\cos^3 x + c$

c) $3x + 2$; $\frac{1}{171}(3x+2)^{19} + \frac{7}{162}(3x+2)^{18} + c$

d) $1 - 4x$; $-\frac{1}{4}e^{1-4x} + c$ e) $\sin x$; $e^{\sin x} + c$ f) $e^x + 3$; $-\cos(e^x + 3) + c$

2 a) $-\frac{1}{2}\cos x^2 + c$ b) $\frac{1}{3}\sin x^3 + c$ c) $\frac{2}{3}(\sin x)^{3/2} + c$

d) $\ln(e^x + 2) + c$ e) $e^{\tan x} + c$ f) $\frac{1}{2}\ln(x^2 + 4x) + c$

Exercise 26C page 442

1 a) $e^x(x-1) + c$ b) $\sin x - x \cos x + c$ c) $\frac{1}{3}x^3 \ln x - \frac{1}{9}x^3 + c$

d) $-e^{-x}(x+1) + c$ e) $e^x(x^2 - 2x + 2) + c$ f) $(x^2 - 2)\sin x + 2x \cos x + c$

g) $-\frac{1}{x}\ln x - \frac{1}{x} + c$ h) $x \tan x + \ln \cos x + c$ i) $(x+2)\ln(x+2) - x + c$

2 a) 41.2   b) $\frac{\pi}{2} - 1$ c) 5130

3 π 4 0.537   5 0.467   6 208.6

Exercise 26D page 444

1 a) $-\frac{1}{8}\cos 4x - \frac{1}{4}\cos 2x + c$ b) $\frac{1}{6}\cos 3x - \frac{1}{14}\cos 7x + c$

c) $\frac{1}{8}\sin 4x + \frac{1}{4}\sin 2x + c$ d) $\frac{1}{20}\sin 10x + \frac{1}{4}\sin 2x + c$

e) $\frac{1}{4}\sin 2x - \frac{1}{8}\sin 4x + c$ f) $\frac{1}{2}\sin x - \frac{1}{26}\sin 13x + c$

g) $\frac{1}{3}\cos^3 x - \cos x + c$ h) $\sin x - \frac{1}{3}\sin^3 x + c$

i) $\sin x - \frac{2}{3}\sin^3 x + \frac{1}{5}\sin^5 x + c$ j) $\frac{1}{5}\cos^5 x - \frac{1}{3}\cos^3 x + c$

k) $\frac{1}{4}\sin^4 x - \frac{1}{6}\sin^6 x + c$ l) $\cos x + \frac{1}{\cos x} + c$

m) $\frac{1}{2}x - \frac{1}{4}\sin 2x + c$ n) $\frac{1}{2}x - \frac{1}{8}\sin 4x + c$ o) $\frac{1}{2}x + \frac{1}{16}\sin 8x + c$

2 $x - \frac{1}{4}\sin 4x + c$ 3 $\frac{1}{32}\sin 4x + \frac{1}{4}\sin 2x + \frac{3}{8}x + c$

4 $\frac{1}{36}\cos 9x - \frac{1}{28}\cos 7x - \frac{1}{12}\cos 3x + \frac{1}{4}\cos x + c$

5 a) $\dfrac{\cos(m-n)x}{2(m-n)} - \dfrac{\cos(m+n)x}{2(m+n)} + c$ b) $-\frac{1}{4n}\cos 2nx + c$

Exercise 26E page 446

1 a) $\sin^{-1}\frac{1}{3}x + c$ **b)** $\frac{1}{2}\sin^{-1}2x + c$ **c)** $\frac{1}{3}\sin^{-1}\frac{3}{2}x + c$

d) $\frac{1}{2}\tan^{-1}\frac{1}{2}x + c$ **e)** $\frac{1}{3}\tan^{-1}3x + c$ **f)** $\frac{1}{6}\tan^{-1}\frac{2}{3}x + c$

2 $\sin^{-1}(x+2) + c$ **3** $\tan^{-1}(x+2) + c$

Exercise 26F page 447

1 $\ln(x+1) - \ln(x+2) + c$ **2** $\ln(x+1) + 2\ln(x-2) + c$

3 $\ln x - \ln(x+1) + \dfrac{1}{x+1} + c$ **4** $\dfrac{1}{2(1-x)} + \frac{1}{4}\ln(x+1) - \frac{1}{4}\ln(x-1) + c$

5 $-\dfrac{1}{x} + \ln(x+1) - \ln x + c$ **6** $\frac{1}{10}\tan^{-1}\frac{1}{2}x - \frac{1}{10}\ln(x^2+4) + \frac{1}{5}\ln(x+1) + c$

Exercise 26G page 448

1 a) $\frac{3}{7}(x+3)^{7/3} - 3(x+3)^{4/3} + c$ **b)** $-\frac{1}{4}\cos^4 x + c$ **c)** $\frac{1}{6}\cos^3 2x - \frac{1}{2}\cos 2x + c$

d) $\sin x - \frac{2}{3}\sin^3 x + \frac{1}{5}\sin^5 x + c$

e) $\frac{1}{4}\sin 2x - \frac{1}{16}\sin 8x + c$ **f)** $\frac{1}{8}\sin 4x + \frac{1}{4}\sin 2x + c$

g) $\frac{2}{5}\tan^{-1}x - \frac{1}{10}\ln(x^2+1) + \frac{1}{5}\ln(x+2) + c$ **h)** $\dfrac{1}{x} - \ln x + \ln(x-1) + c$

i) $x - \frac{1}{2}\cos 2x + c$

2 $\frac{1}{2}x - \frac{1}{4}\sin 2x + c$ **3** $\frac{1}{2}\ln(x^2-1) + c$ **4** $2\ln(x+1) + \dfrac{2}{x+1} + c$

Exercise 26H page 449

1 a) $x\sin^{-1}x + \sqrt{1-x^2} + c$ **b)** $(x^2-1)\ln(x+1) + \frac{1}{2}x(2-x) + c$

c) $\frac{1}{2}\cos 2x + x\sin 2x + x^2 + c$ **d)** $\frac{1}{2}(x^2+1)\tan^{-1}x - \frac{1}{2}x + c$

e) $\frac{1}{2}\tan^{-1}x^2 + c$ **f)** $\frac{1}{2}\sin^{-1}x^2 + c$ **g)** $\frac{1}{3}\tan^3 x + \tan x + c$

h) $\frac{1}{3}\ln(x^2+1) - \frac{1}{3}\ln(x^2+4) + c$ **i)** $\dfrac{1}{\sqrt{2}}\tan^{-1}\dfrac{e^x}{\sqrt{2}} + c$

2 a) $(x+1)\tan^{-1}\sqrt{x} - \sqrt{x} + c$ **b)** $(x-\frac{1}{2})\sin^{-1}\sqrt{x} + \frac{1}{2}\sqrt{x}\sqrt{1-x} + c$

c) $\frac{1}{2}e^{x^2}(x^2-1) + c$ **d)** $x\ln(1-x^2) - \ln(x-1) + \ln(x+1) - 2x + c$

e) $\frac{1}{7}(x^2-2)^{7/2} + \frac{2}{5}(x^2-2)^{5/2} + c$ **f)** $\frac{3}{2}x^{2/3} - \frac{3}{2}\ln(x^{2/3}+1) + c$

Examination questions page 450

1 **1 a)** $\frac{1}{2}$ **b)** $\frac{1}{3}\pi^2$

3 a) $-\dfrac{2}{x} - \dfrac{1}{x^2} + \dfrac{4}{2x-1} + c$; $2\ln(\frac{3}{2}) - \frac{1}{2}$

b) $\frac{1}{4}x^4\ln 4x - \frac{1}{16}x^4 + c$ **c)** $\dfrac{\pi}{216} + \dfrac{1}{108}$

4 $\frac{2}{5}(1+e^{-2\pi})$; $\frac{1}{5}(1+e^{-2\pi})$; $\dfrac{9\pi}{20} - \dfrac{\pi}{20}e^{-2\pi}$

Consolidation section F

Extra questions page 452

1 128 cubic feet **2** 13.8 **3** 1.115

4 15.05 **5** 126 **6** 15.1

7 1.111 **8** 13.2, 43.1°; 56.3°, 50.2°

9 34.4° **10** 92.3°

11 $4\mathbf{i} + 5\mathbf{j} + 3\mathbf{k}$, $4\mathbf{i} + 5\mathbf{j} - 3\mathbf{k}$, $5\mathbf{j} - 3\mathbf{k}$; 34.4°

12 a) $\dfrac{4}{x-2} - \dfrac{2}{x-1}$ **b)** $\dfrac{4}{x-1} - \dfrac{3}{x^2} - \dfrac{4}{x}$

c) $\dfrac{\frac{4}{3}x - \frac{1}{3}}{x^2+2} - \dfrac{1}{3(x+1)}$ **d)** $1 + \dfrac{3}{x-2} - \dfrac{1}{x-1}$

13 a) $x + 1.5x^2 + 1.75x^3$ **c)** $-0.5 + x - 0.25x^2$

14 $\sec^2(x+\frac{\pi}{4})$, $2\sec^2(x+\frac{\pi}{4})\tan(x+\frac{\pi}{4})$, $2\sec^4(x+\frac{\pi}{4}) + 4\sec^2(x+\frac{\pi}{4})\tan^2(x+\frac{\pi}{4})$;

$1 + 2x + 2x^2 + \frac{8}{3}x^3$

15 a) $1 - 3x + \frac{9}{2}x^2$; all x **b)** $\frac{1}{3}x - \frac{1}{18}x^2$; $-3 < x < 3$ **c)** $1 - 2x$; all x

16 $k = -4$ or $k = 5$

17 a) $\frac{1}{11}(x-1)^{11} + \frac{1}{5}(x-1)^{10} + c$ **b)** $\frac{1}{2}\tan x^2 + c$

 c) $(x-3)\ln(x-3) - x + c$ **d)** $-\frac{1}{6}\cos 3x - \frac{1}{2}\cos x + c$

 e) $\frac{1}{6}\cos^3 2x - \frac{1}{2}\cos 2x + c$ **f)** $\frac{1}{2}x + \frac{1}{20}\sin 10x + c$

 g) $\frac{1}{2}\sin^{-1}\frac{2}{3}x + c$ **h)** $2\ln(x-3) - 2\ln(x-2) + c$

 i) $\frac{1}{4}\ln(x+2) - \frac{1}{4}\ln x - \dfrac{3}{2x} + c$

Mixed exercise page 453

1 5.31 cm **2** 1, −2 **3** $b = 30$, $a = 5$

4 a) 0.460 **b)** 0.461 **5** $\frac{1}{2}\pi k^2$; $\frac{16}{5}\pi$ **6** $\frac{1}{2}\pi^2$

9 $\dfrac{\pi}{3}(k^3 - 3k + 2)$; 0.226

Longer exercise page 454

Painting strips

1 $\displaystyle\int_1^\infty x^{-0.75}\,dx$ **2** $\displaystyle\int_1^\infty \pi x^{-1.5}\,dx$; 2π

Examination questions page 456

1 a) $\dfrac{2}{2x+1} - \dfrac{1}{x+1}$ **b)** $(-\frac{3}{4}, -8)$, maximum

2 77.2° **3** $p < 4$ or $p > 20$

4 744 000 m³; under **5 b)** $1 + \frac{3}{4}\pi$

27 Functions and graphs

Exercise 27A page 461

1 a) all x, $y \geq 0$
b) all x, $y \geq 0$
c) all x, $y \leq 1$
d) all x, $y \geq 1$
e) all x, $y \geq -2$
f) all x, $y \leq 2$
g) $x \neq 0$, $y \neq 0$
h) $x \neq 0$, $y \neq 1$
i) $x \neq 0$, $y \neq 3$
j) $x \neq 2$, $y \neq 0$
k) $x \neq -2$, $y \neq 1$
l) $x \neq \frac{1}{2}$, $y \neq 1$
m) $x \geq -3$, $y \geq 0$
n) $x \geq 0$, $y \geq 1$
o) $x \leq 1$, $y \geq 0$
p) $x \geq 0$, $y \leq 1$
q) all x, $y \geq 0$
r) all x, $y \geq 1$
s) all x, $y > 0$
t) all x, $y < 2$
u) $x > 0$, all y
v) $x > -3$, all y
w) all x, $-1 \leq y \leq 1$
x) $x \neq \frac{\pi}{2} + n\pi$, all y

2 $x \neq 1$, $y \neq -3$

3 a) $y > 1$
b) $0 < y \leq \frac{1}{2}$
c) $0 \leq y < 4$
d) $y \leq 1$
e) $0 \leq y \leq 1$
f) $y > 1$

4 $0 \leq t \leq 8$; $0 \leq h \leq 80$

Exercise 27B page 463

1 (a), (c), (d) and (f) are one to one functions
2 $x \geq 0$
3 $-\frac{\pi}{2} \leq x \leq \frac{\pi}{2}$
4 (b)

Exercise 27C page 464

1 a) 23
b) 13
c) $8x - 11$

2 a) 9
b) 3
c) $(1 + 2x)^2$
d) $1 + 2x^2$

3 $x = 0$ or $x = -2$

4 a) 0
b) 1
c) $1 - e^x$
d) e^{1-x}
e) x

5 a) 1
b) $-\frac{1}{2}$
c) $\frac{1}{x-1}$
d) $\frac{1}{x} - 1$

6 a) $acx + ad + b$
b) $acx + cb + d$
c) $a^2x + ab + b$
d) $c^2x + cd + d$

7 $c = 1$ or $b = 0$

Exercise 27D page 466

1 a) $\frac{1}{3}(x - 5)$
b) $\frac{1}{3}(2 - x)$
c) $2(x + 2)$
d) $\frac{1}{2}x + 1$
e) $3x - 2$
f) $2 - 2x$

2 $\sqrt[3]{x + 1}$
3 $0 < x$; $\ln x$; $0 < x$

4 $y \geq -2$; $\sqrt{x + 2} + 1$, $x \geq -2$, $y \geq 1$

5 $y \geq -2$; $f^{-1}(x) = (x + 2)^2 - 1$, $x \geq -2$, $y \geq -1$

6 $k = 1$; $f^{-1}(x) = \sqrt{x + 1} + 1$, $x \geq -1$, $y \geq 1$

Exercise 27E page 468

2 $-\frac{\pi}{2} \leq x \leq \frac{\pi}{2}$
3 $-\pi \leq x \leq 0$
4 $\frac{3\pi}{2} < x < \frac{5\pi}{2}$
5 $\frac{\pi}{2}$

Exercise 27F page 470
1 a) odd **b)** even **c)** none
d) even **e)** none **f)** odd
g) even **h)** odd **i)** even, period 2π
j) odd **k)** even **l)** odd, period $\dfrac{2\pi}{3}$
m) even, period 6π **n)** period 2 **o)** odd, period $\dfrac{\pi}{2}$

2 a) even **b)** odd **c)** periodic, odd

6 a) even **b)** odd **c)** even **d)** even
e) even **f)** even **g)** odd

9 $f(x) = \frac{1}{2}(f(x) + f(-x)) + \frac{1}{2}(f(x) - f(-x))$ **10 a)** b **b)** ab

Examination questions page 472
1 (i) 8 (ii) ± 2 (iii) -1

2 a) $y > 1$ **b)** $x > 1,\ y > 2$

3 $y > 1;\ f^{-1}(x) = \dfrac{1}{x-1}$ (ii) $1 < y < 2$

28 Applications of the chain rule

Exercise 28A page 475
1 $0.0318\ \text{cm s}^{-1}$ **2** $25.1\ \text{cm}^2\ \text{s}^{-1}$ **3** $0.8\ \text{cm s}^{-1}$

4 $0.5\ \text{cm}^2\ \text{s}^{-1}$ **5** $30\ \text{cm}^3\ \text{s}^{-1},\ 12\ \text{cm}^2\ \text{s}^{-1}$ **6** $\dfrac{\sqrt{3}}{10}\ \text{cm s}^{-1}$

7 $\dfrac{2q}{x}$ **8** $\frac{1}{2}r^2\theta = A,\ -\dfrac{4Aq}{r^3}$

9 $8\pi\ \text{cm}^3$ **a)** $\dfrac{5}{\pi h}\ \text{cm s}^{-1}$ **b)** $\dfrac{5}{h}\ \text{cm}^2\ \text{s}^{-1}$

10 $0.015\,625\ \text{m s}^{-1}$ **11** $0.118\ \text{cm s}^{-1}$ **12** $\dfrac{1}{175\pi}\ \text{cm s}^{-1}$

Exercise 28B page 478
1 a) $3y^2\dfrac{dy}{dx}$ **b)** $e^y\dfrac{dy}{dx}$ **c)** $3\cos 3y\dfrac{dy}{dx}$

d) $\frac{1}{2}y^{-\frac{1}{2}}\dfrac{dy}{dx}$ **e)** $2xy + x^2\dfrac{dy}{dx}$ **f)** $\sqrt{x}\dfrac{dy}{dx} + \frac{1}{2}yx^{-\frac{1}{2}}$

g) $y^2 + 2xy\dfrac{dy}{dx}$ **h)** $3x^2y^2 + 2x^3y\dfrac{dy}{dx}$ **i)** $e^y\cos x + e^y\sin x\dfrac{dy}{dx}$

2 a) $-\frac{1}{3}$ **b)** $\frac{8}{3}$ **c)** $-\frac{1}{2}$ **d)** infinite
e) $-\frac{5}{2}$ **f)** $\frac{7}{4}$ **g)** -1 **h)** -1

3 a) $-\dfrac{x}{3y}$ 　　**b)** $\dfrac{4x}{3y}$ 　　**c)** $-\sqrt{\dfrac{y}{x}}$ 　　**d)** $-\dfrac{\cos x}{\cos y}$

e) $\dfrac{-4x-3y}{3x+4y}$ 　**f)** $\dfrac{2x+3y}{10y-3x}$ 　**g)** $-\dfrac{1}{x}$ 　　**h)** $-\dfrac{\cos x \sin y}{\cos y \sin x}$

6 $3y+2x=7$ 　　**7** $13y+18x=-28$ 　　**8** $2y=3x-4,\ 18y-13x=75$

9 $\dfrac{-2x-y}{x+4y};\ (\sqrt{2},-\sqrt{2}),\ (-\sqrt{2},\ \sqrt{2})$ 　　**10** $(\pm\dfrac{2}{\sqrt{7}},\ \mp\dfrac{4}{\sqrt{7}})$

11 $\dfrac{4\sqrt{2}}{\sqrt{7}}$ 　　**12 a)** $\dfrac{5\sqrt{2}}{3}$ 　　**b)** $\dfrac{2\sqrt{5}}{3}$

13 $-\dfrac{2y}{3x}$ 　　**14** $(1+\ln x)x^x$ 　　**15** $(\dfrac{1}{x}-\ln y)\dfrac{y}{x}$

Exercise 28C page 481

1 a) $\dfrac{2}{3t}$ 　　　**b)** $-\tfrac{1}{2}$ 　　　**c)** $\tfrac{2}{3}$

d) $\dfrac{1}{2\sqrt{t}}$ 　　**e)** $\dfrac{\sqrt{t+1}}{\sqrt{t-1}}$ 　　**f)** $-\dfrac{t^2}{5}$

g) $\dfrac{(t+1)^2}{(t+2)^2}$ 　**h)** $-\tfrac{4}{3}\cot\theta$ 　　**i)** $-\tfrac{1}{2}\cot\theta$

j) $\operatorname{cosec}\theta$ 　　**k)** $\tfrac{4}{3}\sin\theta$ 　　**l)** $-\dfrac{2}{e^{3t}}$

m) $\dfrac{e^t+e^{-t}}{e^t-e^{-t}}$

2 a) $y=\tfrac{2}{3}x+\tfrac{1}{3}$ 　　**b)** $4y-x=1$ 　　**c)** $4y-x=1$
　d) $3\sqrt{3}y+4x=24$ 　**e)** $y-\sqrt{2}x=-1$ 　**f)** $x=1$

3 a) $2y+3x=5$ 　　　**b)** $y+4x=30$ 　　**c)** $y+4x=4.5$
　d) $4y-3\sqrt{3}x=\dfrac{7\sqrt{3}}{2}$ 　**e)** $\sqrt{2}y+x=2\sqrt{2}$ 　**f)** $y=0$

4 $45°$ to horizontal 　　**5** $75°$ to horizontal 　　**6** $y=x+2,\ x=y^2-3y+2$

7 $4y-3x=14,\ y=\tfrac{2}{9}(x^2-2x+1)$ 　　**8** $3y+x=28$ 　　**9** $y^3=x^2+3xy+2y^2$

10 $y+2t^3x=3t^2$ 　　**11** $x^2+\tfrac{1}{4}y^2=1$ 　　**12** $\tfrac{1}{9}x^2+\tfrac{1}{16}y^2=1$ 　　**13** $\dfrac{1}{\sqrt{2}-1}$, infinity

Exercise 28D page 484

1 $0.03\,\text{cm}$ 　　　　　**2** 1.4 　　　　　　**3** -0.06

4 -0.05 　　　　　　**5** $0.07\,\text{m}$ 　　　　　**6** 1.25δ

7 $0.0027\,\text{cm}$ 　　　　**8** $0.0035\,\text{cm}$ 　　　**9** 2%

10 1.2% 　　　　　　**11** 0.2% 　　　　　　**12 a)** 0.1 　　**b)** 0.5%

13 $3x\%$ 　　　　　　**14** $2x\%$

15 a) 3.05 　　**b)** 9.95 　　　　**c)** 2.008 　　**16** 8.0078 　　　　**17** 0.24p

Longer exercise page 485
2 $\delta x = 0.0488$; 2.0488   **3** 2.046 615 411   **4** 2.046 610 92   **5** 0.739 085 13

Examination questions page 485
1 $0.22\,\text{m s}^{-1}$

2 $n\%$

3 (i) $\delta V = 4\pi r^2 \delta r$ (ii) 0.07 cm   (iv) $0.707\,\text{cm s}^{-1}$

4 $\dfrac{\mathrm{d}y}{\mathrm{d}x} = \dfrac{3t}{2}$; tangent $2y - 3tx + t^3 = 0$

5 (i) $(0,\ \pm\tfrac{3}{2})$ (ii) $\dfrac{-(x+y)}{x+4y}$ (iii) $x + y = 0$ (iv) $x + 4y = 0$

29 Differential equations

Exercise 29A page 489

1 $\dfrac{\mathrm{d}P}{\mathrm{d}t} = kP$

2 $\dfrac{\mathrm{d}M}{\mathrm{d}t} = -kM$

3 $\dfrac{\mathrm{d}v}{\mathrm{d}t} = kv^2$

4 $\dfrac{\mathrm{d}F}{\mathrm{d}t} = k(1000 - F)$

5 $\dfrac{\mathrm{d}P}{\mathrm{d}t} = kP(1 - P)$

6 $\dfrac{\mathrm{d}V}{\mathrm{d}t} = p - qV$

7 $\dfrac{\mathrm{d}y}{\mathrm{d}x} = kx^2$

8 $\dfrac{\mathrm{d}y}{\mathrm{d}x} = kyx$

9 $\dfrac{\mathrm{d}V}{\mathrm{d}t} = -kV$

10 $\dfrac{\mathrm{d}V}{\mathrm{d}t} = k' - \dfrac{V}{100}(k + k')$

11 $\dfrac{\mathrm{d}P}{\mathrm{d}t} = kP - k'P^2$

Exercise 29B page 493
1 a) $y = A\mathrm{e}^x$

b) $y = (x^2 + c)^2$

c) $y = k(x + 3)$

d) $y = \sin(x^2 + c)$

e) $y = -\ln(c - \mathrm{e}^x)$

f) $y = \dfrac{1}{c - \tan^{-1}x}$

g) $y = A\mathrm{e}^{\frac{1}{2}x^2} - 1$

h) $y = A(x + 2) - 3$

i) $y = \sqrt{Ax - 1}$

j) $y = -2 - \ln(\mathrm{e}^{-x} + c)$

k) $y = \tan^{-1}(c - \cos x)$

l) $y = -\ln(c - \sin x)$

m) $y = \dfrac{A\mathrm{e}^{2x}}{1 - A\mathrm{e}^{2x}}$

n) $y = \dfrac{Ax - 1}{Ax + 1}$

o) $y = \dfrac{A\mathrm{e}^{\frac{1}{2}x^2}}{1 + A\mathrm{e}^{\frac{1}{2}x^2}}$

2 $\dfrac{\mathrm{d}P}{\mathrm{d}t} = 0.02P$; $P = A\mathrm{e}^{0.02t}$; 34.7 years

3 $\dfrac{\mathrm{d}h}{\mathrm{d}t} = k(H - h)$; $h = H - A\mathrm{e}^{-kt}$

4 $\dfrac{\mathrm{d}y}{\mathrm{d}x} = A\mathrm{e}^{5x}$; $y = B\mathrm{e}^{5x} + c$

5 a) $y = Ax^3 + c$

b) $y = \tfrac{1}{80}x^5 + \tfrac{1}{6}cx^3 + c^2x + d$

c) $y = -A\cos x + c$

Exercise 29C page 495
1 a) $y = 3\mathrm{e}^x$

b) $y = (x^2 + 2)^2$

c) $y = 2(x + 3)$

d) $y = \sin\left(x^2 + \dfrac{\pi}{2}\right)$

e) $y = -\ln(2 - e^x)$

f) $y = \dfrac{1}{\frac{1}{2} - \tan^{-1}x}$

g) $y = 4e^{\frac{1}{2}x^2} - 1$

h) $y = 2(x + 2) - 3$

i) $y = \sqrt{2.5x - 1}$

j) $y = -2 - \ln(e^{-x} + e^{-7} - 1)$

k) $y = \tan^{-1}(1 - \cos x)$

l) $y = -\ln(e^{-1.5} + 0.5 - \sin x)$

m) $y = \dfrac{\frac{1}{3}e^{2x}}{1 - \frac{1}{3}e^{2x}}$

n) $y = \dfrac{-3x - 1}{-3x + 1}$

o) $y = \dfrac{e^{\frac{1}{2}x^2}}{1 + e^{\frac{1}{2}x^2}}$

2 $50e^{0.03t}$; 13.5 years

3 $h = 5(1 - e^{-0.0002t})$; 1116 seconds

4 $y = \dfrac{1000e^{t/10}}{1 + e^{t/10}}$; $t = 22$

5 $\dfrac{dP}{dt} = kP(1 - P)$; after 1.47 years

6 $P = \dfrac{1\,000\,000}{1 + 3^{-t}}$

7 $M = 60(1 - e^{-0.05t})$; 23.6 litres; $\frac{3}{5}$

Exercise 29D page 498

1 a) $\dfrac{dy}{dx} = 3x^2$

b) $\dfrac{dy}{dx} = 2y$

c) $\dfrac{dy}{dx} = \dfrac{y}{x}$

2 a) $\dfrac{d^2y}{dx^2} = y$

b) $\dfrac{d^2y}{dx^2} - 5\dfrac{dy}{dx} + 6y = 0$

c) $\dfrac{d^2y}{dx^2} = \dfrac{dy}{dx}$

d) $\dfrac{d^2y}{dx^2} = -y$

e) $\dfrac{d^2y}{dx^2} - 2\dfrac{dy}{dx} + 2y = 0$

3 a) $x + y\dfrac{dy}{dx} = 0$

b) $\dfrac{dy}{dx}(2y + x) + y = 0$

c) $\dfrac{dy}{dx} = -e^{x-y}$

4 $\dfrac{dy}{dx} = \dfrac{y \ln y}{x \ln x}$

Exercise 29E page 499

1 $y = Ae^{5x}$ ⟶ **a)** $y = 5e^{5x}$ **b)** $y = -3e^{5x}$ **c)** $y = \dfrac{4}{e^5}e^{5x}$

2 $y = (x^2 + c)^2$ ⟶ **a)** $y = (x^2 + 1)^2$ **b)** $y = (x^2 - 15)^2$ **c)** $y = x^4$

3 $y = \sin(x + c)$ ⟶ **a)** $y = \sin x$ **b)** $y = \sin\left(x - \dfrac{\pi}{2}\right)$ **c)** $y = \sin\left(x + \dfrac{\pi}{2}\right)$

4 $y = Axe^{-x}$; Ae^{-1} **5** $y = A + e^x$ or $y = B + e^{-x}$

Exercise 29F page 501

1 a) 1.46   **b)** 5.58   **c)** 1.15   **d)** 5.90

2 a) 1.87   **b)** 1.76; 6%

3 $\dfrac{dh}{dt} = -0.01h$ **a)** 0.9512 m   **b)** 0.9510   **4** 10 131

Examination questions page 503

1 $\dfrac{dh}{dt} = -kh$; 10 minutes; $h = H2^{-\frac{1}{3}t}$

2 $y = \tan\left(x + \frac{1}{3}x^3 + \dfrac{\pi}{4}\right)$

3 a) $\dfrac{dy}{dx} = -4xe^{-2x}$; (0, 1), maximum

4 $y = 1.108$ when $x = 0.5$; greater

Consolidation section G

Extra questions page 509

1 a) all x, $y \geq -5$ **b)** $x \geq \frac{3}{2}$, $y \geq 0$

   c) $x \neq 2n\pi \pm \dfrac{\pi}{2}$, $y \leq -1$ and $y \geq 1$ **d)** $x \neq n\pi$, $y \leq -1$ and $y \geq 1$

2 $y \geq 3$ **3** **(b)** and **(c)**

4 a) -5 **b)** -9 **c)** $3 - 4x$ **d)** $-4x - 1$ **e)** $4x + 3$ **f)** $\frac{1}{2}(x - 1)$

5 **(b)** is even, **(c)** is odd, with period 2π

6 $\dfrac{4}{\sqrt{3x}}$ cm s^{-1} **7** $-\frac{3}{2}$ **8** $2y + 3x = -4$

9 $\left(\dfrac{-4}{\sqrt{5}}, \dfrac{1}{\sqrt{5}}\right)$, $\left(\dfrac{4}{\sqrt{5}}, \dfrac{-1}{\sqrt{5}}\right)$

10 a) $\dfrac{1}{4t^{1.5}}$ **b)** $y = \frac{1}{32}x + 1\frac{15}{32}$ **c)** $x = 1 + y^4$

11 0.023 cm **12** $\dfrac{dN}{dt} = kN(p - N)$ **13** $y = Ae^{-1/x} - 1$

14 $y = 4e^{1 - 1/x} - 1$ **15** $A = 5^{t/10}$

16 $y = A \sin x$ **a)** $y = \sin x$ **b)** $y = -2 \sin x$ **c)** $y = 0$

17 2.212

Mixed exercise page 510

1 Total of 4.85 years

2 2π **3** after total of 5.27 years **4** $y = Ae^{3x} + c$; $y = c$

7 $6\frac{2}{3}$ mins before 1 **8** $\dfrac{dV}{dt} = 1 - kV - kV^2$; 11.09 mins

9 all x **10** 0.000 06 cm s^{-1} **11** 0.013 m s^{-1}

12 0.0065 feet per second **13** $-1 \leq y$

Longer exercise page 511

2 a) $v = e^{-t}(10 + u) - 10$; $t = \ln\left(1 + \dfrac{u}{10}\right)$

b) $s = -10 \ln(u + 10) + 10 \ln(v + 10) + u - v$; height $= u - 10 \ln\left(1 + \dfrac{u}{10}\right)$

3 $u > 11300$ m s^{-1}

Examination questions page 512

1 Compress by factor 4, parallel to y-axis; reflect in x-axis; translate up by 1.
Reflect in $y = x$; g^{-1} has domain $x \leq 1$, range $y \geq 0.83$

2 (iii) $1 - \dfrac{1}{x - 1}$

3 (i) -0.395 (ii) -0.390 (iii) $(0.766, \ 0)$ (iv) $\left(\dfrac{\pi}{2}, \dfrac{1}{2}(e - 2)\right)$

4 $\dfrac{5\pi}{4}, \dfrac{7\pi}{4}$; $y = x - 4$, $y = -x - 2$

Mock examinations

Examination 1 page 514

1 $(0, 0)$, $(1, 3)$, $(2.25, -0.75)$; $(1, 3)$, $\sqrt{10}$

2 $x \geq 0$, all x **a)** $fg: x \rightarrow (x + 1)^2$ **b)** $gf: x \rightarrow x^2 + 1$ **c)** $ff: x \rightarrow x^4$; g has inverse

3 c) $33.7°$ **d)** $3\frac{4}{7}\mathbf{i} + 2\frac{6}{7}\mathbf{j}$

4 a) $2R \cos\theta$ **b)** $4R \cos\theta \sin\theta$ **c)** $4R^2\theta \cos^2\theta$
   d) $\frac{1}{2}R^2(2\pi - 4\theta + \sin 4\theta) + 2R^2\cos^2\theta(2\theta - \sin 2\theta)$

5 $10e^{-0.1t} - 5$; maximum distance $= 15.3$; returns when $t = 15.94$

6 a) $x \tan x + \ln \cos x + c$ **b)** $\tan x - x + c$ **c)** $\frac{1}{2}\tan x + c$

7 a) $\sqrt{2}$ **b)** $4\sqrt{3}, \frac{4}{3}$ **c)** $2\sqrt{3}, \dfrac{\sqrt{2}}{3}$ **d)** $\mathbf{i} + \mathbf{j} - \mathbf{k}$; $70.5°$ or $109.5°$

8 $1 + 2x + 3x^2 + 4x^3 + \cdots$ **a)** $\frac{1}{6}$ **b)** $\frac{5}{36}$ **c)** $\frac{1}{6}(\frac{5}{6})^{n-1}$; $\frac{1}{6}[1 + 2 \times \frac{5}{6} + 3 \times (\frac{5}{6})^2 + \cdots]$; £6

9 $\dfrac{dp}{dt} = kp(1 - p)$; total of 55.4 years

Examination 2 page 516

1 b) $-3 < x < -2\frac{3}{4}$ **2** $2\pi x^2$; $0.006\,37$ cm s^{-1}; 1.6 cm^2 s^{-1}

3 $\left(\dfrac{\pi}{6}, \dfrac{1}{\sqrt{3}}\right)$, maximum; $\left(\dfrac{5\pi}{6}, \dfrac{-1}{\sqrt{3}}\right)$, minimum; $-\dfrac{1}{\sqrt{3}} < f(x) < \dfrac{1}{\sqrt{3}}$; $\ln 2$

4 120; at $p = 8$; $\displaystyle\int_{p=2}^{p=4.4} pf(p) \, dt$; 6518 **5** 80.5 m; underestimate

6 5000×1.1^n; $10\,000(1.05^n - 1)$

7 a) is odd, period $2a$ **b)** period $4a$; $k = 2a$ **8** 183.6 m^2

9 a) $\frac{7}{55}$ **c)** e.g. $\sqrt{2.5}$ **10** $x = 32$, $y = 42$

Examination 3 page 518

1 **a)** $e^x(x^{3/2} + \frac{3}{2}x^{1/2})$ **b)** $\dfrac{-(x^2 + 14x + 17)}{(x^2 + 3x + 4)^2}$ **c)** $\frac{2}{3}\cos 2x(1 + \sin 2x)^{-2/3}$

2 $2\pi rh\delta r + \pi r^2\delta h$; $\dfrac{100\delta r}{r}$; 2(% change in r) + (% change in h)

3 $(\frac{1}{3}, 6)$, minimum; $(-\frac{1}{3}, -6)$, maximum; area $= \ln 2 + 13.5$; volume $= 207.5\pi$

4 14 142 m **5** $\dfrac{1 + 2x}{1 + 2x^2} - \dfrac{1}{1 + x}$; $3x - 3x^2 - 3x^3$; $\dfrac{-1}{\sqrt{2}} < x < \dfrac{1}{\sqrt{2}}$; 0.870

6 250 000 cm³ **7** $\dfrac{-1}{3t^2}$; $3y + x = 8$; $y = \dfrac{3}{x - 2}$; $\ln 27$

8 $\dfrac{\mathrm{d}M}{\mathrm{d}t} = -kM$; $M_0\, 2^{-t/5600}$; 7009 years ago **9** bearing of 61.9°; 16.2 seconds

10 971.8 tonnes; 16p; £1.80 **11** **b)** 11.08 Ω; 0.46 **c)** 0.695; 0.87

Index